NON-WASTE TECHNOLOGY AND PRODUCTION

NON-WASTE TECHNOLOGY AND PRODUCTION

Proceedings of an international seminar organized by the Senior Advisers to ECE Governments on Environmental Problems on the Principles and Creation of Non-waste Technology and Production
Paris, 29 November - 4 December 1976

Published by
PERGAMON PRESS
for the
UNITED NATIONS

U.K.	Pergamon Press Ltd., Headington Hill Hall, Oxford OX3 0BW, England
U.S.A.	Pergamon Press Inc., Maxwell House, Fairview Park, Elmsford, New York 10523, U.S.A.
CANADA	Pergamon of Canada Ltd., 75 The East Mall, Toronto, Ontario, Canada
AUSTRALIA	Pergamon Press (Aust.) Pty. Ltd., 19a Boundary Street, Rushcutters Bay, N.S.W. 2011, Australia
FRANCE	Pergamon Press SARL, 24 rue des Ecoles, 75240 Paris, Cedex 05, France
FEDERAL REPUBLIC OF GERMANY	Pergamon Press GmbH, 6242 Kronberg-Taunus, Pferdstrasse 1, Federal Republic of Germany

First edition 1978

British Library Cataloguing in Publication Data

Non-waste technology and production.
1. Non-waste technology - Congresses
I. Economic Commission for Europe
600 TS149 77-30186
ISBN 0-08-022028-2

In order to make this volume available as economically and as rapidly as possible the authors' typescripts have been reproduced in their original forms. This method unfortunately has its typographical limitations but it is hoped that they in no way distract the reader.

Printed in Great Britain by Page Bros (Norwich) Ltd., Norwich and London

Contents

Introduction

The Senior Advisers to ECE Governments on Environmental Problems, a principal subsidiary body of the United Nations Economic Commission for Europe, provides a means for the thirty-four ECE members to exchange their experience and consult with one another on their environmental policy plans and intentions, to study the various environmental options and methods open to them, to sponsor joint studies on matters of common concern, and to give a lead for the development of policies and projects which aim to protect and improve the environment.

This body was established by the Economic Commission for Europe in 1971—after many years of activity by ECE in various environmental fields. At their first session, in 1973, the Senior Advisers decided to include, among other subjects, the principles and creation of non-waste production systems in their work programme.

An informal meeting, organized by the Senior Advisers in Geneva in October 1974, defined non-waste technology as "the practical application of knowledge, methods and means so as, within the needs of man, to provide the most rational use of natural resources and energy and to protect the environment". Non-waste technology, it was stressed, should be seen as a long-term strategy, as a philosophy of the evaluation of the environmental complex. Its concepts and principles should be reflected in national economic policies and development strategy, in the production and consumption activities of man.

A meeting of experts, held in Geneva in January 1975, discussed the principles of non-waste technology, the structure of non-waste systems, and ways to transform contemporary technology into non-waste technology.

The subject was discussed in depth at the third session of the Senior Advisers immediately afterwards, and the decision was taken to hold a seminar. The main aims of the seminar would be to exchange experience and information on the development and application of non-waste technology, to demonstrate that the introduction of its concepts into economic, technical, scientific and social activities could lead to the solution of many environmental problems, and to make conclusions and recommendations based on the issues discussed.

The Government of France offered to act as host for the seminar, which was held in Paris from 29 November to 4 December 1976. More than 150 representatives of thirty countries and nine international inter-governmental and non-governmental organizations took part.

Jean-François Saglio (France) and Czeslaw Cala (Poland) were elected, respectively, chairman and vice-chairman of the seminar.

The inaugural address was given by Vincent Ansquer, Minister for the Quality of Life in the French Government, and participants were welcomed by Janez Stanovnik, Executive Secretary of ECE.

Discussions at the seminar were based on six introductory reports by national rapporteurs, and on case studies. The subjects, and the countries providing the introductory reports, were as follows:

Concepts and principles of non-waste technology (USSR)

State of non-waste technology

National experience and policy (Canada)

Industrial experience (France, Hungary, USSR)

Case studies from the iron and steel industry, pulp and paper industry, packaging, and the tyre industry

Cost/benefit aspects of non-waste technology (USA)

Ways and means of implementing non-waste technology (German Democratic Republic)

Methodological and strategic aspects of non-waste technology (France)

On a day reserved for study tours, the host government offered participants the choice of visits to a refinery, a paper mill, and plants dealing with the recycling of plastics and the recovery and sorting of textiles.

Conclusions and Recommendations

PREAMBLE

Man has continually sought to improve the quality of life, transforming nature to provide more food, better living conditions, and a longer life. Technology has helped to accomplish this transformation and to achieve many of man's goals. It has, however, left a profusion of environmental problems in its wake. Indeed, in many respects modern technology is becoming too costly to human society in terms of economic and social values and in terms of future implications. The question today is whether technology can solve the environmental problems which technology has helped to cause. There is widespread belief that this question can be answered positively. But in order to answer it in a more practical way, technological evaluation activities should be expanded and more thought devoted to determining what technology is needed to solve today's pressing problems.

Awareness of negative side effects of modern technology has, in recent years, brought about new economic and legislative measures which are fostering new industrial attitudes and approaches. Attention has been mainly focused on problems connected with treatment of wastes at the end of the production line, once the product (and its consequent wastes) has been produced. But more and more frequently it is being asked whether it would not be economically and socially less costly to minimize all along the line the creation of wastes that need to be treated - from the extraction of raw materials to the end of life of final consumer goods. The essence of non-waste technology is in the answer to this question.

TOPIC 1. CONCEPTS AND PRINCIPLES OF NON-WASTE TECHNOLOGY

The generally agreed definition of non-waste technology, as determined during an inter-governmental meeting at the outset of this ECE project, is:

> "the practical application of knowledge, methods and means, so as—within the needs of man—to provide the most rational use of natural resources and energy, and to protect the environment."

Thus, in its essence, non-waste technology is the planning and management of human activities so as to provide the minimum waste of materials and energy.

In this context, the term "non-waste" is to be understood as a kind of technological optimum which can be used as a norm for measuring the final degree of efficiency, taking into account human needs—which is the object of technological development. As non-waste technology replaces conventional technology, wastes must be considered as a loss of potential resources, and efforts aimed at reducing waste produce numerous positive results.

The problem of non-waste technology is of great social importance both for the level of the national economy and the creation of industrial potential, and also for the related task of ensuring the optimum ecological interaction between a production process and the environment. Scientific and technological progress is unthinkable without taking into account its effect on the environment and the attendant social consequences.

The above concept of what might be called "conventional" non-waste technology addresses itself to reducing or eliminating waste in production processes and in consumption as well. A broader definition will question the entire principle of developing a given product; it will question whether we need that product to satisfy a given objective, or whether we could rather use the same resources to develop another product with more potentialities and more benefit to the human race.

Non-waste technology, in this broader sense, becomes a problem that no single country can handle in isolation from the rest of the world. Since it could result in a new appreciation of our social values, our life styles, our behaviour and our trading and economic patterns throughout the world, nothing less than a consolidated effort by the entire international community is required to provide a comprehensive framework for non-waste technology.

The need for the renewal and most rational use of natural resources has already reached proportions which transcend the limits of development of individual countries and raise vast new problems for co-operation in the conservation of the environment. The development of non-waste technology is, therefore, a long-term endeavour of the greatest importance, whose realization will bring about positive changes in all sectors of industry and in ecology and will help to raise living standards throughout the entire world. Indeed, to make an enhanced quality of life sustainable on this planet depends upon reorienting ourselves towards non-waste technology.

Among the methodologies for analysing and comparing different technological processes, one of the more important is systems analysis (also known as resource analysis and energy analysis). This quantitative analytical tool offers the potentiality of optimizing technological chains and production lines, i.e. of determining those production processes with the maximum economic and energetic efficiency. It enables the problem of non-waste technology as a whole to be considered as comprising two interrelated sub-systems: a production sub-system (model) and an environmental sub-system (model).

To measure quantitatively the efficiency of a non-waste technology considered as a possible solution, a general criterion, known as "exergo-economic", can be used. The measure of the utilization of resources contained in the production of thermal and material flows defines what is known as "exerty". The difference between the total energy introduced into the system and the energy going out determines the total loss caused by the irreversibility of the system, i.e. the possibility of the utilization of secondary resources. In order to establish the most profitable means for improvement of production, it is necessary to compare the energy losses in the various elements and then to study the possibility of reducing the losses which influence most substantially the improvement of production and the functioning of its various units. This provides a method for the most effective means of estimating production wastes.

On the other hand, exergetic analysis describes and examines the general orientation of a process, the regularities of the transfer of mass and energy, and also establishes general alternatives for the realization of a process—i.e. effectiveness, productivity, entropy—and is tied to thermodynamic preferable alternatives and capital investments to achieve the minimum cost for a production unit. Thus, the criteria for optimization of non-waste production can be a composition of additive functions which assess energy, equipment and other investments in monetary terms.

TOPIC II. (a) NATIONAL EXPERIENCE AND POLICY

An examination of the papers submitted on this topic has made it clear that there are many different points of view as to how to promote non-waste technology and to what degree it should be promoted. Even though the range of ideas was very wide, the need for a technology that reduces or avoids waste was universally recognized. Thus, even though the various countries demonstrated their unique problems, they all supported the promotion of non-waste technology and agreed on the possibility of discussion of the common themes.

The first step in the promotion of non-waste technology is the identification of opportunities. General techniques are available, one of which is to analyse material and energy balances throughout the system so as to detect points of inefficiency, leakage and waste. Another approach, to be used in combination, is dialogue between informed sources in government, industry and the research community. There will be many cases where the innovations and practices of one industry can be found useful to other industries. There are cases where it was found to be very productive for governments to gather carefully selected and detailed information and data, so as to assist in the analysis.

In the analysis of systems there is always the problem of establishing boundaries. It seems clear that the boundaries should embrace larger rather than smaller areas. Only in this way will it be possible to include all the costs and benefits and thus to demonstrate that the non-waste approach is for the benefit of society.

These problems and opportunities, once identified, should be assigned some indication of importance or priority. Comments were made in some papers about social costs/benefits, and this aspect constitutes a constraint in that these costs and benefits are often not yet quantifiable. Nevertheless, they must not be excluded from consideration because of that.

The papers indicated a wide range of policy for the promotion of non-waste technology. It is clear, for instance, that anti-pollution legislation can have the effect of promoting waste reduction but it is equally clear that most countries recognize a need to go beyond this point. Some countries, for instance, have instituted active programmes dedicated to the conservation of energy.

The government options for influence that are being used range from legal regulations, financial and economic incentives (both positive and negative) to the encouragement of research and development. The choice of options will always be a matter of national selection to suit domestic conditions, but there does seem to be potential benefit in continuing dialogue among countries. The same dialogue would serve the need to discuss and prevent the export of unwanted industrial wastes and/or polluting technology from one country to another. The exchange of experience and insights cannot but be productive in the promotion of "non-waste technology".

TOPIC II. (b) INDUSTRIAL EXPERIENCE

The biosphere is in a state of dynamic equilibrium which human activities and, in particular, the growth of industry, tend to shift. The role of non-waste technology is to reduce appreciably the negative effect of industrialization on the consumption of natural resources and on natural conditions due to noxious emissions, and not to overstep the bounds of biosphere elasticity.

The reports and oral statements presented under Topic II. (b) of the seminar have shown that it is indeed possible to create such a technology. Numerous discussion papers received for this topic provide information on many industrial applications of this technology. It was noted that different methods could be used to eliminate or significantly reduce wastes:

(a) by improving existing technologies: recycling, increasing yields, development of recovery processes, and waste transformation;

(b) by creating new techniques or by radically modifying existing techniques, in order to obtain production processes which produce less wastes and noxious pollutants.

It is clear that the research work necessary to promote non-waste technology has not attained a desirable level. Countries must develop multi-disciplined research in order to improve non-waste technology for all branches of industry. The economic aspects of the rational utilization of raw materials and energy must be tackled simultaneously. Where necessary, this research should be co-ordinated between the various countries by means of bilateral or multilateral agreements and/or exchange of information.

There is an urgent need for research at the micro, as well as macro, level. The criteria for economic evaluation of non-waste production processes are at present insufficient. It is necessary to develop economic research—in particular, the elaboration of economic criteria—for the evaluation of the transformation of wastes and creation of non-waste production processes, taking into account ecological factors.

The difference in the permissible national standards of pollution in the atmosphere, as well as the hydrosphere and lithosphere, poses significant problems. Harmonization of the various national regulations, taking into account the opinion of the World Health Organization, should be sought. In the course of the elaboration of these standards, all technical, economic and ecological aspects should be considered.

The introduction of non-waste technology in industry cannot be accomplished without the active participation of everyone concerned. It is therefore necessary that educational institutions (particularly for technical staff) take practical measures to ensure that their courses take into account the impact on the environment of the technologies which are being taught and that the ideas relative to non-waste technology are propagated. Moreover, it is necessary that, in the course of their education, young people are familiarized with environmental problems, such as the use of natural resources, protection of the countryside, etc.

It would be important for meetings, conferences and seminars to be devoted to relevant problems in well-defined industrial and agricultural sectors.

TOPIC II. (C) CASE STUDIES

Iron and Steel Industry

It was recognized that the iron and steel industry is one of the most polluting sectors with respect to water and air pollution. In addition, it is an important source of solid waste. Nevertheless, efforts already undertaken in all countries have permitted large reductions in the emission of these pollutants. One can cite that with regard to water, a plant with a closed-circuit discharges significantly less waste than a plant with an open-circuit. As regards air, the levels achieved in the emission of dust (red smoke) from steelworks using a Thomas converter and one using oxygen are of the same order. However, important air emissions from coke-making and sinter strands are not as readily controlled. Finally, concerning solid wastes, out of 600 to 800 kg of water per ton of finished steel, only sludge and very fine dust coming from various steelmaking processes, i.e. 60 to 80 kg, are not recycled or used for other purposes.

In spite of the above, a satisfactory and economically viable solution has not yet been found for certain important pollutants, such as sulphur oxides emitted into the air from numerous types of plants (coke, sintering, steel) and chemically polluted waste water (coke, cold rolling mill).

New technologies, such as direct reduction, must still be studied actively in order to reduce the emission of pollutants from the iron and steel industry which, after all the efforts undertaken, still generate considerable pollutants.

Increasing the yields of steel mills (in terms of the ratio of steel shipped to steel poured) can have a major effect in reducing steel mill pollutants. Continuous casting is a major development here, and the control of steel mill processes by computer will help in the future.

Continued emphasis should be placed on scrap recycling.

The analyses preceding adoption of non-waste technology for the steel industry must consider all factors, including energy conservation and increasing yields, as well as pollution control. Particular emphasis should be placed on the very significant area of waste heat recovery from steel-making processes.

The Pulp and Paper Industry

The traditional technologies to transform wood and vegetable fibres into pulp and paper generate various kinds of waste:

wood and bark waste (while cutting trees);

organic and chemical material (in the production of wood-pulp);

mineral and fibrous material (in the production of paper).

For ten years, great progress has been made to reduce this waste by:

trying to utilize the whole tree;

using closed circuits in pulp and paper production;

utilization of oxygen instead of chlorine as a bleaching agent;

systems of recovery of wood fibres in paper production.

But further technological innovations are required to:

improve efficiency in producing wood-pulp and trying to reach yields of 75 to 80 per cent, or higher;

introduce new pulping solvents which can be easily recoverable;

obtain by-products of cellulose in a state which would permit them to be used.

These objectives can only be attained through considerable research and development efforts and by continued association of the paper industry with the mechanical and chemical industries.

Packaging

The non-waste technology of a package type must be examined in all its aspects before definitive conclusions may be drawn. These aspects include the stages of design, production, distribution, transport, consumption, recycling and waste management and environmental impact.

At the present time, sufficient data are not yet available concerning environmental pollution and resource consumption, thus a judgement of a package or a product nowadays must be made in the knowledge that some factors are not known. Moreover, known factors are interpreted in different ways, due to the lack of an international unified system for evaluation.

An examination of the physical impacts alone of alternative product systems will not necessarily determine the "optimum". One option may require the use of more of one resource, while others may use more of a different resource. In order to determine the "optimum" (with respect to some objective function which itself must be agreed upon and might vary from country to country) it is necessary to make trade-offs based on social and economic impacts and value judgements.

The trade-off analysis should also take into account the possibility of varying the specifications of the function served by the package. For example, cans and re-usable bottles both deliver liquids, but one provides a different level of convenience to the consumer.

Wherever possible, changes should be made to packaging that are less harmful to the environment.

Tyres

With regard to non-waste technology for tyres, industrial research continues for the development, without prejudice to safety, of longer-lived tyres (160,000 km per tyre). The main problem of durability seems to be that often industry does not have sufficient incentive to push this development more rapidly; nor does it have sufficient incentive to increase tyre durability in general in order to facilitate retreading. As a result, a method has to be found to induce consumers to purchase longer-lived tyres and to retread them. The latter would significantly reduce the number of waste tyres thrown indiscriminately into the environment. The immediate needs for research for longer life are thus in this area of incentives.

In addition, more research is needed on methods to re-utilize tyres in other applications once they have been irrevocably discarded. There does appear promise (both technically and economically) in the technologies of rubberized asphalt. Energy recovery from tyres will generally be a desirable strategy under varying local conditions and is, in general, also an economic short-term strategy until more beneficial uses can be developed.

TOPIC III. COST-BENEFIT ASPECTS OF NON-WASTE TECHNOLOGY

There are many benefits associated with achieving new technologies, developing new products, eliminating old products and production techniques, so as to reduce the accumulation of waste that pollutes the environment and diminishes the stock of non-renewable resources. The benefit/cost framework is a useful method for comparing these benefits with any additional costs that moving in this direction entails. All nations have limited budgets for environmental expenditure. Benefit/cost analysis, along with other evaluation methods, can be used to help select those non-waste technologies that should be given high priority, and thereby assist in making the environment as clean as possible.

Decisions concerning the adoption of a particular non-waste technology are made at various levels of decision-making in different countries. There are two important factors to consider. First, some governments will impose regulations, taxes, subsidies or policies that will internalize social costs and concerns so as to reduce waste. The benefit/cost analysis for the plant manager is restricted to the matters that will implement this new policy along with other cost minimization and output maximization objectives. Second, for other governments, decisions are made at a level that is more centralized. These are macro in nature. At this stage, benefits and costs must also be assessed to formulate policy at this more centralized level. The analysis is more difficult because more factors must be considered and each factor may have greater variability. The important conclusion is that the method of benefit/cost analysis is highly dependent upon the specification of the objectives and the definition of constraints.

Often benefits are not easily quantifiable. Two methods are suggested for dealing with this problem. First, the problem can often be specified in terms of the costs avoided by reducing wastes. These can often be more readily quantified than benefits. Second, sometimes neither the benefits of reducing wastes nor the costs saved from reducing wastes can be quantified in monetary terms. An alternative approach is recommended. The quantifiable costs can be substracted from the quantifiable benefits and this amount can be compared with the remaining unquantified social costs and/or benefits to the environment. Both approaches are equivalent to calculating "shadow prices" or, to state it less elegantly, to determining how significant the unquantifiable benefits must be in order to justify a specific decision. Benefit/cost analysis is an important tool, but decisionmakers must still exercise their human judgement to make the important choice that the future demands.

There is an additional matter that must be understood. There are sometimes several ways of reducing pollution. The benefits of each method may exceed their respective costs. But the appropriate method is to select the approach which can achieve the objective in the lowest-cost manner, in order to honour the true spirit of non-waste technology.

If a nation's immediate goal is pollution reduction, waste-treatment versus non-waste technology may be seen as competing but in reality as complementary alternatives. At present, this question needs further investigation and research. It must be noted that "non-waste technology" reduces pollution and conserves non-renewable resources, while waste-management merely reduces the problems of pollution. Over a period of time, it is likely that waste-treatment will become increasingly costly and that non-waste technology will become less costly. This reality must start to be included in present decisions.

TOPIC IV. WAYS AND MEANS OF IMPLEMENTING NON-WASTE TECHNOLOGY

To help ensure conservation of resources and minimization of waste, it would be desirable and practicable to impose the following economic incentives:

The maintenance of an appropriate valuation for raw materials which are rapidly being depleted, thereby stimulating economical use of such materials. Prices for secondary raw materials should be such as to stimulate firms to use them.

Fees for the pollution of natural resources such as air, soil and water. The amount of such fees should depend upon regional and quantitative conditions.

Sanctions imposed on firms when laws, in the form of regulations for keeping the biosphere clean, are not observed.

Financial aid (lower taxes, depreciation) to firms which observe environmental protection. Additional taxes for products and plants which pollute the environment more than is allowed by regulations.

Governmental financial support and credits for environmentally appropriate research, development and demonstration facilities.

The long-term programmes for science and technology have to include further projects on environmental research and development, on better use of resources, on optimization of processes, and on maintenance of a clean biosphere.

Within the concept of non-waste technology, the following should be identified:

The optimization of industrial production processes. To solve the problem of minimizing the use of materials and energy and preventing pollution caused by industrial production processes, special attention should be given to the design of new production processes. If such modifications are not feasible, existing processes should aim at a decrease in resource consumption and in abatement of their environmental pollution.

The optimum design of final products. To ensure optimal design by individual producers, special design methods should be defined in order to minimize the use of raw materials and energy and minimize environmental pollution during the total lifetime of the products. Total lifetime not only includes the manufacture and normal use of products but also their maintenance and repair and their re-use and/or recycling in order to arrive at non-waste design for products.

The development of national policies or careful and integrated planning of all aspects of industrial activities (including mining of ores, separation of basic materials, manufacture of products, recycling or disposal of wastes, etc.) is required keeping in mind the overall concept of non-waste technology. It is important that the concept and principles of non-waste technology should be involved in new technological procedures from the very outset, i.e. at the conceptual research and development stages.

Government incentives or regulations should encourage the prevention or the re-use of unavoidable wastes, through recycling, recovery and treatment.

The use of wastes is not yet satisfactory because of a lack of:

- efficient organization of collection and disposal of wastes;
- existence of a sufficient demand for wastes susceptible of being secondary raw materials;
- existence of technical methods and facilities for waste treatment.

A commonly stated, but incorrect, opinion is that the use of non-waste technology is too expensive in general, that there is a lack of good solutions for such technologies, and that the costs for research and development are too high. This prejudice must be overcome by comprehensive information about the general significance of non-waste technology, and possibilities for its use. Every engineer, scientist, economist, socio-economist and the general public must be encouraged to help. The results of major international conferences on the subject must be publicized.

There is still a very significant lack of knowledge concerning the concepts, potentialities and technical aspects of non-waste technology. Therefore, more attention must be given to education and training in this field. Executives, managers, students and the public must be educated as to the long-term desirability of achieving non-waste technology. Their basic and special knowledge will help the nations of the world to move in the direction of that goal.

The far-reaching and difficult tasks of the development of non-waste technologies and their implementation need further bilateral and multilateral co-operation, in accordance with the recommendations of the Final Act of the Conference on Security and Co-operation in Europe.

TOPIC V. METHODOLOGICAL AND STRATEGIC ASPECTS OF NON-WASTE TECHNOLOGY

In the short term, non-waste technology offers an economic alternative to conventional technology. In the mid-term and long-term, however, it will appear to be a necessity that is consistent with the exigencies concerning the protection of nature and the environment and with the economic choices.

The elements relating to a strategy of non-waste production are the following:

Objectives: the production of wastes, especially wastes noxious to the environment, must be reduced in all cycles of production and transformation which relate to human activities as a whole—and, in particular, to agricultural, industrial and domestic activities;

Motivations: the common good is the basis of this objective. More specifically, its aim is to improve the quality of life for mankind in his environment;

Methods: after complete evaluation of all factors which enter into the cycle of production or transformation—whether technical, economic, ecological or social—one can proceed by optimization of the various sub-systems involved;

Means: action is possible at the level of government policy, in particular by means of laws and regulations. In any case, action must be taken at the level of the economy by setting up statutory systems (taxes) or incentives (fees, state participation in financing, etc.). Research will have to be the first focus of state intervention, the objective being the selection and promotion of research activities. When the concepts, principles and a number of specific examples of non-waste technology have been elaborated, the state should promote its development, keeping in mind the need not to perturb excessively the economy and the industrial structures.

Measuring standards: it will be useful to indicate the quantities of raw materials and energy, together with the amount of encroachment on the environment, required for the manufacture of each product so that objective measuring standards can be developed for products.

Practical methods of implementation: it is necessary to have long-term programmes of research and development because tomorrow's technology will be the reflection of today's efforts. It is necessary to approach the technical realization of non-waste production processes at the regional level and, particularly in the industrial field, at the level of the whole of the sector considered. The ecological unity (ecosystem) must guide the establishment of geographically coherent contours for a given sub-system. There is no *a priori* antagonism between this approach and a centralized management of territory. The two can and must be complementary.

It is clear that the development of non-waste technology requires close co-operation between all nations, all being equally concerned by this issue. This field is particularly suitable for technology transfers, as well as for bilateral and multilateral co-operation. In this respect, appropriate international organizations should play a useful, indeed, essential role in collecting and disseminating data concerning experience and achievements in the field of non-waste technology. Indeed, in spite of the efforts made to date, the field is still in an embryonic stage.

The strategy that should be developed and promoted must consist in choosing the most efficient methods and means of developing and implementing non-waste technology. It is on these criteria of choice (from among the various possibilities that are offered) that one must initiate an objective evaluation.

RECOMMENDATIONS

Preamble

As a result of the papers which were developed from this seminar, and of the discussions which took place, the following recommendations have been formulated and approved by the participants. These recommendations are directed (a) to the ECE Senior Advisers on Environmental Problems for possible inclusion in the future programme of work of ECE or (b) through the Senior Advisers, to the ECE governments themselves for their consideration and implementation as appropriate.

Recommendations to the ECE Senior Advisers on Environmental Problems (SAEP)

Under the catalytic role and support of UNEP, the SAEP should undertake: (a) to assemble a compendium of all available knowledge (or references to knowledge) concerning low- or non-waste technologies which have been or are being developed in the ECE region. This compendium should be updated on a periodic basis and be made available through ECE to all ECE countries, and through UNEP to other interested countries of the world; (b) to stimulate the development of various quantitative, semi-quantitative and qualitative methodologies which could facilitate a comparative analysis and evaluation of various competing technologies—taking into account economic, technological and social (including environmental) considerations. These methodologies include: benefit - cost analysis, systems analysis, resource analysis, technological assessment, environmental impact assessment, and input-output analysis; (c) to stimulate the exchange of information among member governments on the type and magnitude of incentives utilized to encourage non-waste technology and on the methods used in reaching decisions whether (or not) to proceed with the development of non-waste technology for various types of industries.

It is recommended that the Senior Advisers on Environmental Problems envisage wide consideration of the problems of non-waste technology in the chemical and petro-chemical industries and possibilities for the creation of energo-technological complexes with no harmful discharges into the environment.

It is recommended that the Senior Advisers on Environmental Problems encourage the study of:

1. methods and techniques for the identification of waste flows;
2. methods and techniques for dealing with non-quantifiable factors in problem analysis;
3. relative effectiveness of policy instruments as applied by ECE countries;
4. development of innovative policy instruments;
5. resource consumption and environmental pollution associated with packaging, performance of an individual package. This could also be applicable to the design of products;
6. incentives which will lead to uses of packaging consistent with the concepts of non-waste technology;
7. natural resources management based on long-term perspective planning aimed at the creation of non-waste technology.

The United Nations, including its specialized agencies, is one of the largest international organizations in the world and consumes huge quantities of paper in the preparation of its many documents. It is recommended that, as part of its desire to reduce the consumption of natural resources, the ECE should explore the possibility of using, wherever appropriate, only recycled paper or paper with a high content of recycled fibre. In this way the ECE—and, in the footsteps of ECE, the United Nations as a whole—could set an example to the world of its broad concern for problems of the environment.

Recommendations to ECE Governments

It is recommended that governments support the promotion of non-waste technology by:

1. Actively seeking out opportunities to avoid or prevent waste, by collecting information on the flows of materials in their economies, by analysing the data and by encouraging dialogue between interested and competent sources of information within governments, industry and the research community.
2. Assigning priorities to these problems using as broad and inclusive a system as possible, including both quantifiable and non-quantifiable factors.
3. Implementing policy that conserves resources, avoids waste, stops pollution and encourages research and development of appropriate technology.

It is recommended that governments:

1. Develop multi-discipline research, in order to improve non-waste technology for all branches of industry. The economic aspects of the rational use of raw materials and energy must be tackled simultaneously. Where necessary, this research should be co-ordinated among the various countries by means of bilateral or multilateral agreements.
2. Pursue economic research—in particular, the elaboration of economic criteria—for the evaluation of the transformation of wastes and the creation of non-waste production processes, taking into account ecological factors.
3. Seek to harmonize national standards of pollution in the atmosphere, the hydrosphere and the lithosphere.
4. Encourage their educational institutes to include in their curricula courses on the principles and implementation of non-waste technology.

It is recommended that governments co-ordinate economic and legislative activities in the sphere of the rational use of resources. These could include in particular:

the maintenance of an appropriate valuation for raw materials which are rapidly being depleted and low prices for secondary raw materials and products made from them;

fees for the pollution of natural resources;

sanctions when the regulations are not observed;

financial aid (lower taxes, depreciation) to firms which observe environmental protection regulations;

governmental financial support and credits for research in non-waste technology; and

general financial support to activities aimed at improving the environment.

It is recommended that governments stimulate education and training in the field of non-waste technology, both in production and consumption sectors, as well as the dissemination of relevant information to the public.

It is recommended that closer international co-operation be developed—through ECE, UNEP, UNESCO or other appropriate international bodies—regarding the exchange of knowledge and experience in standardization as well as on education and training of personnel, relevant to non-waste technology. Assistance given by governments to other countries should seek to ensure that, to the extent appropriate, the concepts of non-waste technology are applied under such assistance.

With regard to the industries considered as case studies for the seminar, it is recommended that in:

Iron and Steel Industry

(a) there be further research and development in closed-circuit systems;

(b) more emphasis be placed on scrap recycling;

(c) all factors, including energy conservation and increasing yields, as well as pollution control, be considered.

Pulp and Paper Industry

(a) more advanced techniques be developed for the separation of the various types of waste paper, and on the methods of recycling of waste paper;

(b) further research and development be undertaken in the paper industry in association with the mechanical and chemical industries, with the following objectives:

to increase the basic yield in the hope of achieving a minimum objective of 75 to 80 per cent;

to make use of new solvents which are easily recovered;

to obtain by-products of cellulose in a state which would permit their use.

Tyres

(a) A continuing effort be maintained to disseminate research results for member countries' efforts to resolve scrap tyre problems, with respect to the solution of technological problems, and to develop effective incentive mechanism to encourage adoption of tyre waste-reducing practices;

(b) efforts be continued to test and develop candidate technologies for tyre re-use in whatever application.

List of Contributors

David Berg, Adviser for Energy Conservation, Environmental Protection Agency, Washington, D.C. 20460, USA

R. Berry, Department of Industry, Room 2709, Millbank Tower, London SW1P 4QU, United Kingdom

F. Bertrand, Centre Technique du Génie Rural des Eaux et des Forêts, Ministère de l'Agriculture, Parc de Tourvoie, 92160, Antony, France

Tibor Blickle, Research Institute for Technical Chemistry of the Hungarian Academy of Sciences, Schönherz Z.u. 10, Veszprém, Hungary

A. G. Buekens, Professeur, Vrije Universiteit, Pleinlaan 2, 1050 Bruxelles, Belgique

Czeslaw Cala, Ministry of Science, Higher Education and Technology, ul. Miodowa 6/8, Warsaw, Poland

P. Chassande, Secrétariat d'Etat à l'Environnement, Ministère de la Qualité de la Vie, 14, Bd. du Général Leclerc, 92521 Neuilly-sur-Seine, France

Charles J. Cicchetti, University of Wisconsin, 1930 Regent, Madison, Wisconsin, USA

Jerome F. Collins, Chief, Materials Optimisation Branch, Division of Industrial Energy Conservation, US Energy Research and Development Administration, Washington, D.C., USA

W. David Conn, Assistant Professor of Environmental Planning, University of California, Los Angeles, California 90024, USA

V. Cziglina, Collieries of Tatabanya, Vértanuk tere 4, 2800 Tatabanya, Hungary

J. C. Deloy, Editor-in-chief, "Nuisances et Environnement", 40, rue du Colisée, 75008 Paris, France

L. Dzsida, "Vidus" Enterprise, Tóth-Buesoki u.6, 2800 Tatabanya, Hungary

B. Finnern, Fa. Degussa, Postfach 602, 6450 Hanau 1, Federal Republic of Germany

M. Fischer, Institut für Systemtechnik und Innovationsforschung Fhg, Breslauer Str. 48, 75 Karlsruhe 1, Federal Republic of Germany

W. P. Fornerod, Director, Institute TNO for Packaging Research, Landsteinerbocht 12, Delft, Netherlands

J. Francia, Head of section, State Committee for Technical Development, Akademia u.17, 1054 Budapest, Hungary

Haynes C. Goddard, Solid and Hazardous Waste Research Division, Environmental Research Center, Environmental Protection Agency, University of Cincinnati, Cincinnati, Ohio 45268, USA

P. Grau, Professor, Institute of Chemical Technology, Suchbatarova 5, 16628 Prague 6, Czechoslovakia

B. Gromov, State Committee for Science and Technology, Moscow, K-9, Gorky St., 11, USSR

Johan Gullichsen, Arhippainen, Gullichsen & Co., Kasarminkatu 46-48 B. 00130 Helsinki 13, Finland

Stefan Gustkowicz, Committee of Science and Technology, Warsaw, Poland

W. Gutt, Department of the Environment, Building Research Establishment, Building Research Station, Garston, Watford, WD2 7JR, UK

Vollmar Hallensleben, Heye Glas-Fabrik, 4962 Obernkirchen, FRG

Seppo Härkki, Outokumpu Oy, Töölönkatu, 4, 00100 Helsinki 10, Finland

M. Haucke, Verein Deutscher, Eisenhüttenleute, Breite Str. 27, 4000 Düsseldorf, FRG

Y. Hellot, Délégué Arrondissement Minéralogique, Ministère de l'Industrie et de la Recherche, Paris, France

M. Huchette, Etablissements Roquette Frères, 62136 Lestrem, France

R. Huissoud, Conseil National du Patronat Français, 31, Avenue Pierre de Serbie, 75016 Paris, France

E. Jochem, Fraunhoffer-Gesellschaft, Hirschstr. 17, 7500 Karlsruhe 1, FRG

V. V. Kafarov, Corresponding member, USSR Academy of Sciences, Mendeleev Institute of Chemical Technology, Miusskaia Place 9, Moscow A-47, USSR

Dénes Kalló, Head of Department for Hydrocarbon Catalysis, Central Research Institute for Chemistry, Academy of Sciences, Pusztaszeri ut 59-67, 1025 Budapest, Hungary

Rudolf Kauders, Wiesinger strasse 6/9, A-1010 Vienna, Austria

I. Kievsky, Ministry of Chemical Industry, 20 Kirov St., Moscow, USSR

W. Kluge, Institute of Energetics, Torgauer Str. 114, Leipzig, GDR

K. E. Kulander, Sveriges Industriförbund, Storgatan 19, PO Box 5501, 11485 Stockholm, Sweden

C. Lembit Kusik, Arthur D. Little Inc., Acorn Park, Cambridge, Mass. 02140, USA

B. Laskorin, State Committee for Science and Technology, Gorky St., 11, Moscow, K-9, USSR

M. Laurent, France

D. Lefort, Centre de Recherches de Pont-à-Mousson, Maidières, B.P. no. 28, 54700 Pont-à-Mousson, France

P. Lieben, Environment Directorate, OECD, 2, rue André-Pascal, 75016 Paris, France

L.-G. Lindfors, Sveriges Industriförbund, Storgatan 19, PO Box 5501, 11485 Stockholm, Sweden

Joseph T. Ling, Vice-President, 3 M Company, 9 av Bush Avenue, St Paul, Minnesota, USA

E. Lohrden, Sveriges Industriförbund, Storgatan 19, PO Box 5501, 11485 Stockholm, Sweden

T. V. Long, Resource Analysis Group, University of Chicago, 5735 S. Ellis, Chicago, Illinois 60637, USA

Miklos Machacs, Ministry of Heavy Industries, PO Box 35, 1363 Budapest, Hungary

Pentti Malaska, Professor, Turku School of Economics, Rehtorinpellontie 5, 20500 Turku 50, Finland

B. Marechal, Isorel, Tour Roussel-Nobel, 3, Avenue du Général de Gaulle, 92800 Puteaux, France

László Markó, Professor of Organic Chemistry, University of Chemical Engineering, PO Box 28, 8201 Vészprém, Hungary

Yves Maystre, Director, Institute of Environmental Sciences, Federal Polytechnic of Lausanne, Av. Chandieu 3, 1006 Lausanne, Switzerland

A. J. McIntyre, Policy Analyst, Environment Canada, Ottawa, Ontario, Canada K1A OH3

Z. Meleg, Hungarian National Council for Environmental Protection, PO Box 613, 1370 Budapest, Hungary

P. Monzie, Chef de département "pâtes, celluloses, polymères", Centre technique du Papier, B.P. no. 175 Centre de Tri, 38042 Grenoble Cedex, France

D. Moyen, Assistant Director, Institut National de la Recherche sur la Sécurité, 30, rue Olivier Noyer, 75014 Paris, France

Martin Neddens, Scientific Adviser, Rat von Sachverstandigen für Umweltfragen, Postfach 5528, D-62 Wiesbaden, FRG

David W. Nunn, Research Fellow, Chr. Michelsen Institute, Nygaardsgt 114, N-5000 Bergen, Norway

J. Orlich, Federal Republic of Germany

Udo Ousko-Oberhoffer, Wiesinger strasse 6/9, A-1010, Vienna, Austria

J. Picard, Director, Agence Financière de Bassin Moire-Bretagne, Rue de Buffon, 45018 Orléans Cedex, France

Arthur H. Purcell, Director of Research, T.I.P. Inc., 217-1346 Connecticut Avenue NW, DuPont Circle Building, Washington D.C. 20036, USA

W. Howard Rapson, University of Toronto, Toronto, Canada

Douglas W. Reeve, ERCO Envirotech Ltd., Toronto, Canada

Robert Reid, Vice-President, Energy and Environmental Analysis Inc., Suite 1213, 1701 N. Fort Myer Drive, Arlington, Virginia 22201, USA

Rudolf Roth, Chemist, Mannesmann AG Hüttenwerk, D-4100 Duisburg, FRG

M. G. Royston, Centre d'Etudes Industrielles, Chemin de Conches, 4, Conches, Geneva, Switzerland

Jali M. Ruuskanen, Finnish National Fund for Research and Development (SITRA), Udenmaank 16 B, Helsinki 12, Finland

Jean-François Saglio, Directeur de la Prévention des Pollutions et Nuisances, 14, Bld du Général Leclerc, F-92521 Neuilly-s/Seine, France

Hussein Saleh, Senior Consultant, Advanced Concepts, Environment Canada, Ottawa, Ontario, K1A OH3, Canada

J. D. Schmitt-Tegge, Federal Republic of Germany

M. Schubert, Technische Universität, Mommsenstr. 13, D-8027 Dresden, GDR

V. Senin, State Committee for Science and Technology, Gorky St., 11, Moscow, K-9, USSR

Fred Lee Smith Jr., Policy Analyst, Environmental Protection Agency, AW-463, RRD-EPA, Washington, D.C. 20460, USA

Walter Sprenger, Schmalbach-Lubeca GmbH, Braunschweig, FRG

A. Takats, Hungarian National Council for Environment Protection, PO Box 613, 1370 Budapest, Hungary

J. Talabér, Directeur-général, Central Research and Designing Institute for Silicate Industry, H-1300 Budapest, Hungary

W. Theobald, Verein Deutscher, Eisenhüttenleute, Breite Str. 27, 4000 Düsseldorf, FRG

N. Torocheshnikov, Mendeleev Institute of Chemical Technology, Miusskaja Square 9, Moskow A-45, USSR

A. W. F. Van Alphen, Ministry of Health and Environmental Protection, Dr. Reyersstraat 10, Leidschendam, Netherlands

H. H. van den Kroonenberg, Twente University of Technology, PO Box 217, Enschede, Netherlands

I. Van Vaerenbergh, Prime Minister's Office, Scientific Policy Planning, Rue de la Science 8, B-1010 Bruxelles, Belgium

Matti Vehkalahti, Ministry of the Interior, Pääskylänrinne 8, 00500 Helsinki 50, Finland

Ilpo Vettenranta, Enso-Gutzeit Osakeyhtiö, Paper Division, 55800 Imatra 80, Finland

Bengt Westerholm, Upo Osakeyhtiö, Metal Division, Metal Finishing Machines, 15100 Lahti, 10, Finland

J. Wieckowski, Professeur, Directeur de l'Institut de Gestion à l'Université de Varsovie, 75/29 rue Koszykowa, Varsovie, Poland

V. N. Yevstratov, Ministry of Chemical Industry, 20 Kirov St., Moscow, USSR

A. Zygankov, State Committee for Science and Technology, Gorky St., 11, Moscow, K-9, USSR

Concepts and Principles of Non-waste Technology

Introductory Report

V.V. Kafarov (Rapporteur)

Corresponding member of the Academy of Sciences of the USSR

A. SOCIAL ASPECTS OF THE PROBLEM OF NON-WASTE TECHNOLOGY

1. It is well known that the problem of non-waste technology is of great social importance both for the level of the national economy and the creation of industrial potential, and also for the related task of ensuring the optimum ecological interaction between a production process and the environment. Scientific and technological progress is unthinkable without taking into account its effect on the environment and the attendant social consequences. Consequently, new technological processes can be developed only by applying new principles for the creation of technological systems which do not involve noxious emissions into the atmosphere. The elaboration of general concepts for the creation of non-waste technology therefore constitutes a highly important contemporary social problem, to which urgent attention must be given. Otherwise, apart from the obvious impairment of the ecological balance and the adverse consequences for human health, the effect on the environment of a rapid growth of industrial production may be critical even for industry itself and may limit its further development.

2. The need for the renewal and rational use of natural resources has already reached proportions which transcend the limits of the development of individual countries, and raise vast new problems for international co-operation in the conservation of the environment. The development of non-waste technology is therefore a long-term endeavour of the greatest importance, whose realization will bring about positive changes in all sectors of industry and will help to raise the living standards of the population.

3. In the following paragraphs, we shall mention some concepts which are essential for the further consideration of the problem of creating non-waste technology production processes.

4. *Production waste:* the residuals of raw materials, other materials and semi-manufactures which are formed in the production process, but have partly or wholly lost their qualities and do not conform to the relevant standards (technical specifications). After preliminary processing, and sometimes without such processing (for example, when used in the manufacture of consumer goods), these residuals may be used in the field of production or consumption.

5. *Consumption waste:* various products and substances which have been used but which can economically be recovered. These items are suitable for use as basic raw materials in production. Some examples are completely worn-out and non-operational machines, usable glass, rubber or plastics products, spent reagents, catalysts, etc. (producers' consumption waste), or worn-out household articles and articles of personal consumption (domestic consumption waste).

6. *By-products:* these are produced in the physical and chemical processing of raw materials along with the main products, but not the prime object of the production process. They are mostly marketable; there are standards (technical specifications) for them, they have a fixed price, and their production is planned. As a rule, by-products may be used as finished products. Generally speaking, they consist of components of the raw material which are not used in the production process in question, or of products formed as a result of chemical transformations. By-products obtained in the extraction or enrichment of basic raw materials are usually called associated products (for example, associated gas).

7. *Secondary material resources.* "Secondary material resources" (SMRs) is the term used to describe all types of waste from production and consumption which can be used as a basic or auxiliary material for production. Depending on the possibilities of using them, SMRs may be divided into actual and potential resources. Actual SMRs are resources for which efficient methods of use and processing capacity have already been created, and market outlets ensured. Potential SMRs are all types of SMRs other than actual SMRs. Secondary material resources sometimes also include by-products and associated products, which are at present insufficiently used and constitute a potential reserve of material resources for industry.

8. The chemical industry makes extensive use of wastes and by-products. Many of its production processes have been developed as a result of the processing of wastes, whose use as secondary material resources helps to solve a number of important economic problems, such as the problem of saving basic raw materials, the prevention of water, soil and atmospheric pollution, the problem of increasing production of components and articles made from man-made materials, and the production of new types of goods for popular consumption.

9. *Non-waste technology:* the organization of production in such a way that production wastes are reduced to a minimum and completely transformed into secondary material resources. Non-waste technology calls for the establishment of optimum flow sheets with closed material and energy flows.

B. SYSTEMS ANALYSIS AS A BASIS FOR THE CREATION OF NON-WASTE TECHNOLOGY

10. The theoretical basis for the creation of non-waste production processes is production systems analysis, including analysis of the interaction between chemical and technological production systems (CTS) and the environment.

11. The structure of the system consists of the channels of communication between the production process and the environment, including:

(1) the development of closed-circuit water-supply systems, in which highly purified waste waters are re-used for industrial and general needs, including ideally also the drinking water supply;

(2) intake of air and its return to the atmosphere;

(3) closed energy links—i.e. the use of internal energy resources with minimum consumption of energy from outside, in other words, the development of production processes with maximum energy efficiency.

12. In order to permit a comprehensive review of all these questions, the entire global production-environment system is first broken down into sub-systems.

C. THE ESSENCE OF SYSTEMS ANALYSIS

13. The principles for the creation of non-waste technology call for the examination of each chemical, petrochemical, biochemical and other enterprise as a sub-system of the global production - environment system which in turn consists, horizontally, of three interacting blocks:

(1) the raw material preparation block;

(2) the chemical (biochemical) transformation block;

(3) the main product separation block

and, vertically, of three hierarchical levels. At the lowest level in the hierarchy, there are the technological operations typical of any chemical, biochemical or petrochemical production process, e.g. the treatment of liquids and gases, heat transfer between products to be treated and the heating agents, chemical transformations in reactors, and the separation processes—absorption, rectification, ion exchange, etc.

14. At the second level of the hierarchy, the aforementioned processes are combined into units or blocks. Already, the individual processes are interacting and exerting an effect on each other; and between the individual processes one can introduce recycling which will make it possible to convert more of the recovered products into end products, will open the way to a fuller use of the materials processed and will reduce waste to a minimum. It is at this stage that the question of the optimum organization of flows arises.

15. At the upper level in the hierarchy, the units and blocks are combined into a factory or combine. The problems which arise at this stage are: operational control of the production process as a whole, the organization of production cycles, the planning of raw material supplies and of water and energy resources, and the manufacture of finished products and semi-finished products.

16. Existing production processes, which were not developed on the basis of the systems analysis principle, were usually based on the parallel or consecutive principle. This resulted in duplication of the basic equipment and consequently in a lengthening of the main communications; it made it more difficult to achieve full use of the raw materials and energy, and led to increased emissions from each individual process. The over-all economic efficiency of the production process was reduced, and environmental pollution was increased. From the technological standpoint, the lengthening of communications leads to losses of raw material and energy, makes it impossible to create closed energy systems, increases the difficulty of recycling, leads to increased emissions into the atmosphere and reduces the general reliability of the flow sheet.

17. Systems analysis of chemical production processes has led to a tendency towards the development of high unit capacities.

18. As an example, mention may be made of high-capacity ammonia plants. In a 1360 t/day ammonia unit, organized on a closed energy technology system,

water consumption was reduced by 90 per cent and there was a considerable reduction in the volume of waste water, particularly as a result of using air instead of water cooling. Consumption of outside energy was reduced from 1200-1600 kWh/t NH_3 to 50-100 kWh/t NH_3, thus obviating the need to construct a 70,000 kW power plant. This, in turn, reduced pollution of the atmosphere by flue gases. The energy-saving, in terms of natural gas burnt at thermal power stations, is 200 m^3 CH_4 per ton of NH_3. The reduction in atmospheric pollution is attributable to the fact that, apart from the drop in the load from the power plant, gas losses are reduced from 5 per cent under the conventional system to 3.5 per cent in the case of large-capacity units. The cost of the ammonia obtained by the closed energy system is 40-45 roubles/t NH_3, i.e. roughly three times less than by the conventional method.

19. The most important aspect of systems analysis in the consideration of non-waste technology is the compilation of material and energy balances for the production process. This is particularly important since, by using mathematical modelling at the design stage, it is possible to make an objective assessment of, and to exclude, potential emissions in the production process. This has been clearly demonstrated in the USSR in a study of emissions at chemical plants. Attempts to assess emissions on the basis of measurements on models or laboratory analyses have shown that this approach is altogether inadequate. Frequently, the total volume of pollutants turns out to be some ten times less than the quantity of raw material which has disappeared without trace. Ultimately, the conclusion was reached that it was essential to use an objective criterion, i.e. to establish actual material balances for all operational workshops and production processes, with subsequent analysis of the data obtained.

20. The total volume of pollutants discharged by plants (in the form of effluent, gas emissions and solid wastes) was defined as the difference between the amount of raw material used and the products obtained. Laboratory analyses and measurements with models were used solely to obtain a more accurate picture of the distribution of pollution among each of the channels where the losses occurred. The absolute figures obtained for the amount of emissions were compared, in the light of the existing situation, with the values calculated for the maximum emissions permitted at the given point without violating the established standards (emission limit). The final stage of the exercise was to devise a set of measures to eliminate any emissions over and above the limit.

21. Modern methods of compiling material and energy balances, using computers, have already been instrumental in reducing emissions to a level ensuring overall observance of the maximum permissible concentration limits.

22. A program has been devised for the automatic calculation of material and thermal balances and for optimizing chemical and technological systems (CTS), with the object of minimizing waste.

23. By means of this program, it is possible:

(1) to analyse CTS, i.e. to study the properties and operational efficiency of a system, depending on the structure of the technological links between the various elements and subsystems, and also depending on the values of the design and technological parameters of the system and the parameters of the technological régimes of the various elements.

(2) to synthesize a CTS, i.e. to choose the structure of the technological links, and the values of the parameters of the system and of the

technological régimes of the various elements, on the basis of the properties and efficiency indicators of a system which has the optimum values from the point of view of minimizing waste.

(3) to optimize a CTS. One method of optimizing integrated production processes is to co-ordinate the local objectives of the subsystems by seeking an optimum structure for the interrelation of the sub-systems in regard to material and energy flows. This problem is solved by structural optimization methods. Each sub-system i ($i = 1,2, \ldots, N$) producing products and utilizable energy from chemical transformations is capable of transforming the initial material or energy flows $\vec{x}_i$ into terminal flows $\vec{y}_i$.

$$\vec{y}_{ij} = \alpha ij(\vec{x}_{ik}) \qquad \begin{array}{l} i = 1,2, \ldots, N \\ j = 1,2, \ldots, N_i \\ k = 1,2, \ldots, n_i \end{array}$$

where $\vec{y}_{ij}$ is the jth output flow of the ith sub-systems,

$\vec{x}_{ik}$ is the kth input flow of the ith sub-systems,

N_i, n_i are the number of output and input flows of the sub-systems, respectively.

24. The problem of finding the optimum structure for the chemical and technological system (CTS) of a chemical production process may be formulated as follows: in a given set of CTS {M}, capable of effecting the transformation

$$y = \phi(\vec{x})$$

and satisfying the limitations imposed on a chemical and technological system by the "external mean" in regard to the quantity and composition of the raw material

$$\vec{x} \leqslant \vec{x}^*$$

and also in regard to the required quantity of the product to be manufactured and the health standards to be complied with:

$$\vec{y} \geqslant \vec{y}^*$$

$$\vec{y} \text{ health standard} \leqslant \vec{y}^* \text{health standard},$$

find the {M} for which the extreme values of the local economic criteria of the individual sub-systems are matched.

25. The use of the structural optimization method is based on a general description of the structure of the CTS using structural parameters {K} as follows:

$$\vec{x}_{pl} = \sum_{i=1}^{N} \sum_{j=1}^{N_i} K_{ij}^{pl} \vec{y}_{ij} + \sum_{g=1}^{N_v} K_{og}^{pl} \vec{y}_{og}$$

where K_{ij}^{pl} are the structural parameters characterizing the proportion of the jth output flow of the ith sub-system which is fed into the ith input of the pth sub-system;

K_{og}^{pl} are the structural parameters showing what proportion of the input flow of the CTS $\vec{y}_{og}$ is fed into the pth input of the lth sub-system;

N_v = the number of input flows in the CTS. The general description of the structure of the CTS is used for seeking the optimum structure of the integrated production process.

26. The above methodology can be used for optimizing the structure of integrated production processes of whatever complexity, and has been used for analysing and planning a number of chemical production processes. In particular it has been used for optimizing acetylene production, with the result that the economic efficiency of the process has been considerably improved and the amount of noxious emissions into the atmosphere reduced.

D. INTRODUCTION OF RECYCLING IN CTS AS A MEANS OF REDUCING WASTE

27. In principle, the use of recycling or the return of a part of the products being processed into the production cycle reduces waste arising from the incomplete use of the basic reagents. However, the use of the block principle and mathematical modelling for the calculation of individual blocks has shown that there are limit régimes for carrying out recycling in reactor systems for the separation of specific products (absorption, rectification, etc.).

28. For example, in the case of a "reactor-rectification column" system with a reaction of the type $A + B \rightarrow 2C$ and equimolar feed of the initial reagents, two operating régimes are possible with a recycling value higher than the critical value R_{kp}

$$R_{kp} = \frac{G}{\left(\frac{V_r k}{2G} - 1 \right)}$$

which depends on the load on the system G, the volume of the reactor V_r and the reaction-speed constants k.

29. In régime I, the initial reagents A and B are almost fully used, whereas in régime II, with identical technological parameters (load on the system, column fractionating capacity, recycling value), complete transformation is not achieved. An analysis shows that the first régime is astatic as regards its stability, and that in the absence of means for stabilizing it, the accumulation of random disturbances may cause it spontaneously to transform itself into régime II, in which the initial reagents are not completely transformed.

30. Accordingly, a given recycling value does not yet determine the efficient functioning of the system as a whole, and special means are required to achieve the desired operating régime.

E. METHODS OF PROCESSING WASTE INTO SECONDARY MATERIAL RESOURCES

Development of Closed-circuit Water-supply Systems

31. In these systems, provision is made for the maximum possible water circulation and the use of purified water in technological processes. Roughly 97 per cent of the industrial water supply is re-used, thus sharply reducing fresh river water consumption.

32. The basic features of the integrated system are:

independent recycling of the used water from each production process. The construction of local purification installations, and use of the purified effluent in production;

construction of drainage systems with separation of effluents;

establishment of special waste-water treatment installations for the entire plant (biochemical treatment of organically polluted effluents and waste water from the industrial network, distillation of highly mineralized waste waters);

construction of installations for the final purification of biochemically treated, slightly mineralized effluents, and their return to the industrial water circuit;

construction of installations to produce commercial products, using waste from the effluent treatment plant (albumin-vitamin concentrate, nitrogenous and organic fertilizers, chlorine, polyvinyl materials, commercial sodium sulphate, etc.);

use of air cooling in heat exchange equipment.

33. The obvious advantages of physical and chemical methods for the purification of waste waters are illustrated by the trend, which is discernible in a number of countries, and above all, in the United States of America, to substitute this form of treatment entirely for the biological purification of industrial effluent and town sewage. Some other countries, where town sewage is more highly polluted, for example the United Kingdom, are proposing in the near future to adopt town-sewage purification systems based on combined physico-chemical and biological methods of treatment.

34. In the chemical treatment of effluent, increasingly wide use has been made in recent years of new and promising methods of separating the various phases: flotation, foam fractionation and magnetic separation. At the same time, research is being undertaken and new reagents are being introduced (particularly polyelectrolytes) for the chemical treatment of waste waters, with a view to obtaining precipitates with improved structural and mechanical properties. Centrifugal separation is the most advanced method of treating precipitates obtained by chemical purification. The leading countries in the mechanical and chemical treatment of waste waters and precipitates are the United States of America, Japan, Sweden, Canada, the Federal Republic of Germany and the United Kingdom.

35. For the adsorption treatment of waste waters, the most promising materials, apart from activated carbon, are the new synthetic polymer high-porosity "Amberlit KHAD-4" adsorbants, which have many advantages over activated carbon. Experts believe that activated carbon in powder form will not come to be so

widely used as granulated carbon. The leading countries in the development and production of adsorbents for cleaning waste waters are the United States of America and the Netherlands. The adsorbent method has been adopted most extensively in the United States.

36. The most promising of the other physico-chemical methods for treating waste waters are the electro-chemical processes, mainly electro-coagulation and electro-flotation (leading countries: France, United States of America, Canada, Australia, USSR); reverse osmosis (leading countries: France, United States of America, United Kingdom) and ozonation (leading countries: Japan, France, United States of America, USSR, United Kingdom, Federal Republic of Germany). Some new methods of disinfecting waste waters are also promising—for example, gamma radiation and ultra-violet radiation which, however, have not hitherto been widely used owing to the high cost of treatment.

Creation of Systems for the Purification of Gaseous Emissions

37. The following are the main methods used for prevention of atmospheric pollution by oil refineries and petrochemical plants:

(1) preparation of the raw material for processing (desulphurization of the raw material for catalytic cracking, catalytic reforming, etc.);

(2) lengthening the operation and increasing the activity and selectivity of catalysts (catalytic reforming, hydrorefining, catalytic cracking, etc.);

(3) improvements in the mechanical properties of catalysts;

(4) use of regenerating reagents (with removal of sulphur compounds, etc.);

(5) catalytic burning of gases produced in various production processes (bitumen production, paraffin oxidation, etc.);

(6) use of hot gases in recovery boilers, followed by processing of the slag;

(7) use of non-blown water-circulation systems;

(8) increases in process capacity by creating single flow lines with minimum accumulation capacity;

(9) combination of individual processes using rigid connections and minimizing intermediate storage of raw materials and products;

(10) introduction of higher requirements for equipment and development of new equipment, fittings and apparatus, and also sealing materials for ensuring a high degree of gas tightness.

38. In conclusion, I would mention that of the papers proposed for discussion under topic I, I have received the following contributions:

a paper from the Hungarian People's Republic, which has been prepared by Messrs. Tibor Blickle and Miklós Machács, deals with aspects of the problem as a whole, and mentions some results achieved in non-waste technology in such processes as absorption, desorption, granulation and adsorption;

a paper from Finland, prepared by Mr. Pentti Malaska, on general aspects of non-waste production processes, scientific and technological links, etc. The same author has submitted a discussion paper on "Concepts and principles of non-waste technology" in which the main concepts of non-waste technology are defined; and a paper from Canada, prepared by Mr. Hussein Saleh, on the social aspects of non-waste technology.

39. Since the papers concentrate mainly on the terminology relating to the concept of non-waste technology and the social implications of this problem, it would seem to be advisable in the course of the discussion to examine specific proposals concerning principles for the creation of non-waste technology.

40. Finally, I should state that the concepts and principles of non-waste technology which I have outlined here do not, of course, exhaust all the possible methods of developing non-waste production processes, since the methods to be used always relate to the specific technology concerned. This report should therefore be regarded solely as a statement of the terms of the problem and of the methodology for its solution.

Main Results of the Symposium of the CMEA Countries on the Theoretical, Technical and Economic Aspects of Low-Waste and Non-Waste Technology*

Prepared by the Organizational Committee of the CMEA Symposium for the ECE Seminar on the Principles and Creation of Non-Waste Technology and Production

(16-19 March 1976, Dresden, German Democratic Republic)

I. INTRODUCTION

1. The economic policy of the CMEA member countries is designed to raise the population's material and cultural standards by expanding social production rapidly and in a balanced manner, and increasing its efficiency, speeding up scientific and technological progress, and improving labour productivity in all sectors of the national economy. This is the basic long-term goal of the policy of the communist and workers' parties and governments of the countries constituting the socialist community, as may be seen from the decisions adopted recently at the congresses of these parties.

2. In this connection attention is being concentrated to an increasing extent on the direct application of scientific and technological advances to production and on making more effective use of the advantages offered by socialist integration. Only in this way can an organic link be established between rapidly developing production forces and the advantages of the socialist economic system.

3. According to forecasts of the development of the CMEA member countries, production in basic industrial sectors such as power, chemicals and iron and steel will expand by four to six times during the next 15 to 20 years. The fulfilment of the 5-year economic development plans of these countries indicates that these forecasts are quite realistic.

4. The continued use of current production technologies would, of course, result in a commensurate increase in waste and the depletion of non-renewable natural resources. Even the planned construction of treatment plants on a massive scale would obviously not fully protect the biosphere, nor would it solve the question of the recycling of raw materials.

5. Of decisive importance in any radically new approach to environmental protection problems is the reorganization of traditional technological processes and methods with a view to developing low-waste and virtually non-waste production that is highly efficient from the technical and economic standpoint, and solving the problem of ensuring the more complete and rational use of natural

*People's Republic of Bulgaria, Hungarian People's Republic, German Democratic Republic, Polish People's Republic, Union of Soviet Socialist Republics and Czechoslovak Socialist Republic.

resources. The socialist countries are therefore vitally interested in stimulating work on non-waste technology.

6. A proportion of the requirements for many types of raw materials, such as steel, non-ferrous metals, plastics, cellulose, textiles, glass and building materials, are already being met by the organized recycling and reprocessing of raw materials in the context of the planned socialist economy.

7. However, fully satisfactory technical or economic methods and technologies have not yet been developed in respect of other types of raw-material resources that could well be recycled. For this reason, the socialist countries are aiming at a gradual conversion to closed-cycle production flow diagrams for the repeated use of material inputs in various branches, and at a maximum reduction of atmospheric emissions.

8. We are convinced that this approach will enable us to put an end to the situation created over a number of decades by the technological methods used at the present time as a result of which emissions of harmful substances and effluents containing mineral and organic compounds as well as solids have exhausted the possibilities of the self-purification process.

9. We are also certain that general conditions in a socialist society, namely, the collective ownership of the implements and means of production and a policy aimed at ensuring the welfare of all members of society, favour the very rapid introduction of non-waste technological processes.

10. These were the problems discussed at the first scientific and technical Symposium of CMEA member countries on the Theoretical, Technical and Economic Aspects of Low-Waste and Non-Waste Technology, held in the German Democratic Republic (Dresden) from 16 to 19 March 1976.

II. THE SOCIAL AND POLITICAL IMPORTANCE OF NON-WASTE TECHNOLOGY

11. The purpose of introducing non-waste technology on a large scale is to expand production in step with the growing requirements of society and, at the same time, to make rational use of natural resources and energy in the context of qualitatively new interrelationships between nature and society.

12. The introduction of non-waste technology ensures that production wastes are converted into reusable resources and thereby preserves the quality of the environment. In a broad interpretation of the economic laws of socialism, non-waste technologies create increasingly close interrelationships between scientific and technical, political and economic, and organizational and social factors. In terms of the intensification process and its various implications, the recycling or harmless disposal of wastes and their recovery, as well as the substitution of natural materials, implies a transition to non-waste technology, namely, a comprehensive process aimed at achieving a closed "production - consumption" system and, therefore, the creation of a non-waste economy.

13. Non-waste technology must therefore be regarded as an important component of the production process which at the same time resolves the problem of preserving the quality of the environment.

14. The implications of production requirements and of the need to make rational use of natural resources are no longer confined to the economies of

individual countries but give rise to new and increasingly serious problems of international co-operation, not only between countries of the socialist community but also with other countries of the world.

15. Co-operation in this field is an integral part of the efforts being made by the socialist countries to realize the principles set forth in the Final Act of the Conference on Security and Co-operation in Europe.

16. The socialist community's increasingly stringent requirements as regards the quality of the environment can be satisfied at the present stage of the development of productive forces, production capacity and social needs only if the entire cycle of materials between man and nature is rationalized. This must be regarded as one of the main ways of achieving intensive national economic development.

17. Natural resources and energy, which constitute an important element of the material and technical base, have a considerable influence on economic development and particularly on rates of production and consumption growth, labour productivity and the efficient use of public funds.

18. The growing demand for raw materials can be satisfied and the environment protected mainly through the more complete, integrated, efficient, economically rational and scientific use of natural material and energy resources in a manner aimed at meeting ecological requirements. For this purpose, economic principles must be drawn up concerning the use of material and energy resources and steps taken to make better use of social labour in the extraction and use of all materials and sources of energy entering the production process.

19. Satisfaction of the growing material and cultural requirements of society which, according to K. Marx, originate in production, calls for the planned expansion of production and services.

20. What has been said above reveals that non-waste technology is necessary for a number of reasons and that it can be introduced only if the necessary guidance and control is well organized. The most important problems relating to the development of the national economy of the socialist countries are resolved by state plans whose fulfilment is monitored by the State.

21. Appropriate regulations, which are constantly being improved and which are enforced by the competent state bodies, have been laid down for the protection of the environment and the rational utilization of natural resources. The CMEA member countries engage in broad scientific and technical co-operation in matters relating to the protection and improvement of the environment and the related rational utilization of natural resources in accordance with the common expanded programme which they have drawn up on the subject and which covers the period up to 1980. This programme comprises a wide variety of topics, including the development of non-waste technology.

22. Co-operation in respect of this common expanded programme is co-ordinated by the Council for Environmental Protection and Improvement that was set up in 1972 under CMEA's Committee for Scientific and Technical Co-operation. Countries are represented on this Council by deputy ministers and eminent scientists having the powers necessary to organize the implementation of co-operation programmes in their respective countries.

23. The work of the Council during the past few years on matters relating to the protection and improvement of the environment, as well as that of other

CMEA bodies, shows that the planned development of the economies of the socialist countries enables them, in the course of such co-operation, to develop technological processes and methods that ensure maximum use of raw materials and at the same time satisfy environmental protection requirements.

III. ECONOMIC ASPECTS OF THE DEVELOPMENT AND INTRODUCTION OF LOW-WASTE AND NON-WASTE TECHNOLOGY

24. The development and introduction of non-waste technology and methods of using material resources and wastes is necessary from the economic standpoint mainly because of the increasing cost of this form of production. This trend is basically a reflection of the following:

1. Natural resources are being depleted, so that deposits of raw materials that are less promising from the geological and hydrogeological standpoint have to be developed and worked.
2. The volume of waste is growing because production is steadily expanding and because an increasing proportion of the gaseous, liquid and solid substances as well as energy that enter the production process do not end up in the finished product but are released into the environment in the form of partially processed wastes.

25. As the introduction of non-waste technology is also an economic problem, it must be approached from both its macro and micro aspects. The micro-economic problem is one of the interests of individual production enterprises in various branches of industry. The macro-economic problem calls for an inter-sectoral approach at the national economic and international levels. From this standpoint, what may well be good for a specific branch of industry is not necessarily good for the national economy. And conversely what might not be good for a specific branch may well be necessary for the national economy as a whole, regardless of the greater expenditure it implies.

26. The question of the profitability of introducing non-waste technology and the rate at which it should be introduced is approached in different ways in different countries. Decisive importance may be assumed by various factors such as the local situation as regards raw materials resources, the relative level of production techniques in various branches, the situation on the labour market and even differences in social conditions.

27. Another important factor that influences decisions on the introduction of non-waste technology is the social division of labour in the context of co-operation between the CMEA member countries. Specialization by individual countries in certain types of production also offers possibilities of introducing the most advanced technological systems based on principles of non-waste technology.

28. Lastly, it must be borne in mind in tackling the economic problems involved that the economy of each country, as well as that of the socialist community as a whole, is part of the world economy. This situation creates specific mutual relationships between various social and economic systems, and they too will influence the way in which non-waste technology is introduced as well as the speed with which this is done.

29. It follows from what has been said above that the economic calculations necessary in connection with the introduction of non-waste technology on a

large scale in the production sector are not simple or easy. Suitable scientific methods for the calculation of efficiency have yet to be developed, and sufficient experience has not yet been acquired in the introduction of non-waste technology on a massive scale. For this reason, the economic problem must be tackled at the same time as research is being done on the technical aspects of non-waste technology. Such research must be carried out at various levels in respect of individual areas of technology on a national basis as well as through the joint efforts of the CMEA member countries.

30. Moreover, studies should also be made of especially important matters such as:

the expression of non-waste economic activities (in micro-economic units of measurement) in terms of effectiveness; and

aspects of the geographical distribution of enterprises in a non-waste economy.

31. The rationalization of a non-waste economy should be achieved not only by means of technical, organizational and legal standards but also by economic methods (devices). These devices can be used to stimulate macro-economic units such as organizations in various industrial branches, administrative regions (urban and rural centres) as well as financially independent units. Discrete methods should be used in each of these cases to promote non-waste economic activities.

32. Macro-economic decisions taken in respect of major economic and administrative systems should be based on comprehensive calculations of the effectiveness of non-waste technology. In such cases, all the various costs and benefits connected with the rationalization of natural resource management must be studied from every point of view—particularly in the case of scarce natural resources. The problem of maintaining the biological equilibrium of the environment must also be taken into account in these calculations. Major industrial centres as well as conurbations constitute a potential danger to this equilibrium as it has so far existed. The development of industry and the remoteness of production units from sources of raw materials and energy as well as from potential consumers make enormous demands on transport facilities (carriage of goods, passengers, etc.) and therefore pose an additional threat to the environment.

33. The solution of this broad spectrum of problems also calls for the joint consideration of this subject and particularly of its various aspects:

new investments in industry and their distribution;

new conurbations or the expansion of existing towns;

the creation or reorganization of land,water and air transport facilities;

the introduction of systems for the regulation of water supplies as well as the drainage or irrigation of certain areas; and

modification of agricultural structures, cultivation methods and fertilization practices.

34. Decisions of this nature should be based on calculations of direct expenditure as well as the outlay necessary to ensure the least possible waste in every area of economic activity. The most difficult problem in this respect is that of the evaluation of possible solutions. Evaluation of the cost and

the implications of these comprehensive decisions may call for technical, medical, physiological or natural science information, and the data collected will have to be expressed in cost/benefit terms by economists.

35. The task of reflecting the results of non-waste technology in economic indicators is an easier matter for organizations that use profit and loss accounting methods; conventional cost/benefit calculations can be used for this purpose. At the outset, however, it must be decided which aspects of non-waste technology can be related to the economic activities of these organizations and how the cost and benefits of these activities are to be evaluated.

36. Only then should ways of including the cost and benefits of non-waste technology in micro-economic calculations of profitability be determined. The problem is one of inducing the management of industrial organizations to adopt non-waste technology by the establishment of a link between these costs and benefits and calculations of profitability. Needless to say, this approach is effective only when profitability relates to the generally accepted criterion of efficiency.

37. The economics of non-waste systems may be considered from the following standpoints:

as a measure of the social losses caused by failure to comply with non-waste production and consumption principles;

as a means of calculating the effectiveness of the cost and benefits of the wider application of non-waste production and consumption techniques; and

as a possible inducement, if economic methods (incentives) are used, for various economic units to raise the level of their non-waste technology.

38. The development of new products will call for particularly broad participation by economists, especially to evaluate the effectiveness of measures aimed at extending the application of non-waste technology. At the same time, the following will be necessary in connection with the adoption of incentives to promote the use of non-waste technology in this phase of the production process:

the development and use of methods of calculating the effectiveness of the cost structure selected for the product that has been created;

the effectiveness of the design variants in question should be analysed from the standpoint of the amount of waste that is likely to be generated when the raw material inputs are being processed;

the analysis of the effectiveness of individual design variants should take into account the expenditure on raw materials and energy that will inevitably be incurred at a subsequent stage in the processing of the product, as well as the waste generated in the course of such processing; and

the final stage in the life of the product, namely, its disposal, should be taken into account in the analysis of the effectiveness of different design variants. In this connection, thought should be given to the development of a method for the disposal of the worn-out product.

39. Policies relating to the rational geographical distribution of the forces of production play an important role in current socialist economic thinking. Industrial enterprises are best grouped into territorial units. Such enterprises can be operated on the basis of a specific organic relationship and thus

constitute a single production unit. Low-waste and non-waste technological processes can be introduced on a large scale in this type of industrial complex. This particular aspect of non-waste technology makes it possible to derive even greater benefit from a number of economic advantages which can be secured only with difficulty if other siting arrangements are used. These advantages include:

1. The rational interrelationship created between various types of production units situated immediately next to or near one another which process one and the same raw material in succession enables a vertical relationship to be established between the various stages of the industrial process in question. This arrangement considerably reduces capital outlay and operating costs, as well as expenditure on the transfer of the product from one processing stage to the next because production forces are concentrated in one place. It offers possibilities of labour savings and reduces production costs because outlay on transport declines. Indeed, in some branches such costs for industrial units that are scattered over a large area account for 20 to 40 per cent of the cost of the finished article (foundry and chemical production, building materials, etc.).

2. Problems of organization and management are solved by refinements in specialization and co-operation.

3. The efficient use of the waste products of certain enterprises by others, including the use of secondary sources of energy (combustion products with a residual fuel value, hot water, spent steam, etc.).

4. Economic advantages are offered by phasing (daily and seasonal) the electric and thermal energy consumption of enterprises with different production regimes (at different times during the 24-hour period).

5. The advantages offered by the joint use of major water preparation and effluent treatment installations, as well as environmental protection installations. In addition to these advantages which are inherent in the concept of non-waste technology, other benefits can be achieved through the rational management of manpower resources, the training of higher level staff and workers, joint public utilities, etc.

IV. SCIENTIFIC AND TECHNICAL ASPECTS OF THE INTRODUCTION OF NON-WASTE TECHNOLOGICAL PROCESSES AND FLOW DIAGRAMS IN VARIOUS BRANCHES OF INDUSTRY

40. Non-waste technology is the theoretical limit to which low-waste technology can be carried. The very term "technology" implies the knowledge, methods and means of obtaining material wealth from natural resources, as well as a knowledge of the processes by which raw materials are converted into a finished product. An integral part of technology is the knowledge, machinery and equipment used in production; and an integral part of non-waste technology is the direct and non-waste production of energy.

41. The introduction of non-waste technology consists in the application of a long-term priority programme on a very broad scale to bring about changes in the industrial and raw materials sectors, in consumption interrelationships, in the structure of levels of living, etc.

42. The concept of non-waste technology viewed in the broad sense covers not

only production but also consumption, energy and raw materials, as well as matters relating to the protection and formation of the environment.

43. An analysis of the achievements of science and technology and prospects for their application in various branches of the national economy has revealed that the technical arsenal at present available to the socialist countries is adequate to enable them to embark upon the elaboration and implementation of comprehensive programmes relating to non-waste technology.

44. It was agreed at the first scientific Symposium at Dresden to aim at more practical and tangible objectives. For this reason it was decided to concentrate attention on industrial production, which constitutes the greatest ecological danger owing to the harmful wastes it generates. This problem was given top priority.

45. The question of the development of non-waste technology for the comprehensive utilization of raw materials and its application as soon as possible to industrial production processes is extremely important and topical because it must be borne in mind that the rapid expansion of industrial production not only upsets the ecological equilibrium and has an adverse effect on human health, but also that its impact on nature may be of critical importance for industry itself in that it might limit possibilities of further development.

46. A large number of draft solutions and technical proposals, reflecting the principles of non-waste technology, were submitted to the Symposium. Attention was concentrated on four selected areas, namely, the chemical industry, the coal and power industries, ferrous and non-ferrous metallurgy, and the building materials industry.

47. *The chemical industry* offers enormous possibilities of drastically reducing waste in a wide variety of production processes, using waste products, and considerably mitigating the effects of waste disposal on the natural environment. Conditions in a number of branches of the chemical industry are already favourable for conversion to virtually non-waste technology.

48. The industry as a whole can be sub-divided on the basis of inorganic and organic technology and, on the basis of production capacity into large units with outputs of 10^5 to 10^8 t/year, average units with outputs of 10^3 to 10^5 t/year and small units with outputs of under 10^3 t/year. Large capacity units in the inorganic branch include plants producing sulphuric acid, superphosphates, phosphorous acid, ammonia, nitric acid, ammonium nitrate, caustic soda and salts.

49. The various problems that have to be solved in the chemical industry with a view to the protection of the environment include:

1. Development of highly selective high-quality catalysts.
2. Development of efficient counterflow separation apparatus (high-performance extraction columns, absorption apparatus with large gas throughputs, absorption and ion-exchange apparatus).
3. Development of instruments for the transport, mixing and separation of powdery, viscous and highly non-newtonian materials as well as materials of irregular shape.
4. Study of the physical properties of foams, aerosols and emulsions.

5. Study of reactions that take place with particles of under 0.1 micron in size and development of instruments for such reactions.

6. Study of unconventional methods of separation, i.e. reverse osmosis, ultra-filtration, dialysis, electrodialysis, etc.

50. The steady growth of the *fuel and energy* economy throughout the world as an important prerequisite for expanding the production of various industrial items and improving material well-being creates environmental protection problems. At the present time, energy is still produced mainly from mineral sources, particularly hard and brown coal, and this situation is giving rise to major problems which a number of countries are trying to solve.

51. Whereas the electric power stations of the 1920s had an average capacity of 100 to 200 MW, some 20 years later stations of up to 400 MW were being built. At the present time, thermal power stations of up to 3000 MW are under construction and, according to the latest information, stations of up to 6400 MW fired by fossil fuels are being designed.

52. This steady increase in capacity has given rise to problems of sulphur dioxide, dust, nitric oxide and ash emissions, as well as heat emissions into the air and water. Dust removal by modern installations in amounts up to 180 kg/m^3 represents a considerable improvement.

53. Notwithstanding the major efforts made to introduce new technical processes and equipment, sulphur dioxide emissions are still an important economic problem. Emissions of nitric oxide and other harmful substances are a source of similar concern. Research on the use of thermal power-station ash is being carried out in a number of countries and the results are applied in practice. Nevertheless, the present situation as regards ash utilization must still be viewed as unsatisfactory.

54. The steady depletion of oil resources inevitably implies a return to coal as the basic source of energy. As in the case of petroleum, coal deposits of lower grade are being tapped to an increasing extent, and the cost of extracting these fuel resources is steadily increasing. Such deposits cannot be fully worked out by conventional methods, and most of the coal they contain remains unused in the bowels of the earth.

55. A typical example of non-waste technology in this context is underground coal gasification which can be used to produce heating and synthetic gases; in this method the gas can be treated underground and actual extraction confined to its energy and synthetic components which, when used for energy production and in the chemical industry, do not produce any pollutants.

56. The *non-ferrous metallurgical and metallurgical sector* offers enormous possibilities, which have by no means been fully explored, for the organization of comprehensive non-waste technological processes. In this connection, a decisive role is played by the introduction of new highly intensive advanced technological processes and flow diagrams.

57. Considerable interest was aroused at the Dresden Symposium by documents submitted on the experience of enterprises in the German Democratic Republic and the USSR concerning the successful comprehensive use of waste generated during the pig-iron production process.

58. The variety of problems connected with the production of non-ferrous metals, the complex composition of raw materials and the complexity of techno-

logical processes call for the development of various types of non-waste technology taking into account specific requirements reflecting the composition of raw materials, geography, transport, etc., and based on the more complete recovery of the useful ingredients of raw materials and their comprehensive processing.

59. Special attention must be paid to the rational utilization of red mud, namely, the waste generated by the alumina production process in the bauxite-based aluminium industry.

60. As regards the metallurgy of heavy non-ferrous metals (copper, lead, zinc, nickel, cobalt, molybdenum, etc.), a recommendation should be adopted for the introduction, as rapidly as possible and on the widest possible scale, of advanced technology based on autogenous processes (oxygen jet smelting, vacuum technology, electrometallurgy, etc.), that can considerably reduce the pollution of the atmosphere by sulphur dioxide. This would constitute a major step in the direction of non-waste technology in the most important branches of non-ferrous metallurgy.

61. It is recommended that greater use should be made of hydrometallurgical flow diagrams based on new advanced processes, such as autoclave technology for the treatment of ores and concentrates of non-ferrous metals, sorption-extraction technology for the selective extraction and efficient separation of metals, as well as the treatment of liquid effluents and water recycling. Hydrometallurgical flow diagrams using extraction and hydro-exchange sorption offer a sound basis for the development of non-waste technology in non-ferrous metallurgy and open up further possibilities of resolving problems relating to environmental protection.

62. The following will also be necessary:

further research on and introduction of new advanced metallurgical technology that has fewer process steps (i.e. technology consisting of a small number of separate operations) and better (as compared with conventional technological flow diagrams) technical and economic indicators and that, at the same time, considerably reduces emissions of harmful substances into the biosphere and ensures the complete utilization of production waste;

maximum improvement and refinement of existing production processes with a view to reducing harmful emissions at all stages and ensuring that the waste produced is fully used; and

more extensive use of closed stabilized recycled water supply systems and reduction of process water consumption to the minimum.

63. Indicators reflecting quantitative and qualitative trends and various methods of using wastes in the *building materials industry* reveal that possibilities in this sector are far from exhausted. The proportion of waste used in comparison with the amount that is not used as a secondary raw material is not yet very high.

64. In this connection, the Symposium proposed the adoption of measures to:

speed up research with a view to the development of economic accounting procedures and the selection of profitability indicators for enterprises that would promote the use of secondary raw materials in the building and building materials industry;

offer non-economic incentives to promote the use of secondary raw materials;

speed up work connected with the design of machines and equipment for the conversion of waste into raw materials or products which could be used directly in building;

bearing in mind that the waste products used in building are extremely bulky and heavy and that transport costs are of decisive importance to economic effectiveness, reduce such costs by siting processing plants in a way that would ensure optimum conditions for the carriage of bulky loads;

draw up a classification of and standards for waste products and by-products used in the manufacture of building materials and in building;

improve and speed up exchanges of scientific and technical information and, for this purpose, create something in the nature of a data bank (patents, licences, know-how, etc.).

65. Existing technologies which have proved their worth in individual countries can, even at the present time, be applied to achieve a considerable improvement in the use of waste—and particularly ash, slag and dust—in the building materials industry simply through exchanges of experience. For example, one such technology is used to obtain alumina from ash by the Grzymek process in the Polish People's Republic.

66. New, more advanced technologies, more in line with the current concept of low-waste and non-waste technology, must be developed more rapidly. The various problems that arise in this field should be included in the programme of work of the Senior Advisers to ECE Governments on Science and Technology.

V. CONCLUSION

67. The Symposium of the CMEA countries on the Theoretical, Technical and Economic Aspects of Low-waste and Non-waste Technology held at Dresden served to draw attention to the great importance and role of non-waste technology and economics in the further social and economic development of all countries.

68. The adoption of non-waste technology on a wide scale as rapidly as possible is obviously bound up with the identification of problems and arrangements for exchanges of information on this technology between individual branches of industry and especially between countries. The development of non-waste technology must, of course, be based on broad international co-operation in the implementation of research programmes in this area.

69. The rapid development of non-waste economics at the international level depends to a great extent on international co-operation. The following topics might well be examined with a view to the further development of scientific and technical co-operation and exchanges of information on various aspects of non-waste technology:

Development of methods and arrangements for the processing of raw materials that considerably reduce the creation of wastes.

The development of virtually new production processes in which waste is sharply reduced or eliminated.

Development of methods of using and processing industrial wastes to obtain a commercial product.

Development of effluent-free and emission-free production processes in which valuable components are recovered.

Development of processes based on energy and technological principles.

Development of processes and equipment to render particularly toxic and poisonous wastes harmless.

Development of systems models for the optimization of material and energy flows in the creation of non-waste processes.

Analysis of patents and information documents on non-waste production processes and the preparation of manuals on industrial wastes.

70. Consideration of the above topics at the forthcoming ECE Seminar on the Principles and Creation of Non-waste Technology and Production, to be held in Paris in the fourth quarter of 1976, would, in our view, be extremely valuable and would contribute to the development of co-operation between CMEA and ECE in this sphere.

A Broader Definition of Non-Waste Technology

Hussein Saleh

Senior Consultant, Advanced Concepts Centre, Environment Canada, Ottawa, Canada

In recent years there has been a growing realization that the contribution of the industrial revolution to the human experience is threatened by wasteful practices. The Club of Rome sparked debate on the limits implied in the wasteful use of resources. The growing concern for the finite capabilities of the environment has led to programs for pollution control, recycling, waste reclamation and conservation. But recycling and waste reclamation imply technologies for handling wastes. There are important and indeed necessary components in a strategy for adjustment to a less wasteful course of development. This paper submits, however, that it is now time to begin to look beyond the management of wastes to the development of technologies that will meet the needs of society without waste.

Perhaps it is well to begin an exploration of "non-waste technology" with a review of the meaning of the word. "Waste" draws images of destruction and despoliation: the squandering of resources, energy and time; the desolation of a ravaged countryside; the effect of wear, decay and spoilage. "Waste" encompasses much more than effluents, refuse and the by-products of industrial processes. Wastefulness is the irrational use of more than necessary resources to fulfil human needs and objectives. And if this broader definition of waste is accepted, then non-waste technology becomes a positive image for the application of technology to fulfil human needs and aspirations without destructive impact on the environment.

A serious consideration of non-waste technology in these terms leads to fundamental questions about our production and consumption patterns. We should be able to stop for a while and think whether we need a certain line of product to fulfil a given need. We must be able to shift from our traditional approach to satisfying our needs in terms of *products* to the more basic approach of *functions*. In this way a new direction is provided for reconsidering traditional patterns of development. The Prime Minister has urged Canadians to shift their aspirations to self-fulfilment through what they do rather than looking for progress measured in the possession of material goods.

The family car provides a good example of a wasteful technology. Is the proliferate use of the automobile really required for transportation when we see the impact of freeway expansion, suburban sprawl, the boredom of the automobile assembly line, and many other impacts? If we want to minimize the wastage of fossil fuels while meeting objectives for moving from one place to another, then our thinking should go well beyond the car. We would move closer to a non-waste technology by promoting transportation and communication systems where less energy is required, or where renewable energy sources could provide the motive power. But we should not stop there. If people can provide for their needs and wants without having to move so far, so frequently, that would surely reduce the consumption of energy and resources. Perhaps technology that gives us the means to communicate without having to move could justifiably be called a non-waste technology. Is the telephone non-waste technology? Certainly it is closer to the definition than pollution-control devices on the automobile exhaust.

The developing technologies of solar heating, wind energy, and other environment-derived energy systems suggest non-waste technology. The manufacture of these products admittedly will use resources and can lead to waste, but these technologies to produce warmth and energy without consuming oil or coal present avenues for developing effective non-waste technologies. Technologies that will support "environmentally appropriate development", that will assist producing food without depleting the soil, that will heat our homes without depleting resources, that suggest permanence and continuity; all are non-waste technologies.

This broader view of non-waste technology could and should have dramatic impact upon our entire approach to resource management. It will require drastic shift in our planning exercises because having to deal with the functions and not the products, the ends and not the means would require the adoption of long-range planning principles where the years 2050 are perceived and considered. Conventional non-waste technology addresses itself to reducing or cutting waste in production process and in the consumption as well. A broader definition will question the entire principle of developing a given product and whether we really need it to satisfy a given objective, could we instead use the same resources to develop another product with more potentialities and more benefit to the human race.

The process of developing such technology addressing itself to the optimum utilization of our natural resources could only function properly through long-range planning. This non-waste technology policy could put an end or at least reduce the terrible waste in human material and financial resource on products that become obsolete before the end of the year, we all know of some products that become obsolete even before they leave the production line. The short-range expedient must yield to the long-range necessity. But as petroleum fuels are depleted, as material resources become more expensive and as the requirements for stable food production becomes more essential, positive shifts must be made towards life styles and technologies that put no demands upon the environment.

The Canadian Government is increasingly concerned with the development of alternatives to highly wasteful courses of development. The Science Council of Canada, the Department of Environment and other agencies have undertaken a number of research projects under the theme of "Conserver Society". These have attempted to explore non-waste technologies and approaches to development planning that are in balance with sustained production from the environment. These agencies are collectively supporting an interagency, interdisciplinary research project on this theme through a university-based consortium called GAMMA. This study is attempting to develop "Conserver Society" models that range from resource conservation to substantial changes in life styles, and to bring these proposals forward for public discussion. Non-waste technology would undoubtedly be one aspect of a Conserver Society.

As one concrete example of non-waste technology, Canada is sponsoring the development of a structure that is called an Ark for Prince Edward Island. It is a solar-heated, wind-powered structure for family living which produces food for the family from a greenhouse and fish culture system. It creates a miniature ecosystem in which water and nutrients are recycled and the energy is provided from renewable sources. It is not put forward as a panacea; only as a demonstration of a practical outer limit of self-sufficiency with present technology. It hopefully will be a spark for considering new models for environmentally appropriate development in Canada and other developed countries. It may be a signal to less-developed countries to look to models for appropriate development within their indigenous capabilities and to avoid highly "wasteful" development schemes.

Non-waste technology in its broader definition becomes a problem that not one single country can handle in isolation from the rest of the world. As the new definition would result in questioning, and most probably affecting our social values, our life styles, our behaviour and the world's trade and economic patterns, nothing less than the international community's consolidated efforts could provide a comprehensive framework for non-waste technology. A global approach to managing our resources should achieve broader social needs and not individual demands of those countries who own most of the resources and most of the knowhow. The technology that provides such global approach where the third world basic needs are given preference to the developed countries extravaganza is a non-waste technology.

New Ways of Developing Chemical and Related Procedures Free of Wastes or Low in Wastes in Hungary

Tibor Blickle* and Micklós Machács**

**Research Institute for Technical Chemistry of the Hungarian Academy of Sciences, Veszprém, Hungary*
***Ministry of Heavy Industries, Budapest, Hungary*

In all industrial societies, industrial plants emitting by-products damaging to the human organism or to living organisms important to human life represent a quite considerable source of danger. Among the industrial plants especially hazardous are those connected with the chemical industry and related industries processing similar materials, because wastes of these plants represent a danger to the natural surroundings.

Another approach to the problem is that wastes of production are valuable materials and may have significant economic interest.

In order to protect constitutionally the human environment a law was enacted in Hungary and in recent years numerous official and administrative provisions have been taken to enforce the law. Research and development bases for protection of environment were organized and developed by the economic branches and soon produced the first results. The application of a progressive penalty system forced factories contaminating the environment to stop this activity. As a consequence of rapidly increasing prices of raw materials, the need to process materials with the least possible loss and to re-utilize wastes became insistent.

In the elaboration and realization of a research programme a first task was to develop chemical and related* technologies in such a way that the emission of damaging wastes would be completely prevented or decreased considerably. In formulating our research programme we had to start from our national situation while taking into account international experiences and trends. The results might be expected to contribute to international research as well as finding use at home.

In accordance with the above aim of our research programme, research into new chemical industry and related industrial processes as well as the technical and development methods leading to these processes takes into account:

demands on chemical industry production after 1990;

requirements for the economic and effective utilization of raw materials and sources of energy;

*The term "related" refers to technologies similar to those employed in the chemical industry, e.g. mechanical, building materials, food industry technologies.

expected requirements for the protection of environment;

expected technical and technological levels and demands for manpower;

expected national income and living standards;

possibilities of international co-operation in production, domestic possibilities of industrial situation and concentration, economic size of plants in line with market demands;

the supply of basic raw materials and cost/price ratios.

This research may lead to the development of basically new or modified chemical and related technologies which are free of wastes or low in wastes. The aim of the programme is that the results gained in the period from 1976 to 1990 must give theoretical and practical guidelines on developing the chemical industry after 1990, in addition to the possible immediate rise of applicable research.

Research into the development of technologies which are free of wastes or low in wastes serve the following purposes:

development of basically new methods of production, the organization of production, technology and siting as well as the development of new materials;

transformation of existing, well-known or prospective procedures to ones being free of wastes or low in wastes and the additional application of such transformation methods.

Research based on present scientific results so must be comprehensive taking into account the theoretical, organizational, economical and technological production aspects of the problem and their interrelationships. The possible results and the real scope and limits of their application must be indicated.

During the preparation of the concept of our research we had to determine—in the absence of a more accurate definition—what we meant by "waste" from the viewpoint of the programme. *Waste materials* are specified as follows:

loss of material, contaminated materials, by-products or heat formed in different operational stages of production, which are not fed to the next stage and are not utilized as intermediate or final products;

material or heat released into the environment during processing of material and product (intermediate product, by-product) outside technological operations, e.g. during transport, loading and analogous physical manipulations;

materials or heat formed under the above circumstances, if they are

released into the environment without control or without possibility of control and appear as wastes, or change the equilibrium of the environment;

released into the environment under control but modify the environment locally or over a larger area, for a longer period or permanently;

stored in a closed system but the problems of utilization or destruction are not solved

the products, if

in such a form that if carefully and properly used, certain parts become wastes;

their use has a guaranteed time-span and complete regeneration is not possible, or is not being undertaken;

they are partially used or limited in function and regeneration, destruction and re-utilization are not practicable.

As can be seen from the definition of the term "waste", the research programme sought to assist the choice of main directions and themes in research as well as systems principles in a research programme which could not cover the problems completely, priority projects had to be developed, taking into account the urgency of the problems, research capacities and financial possibilities.

Studies of the creation of waste-free procedures cover the following fields:

application of known and new methods of physical chemistry and chemical technology and their possible combinations on the basis of systems theory;

investigation of constructional possibilities and suitable linking of new procedures and new operational units;

construction of closed systems of main and auxiliary material flows with special regard to the application of blocks of operation and possible connection of recirculation;

construction of hermetic systems;

investigation of new raw materials, auxiliary, technological constructional materials, and realizable states of operation;

investigation of possibilities to develop new forms of products ensuring waste-free processing and application of materials.

In this field valuable research has been carried out in Hungary and applicable results have been obtained. Staff of the Research Institute for Technical Chemistry of the Hungarian Academy of Sciences, the Veszprém University of Chemical Engineering and research organizations grouped around them, having investigated new procedures, new structural and auxiliary materials, and technical chemical systems theory, have developed practical procedures which satisfy the demand for waste-free technologies. Some of the results are as follows:

Absorption - desorption realized in one operational unit simultaneously. The processes take place only on two sides of a selective separation wall through which only the absorbent liquid permeates. With this procedure, the content of contaminating material in exhaust gases of different chemical composition can be practically eliminated. The contaminating material is drawn off in a form which does not add to the contamination of the environment but on the contrary: feeds back into the technological process for use as raw material. Losses and production costs are thus reduced.

The procedure separating out the solid material content of gases containing dust formed in chemical processes has similar importance from the viewpoint of protection of the environment. In the procedure, directly utilizable granules are formed in one operation.

In the Research Institute for Technical Chemistry of the Hungarian Academy of Sciences new and effective procedures which avoid the creation of contaminating materials and prevent their escape to the environment were developed by improvement of the final product form. These procedures—mainly by the forming of granulates—besides the creation of a more satisfactory final product are more economical when compared with "traditional" procedures.

At the Veszprém University of Chemical Engineering development of procedures applicable to the absorption of certain wastes as intermediate technological element is being investigated. The active material of these operations is a special zeolit which can be found in Hungary under natural conditions.

Investigations tending to low-waste procedures take into account the following considerations:

all non-waste technology considerations and, in addition;

procedures in which waste can be destroyed where it is created, or can be used, or can be transported in a closed system in order to be destroyed or utilized;

methods which give reasonable protection against accidental formation of waste as a consequence, for example, of handling errors, operating abnormalities or other reasons.

The completeness of the programme and the extensiveness of its content call for co-ordination among the economic branches—and, in some cases, international co-ordination—and broad professional direction of research under the leadership of the appropriate research centre or co-ordinating institute.

The professional leadership and co-ordination of the programme have been assumed by the National Council for Environmental Protection and the Ministry of Heavy Industries. Research is sponsored from state resources.

Simultaneously with the theoretical identification of the problems practical research has been launched on priority topics or on those which offer the hope of new solutions.

The Hungarian concept of the development of procedures and methods which produce little or no waste and the launching of relevant research in the chemical and related industries go back barely 6 months. It is still not possible to determine the scope of the tasks or to gauge the results. Obviously, only some specific results can be achieved within the scope of this programme. The extension of co-operation in research to the international level and the organized exchange of results may offer the prospect of a solid foundation for work which has a most important aim.

Concepts and Principles of Non-waste Technology

Pentti Malaska

Turku School of Economics, Rehtorinpellontie 5, 20500 Turku 50, Finland

NEW TECHNOLOGIES

When the development of new sciences and new scientific knowledge reaches a point of maturity at which it is possible to control phenomena by stating boundary conditions radically new technologies become possible.[1]

Such sciences as ecology, biology, genetics, systems science and cybernetics, together with the study of the matter and with the study of the possibilities of industrial production under extreme land, marine, space conditions, are opening up new highways and novelties for technological development in this sense. Each of these would deserve a discussion and an assessment of its own, but in this context I cannot do more than draw attention to them, and point out that they should be included in the concept of non-waste technology.

NON-WASTE ENGINEERING

The other possibilities of direct material control—as opposed to monetary control—lie in the improving material efficiency of present technologies and techniques, by means of new technical constructs and production systems, based on the principle of minimum use of material and energy resources and minimum degradation of the environment. The term which has been introduced in the ECE for a technology based on such a principle is "non-waste technology", and considerable activity is being shown in the development of this way of thinking and the testing of its applicability. I would like to recognize the content of the concept by following quotations from recent reports of ECE:[2,3]

> Non-waste technology is the practical application of knowledge, methods and means so as to ensure the most rational use of natural resources and energy to provide for the needs of man and to protect environment.[2]

And further from another report:[3]

> Non-waste technology is simply the planning and management of human activities so as to incur the minimum waste of materials and energy. This concept then achieves the twin goal of reduced degradation of the environment and increased conservation of natural resources. Non-waste technology can be applied to all activities of industrial man. Most of these activities can be grouped into the chain of production, thus: 1. Production of raw materials, 2. Transportation of raw materials, 3. Manufacture of finished and semi-finished products, 4. Distribution of products, including

transportation, storage and sale, 5. Disposal and possible recycling of used products. At each stage in the chain the concept of non-waste can be applied at various levels: A. The conception of the product or output, B. The design of manufacturing, C. The design of residuals and re-use of products, D. Design of the centralized waste handling facilities for use of waste as new resources (however, designing to avoid waste before a process begins, rather than handling waste effectively after it has been produced), E. Designing the total system as a whole.

We may note that almost the same ideas and aims have been the point of departure for the activity for the FEANI environmental committee, although here it has taken a more general form which also includes new technologies in the sense I already mentioned.[4,5] The work of the ECE, however, has operated on a more concrete level, as appears from the individual case studies reported in more or less detail so far. The term "non-waste technology" itself is perhaps too narrow to describe the whole extent of the goal in its most general form; on the other hand, this term is to be understood as a kind of technological limit, which can be used as a norm and a criterion of efficiency but which cannot be actually achieved by any technology realizable in practice. In this sense, and from the engineers' point of view, an activity similar to the concept of the non-waste technology should be considered as a stimulus of crucial importance to technological development and as a mean of real control for material development.

WASTE ENGINEERING

One of the most important reasons for the disproportionate magnitude of waste problems and environmental degradation in the west—disproportionate to production, that is—is, I think, the fact that these problems and the possibilities of their solution have not involved any incentive which would stimulate the market mechanism. The market mechanism is always channelled in the direction of effective demand, and up to the 1970s, despite the need, there has been no such market activity for solutions to these problems.

It is evident that the market mechanism really can be exploited effectively along with other means. This might be made possible by means of an arrangement whereby the primary responsibility for solving a waste problem would be borne by the producer or marketer of the product in question. This would probably lead very soon to the organization of producers for the purpose of solving waste problems, to the implementation of the technical and other knowledge necessary for the treatment of wastes, and to a corresponding change in product prices. Only in some such way can the rapid and effective exploitation of scientific and technological means be harnessed for the purposes of the—at present—unorganized field of waste-processing. Furthermore, it would open up new and positive business opportunities; this, of course, has already begun to take place, though not yet to a sufficient extent. And—what seems to be important—the market mechanism itself is not capable of creating these conditions necessary for its functioning. It will be left to governmental activities to take the initiative.

RESOURCE ENGINEERING

A basic prerequisite for any technologically based development is the sufficient availability of new material resources. Furthermore, this condition cannot be ensured by any increase in efficiency, not even by recycling; even in a situation of zero growth entirely new resources would constantly be needed. As Georgescu-Roegen has rightly pointed out, the only type of economy which would have a durable ecological foundation would be one in which economic activity and production were constantly falling.[6] The development of resource engineering will thus be subject to constantly growing interest, and repeated estimates of the sufficiency of raw materials globally, their technological and economic exploitability and their commercial availability, form an important part of every country's security for the future. It may be useful here again to recall some of the recent estimates as to the sufficiency of some of the important raw materials. I refer here to the estimates made by the working group of Dennis Gabor and Umberto Colombo.[7] This working group has divided raw materials into three different categories according to their sufficiency in global terms: those which are available to an unlimited or otherwise sufficient extent at the prices approaching the present ones, those available to a limited extent, and those which may be considered scarce resources.

For example, oxygen and nitrogen are available without limit from the air, magnesium and bromine from the sea water, silicon from sand, calcium from limestone, sodium and chlorine from rock salt. Sea water itself is, of course, available without limitation. The reserves of aluminium, iron, and sulphur are considered adequate with slower increasing prices. The reserves of phosphor, essential to all life, are considered adequate in quantity, although there are technical difficulties involved due to the low grade of the deposits, and thus price increases could be expected. The available reserves are scanty in the case of silver, mercury, platinum and helium, as well as, of course, food, and fuels such as oil and gas.[8] Other raw materials, the working group sees as available to a limited but not yet scarce extent.

There is only one response to the challenge of raw material scarcity: modifying the technology so as to be able to exploit as substitutes other materials, which are yet not so scarce. In the case of raw materials of limited amounts, we can try to utilize inferior deposits; it is in fact precisely here that the non-waste technology, with its efforts at increased efficiency, can lead to good results in shifting the boundaries.

Many, however, put their hopes in the possibility of discovering new and rich deposits, thus trying to eliminate the problem of scarcity altogether. It is certainly possible that good deposits may be found, for example at the boundary regions of the continental shelf or in heretofore unexplored regions. However, we must take into account that there has been not only constant quantitative growth in the use of raw materials but also qualitative change at the same time. If in their sufficiency, resources were to keep up to the level of previous years, due to this qualitative change new and rich deposits would be needed to cover in an even higher proportion relative to yearly consumption than in earlier years. This process of qualitative change can be briefly outlined by means of the following three components:

the end consumption of raw materials has increased greatly in extent, in the case both of self-renewing and of non-renewable materials;

at the same time, there has been a change in the relative proportion of renewable and non-renewable materials; the share of renewable materials has decreased and that of non-renewable resources has increased in end consumption;

finally, the primary production of both renewable and non-renewable materials has shifted to a relatively more intensive use of non-renewable input materials.[9]

One consequence of these factors has been and still is the disproportionately large additional need for non-renewable resources relative to the growth in production, and thus an increased dependency of all production on the reserves of these materials and on those who can control them. I should in fact emphasize that in the future the actual commercial availability of raw materials will be a quite different matter from their actual sufficiency, and technical exploitability, as has been clearly shown by recent developments in the case of oil. In this sense, I think that it would be quite useful for engineers, as for others, to take a look at global development programs such as the United Nations' New Economic Order.[10] Since I do not think that we can trust in new and rich deposits to solve our problems of sufficiency, technological development must be directed toward other alternatives. This re-orientation may be called that of low-grade technology, and its application will be needed both on land and on and in the ocean.

The solution of resource problems by this means, however, implies that the solution of environmental problems and especially of energy problems will become much more severe, at least in relation to the lowering of the grade of deposits exploited. I shall, however, leave the discussion of energy problems at this point, and trust that these problems are better conveyed by picture.[11]

An assessment of energy problems

	Energy supply	Feasibility						
		A	B	C	D	E	F	
Long-term options	Solar	Solved	In part	In part	In part	In part	Solved	Flow
	Geothermal	Solved	In part				Solved	Flow
	Fusion	Solved					In part	Flow
	Breeders	Solved	Solved	In part				Stock
	Hydrogen fuel	Solved	Solved			In part	Solved	?
Short and medium-term options	Conventional fission	Solved	Solved	In part	Solved	Solved	In part	?
	Fossil fuels	Solved	Solved	In part	Solved	Solved	In part	?
	Low-grade heat	Solved	In part	In part	In part	Solved	Solved	Flow
	Wave and air motions	Solved	In part	In part		In part	Solved	Flow
Science fiction	Gravitation waves							Flow
	Black holes							?
	Granite as fuel							Stock

A Scientific feasibility
B Technological feasibility
C Practical feasibility
D Economic feasibility
E Social feasibility
F Ecological feasibility

|——►| Solved
|► | In part
| | Unsolved

REFERENCES

1. P. Malaska, *Mankind's Dowry and Technology*, FEANI seminar on Engineering and Education for Environment, Stockholm (1976).

2. Report of the *ad hoc* meeting of experts on the principles, concepts and practices of non-waste technology and production systems, Economic Commission for Europe (ECE) 7 (1975).

3. Some examples to illustrate the concepts of non-waste technology. Note by the secretariat, Economic Commission for Europe (ECE), 17 (1975).

4. P. Malaska, *Prospects of the Future of Technical Man*, Helsinki 8 (1971).

5. P. Malaska, Toward engineering conclusions in human problems of environment. Annual meeting of den Norske Ingenjörförening, Oslo, 20 (1972).

6. N. Georgescu-Roegen, Energy and economics I. *Ecologist*, 5, 5, pp. 164-174 (1975); Energy and economics II. *Ibid*. 5, 7, pp. 242-252 (1975);

7. V. Colombo, Research and the future management of natural resources, lecture VOF-seminar, London, 43 pp. (1974).

8. M. Mesarovic and E. Pestel, *Mankind at the Turning Point*.

9. P. Seiskari, Ihminen ekosysteemin häiriötekijänä. Helsingin Sanomat 23 marrask. (1974) (Finnish).

10. Reviewing the international order (RIO), interim report 17-20 June 1976, Rotterdam (J. Tinbergen as coordinator), 29 pp.

11. P. Malaska, Analysis of the World energy game. *Turun Kauppakorkeakoulu*, Serie AII-1: 1975, pp. 221-241 (1975).

Eco-Productivity: A Positive Approach to Non-waste Technology

M. G. Royston

Centre d'Etudes Industrielles, Geneva

INTRODUCTION

Pollution is waste. Waste today leads to shortages tomorrow. "Waste not want not" is a motto as true now as it was for all those generations before the brief flowering and decaying of the affluent/effluent society. The very sustainability of dignified life on this planet earth must depend on re-establishing a non-waste society—a non-waste economy—a non-waste technology—and, above all, a non-waste value system.

If we talk about recycling wastes back into useful products, and if we do not have abundant cheap labour (and time) available as did our thrifty* forefathers or as might be available in some developing countries, then we may have to talk not about non-waste but rather about low-waste, where the "low" is defined by some minimum level of net energy usage.

But is this unfortunate situation when we have to "waste" in order to avoid waste due to our approach to non-waste technology?

It would be the case if non-waste technology were concerned with the technology of making useful products out of the wastes from some other technology which took no account of the value or cost (economic or social) of its residuals. However, this is not the case—despite all temptations for it to be so. Non-waste technology is the technology based on the conceptualization of the total system of raw material supply—production—consumption—disposal and recycling, viewed in an integrated and a systematic fashion so that no waste occurs.

So, if our perception of the possibilities of achieving a true non-waste system are limited, perhaps we can look at the definition from a different viewpoint. And by doing so, avoid the other trap which is that of a world without waste, achieved by eliminating production, industry, agriculture and even people.

What then is the positive version of "non-waste"? It is, in fact, something like "all-product", i.e. if our technology produces no wastes, it produces instead only products (which are by definition useful). The definition is basically the same, the technology should be basically the same, but the process of technology selection will be different since it would set out to seek new sets of products which could be produced in a technological system without

---

*Thrift—saving ways; sparing expenditure } *The little Oxford Dictionary*, 1945
Thrifty—economical
Thrive—prosper, grow vigorously
Common root "Thrifa", Old Norse

engendering waste. Such an approach would automatically generate the alternatives which we lack so much at the moment. A world of maximum (eco-) product (-ivity), i.e. without waste, would be a world with a sustainable yield of food and drink, goods and services to meet the needs and provide health and well-being for this and future generations.

Let us remember the case of the thrifty Scots. That large distillery in Scotland which was dumping its wastes into the nearby river was threatened by the law either to clean up or to close down. It did clean up, by concentrating and drying its wastes and selling the product (waste) as highly nutritious animal feed. The plant paid for itself in 6 months and generates over £1 million additional revenue for the company each year.

What might the world look like then, if all waste were prohibited by law and if all sectors of the economy were enjoined to search for and then produce products which would meet the basic needs of man for physical and mental health, i.e. the food, the water, the air to breathe, the shelter, the transportation, the recreation, the protection from disease and from the climate which form the basic elements of that health, and the energy, raw materials, land, oceans and technology which provide those elements? Let us, then, develop this particular eco-productivity scenario a little.

SCIENCE AND TECHNOLOGY

In a finite world, the one resource which is unlimited is the human spirit and the love, sense of purpose and quest for knowledge that flows from it. Indeed, the one resource in this world which grows is this resultant knowledge and from which human understanding, human wisdom and, hopefully, human institutions and technology spring. Thus, one key to the new 'product-not-waste' society is the liberation of the human spirit, the encouragement of new scientific research and the application of the new insights to develop the new systems which meet human needs without creating waste. Particular areas to be stressed for new research might be the theoretical bases for a systems engineering of the integrated units which will produce a wide range of products without waste, with particular stress on the integration of inputs; secondly, research on the basis for selecting the outputs, i.e. those products which contribute most to meeting human needs; and, finally, research on renewable resources of both energy and raw materials.

ENERGY

What is the "product-not-waste" form of energy? The use of fossil fuels such as coal or oil produce waste heat (hot air and cooling water), ash, carbon dioxide and water vapour, nitrogen oxides and sulphur oxides. In shipping oil, we could avoid the waste from tanker washings, and the oily ballast waters as is done at the Port of Ashkelon, where the oily ballast water and residues are pumped ashore and passed through highly efficient separators to recover the oil. The installed piping and pumping system allows a faster turn-round of the tankers and at a recovered oil price of $12 per barrel, the whole operation is self-financing. Pollution is avoided and a valuable resource is recovered by thinking the system through with the positive view of avoiding waste and maximizing recoverable products. But, if we cannot use as products the carbon dioxide or the sulphur resulting from the combustion, then the application of the strict "product-no-waste" rule means doing without fossil fuels. Nuclear energy might appear to be better in this respect. Application of the "product-not-waste" principle would mean that power plants would be built in an urban - industrial agricultural - mari - cultural complex such that the cooling water passed first to industrial processes for space heating/cooling, then to green-houses and field-crops and, finally, to the fish ponds so that the totality of the energy is productive. Nevertheless, even here, the strict application of the "product-not-waste" principle would mean that the loss of uranium ore, the high-energy consumption of fuel production and the haunting and unresolved problem of radioactive *wastes* would rule out nuclear power also.

Looking at all the various alternative energy sources, the one which conforms best to the "product-not-waste" principle is solar energy—not via panels of photo-electric cells, not even necessarily by the cheap heat-collection panels, but by the better management and increased productivity of the world's forests and arable lands. Suffice it to point out as an example that either those tropical forests whose soils are unsuited for agriculture, or tropical arable land producing starchy roots such as manioc, or the wastes of the world's towns, farms and forest would each meet all the world's energy needs by direct combus-tion or by prior physico-chemical or biochemical processing. There are similar examples of converting wastes to products in many industries. In the public sector, in Europe again, it has been common practice for many years to burn garbage in specially designed plants in order to generate electricity. Such plants exist in Geneva, Zürich, Münich, Stuttgart, Paris and many other cities and can provide around 15 per cent of a city's need in power.

Also in this area, a number of power plants in Europe have for many years used their waste heat to supply hot water and space heating for houses and apartment blocks.

The lack of development of these processes in the U.S. is almost entirely due to the much lower energy costs in the U.S. compared with Europe. Since the oil crisis, however, American engineers and city authorities have made up for this lack of interest.

In the case of using trees and plants to capture solar energy, we have the ideal situation of a useful product (liquid, solid, gaseous fuel or electric power—still with integrated waste" heat utilization) being produced without waste since the residues such as leaf and bark and ash return the fertilizing nitrogen and minerals to the soil and the combustion products of carbon dioxide and water are the raw materials for photo-synthesis to build up new trees and plants. There would be no waste, the global heat balance would be maintained and the economic and social benefits accruing to those rural areas in tropical and semi-tropical areas would be enormous. Increased forestation would even ameliorate the climate, increase rainfall and prevent floods and soil erosion.

With a non-waste or eco-productive energy supply based on green plants, whole new vistas of non-waste technology open up.

RAW MATERIALS

Organic Chemicals

The production alternatives for organic chemicals are somewhat similar to those of energy inasmuch as the use of oil, natural gas and coal (which in the present world should be reserved for chemical synthesis) would in the product-not-waste" world produce carbon dioxide and sulphur wastes, waste heat and would "waste" finite resources. However, the natural process of applying pressure and temperature to rotted algae and wood to produce oil plus gas and coal, respectively, over a few 100 million years, can be the basis of a new technology using higher pressure and temperature to produce solid carbonaceous, oily and gaseous fractions from virtually any natural or synthetic organic material in a matter of a few minutes. Thus the old petro-chemicals and their associated wastes can be replaced by a new range of "bio-chemicals" produced without waste. Nevertheless, within this "scenario", it might be noted that considerable advantages accrue from the integrated use of energy production from biological materials, and organic chemicals production, e.g. from pyrolysis, the solid and gas used as fuel and the oil fraction used as "bio-chemical" feedstock. The organic chemical products thus produced would include polymers which could be designed for recovery and recycle, either integrally, such as is done to some extent with rubber tyres via recycled "crumb", or like polystyrene, such that the application of heat causes the material to break down to its monomer from which "virgin" polymer can be made.

However, it should be remembered that in the "product-not-waste" world, construction materials, clothing and packaging materials would be made from substances such as lignin or cellulose or sugar extracted from the vegetable material or directly from the solid or fibrous material itself. Selection of the route to follow would be by minimum cost (economic plus social plus energetic).

Naturally, the biggest group today of non-waste organic chemicals are the classical bio-chemicals which are those made by biochemical rather than physico-chemical processes, i.e. by fermentation. These range through alcohols, ketones, aldehydes and acids through polysaccharides and polyamides to amino acids, vitamins and antibiotics. It has to be remembered that to these as an energy source and chemical can be added methane, the sewage gas of old, and the "bio-gas" of new, low impact technology.

The "product-not-waste" aspect of these biochemicals is that they are produced from readily available raw materials such as corn steep liquor, molasses, manioc mash, human, animal, farm or food factory "wastes" and that their own "waste" can in turn be used as animal feed. Thus the production of biochemicals should be integrated with the farming, animal husbandry and food processing which can provide its raw material and use its by-product (waste).

This idea of producing a useful by-product instead of an environmentally degrading waste product is not new, of course, although price competition from other cheap products has tended to make it less widespread than formerly. It is ironic that in the 1960s, many Scandinavian pulp mills stopped converting their liquid wastes to protein and industrial alcohol because they could not compete with imported soya and petrochemical alcohol.

Inorganic Chemicals

Many inorganic chemicals are made from materials extracted from land, water or the air. In many cases, the production of the chemicals includes the creation of wastes which are discarded again on land, water or the air. The production of phosphoric acid from phosphate rock involves the creation of a by-product (waste) which contains calcium sulphate from which plaster of paris can be made. Thus, again, integrated phosphoric acid and gypsum industry is eco-productive and not wasteful. Many inorganic chemicals are used in industry, e.g. acids for pickling, alkalis in paper-making, cyanides in electroplating, etc., which can be recovered and reused by rearranging drains and washing procedures and by putting water on a closed cycle with hierarchic use and recovery of chemicals from the relatively concentrated streams. The key to avoidance of waste is segregation so that mixing and dilution are minimized. This leads a tube manufacturer in Australia to recycle pickle acid and make a range of iron-based pigments as a by-product and also enables miniature battery manufacturers to recover the mercury from cells returned to the retail point by the consumer.

In general, the production of inorganic chemicals can be readily achieved without waste due to their readily defined physical properties, e.g. crystallizing conditions, so that internal separation and recycle of mother liquors present few problems. During industrial use, segregation and internal recycle can lead to new "products", e.g. steam and salts from the incineration of black liquor. The domestic use of inorganic salts is likely to end up in the ash from power plants burning garbage and, as such, can be used as aggregate or hard core in construction and road building.

Non-metallic Minerals

One major feature of waste is the devastation of the countryside which results from large-scale mining of chalk, limestone, gravel, sand, clay, etc., as well as for the metals to be discussed in the next section.

To the west of London, England, the floor of the Thames Valley is scarred and pocked with ugly waterlogged holes, left by years of digging for the gravel used as concrete aggregate. In this same area, crowds acclaimed the winner of the 1975 World Water Ski Championships which were held in a newly built water sports complex. The complex was in fact produced as a result of the gravel-winning operations of the Ready Mixed Concrete Company—the company which now owns and operates the recreational centre for its own profit and for the benefit of everyone who lives in the Thames Valley. In the U.S., Amax has also created recreational areas from its mining operations.

Thirty years ago, English Clays, one of the world's largest producers of high-grade china clay, set about developing and marketing a prefabricated home, the Cornish Unit House, in order to consume the ever-increasing quantity of mine wastes which were threatening the surrounding countryside.

The "product-not-waste" principle is clear here as it is for the cut-and-fill operations of road and railway construction. Thus, we need to look at other needs for recreational areas, harbours, etc., and coordinate these projects with the extraction of minerals.

The other aspects of this particular approach are the use of mining operations to improve agricultural land after open-cast mining, for example, creation of recreational hills on coal spoil tips, as in the "Wigan Alps" in the old coal-mining areas of industrial Lancashire, and creation of ski slopes as on Mount Trashmore in the USA.

The other side of this particular coin is the use of waste stone, sand, concrete, etc., as fill for natural and man-made holes, construction of harbour works, erosion and flood control or coast defenses, as well as incorporation in construction works as hard core and aggregate for concrete.

One of the most widely spread minerals is silica, and one of the most widely used products, and one with a high product/waste potential, is glass. Here, the case of present-day Switzerland shows the way in which a "product-not-waste" approach requires not only a whole infrastructure to back it up but also a strong conservation value system.

Thus in Switzerland, more than half of all glass containers are returned and re-used as such because there is standardization of 1-litre bottles for mineral water, soft drinks, fruit juices, beer and wine, and there is a deposit system which operates via semi-automatic receivers of empty bottles in most supermarkets. In addition, half of the non-returnable bottles and jars are recycled via segregated bins in each neighbourhood into which the consumer puts empty green, brown or white bottles.

Thus, we see a need for, and the effect of, a public education process which calls not for the destruction of the present industrial and technological system, but rather for a redirection of technology which will, over time, reduce progressively both pollution and the demand on resources. It is against this background that the use of the returnable bottle, which involves three times less energy than its non-returnable counterpart, is seen not as an isolated instance, but as part of a whole programme of fighting against waste.

Already, manufacturers, at least in Europe, are beginning to realize that there is a vast untapped reservoir of consumer motivation in the non-waste area. The evidence for this new attitude is the outstanding success of other domestic waste segregation schemes which have sprung up all over Europe and which are, surprisingly, continuing to engage the co-operation of the consumer. It appears that the reason why individuals are prepared to go to such personal inconvenience to segregate paper, metal and three different colours of glass at source, is not for the $250 worth of material a ton of it might contain, but because it is felt to be a personal contribution to conserving resources and hence helping the sustainability of our society.

This reaction is particularly due to a growing consciousness of a crisis looming due to the inevitably finite nature of resources of raw materials and fuels. And even more acutely due to our growing awareness of our own wastefulness, not only due to our throw-away habits but also in the very products and services we use.

All this is building up a groundswell of opinion which industry and government are responding to. An opinion that requires that waste is stopped at all stages of our industrial society, from producing and transporting raw materials and fuels, to our manufacturing processes and finally to the products we use and how we use them and what happens when we junk them.

Metallic Minerals

The application of the principle of "produce-not-waste" to metals would also be a severe but salutory lesson. The example of American Metal Climax designing their molybdenum mine in Colorado as a recreational site has already been mentioned. In manufacturing of metals, one will have to adopt product-oriented closed-circuit techniques such as the Peterson process for alumina production which produces iron and cement as sole by-products, followed by smelting with total fluorine retention and regeneration of synthetic cryolite. In steel-making, charcoal would replace coke (as it has in certain South East Asian and Japanese blast furnaces) or even the OSCAT process using chemical extraction of iron followed by forming parts by powder metallurgy. Copper ores can be smelted with charcoal, steam and air which produces ingots and elemental sulphur. Which raises a crucial research question about whether we can find a use for sulphur, e.g. in raising the fertility of sulphur-deficient soils.

However, the basic "product-not-waste" aspect of metals is, of course, designing metal products for recycle. Since recycling aluminium uses only 5 per cent of the energy required for making primary aluminium from bauxite, and recycling iron only 28 per cent of the energy needed to make it from ore, the lesson is clear.

But we will have to design our products, from our beer cans to our automobiles, so that we can easily segregate and recycle relatively homogeneous streams of aluminium, steel, copper, zinc, etc.

Where technology fails, substitution by other materials can step in. Thus, the selection of materials and the design of products will be for their ease of recycling rather than for their ease of manufacture or marketability.

AIR

Animals need the oxygen of air, plants need its oxygen, its nitrogen and its carbon dioxide, all need this canopy to protect them from the ultra-violet of the sun, to cool in summer and warm in winter, to carry the life-giving rains. However, as Shakespeare said,

> "This most excellent canopy,
> The air,
> Why, it appears no other thing to me,
> But a foul and pestilent
> Congregation of vapours!"
> Shakespeare, *Hamlet*

We waste "air" not only quantitatively, but also by degrading its quality when we use it in our power plants, in our chemical and steel plants and in our fertilizer plants. The composition of the air is changed quantitatively and the quality of the air changes also by the innumerable rejects it receives from hydrocarbons to nitrogen oxides or dust. The "waste" in human terms alone is the toll of air-pollution-induced diseases and death.

Eco-productivity means using the components of air more effectively without changing it quantitatively or qualitatively. This means growing more plants and, when we burn plants, returning the phosphorous, nitrogen and potash to the soil, shifting from inorganic nitrogenous fertilizers which "waste" air to recycled organic "wastes" and leguminous crops.

It also means changing the quality of air only insofar as that changed quality can be a useful product. This means limiting the production of carbon dioxide to that which the green plants can absorb. Otherwise, the "greenhouse" effect will cause disastrous consequences to our delicately balanced climate. At the same time, we must limit sulphur and nitrogen oxides which might cause a haze of ammonium sulphate aerosols which, in turn, would affect solar reflectivity and hence the global heat balance. Finally, we must limit the generation of excess heat itself, for fear of over-heating the planet, melting the thin layer of Arctic sea ice and precipitating a major climatic swing.

All of these misuses of air can be avoided if we use air productively to carry the carbon dioxide and water vapour from our renewable fuels back to the green plant "fuel forests" from which they came.

Air is so essential for the planet and for human health that we must cease to look at it as a waste bin and start looking at it as a precious and valuable product. Then we would not use the air to receive fluoro-carbon from spray cans, putting at risk the protective ozone layer and hence the possibility of problems from increased skin cancer and increased destruction of soil micro-organisms and, hence, reduced soil fertility. Instead, we might use air productively as the propellant in a pump-spray.

Thus, we need to "produce" clean air with more trees and plants and we need to use air more effectively so that all its uses are productive and beneficial.

LAND

John Donne said "Everyman is a piece of the continent, a part of the Main. If a clod be washed away by the sea, Europe is the less", and ended: "therefore do not ask for whom the bell tolls, it tolls for thee". The future of man on this planet is the future of man's care for land.

If man loses the environmental battle and destroys himself, it is likely not to be due to pollution, nor even because he wasted land—that limited and precious resource—but because with a few exceptions, such as the Dutch and the Balinese, he did nothing to increase its extent nor to increase its yield. Nevertheless, it has been suggested that 6.7 per cent of the land-surface of the world has been lost to cultivation by bad land management. A figure easy to credit if we remember that day on 12 May 1934, when 350 million tons of rich topsoil was lifted off the Great Plains of the USA and deposited 4000 kilometers away in New York, Washington and out into the Atlantic Ocean. This loss affects us all and all will feel the consequences.

Erosion must be reversed, land reclaimed from the sea, deserts pushed back and mountains clothed again. It is a battle that man is losing because he has destroyed the trees—trees which stabilize the soil on coast and mountain, trees which reduce the wind, attract the rain and store it beneath their roots, trees which hold back floods and mulch the ground with their leaves.

The eco-productivity approach is to produce more productive land by planting trees on the coasts, up the mountain slopes, even in the deserts, helped by pre-cultivating their roots in plastic tubes so as to conserve moisture. Once again, just avoiding the wastage or degradation of land is not enough; we must through all our operations, agricultural, industrial, urban, etc., aim to create more land and more productive land, ensuring that the trees themselves provide us with the fuel and food we need.

Another aspect of eco-productivity and land is the composting of garbage. Whereas once, land was lost under vast unsanitary garbage dumps, now, in many countries—around the Mediterranean, for example—wastes are composted and spread on the land to bind it and nourish it.

We waste the land by covering it with concrete and wastes, by letting it erode into the rivers and seas, by allowing it to turn into dust bowl or salt desert. Again, we can try to stop this waste, or we can try to extend and make more productive our land. Does the urban conglomeration with its high rises save land when one takes into account the surface occupied by the commercial services and distribution and transportation systems needed to support it? Is not the cottage with its small-holding more productive, in which every square metre is producing food, visual appeal and a sense of belonging?

Once again, the positive non-waste technology approach is to ensure that when we build roads or cities, we aim to create at the same time more productive land by building on less productive areas, but also using the new infrastructure of transportation, water, electric power and sewerage to expand the productivity of adjacent land areas, to reclaim land from the sea, to drain swamps and hence add to land resources and improve human health at the same time. How much better to enjoin the developer to maintain or increase the area of productive land through his project rather than just to "waste" as little as possible or, worse still, to allow him to destroy land without restriction.

So, if we look at the whole system, we see how the increase in forests for fuel and raw materials will arrest erosion, how the eco-productive non-waste mining, industrial and construction operations will create new land for farming and recreation, how use of organic fertilizers, maintaining hedgerows and a diversity of crops all reduce the tendency to form dust bowls. Eco-productivity in land use will affect the quality of water and air and health and life itself. A thread which runs through eco-productivity is that single solutions using this principle yield multiple benefits and integrated multiple approaches reinforce each other even more strongly.

WATER

As St. Exupéry said, water is "not necessary for life but rather is life itself". Unlike air which can be wasted quantitatively globally, water is only wasted quantitatively locally with drought and desertification induced by man's shortsighted stupidity. Globally, however, we do "waste" water qualitatively by pollution of streams, rivers, lakes and oceans which render it unfit to drink, to support fish life or to support the very plankton of the oceans which produce two-thirds of the oxygen we breathe. Air, land and water are inextricably interrelated. Old pollution-control techniques simply shifted the problem from one area to the other and, in the end, the poisonous sink is the ocean from which all life came and, ultimately, on which all life depends: this is the so-called "high-stack" syndrome.

But more, the trees we advocate as a major instrument of eco-productivity halt the deserts and induce rainfall as well as holding back floods. The management of land by contour ridging conserves the water. Our eco-productive industries prevent degradation of water by waste materials and waste heat and, in fact, would enhance its currently degraded quality, leading in turn to less "waste" in terms of human disease and death and "waste" of a major recreational and food resource.

However, the most striking feature of water is simply the way we waste it. From the way we let it run through deserts and flood, unused and unproductive, into the seas, to the way we use such vast quantities to flush our lavatories,, the water scene is one of universal waste.

Eco-productivity means that we need to recharge the aquifers, build more reservoirs and barrages, to plant more trees, to irrigate more with low-cost drip irrigation systems, redesign our water systems so that water is used in a hierarchic fashion with the water passing from one use to the next—first cleanest water or coolest water for the cleanest and coolest need, passing from power plant to industry, to houses, to agriculture, to fish ponds, where the plants and fish thrive on the tepid, rich waste waters. Water re-use is already a fact with the citizens of Rotterdam drinking water which reportedly has already passed 40 times through the human stomach.

In industry, water use can be put on complete closed cycle with conservation through hierarchic use, pulsed rinsing, countercurrent flow, etc. In the home, hand washing of clothes, redesigned lavatories, pulsed-mist baths and wash basins, pulsed sprays for car wash and drip irrigation for gardens would cut water consumption to a fraction of the current level and produce a concentrated sewage which could go directly to methane generators with the residue used as fertilizer on the land. Once again, an eco-productivity approach in one sector reinforces the effort in another sector, and maximization of the productive use of water while banning completely waste, would ensure ample water for all.

COMMUNICATIONS AND TRANSPORTATION

Using energy consumption as criteria, we know that in increasing order of eco-productivity and decreasing order of energy consumption, our transportation options for goods are air freight, truck, railway, ship or barge, pipeline and, finally, hand, and for people are air, car, rail, bicycle and, finally, foot.

In communication for business meetings or education, a gramme of material in a satellite handles more communication than kilogrammes of wire in a telephone line or tons of fuel to travel to talk to people. The eco-productivity system of transport and communication will be a strange mixture of canals, bicycles and space satellites.

SHELTER

In the area of our towns and our houses, it is almost as if man has pervertedly decided to turn the precept of "product-not-waste" on its head. Our cities produce little that contributes to the health, safety and well-being of man, but does it in a most costly and wasteful way. We pile buildings on top of each other at great cost, divorcing the work place from the recreation place, from the residence, from the area where foods and goods are produced, thus involving ever more cost. Rates of disease, violence, death, etc., are all higher in cities and demand ever more costly infrastructures to cope.

Our houses are factory made, alien and inefficient, hot in summer, cold in winter, too light in the day, wasting resources to make them liveable.

The "product-not-waste" town might be the small integrated community of no more than ¼ million, with integrated industrial, agricultural, residential, commercial, educational and recreational facilities. The diversity and interrelatedness of the elements makes it a place to stay in, not to migrate from. The size is big enough to provide the necessary services and diversity, but small enough to develop an identity, a spirit of community and a degree of personal involvement and commitment which reduces alienation and violence and promotes opportunity for direct citizen involvement—the only sustainable political system. Modern satellite telecommunications would link this "city-state" to its neighbours and the world.

In such a city, the hierarchic use of water and energy, the use, recycle and re-use of materials would all be facilitated by the short distances and the diversified economy. What can be more eco-productive than the self-supporting community?

Houses would be low cost with basic modules provided and additional modules added by the owner. As is well known "do-it-yourself" solar heating units are cheap compared with bought-in units. The "product-not-waste" house would depend largely on such cheap yet installation labour intensive, devices of the modern alternative living style, such as solar panels, wind mills, heavy insulation, heat pumps operating on a large sump of collected wastes, bio-gas, irrigation systems using digested waste waters, etc.

EDUCATION, POLITICS AND CONFLICT

Billions of dollars every year are lost by conflict outside and inside industrial plants because of the alienation which exists between citizen and technology, between workers and management. This "waste" results from alienation of the basic goodwill of society and can be combated by increasing participation at all levels. Participation requires education and thus another interaction with the "eco-city" is as a setting for "product-not-waste" education. This education could then be the learning of attitudes through interacting with one's fellow citizens, plus the learning of skills through working in the diverse economy as well as the all too conventional acquisition of knowledge by sitting in the classroom. The challenge is clear: If we can't participate and do-it-ourselves in the system, we will opt out of the system and do-it-ourselves all the same.

GOODS AND SERVICES AND EMPLOYMENT

Since Vance Packard wrote *The Waste-Makers* in 1960, one has only been able to remark the further acceleration in the gallop of the waste built into our consumer society. Rather than add to the copious literature which already exists on the subject, suffice it to say that the "product-not-waste" criterion and the total ban on waste that we have previously hypothesized would lead to long-life products, recyclable products, products sold with explicit description of their energy content, e.g. refrigerators, their impact on the human environment, e.g. detergents, and the means to recycle, e.g. lubricating oil. Packaging would be minimum, functional, standardized and re-usable. Emphasis would be on durability and maintenance rather than built-in obsolescence and the throw-away economy. Incidentally, such an approach would increase employment as well as conserve resources, since maintenance and service is more labour-intensive than manufacturing. Here again, we find a link point. Modern industry in the mature as well as in the developing economies "wastes" labour by installing automated equipment, not only in the mines and foundries where it saves human suffering, but also for saving labour in industry generally, while there are jobless on the streets and drawing the dole. Eco-productivity should lead to increased employment and decreased income differentials, and as such, reduces "wasteful" social and industrial conflict.

FOOD

The issue of wasted food production capacity runs through all the above sections. Again, this area has many dimensions which interweave those already examined. Food is wasted by pest and rot after harvest—up to one-third of the crop in some countries—production potential is wasted by growing grain and pulses to feed animals when that food would feed the people of the world, there is waste by destruction of fish stocks by water pollution and bad management, by destruction of animals and plants by air pollution, destruction of good farm land by bad urbanization and industrialization, by water and air erosion, by creation of dust bowls, salt pans and the inexorable march of the deserts. At the same time, let it be remembered that the most extensive pollution is due to those wastes—which choke our dumps and poison our ground waters, which flood our sewers and destroy our rivers, and which run-off our farmlands to eutrophy our lakes and estuaries—which are food wastes from man, industry and farm animals.

Eco-productivity means closing this cycle so that nutritive wastes are used directly as food by man or beast or land, not as a poison for water and fish. We need to develop the use of systems such as rotational pig-raising—muck spreading—barley production—pig feeding, spraying liquid wastes on land, feeding cheese whey to pigs, feeding dried chicken droppings to pigs, feeding solidified food industry waste or human and animal waste to fish, feeding algae with sewage and then fish and chicken with the algae, feeding cattle on their own manure after alkaline and acid digestion, growing yeasts and moulds on starchy wastes to produce highly nutritious human food, i.e. micro mushrooms, or on citrus wastes to supply a nation with an indigenous supply of animal feed, etc. The list of "product-not-waste" opportunities in the food area are endless as we saw with the North British Distilleries in Edinburgh or as we see from the electrostatic precipitator in the Tel-Aviv instant coffee plant which recovered coffee dust and which also paid for itself in 6 months.

Another aspect of the problem is, of course, increased production with less waste and less energy. We know that wet rice culture in Indonesia takes 0.02 calories per calory of food produced, intensive maize 0.5 calories, grass-fed beef 3 calories and feedlot beef 16 calories per calory of food produced. The challenge then is to produce food in quantity and quality and on a small land surface. Here, we can learn more from studying Indonesia than we can from studying Iowa. We can meet the challenge by serial sowing and by intercropping, alternating high and low plants, shallow-and deep-rooted plants, leguminous and nitrogen feeding plants. We need small machines to work single rows of such diverse plants. And in addition, the constant stand of diverse crops holds the rain, prevents drying out on the one hand and soil erosion on the other, and reduces crop damage from animals and insect pests.

Finally, we come to the whole area of increasing food production by better management of land and of water. We can push back the deserts by trees grown in plastic tubes, we can use drip irrigation and hydroponics to cultivate arid lands, we can put "heart" into the sands and rocky soils by "composting" our wastes and spreading them on the inhospitable ground, we can gain land from the desert and from the mountain. We can farm the fish in ponds as well as in lakes, bays and shallow gulfs, we can increase fish productivity many times over, we can manage the oceans more effectively, not only conserving the traditional catches, but by developing the means of catching the other 99 per cent of the oceans' renewable protein resources such as the krill and the giant squid. Again, while we hoe the ground or cast the net, we should not forget the role of the satellite in tracing schools of fish, detecting rain, fire or locusts and so aiding the farmer and fisherman alike to maximize his production and minimize his wasted crop and his wasted effort.

HEALTH AND THE ENVIRONMENT

Finally, we come to the central question of human health. Here, the key question is in "development" of health or how to avoid the astronomic waste in human skill and energy as well as in the health budgets of most nations. Essentially, the point is: do we treat the symptoms of disease or do we prevent it happening in the first place? Obviously, the latter is preferable and likely to be less wasteful. Here, the environment plays the essential role since we now see more and more clearly that the quality of the environment largely determines human health. In industrialized countries, at least half of all pulmonary or liver disease is due to air pollution and water pollution, respectively. Up to 90 per cent of all cancer is due to environmental factors, introduced by man himself. Coronary heart disease is due to our unhealthy life style, too much animal fat, too much worry, not enough exercise. In developing countries, the death toll is due to infectious diseases in the environment induced largely by bad water supply, inadequate sanitation and unsuitable housing. Should we provide modern hospitals and intensive-care units to heal the sick or tackle the root of the problem with local medical aids, simple sanitary measures and education for a healthier life style? The eco-productivity or "product-not-waste" answer is clear. It is better to produce health than to waste precious resources in treating disease. And the best way to "produce" health is to produce a good environment with clean air, clean water, healthy food and proper shelter. And finally, the best way to produce this good environment is to make sure that we concentrate on producing products which do not in their origins, production, use or disposal create waste. This then is the challenge of eco-productivity, "product-not-waste" or the positive aspect of Non-waste Technology.

CONCLUSION

Hence, for sustainable development of human health, safety, happiness and well-being, we need to re-direct ourselves towards this Non-waste Economy and the technology it will use. The great advantage of this new technology should be its low cost, which is due to the fact that it is an integrated technology, involving no double handling of wastes and products. The major obstacles lie in the imaginative effort required to recast the product mix and the diversification of business activities which will result from the new products plus the major problem of financing the new processes. Thus the concrete company finds itself in the leisure business, the whisky company in the feed business, the clay company in the construction business, the municipality in the raw materials and energy business, the port authorities in the oil business, etc.

The common feature of all these processes is that they are designed to produce not a product plus wastes as in conventional processes, but a complete range of products instead. Thus, two major problems are tackled at the same time: pollution due to the dumping of these wastes in the environment and the increasing shortage of raw materials due to wasteful production and consumption practices.

A reaction is setting in about the fact that it can take 10 times as much energy to produce a pound of animal protein as a pound of vegetable protein; that it can take 100 times as much energy to transport by air as by water, that an automobile requires 30 times the energy per passenger mile as a bicycle; that making products from primary aluminium requires 20 times the energy of the same product from recycled aluminium.

In particular, we should stop these wastes ever appearing and we should do all we can to ensure, by pre-planning and design, that more and more we produce only products which are useful to society. This then is where Non-waste Technology enters, a concept, which appears, incidentally, to transcend all geographical and political boundaries, being developed equally enthusiastically in American Metal Climax, the Israeli Department of Shipping and Ports, the Environmental Protection Agency in Washington and the Soviet State Committee for Science and Technology.

It almost appears that avoiding waste is pretty basic to human nature everywhere, as being the right way to grow. Why else does the word "thrift" mean to save and at the same time come from "thrive", which means to prosper? Perhaps after all there is more truth than we recognized in the old addage "Waste not Want not". So let us not waste so that we will not be in want.

Concepts and Principles of Non-waste Technology

J. D. Schmitt-Tegge

1. DEFINITION

Encouragement of "non-waste technologies" is intended to increase application of methods of production in which, compared with other methods:

- less or less dangerous production-waste arises;
- raw materials and energy are used more economically;
- products are manufactured which in their use and on disposal produce less pollution, i.e. which have longer life-expectancy, less energy consumption and are less liable to emit harmful effluents while at the same time permitting simple and cheap disposal.
- Non-waste technology in the narrow sense involves production planning and production processes in the internal works field. Cycle processes in the works fall under this definition which also covers manufacture of products which are purpose-designed to be re-used (e.g. returnable bottles).

Only in a broader sense can the use of externally "imported" secondary raw materials (i.e. materials from outside the works) be considered as application of non-waste technology, i.e. when primary raw materials and energy are saved and/or less waste or less harmful waste arises from the production process (e.g. utilization of waste paper). However, in broadening the definition it must be borne in mind that utilization of materials (*after having been made use of before*) involves other questions than application of non-waste technology in the narrower sense (*before having been otherwise used*). Correspondingly different problems arise, for example, in establishing, processing and newly developing materials and products.

2. AIMS OF NON-WASTE TECHNOLOGIES

The aims of non-waste technologies arise from the definition: The targets are thus

- to reduce production waste;
- to apply environmentally acceptable production processes;
- to reduce or replace use of raw materials and energy during production;
- to manufacture non-pollutant products;
- to increase the life-expectancy of products.

2.1. Reduction of Production Waste

One important aim of non-waste technologies is to reduce waste arising during production, in particular when this involves problematical special-category waste. By use of corresponding technologies, generation of waste should be avoided or reduced "at the source".

A reduction in generation of waste can be arrived at by

- modification of the production process;
- modification of the product manufactured;
- reintroduction of the production waste into the production process (works-internal).

The aim is to minimize the waste leaving the works assessed by quantity, value and degree of danger.

2.2. Application of Environmentally Acceptable Production Processes

Production processes should be designed in such a manner that at maximum product quantity a minimum of harmfulness is achieved for the effluents and pollutants emitted to the environment. In this connection all types of pollution should be considered, i.e. that causing harm to air, water and soil. Shifts from one type of effluent to another are to be avoided. Difficulties are encountered in defining a production process as *umweltfreundlich* (acceptable to the environment) or *umweltfeindlich* (hostile to the environment).

As a rule legally defined standards of effluent emission serve as means of judging various alternative processes. Comparative assessment of various effluents fails because there are insufficient quantified overall economic and ecological parameters. For this reason optimization of various processes with the aid of marginal cost theory is not possible.

2.3. Reduction/Replacement of Raw Materials and Energy Used

Use of raw materials (in particular raw materials in short supply) and consumption of energy must be reduced. Similar to waste reduction, this may be achieved by

- modification of the production process;
- modification of the product (*example:* reduction in wall thickness but retention of other physical characteristics if possible at same level);
- modification of quality criteria by which products are judged (*example:* reduction or repeal of quality criteria for newsprint). By this means the tendency towards use of secondary raw materials to replace primary raw materials will be encouraged;
- examination of raw materials used with respect to end-purpose of products.

As a basic principle reduction in the raw materials used with the aim of decreasing the *growth rates* of consumption is the most effective method of saving resources and preventing waste. The following example may serve as proof:

The recycling quota of copper amounts at present to approximately 40 per cent referred to the amount of available scrap copper. Were this to be raised to 100 percent, then this approach at the present consumption increase of 4.6 per cent per year and present average life-expectancy of products of 22 years would achieve a postponement of the critical point (exhaustion of known reserves) of 3 years. A reduction in the growth rate to 0 per cent would on the other hand at the same recycling quota of 40 per cent and a product life expectancy of likewise 22 years would lead to a postponement of the critical point by 59 years or 195 years.

The conclusion to be drawn here is that reduction in absolute consumption (lowering of consumption growth rates) and corresponding application of raw material saving technologies have much greater repercussions as regards protection of raw material reserves than recycling.

2.4. Manufacture of Environmentally Acceptable Products

Products can be considered as acceptable for the environment if during their use compared with comparable products they cause a minimum of emitted pollution, a minimum of energy consumption and, after termination of their primary utilization life, can either be re-used or can be disposed of by a simple, harmless and inexpensive means.

Assessment of the environmental acceptability of a product or likewise a production process (see sect. 2.2) is only possible to an approximate extent. The inadequacy of exactly testable criteria is also very evident in this field.

2.5. Lengthening of the Product Life Expectancy

From the point of view of definition lengthening of the life-expectancy can be classified under aims 1 to 4. However, on account of its importance it is classified separately. Lengthening of product life-expectancy constitutes an advantage in all of the named fields of non-waste technologies. On account of the reduction in the amount of goods produced per unit time, it also leads to a reduction in production waste. In addition it lowers raw material and energy consumption. This follows from the observation that if the life-expectancy of a product is doubled this does not imply that the material and energy consumption will also be doubled during its production. To an extent higher-quality materials will be employed. Thus a German study dealing with the development of a long-life car showed that the use of higher-quality materials required to double the life-expectancy only led to a cost increase of approximately 30 per cent.

The switch to long-life articles is only proceeding on a long-term basis. In the case of larger articles (e.g. long-life cars) a time horizon of 20 years is estimated. Developments in the direction of long-life goods normally take place motivated by economic criteria. The influence of backup legal measures, for example, in the field of environmental conservation or road safety (protection against corrosion) constitutes an important factor.

Simulation calculations show that manufacture of long-life articles is not bound to lead to the danger of reduction in jobs. In some cases it may even lead to an increase in the number of jobs in other fields of the overall economy.

3. OPTIMIZATION PROBLEMS. INPUT/OUTPUT ANALYSIS

The question as to the manner in which the aims of non-waste technology are to be approached constitutes the basis for an optimization study. In this connection the following repercussions are to be taken into account from an ecological point of view:

- reduction in waste arising during and after production (use and consumption) including all prior stages and intermediary stages;
- reduction in pollution and effluents emitted in the various production processes and by the product;
- reduction in use of raw materials and energy consumption referred to production/consumption cycle.

As already mentioned, assessment of the various effluents emitted is problematical. This also applies for the individual sections of non-waste technologies. Examples: Is reduction of the use of raw materials to be considered as more important than reduction in energy consumption? Is lengthening of product life-expectancy more important than reduction of pollution caused by the production process? Is it more important to reduce the harmful effects or the quantity of waste?

The decision varies from product to product and from process to process. On some occasions reduction in pollution emission will occupy the forefront of considerations and on others energy saving will be more important.

As a rough guide for works planning we can make use of the model in Fig. 1. This scheme, however, only takes into account production waste of the end-process, not preliminary production.

Solutions proposed for input/output analysis are normally based on linear production relationships. Corresponding models permit not only economic but also ecological parameters such as pollution emission, as well as raw material and energy consumption to be taken into account.

Within the scope of such model calculations we can determine the shares of pollution accounted for by end production and preliminary production. Corresponding quantitative repercussions may be represented as a function of a modified end-demand. An example for this is the correlation between a certain demand for furniture and the SO_2 emission resulting from its manufacture.

Application of low-waste technologies can as a basic principle also be taken into consideration in extended systems. The important factor is that the repercussions of preliminary production stages are also included in the equations. Frequently they are of decisive importance for arriving at an assessment.

Fig. 1. Checklist of the ecological effects of a product.[1] Environmental aspects.[2]

	Production					Utilization/Consumption					Disposal		
	Consumption of raw material	Energy consumption		Refuse		Energy consumption			Refuse				
Product	Rare or vital raw materials	per hour or unit	per weight unit	reusable	not reusable	annual	per working life	efficiency of energy conversion	reusable	not reusable	potential of circulation	energy consumption for circulation	non-reusable refuse
1. Automobile	X	X	XX	XX	XX	XX	XX	XX	0	XX	√√	XX	X
2. Refrigerator/ deep freezer	X	X	X	0	X	XX	XX	XX	0	0	√√	XX	X
3. Radio-, television apparatus	X	√√	√√	0	0	√√	√	√	0	0	X	0	X
4. Electric bulb	X	√	√	0	0	√	X	X	0	X	XX	0	0
5. Sodium lamp	X	√	√	0	0	√√	√	√√	0	0	XX	0	0

[1]Reproduced by permission from *The McKinsey Quarterly*, Fall 1972. Copyright 1972 by McKinsey & Company, Inc.

[2]Symbols for ecological effects: Registration illustrative only.

√√ = very positive
√ = positive
0 = neutral
X = negative
XX = very negative

Input/output analysis in principle permits overall economic and simultaneous planning. Corresponding models are, however, hardly practicable because of their complexity and the required volume of data as well as the difficulties of acquisition. For better clarity it is sensible to restrict considerations in the examination to part systems in order to arrive at part solutions.

In addition, certain combinations of part alternatives can be coupled to an "individual measure" which is then analysed in the model so that a transparent choice of alternatives or strategies is obtained.

4. INSTRUMENTS FOR ENCOURAGING APPLICATION OF NON-WASTE TECHNOLOGIES

4.1. *Direct State Intervention*

State intervention for encouragement of non-waste technologies is possible via the following legal provisions:

4.1.1. *Prohibition of manufacture* and/or use of specific products.

4.1.2. *Regulations* for limiting pollution emission.

4.1.3. *Usage regulations* such as the regulations governing the quality and quantity of materials to be used for specific products, in particular regulations for use of secondary raw materials. (Example: regulations for use of waste paper in paper-making.)

4.1.4. *Production regulations,* by means of which certain production processes can be prohibited or made mandatory. (Example: cellulose manufacture by the magnesium bisulphite process instead of the calcium bisulphite method.)

4.1.5. *Usage and consumption regulations.* Prohibiting restrictions have the advantage that they are direct, quick in their effects and clear. Observance of such restrictions is easy to supervise.

Prohibiting restrictions and usage regulations are only used reluctantly, however, since they constitute massive direct intervention. General prohibiting restrictions may also lead to undesirable unequal burdening of those concerned.

4.2. *Indirect State Intervention*

4.2.1. *Environment taxes/charges.* Emission of pollutants into the air, water or soil is subject to a uniform tax imposed per emitted unit of pollutant. This tax can be levied at varying levels depending on the harmfulness of the particular pollutant. This tax must be levied at all production stages. This will guarantee that influences on the environment caused by the preliminary productions stages go into the end product as a cost-raising factor.

Consequences of taxation. Producers and consumers endeavour to avoid the tax burden (costs). The manufacture and consumption of such products thus declines. At the same time there is a switch in the production process or product from hostile to acceptable for the environment and a reduction in pollution emission to the environment.

Advantage. Burdens are borne in accordance with the instigator principle, the means conforms to the market, there is incentive to switch to environmentally acceptable production and reduction of emission of pollutants.

Disadvantage. Slow effect (processes of adaptation). Difficulties are encountered in fixing the level of taxation at which the required effect is achieved. The tax rates also have to be adapted to meet new developments. Side effects and side stepping measures on the part of those concerned can frequently not be predicted.

4.2.2. *Subsidies*. Subsidies to encourage application of non-waste technologies involve granting of direct financial support to prevent environmental pollution and/or to encourage saving of raw materials and energy.

Disadvantage. The burden is borne by the public sector budget, difficulties are encountered in subsequent reductions in subsidies. Subsidies are thus contrary to the instigator principle and thus should as a rule be rejected (Art. 92 of ECC Treaty).

4.2.3. *State demand*. The State guides its demands in the field of public administration in order to deliberately encourage or discourage specific developments.

Example. Regulations on the use of certain paper products (paper from waste paper). Equipping of vehicles with steel belted tyres. Use of returnable vessels in canteens.

This instrument is easy to deal with, it conforms with the market and is politically implementable (exemplary behaviour on the part of the State).

5. SUMMARY-LONG-TERM POLICIES

Non-waste technologies must be seen as only one aim within an overall environmental policy in which competing aims, in particular economic and social policies, have to be included. Implementation of non-waste technologies and their part aims can thus not be considered as an absolute demand.

In the field of environmental conservation and the part sector of non-waste technologies the following developments appear evident:

What used to be "free goods" will be supplied at an increasing price. There will thus be a trend towards a general rise in price levels.

The relative prices of goods (one type relative to another) will change. Environmentally acceptable products will replace environmentally hostile products. The composition of goods will change. As a general trend, the prices of services can drop in comparison to industrial products since services in general cause less pollution.

To assess the interrelationships and effects of measures, etc., the instruments of input - output analysis will have to be employed more sweepingly. In this field development work still has to be carried out. In its 1976 Environment Report the Federal Government makes the following demands in this connection:

"Improvement of the means of understanding systems by compilation of materials balances, revealing of chains of effects and control loops, setting up of ecological models taking into account economic and social factors."

in the field of production, non-waste technologies will continue to be developed. Interlinkage with the overall economic effects will take place for individual works to a sufficiently precise extent via costs and prices because benefit or damage to the environment goes into the prices due to corresponding costs.

In this connection the Federal Government Environment Report for 1976 states the following:

"The Federal Government considers reduction of waste amongst producers and consumers as the priority aim of waste management... Direct prevention of waste is as a rule the most effective method of relieving the environment and also leads to a saving of raw materials and energy. The Federal Government therefore expects that Industry in coming years will give particular attention to development and application of low-waste technologies and to increasing the durability of its products."

State of Non-waste Technology

National Experience and Policy Introductory Report

A. J. McIntyre (Rapporteur)

Policy analyst, Department of the Environment, Environment Canada, Ottawa, Canada

1. Later sessions of this seminar will be dedicated to the generation of recommendations to member countries. The task of this particular session is to explore the papers provided in order to extract information that will be useful in shaping these recommendations. Thus the recommendations that come from this session would, of course, be put forward tentatively.

2. It cannot be assumed that any two of us has seen any given paper in the same way. To provide a basis for discussion, then, I would like to offer my own comments on each of the papers. Hopefully, subsequent comment would correct and refine our understanding of the papers.

3. If common denominators exist, it is hoped that we can provide them. The fundamental issue transcends national boundaries; and if international co-operation is part of the resolution, then these common denominators could possibly provide a basis for such co-operation.

4. The papers will be dealt with in alphabetical order.

AUSTRIA

5. The list of non-waste developments in Austria is extensive and impressive. The authors did a very thorough research job in compiling such a list. It would also appear that there is a great deal of available non-waste technology that may not be recognized since it was not developed under that general heading.

6. This implies an interesting question for all countries. It is possible that a careful search of any given milieu would reveal alternative methods of procedure, some of which are less wasteful than others. Once having exposed this condition, then the question of why one alternative is being used over another might be addressed.

7. There is no question that *economics* is usually the basis for any given choice, but what is needed seems to be a more extended understanding of the economics.

BELGIUM

8. The combination of density of population and high level of industrialization seems to have created an advanced interest in non-waste technology in Belgium. The Government's interest in financing research and development and the response of industry has been very productive indeed. Not only is there a long list of specific developments, there is a waiting list of opportunities for further development.

9. The motivation for these programmes is a mix of raw material saving, energy saving, and pollution abatement. It will be an interesting analysis to attempt to separate these various objectives. It would appear that some of the work being implemented will go a long way toward solving economic problems from both the supply and demand side of the equation and this in turn would indicate, perhaps, entirely new policy options.

CANADA

10. For reasons that are generally conventional, it seems, Canada has some evident interest in non-waste technology. The balancing that goes on between social, economic and political processes is seen to have resulted in some relevant policy and in certain tangible developments.

11. The policy thrusts are in the form of financial support for research and for technology development and also in the form of legislation at both federal and provincial levels. In all cases, the original intention was for the purpose of pollution abatement, but inevitably a certain amount of the result is clearly non-waste in nature.

12. The tangible results are most clearly seen in the Can-Wel project and in the Reeve-Rapson process. There is also a certain amount of activity in the field of municipal waste disposal.

13. A particularly relevant development is the Energy Conservation Programme of the Department of Energy, Mines and Resources. This programme is complementary to the work of the Department of Fisheries and Environment.

14. In view of Canada's population distribution, questions of public information and participation are of considerable interest.

FEDERAL REPUBLIC OF GERMANY

15. The level of state activity in the Federal Republic of Germany is both advanced and extensive. The primary motivation seems to come from pollution abatement needs yet there are programmes devoted to resource conservation alone. In this last regard, mention is made of a law requiring companies to have on staff a waste officer. It would be very interesting to pursue this point in some detail, since it appears to be a very effective mechanism.

16. In the field of water pollution and waste disposal, the use of fees is of special interest and again some further detail would, perhaps, be useful. Lastly, the comprehensive information system mentioned is of special interest.

FINLAND

17. Finland has a number of interesting technical developments that clearly fall into the general category of non-waste technology. With the interest being shown by that Government, it is expected that there will be a flow of useful ideas and developments from this source.

18. The point is made that, being in a colder climate, heating energy is of particular interest. It is likely that this is a particularly fruitful area of investigation for a good many countries.

19. A point, not seen elsewhere, is the interest in waste in the construction industry. Some worthwhile developments seem to be available here, particularly in the area of reconstituting older buildings.

FRANCE

20. This paper focuses on the term "clean technologies" which refers to those technologies that reduce *or evade* waste or pollution. These technologies can be found in basic processes, auxiliary processes or separate processes. Thus, waste can be reduced by altering conventional manufacturing processes, by treating the effluent streams from a process, or by reclaiming discarded materials.

21. It is possible that the linkage of pollution abatement technology to non-waste technology is too direct. The question arises as to whether this approach can expose all the options available for the reduction of waste.

HUNGARY

22. The conflict between *resource conservation and environmental protection* on the one hand and *economic growth* on the other is recognized to the point that it is felt that something more than pollution abatement technology will eventually be required. As in the USSR, it is thought that non-waste technology will be required to solve the problem.

23. It is further postulated that zero waste is not likely to be attained for individual processes. The chance of success improves when *various industries are considered in combinations*. The economics of the situation must be understood at all levels, from individual to national to international, if this is to be done properly.

24. As a first step, the next 5-year plan for economic growth includes measures to maintain present environmental quality. Even this will be difficult considering the growth increment involved and the technology available. Meanwhile, a research programme is being developed to provide the non-waste technology required.

THE NETHERLANDS

25. This country seems to have a considerable interest in non-waste technology and is actively involved in developing approaches that are expected to promote

and encourage industry to innovate in *socially* acceptable ways. It seems clear that this is considered to be a proper and constructive role for government.

26. At the moment the first area of attention appears to be pollution abatement as opposed to resource conservation. Pollution abatement itself can use resources rather than conserve them and this is recognized in the attention directed to *innovative process rather* than to *remedial measures* intended for attachment to conventional processes.

27. Broad government policy is based on dialogue with segments of industry so that the best possible understanding is obtained of the social and economic systems involved. It is recognized that economics is a dominant influence and some special attention is given to measures that are available to government to encourage innovation. It is noted that the rationale leans towards the *sharing of the risks* involved in innovating which neatly bridges the gap that sometimes exists between the goals of society and the goals of industry.

28. It will be interesting to watch the development of these instruments in the Netherlands since it seems that it will be done in such a way that comparative effectiveness of a range of options might be compared within one economy.

29. Another question that might excite our curiosity, is the role of pollution abatement within the larger framework of non-waste technology.

POLAND

30. This paper describes an interesting programme for the recovery of mine wastes and the values they might contain. This focusing of interest is worthy of some attention since there is no doubt that all recovery processes leave some values in the tailings. Thus, we have the choice of improving recovery efficiencies or of recycling tailings and for existing tailings piles, this latter choice can, in many cases, be profitable. A third option of the use of barren tailings is as a building material or aggregate. The basic point is that mine tailings have value.

31. The basis of the programme lies in obtaining detailed information on the nature of ores and tailings throughout Poland. In this way opportunities for further processing, recycling or secondary uses can be identified at various stages in the mineral processes. Poland now has comprehensive plans for the exploitation of mineral process wastes.

32. Some success has already been achieved, notably in coal mining where schist is reprocessed for its residual coal and to produce ceramic clay. Another interesting development has to do with technology for extended exploitation of polyhalite minerals.

33. The benefits derived have been considerable, both in terms of recovered minerals and in terms of environmental quality improvement. The need to develop technology has resulted in much useful research.

34. This unwillingness to accept waste in the mineral industries might well be extended to other areas of industrial activity and some of the lessons learned in Poland could help to point the way.

SWEDEN

35. Sweden has made great strides in reducing pollution which, of course, also has the effect of reducing waste and conserving resources. I was especially interested in the comment on research for ways to keep heavy metals out of municipal waste so that the sludge could be used as fertilizer.

36. Special programmes are in place to increase the volume of recycled paper, to attract scrap automobiles to licensed shredders and to manage the disposal of certain chemicals. The comment concerning the lack of success in the beverage container field might bear further comment.

37. Sweden has a very active energy conservation programme which involves information, training and financial assistance. The financial assistance covers both housing and building innovation and industrial technology.

38. The role of government in controlling land use with water quality and energy among the criteria for allocation is very interesting.

THE UNITED STATES

39. Increasing concern about availability of raw materials is increasing the pace of development of non-waste technology in the US. Pollution law has tended to internalize social costs and this has also raised the level of interest in non-waste technology. Finally energy costs have contributed markedly. The new Energy Policy and Conservation Act contributes directly by focusing attention on energy use at the consumer level and by providing funds for relevant research. Some financial assistance is available in the form of tax relief and in municipal bonds.

40. The policy of requiring deposits on beverage containers was tried in two locations with good results. The amount of the deposit was not mentioned.

41. With respect to policy analysis technique, the US has contributed a great deal of basic development to four useful tools, namely, cost - benefit analysis, input - output analysis, technology assessment and resource analysis.

THE USSR

42. In the USSR it is recognized that man's interference with nature has caused problems. It is further recognized that something more than waste treatment is required if the biosphere is to be protected from the effects of the pollution caused by economic development. That something will have to be the development of new technologies that eliminate or reduce waste. This approach would also conserve resources some of which are wasted needlessly at present.

43. The paper goes on to demonstrate that the various branches of industry have the potential to reduce or eliminate waste. It is suggested that if *social costs* could be included in the *costs of production* the development of these techniques would be accelerated.

44. Meanwhile, the first step toward non-waste technology is thought to be represented by pollution-abatement technology.

THE UNITED KINGDOM

45. Here we are warned that the real aspirations of society are expressed in economic terms and that if this is not recognized we run the risk of being, or appearing to be, idealistic. Thus, full account of impacts on the economic system must be taken in considering non-waste technology developments. We must be realistic in order to be effective.

46. Here, we see industry, government, labour and academe coming together for the purpose of defining issues and assigning priorities to them. As the summary of the 1974 Green Paper indicates, the types of questions raised run the full range of non-waste technology from the nature of demand through design and use to eventual disposal and recycling.

47. It would appear that the first line of attack is municipal solid waste. In this case, it is clearly recognized that there are other facets of the problem that must be attacked if the problem of resource waste is to be solved.

48. It is suggested that one of the *communications problems* we have is to avoid the tendency to think of non-waste technology as only solid-waste disposal or recycling. Those of us who have looked into the matter do not have this problem but a good many of those in industry, whom we hope to influence, do have it.

YUGOSLAVIA

49. Yugoslavia is blessed with a clean rural environment but it is also developing and is aware of the threat to the environment that new technology brings. The plan to evade this problem will be to research ways and means that are environmentally appropriate so as to prevent environmental problems from the outset. This concept is stipulated in the constitution. Longer-term interests will thus be honored even though the short-term costs might be higher.

50. This completes the set of papers submitted for Topic II(a) so now in our discussions let us first attempt to refine our understandings of the papers submitted. It is important that we explore the various points of view and that all information given be used.

When we are confident of our understanding then we can proceed to the development of advice useful to the following sessions of the seminar. In this regard the questions given below might serve as a basis for discussion.

1. Do the papers make a good argument for economic research with particular reference to social costs? (Session III)
2. What is the role of pollution abatement technology within the larger framework of non-waste technology? (Sessions IV and V)
3. What common denominators can be identified?
4. What implements and techniques can be identified for later consideration? (Session IV)

State of Non-Waste Technology in the Netherlands: National Experience and Policy

A.W.F. Van Alphen

Ministry of Health and Environmental Protection, Leidschendam, Netherlands

INTRODUCTION

The idea of "non-waste technology", i.e. to plan human activities with the least waste of raw materials and energy and the maximum protection of the environment, is finding broad acceptance within some public and industrial sectors.

Already before the oil crisis there was a broad concern in the Netherlands over the continuing growth of consumption and production and the accruing technology as an increasing assault upon the world's natural resources. The oil crisis accentuated public awareness of the limits of growth. It also gave additional impetus to the development of a technology making less use of raw materials and energy and affecting less the environment.

The way in which the principles of non-waste technology find acceptance in the Netherlands is also reflected in the attitude of authorities and industries to co-operate tightly, in the interests of producers and consumers, as well as of the environment.

The producers' role is the maximum reduction of waste of energy and raw materials by continuously operating the best available technologies.

The consumers can contribute by minimizing their desires for goods, in particular the more wasteful ones, and by consciously giving preference to goods which do least harm to the environment at the production stage and can easily be recycled.

The authorities concerned can contribute in an active as well as a passive way.

The Government's active approach, in accordance with its environmental objectives, is to support with subsidies and credits the development of less wasteful processes and techniques in conjunction with those industries that have given proof of their willingness to introduce such systems and place them under public control and availability. This is an important addition to the more passive approach of levies and licences to regulate less desirable developments.

A common concern is that not every technological alteration means an improvement beneficial to society: changes brought about by technological innovations could in turn have most undesirable consequences.

PUBLIC INTERFERENCE WITH INDUSTRIAL ACTIVITIES IN FAVOUR OF THE DEVELOPMENT OF NON-WASTE TECHNOLOGY

The Government's policy of encouraging non-waste technology and industrial innovations arises from social and economic objectives, such as the quality of life, protection of the environment and the targets of selective growth. Consequently, innovation should not be limited to technological, industrial, commercial or other improvements leading to the successful introduction of new products and/or to the commercial use of technically new processes and new capital goods without taking into consideration other consequences for society.

Since innovations usually emanate from the recognition of some need, usually resulting in a market, governments can influence to a certain degree the innovation process quantitatively as well as qualitatively in a desired direction by structuring community needs and recurrently emphasizing the interest of society in the most explicit way.

One governmental instrument is the continuous review of the system of norms and standards to protect the environment by approaching as nearly as possible the quality objective (of no-effect levels). Furthermore, the Government can enforce regulations and levies to oblige industries to switch over to less wasteful or harmful systems of production.

In this way the authorities are able to stimulate industry indirectly to the acceptance of a new attitude towards environmental problems, i.e. to the application of low- and non-waste technology and production processes.

Financial reasons may cause delays or prevent a rapid implementation of the desired innovations. Such difficulties can be overcome by long-running financial credits or investments that conform with the "polluter pays" principle.

Another difficulty is the possibility that enterprises covering a substantial portion of the market may delay the introduction of promising innovations because of insufficient social pressure against actual more wasteful products or production processes.

At present discussions take place with industries in order to achieve the development and application of non-waste technology by projects aimed at

I. Reduction of nuisance caused by dust.

II. Reduction of the emissions of nitrogen oxides from big furnaces.

III. Reduction of the emission of nitrogen oxides from the production of nitric acid.

IV. Clean energy production from heavy residual oil gasification-based combined-cycle power stations.

V. Development of low pollution power systems for corporation buses.

VI. Reduction of pollution caused by tanneries.

VII. Reduction of pollution caused by the production of phosphate fertilizers.

VIII. Anaerobic purification of sewage of agro-industries (e.g. the beetroot industry).

IX. Reduction of pollution by the textile industry.

In each of these groups consultations take place with more than one industry (or industrial branch); some individual industries are contacted more frequently when it comes to the actual setting up of non-waste technology projects.

If a project would lead to considerable risks for one industry or industrial branch there is a possibility of support from the Government. This support may consist, firstly, of drawing the attention of industries to the available facilities for the developing of non-waste technology; secondly, of granting subsidies or credits; and thirdly, of guarantees or tax reduction facilities; finally, of the forming of a permanent framework of consultation between government and industries.

When specific situations in the Netherlands demand immediate attention the Government takes an active role. The urgency of introducing an innovation can influence the kind and extent of the support needed. Generally, those projects are favoured that prevent pollution at source rather than those that abate pollution or its effects. However, projects are also eligible when they are directed towards control of already-produced waste as long as this waste is not diffused into the environment.

In such cases, the national and regional authorities have the following instruments of intervention at their disposal:

Research Policy

The Government successfully attempts to involve certain industries, with their research potential, in national research programmes. Two of these national research programmes in which industrial research is involved are on energy research and on environmental protection. The latter programme calls attention to assessment and evaluation of sources and consequences of changes in the environment.

Regulations

By means of prohibitive regulations, licences-under-conditions and product-requirements, the Government can enforce the reduction of pollution and the introduction of best available technology. In practice, however, such regulations are hardly a stimulus for innovations. Additional measures are necessary.

Levies

Levies can be applied on the produced waste. They are a useful supplement or sometimes substitute for regulations, especially when adequate alternatives for the production systems in use seem to be available. Under such circumstances levies are constantly stimulating the adaptation of production and consumption patterns and the development of non-waste technology.

The Active Instruments

Governmental initiatives, inviting branches of industries with similar technological problems to discuss common research projects, are welcomed by these industries. Even if the discussions do not yield immediately applicable results, the exchange of views between representatives of industries, research institutions and the Government is very stimulating for all parties.

The Government endeavours to keep well informed on the latest achievements and present limits of technology (especially those of environmental technology) in order to readjust levies and regulations and to induce industrial innovations and appropriate research. In order to know more about the possibilities of technological change, a more active participation of the Government is required in the research and development activities of industry.

Another criterion is to support projects of which the expected results may be applied on as large as possible scale, i.e. in several industries or industrial branches.

The granting of subsidies and—to a smaller extent—of credits or guarantees, of course only takes place when the project cannot be executed by the industry or branch itself on the desired scale and at the optimal speed, and when the reasons for the delay are really financial or organizational. Under certain circumstances the Government is willing to enter into a joint venture when it is a matter of broad public importance. It goes without saying that government will only take initiatives and responsibility for the consequences when a highly developed technology is required and when risks are too abundant and diversified to be born by a single industry or branch of industry.

Governmental participation can provide a way to prevent the closing down of those branches of which the capital goods are so obsolete that the socially and environmentally desirable non-waste technology cannot be applied economically.

SUMMARY

The authorities and industries in the Netherlands are co-operating to promote the creation and possible application of low- and non-waste technology.

A government policy has been elaborated to ensure rational use of natural resources and minimum generation of waste in the interest of environmental protection. The policy aims at an incentive system supplementary to regulations and levies, to stimulate technological innovations within the industry in the Netherlands in the interests of environmental protection and economically beneficial use of resources.

Inter-disciplinary working groups or study groups combining representatives of industry, the Government and sometimes research institutes are established. Their responsibility is to put forward projects to develop non-waste technology in different branches or types of industry. The Government may support such projects by policy guidance, subsidies, credits, guarantees, etc. A review is made of several working groups that have made specific and detailed proposals for development and application of non-waste technology.

Non-Waste Technology: Comments on the Canadian Scene

A.J. McIntyre
Policy Analyst, Environment Canada, Ottawa, Canada

More by way of explanation than of apology it should be understood from the outset that Canada makes no special claims to expertise in the area termed "non-waste" technology. Any achievements noted below have been the result of conventional economic and political influences rather than of any novel approaches to resource management. So these examples are offered simply as an indication of conditions in Canada as they stand at present.

Before getting to actual cases some comment on Canada's geography might be useful.

Canada is a federal state stretching over some 6000 miles from its Atlantic coast to the Pacific. The topography varies from flat prairie to snow-capped mountains and the climate from mid-temperate to Arctic. There is an abundance of fresh water, much of it harnessed to generate electric power. Minerals are available in quantity including such special endowments as nickel and uranium. Forest and fish resources are large and extensively exploited. Large areas of the country are under cultivation of one kind or another.

By most standards Canada is underpopulated. Much of the population is concentrated along the Southern border with fairly dense concentrations in the Toronto to Montreal area. Vast reaches of the country, especially in the North, are virtually unpopulated except for small numbers of indigenous peoples.

The population itself reflects its immigrant beginning in the wide range of ethnic origins to be found in any populated area of the country.

This diversity of factors makes Canada a difficult country to categorize. It has most, if not all, of the problems of the civilized world. At the moment, inflation and unemployment enjoy the spotlight. In the recent past the language question was dominant and cannot yet be considered a resolved issue. Next year it might be regional disparity or labor-management issues that will pre-empt the attention of the politicians.

Even so, within this mixed and stochastic milieu there are indications that long-term problems can be addressed. For instance, the approach being taken to the proposed McKenzie Valley gas pipeline indicates that conventional economics can be questioned. There are considerable grounds for optimism that the Canadian people can extend their planning horizon so as to include the long-term considerations.

Canada has in place now a number of policies and programs that serve in one way or another to promote the development of non-waste technology.

One of the more successful programs has been the Co-operative Pollution Abatement Research Committee (CPAR). This is a government-industry group who administer the funding of pollution-abatement research for the pulp and paper industry. While the thrust is pollution abatement, some of the work has been on recovery processes which are in themselves non-waste technology. One of the early projects underwritten by this program had to do with the Reeve - Rapson process which is the subject of another paper prepared for this seminar.

Another program is dedicated to the Demonstration of Pollution Abatement Technology (DPAT). The idea is to assist financially the practical demonstration of the feasibility of those technologies or processes which have not been commercially marketed and where there is a high degree of risk involved. Again, the thrust is pollution abatement but some of the work will inevitably involve recovery processes. The large-scale pilot installation of the Reeve - Rapson process was assisted by this program.

Still another program has to do with energy conservation and is managed by the Office of Energy Conservation in the Department of Energy, Mines and Resources. This program has directed its efforts to vigorously educating the public on how to reduce the consumption of energy. They have also made policy proposals to government having to do with taxing the sale of larger automobiles and promoting better gas mileage for any size of automobiles and these proposals are under active consideration. They have held two major conferences with industry representatives to explore the possibilities for conserving energy.

As for legislation, it cannot be claimed that there is any in existence in Canada that is specifically dedicated to non-waste technology. There are, however, a number of Acts on the books that have to do with protecting the environment and many of these have the effect of attracting attention to non-waste technology. Certainly the pressure being applied to polluting industries all across the country to get them to clean-up their effluents has this effect. There are also land use and solid-waste disposal regulations in many areas that encourage a change in attitude toward waste.

In all of the ten provinces and at federal level there is legislation having to do with the protection of the environment or the abatement of pollution. In several cases there is a specific legislation having to do with waste disposal. Virtually all of these government interventions are regulatory in nature but have the effect of internalizing costs so that values shift and motivation for recycling and waste evasion is created.

There are a number of technical developments in Canada that are of interest in the field of non-waste technology.

The Can-Wel project sponsored by Central Mortgage and Housing (CMHC), an agency of the federal government, involves the combining of several proven technologies so as to produce potable water and heat energy from liquid and solid waste from domestic sources. Laboratory pilot models have been operated for the liquid waste phase and water suitable for discharge without further dilution was produced. This also provides input to the second stage designed to produce potable water and pilot trials are now underway. The solid waste phase consists of a special incinerator system which reduces solid wastes (including sludge from the two liquid phases) to a sterile ash and uses the heat to provide hot water.

Further pilot plant work is now underway on apartment block installations and valuable experience is being gained. At this stage of development it seems possible that the system can be closed so as to vastly reduce water intake and solid waste output all without any attendant pollution.

This development is the result of efforts to devise an alternative to conventional sewage systems because of their dependence on expensive underground piping. The Can-Wel system would remove the need for serviced land as we presently understand it and would materially affect the nature of urban growth. Thus, Can-Wel qualifies as non-waste technology not only in that it conserves water and recovers heat but also in that it removes the need to invest in sewage systems and permits a much more efficient use of land in general.

Among the developments in Canada directed at recovering values from solid municipal waste there is an incinerator in Quebec City which produces steam which is sold to an adjacent paper mill. This unit burns approximately 1000 tons/day of refuse and produces up to 300,000 pounds per hour of steam.

Another installation of this type is now under active consideration for the Niagara peninsula area. In this case, the steam output will be maintained at a constant level by supplementing the waste input with conventional fuels as needed. The user for the steam is a large pulp and paper mill which is also participating in the venture financially.

An incinerator is being considered for the city of Ottawa to produce steam for space heating in the city core. There is an installation of this type presently operating in Montreal, although the steam generated is not fully utilized.

Values other than heat are present in municipal waste and one of the more obvious ones is the metal content. The Montreal incinerator is equipped to separate the ferrous portion from the ashes left after incineration. There is an installation in Hamilton designed to separate the metal before going to the furnace. There have been some design problems with the Hamilton plant that have impeded its being operated as intended. These problems will be worked out in due course and the benefits of the approach will be assessed at that time.

So far, the return from metals recovered from municipal waste has been highly variable and generally disappointing. Problems of quality control and costs of de-tinning have complicated the problem. It would seem that the price of virgin material is not quite high enough to make this form of recovery economically attractive.

Some interest is being shown in separating other materials from refuse. The Province of Ontario is funding the construction of a pilot plant to separate municipal waste into main components of ferrous metal, non-ferrous metal, glass, paper and fuel. The interest lies in the fact that values appear to have changed enough to begin to attract attention to solid wastes as a source for recyclable materials.

Another interesting development is the INTERMETCO process which is designed to separate the non-ferrous portion of reclaimed metal from scrapped automobiles into its various components. A pilot plant is presently being built at La Prairie, Quebec. This process depends on the difference in melting points of the various metals.

Waste lubricating oil is being used as a supplemental fuel for a cement kiln in Montreal and the operation has been very encouraging. This particular burning method has the effect of capturing heavy metal additives in the clinker rather than have them emitted into the environment as air pollution. The volumes of waste oil available are large enough to make this form of recycling very attractive.

Canada has a comparatively large forest resource that is not in danger of depletion in the near future but the energy problem, among others, has attracted attention to the potential of forest waste. The Moore gasifier is a case in point. This apparatus provides a low b.t.u. fuel gas from woody wastes thus making the wastes much more useable in that the energy can be made available for a wide variety of uses. Pilot plant operation has been successful and full-scale installations are now under consideration.

The federal government has recently undertaken in two different provinces to assist in full-scale trials to develop an energy self-sufficient community. In both cases the source of energy is waste from lumbering operations and the communities are isolated. These experiments will be interesting but results will not be available for a year or two.

In the face of energy and material shortages, there is little question that forest resources are not exploited efficiently. It would undoubtedly be productive to focus a larger portion of the world's research effort on the identification and reduction of waste in the utilization of forest resources.

To summarize, it is apparent that these are a number of non-waste technology developments in Canada and the list cited above is probably not complete. It is also apparent that these developments are generally the result of environmental legislation (mainly in the area of pollution abatement) and of rising costs of raw materials, most particularly petroleum. There are indications in Canada that the people are ready for some sort of modified approach to resource exploitation. Certainly, the rate of obsolesence of some products seems to have slowed as a function of consumer reaction. There is also the interest being shown in the Conserver Society concept with the Science Council of Canada and the Gamma Institute of Montreal, both conducting major studies in this field. It is still highly problematical as to how far the people are willing to go but a direction does seem to be indicated.

Austrian National Report on Non-Waste Technology

Rudolf Kauders* and Udo Ousko-Oberhoffer**
**Ignatz Pleyel-Gasse 2/Stg. 29, A-1100 Wien*
***Wiesingerstraße 6/9, A-1010 Wien*

AUSTRIAN PROCEDURES OF NON-WASTE TECHNOLOGY

(A) PRACTISED AND APPROVED PROCEDURES

1. The VEW-Electro-Slag-Remelting Process

   Vereinigte Edelstahlwerke AG

2. The VEW-BEST Process

   Vereinigte Edelstahlwerke AG

3. The continuous one-step production of coppered welding wire Vereinigte

   Edelstahlwerke AG

4. Power metallurgical manufacturing process of parts made of sintered iron and sintered steel

   Metallwerke Plansee AG

5. Method of operation to produce contact material on basis of Wolfram-copper alloyances and Walfram-silver alloyances respectively and heavy metal alloyances on Wolfram basis

   Metallwerke Plansee AG

6. Recovery of hydrochloric acid from the pickling process

   Ruthner Industrieanlagen AG

7. Pickling of special steel strips according to the Ruthner neolyte process

   Ruthner Industrieanlagen AG

8. Procedure for reworking caustic sludge, obtained from acid polishing of lead glass, to basic lead carbonate

   Fa. D. Swarowski & Co. Wattens

9. Production of aluminium fluoride from waste gas of phosphoric acid and superphosphate plants

Chemie Linz AG

10. Production of ammonium sulphate and sulphuric acid respectively from phosphoric acid waste

    Chemie Linz AG

11. Procedure for reprocessing of a-tactical polypropylen Petrochemie Schwechat

12. Process for recycling of stretched thermoplastic material in the manufacturing process

    Chemiefaser Lenzing AG

13. The Lenzing magnesiabisulphite recovery process in the magnesia-bisulphite wood-pulping industry

    Chemiefaser Lenzing AG

14. Production of paper and cardboard without environmental pollution Ruthner Industrieanlagen AG

15. Very effective purification of radioactive waste water

    Vereinigte Edelstahlwerke AG

16. Heavy degree fermentors - an integrated component of recycling and environmental protection measures

    Vereinigte Edelstahlwerke AG

17. Flare gas recovery plant

    Österreichische Mineralölverwaltung AG

18. Generation of energy by burning wastes

    Ruthner Industrieanlagen AG

19. Energy recovery at cracking plants

    Österreichische Mineralölverwaltung AG

(B) PROCEDURES STILL IN DEVELOPMENT

1. Environmental favouring process of high efficiency for power stations Vereinigte Edelstahlwerke (VEW) AG

2. Tyres made of cast plastic

   Polyair Maschinenbau Ges.m.b.H.

3. Complete recycling of lees production of tartaric acid

PREFACE

Although Austria is a small country with not too many inhabitants it is highly industrialized. The high standard of Austrian technology has a long tradition.

The Austrians seem to be extremely creative and talented. The car was invented by Siegfried Markus, the incandescent gas-light developed by Auer von Welsbach, and the radio valve by Schrack, who were all Austrians. The development of stainless steel—due to Austrian research—rendered Austria's leading position in the high-quality steel industry possible. The LD-steel process developed in Austria, too, not only gives evidence of the advanced level of Austrian technology but its realization has also directly and indirectly reduced environmental pollution.

Austria's industry is extremely specialized to enable it to compete in Austria as well as abroad in spite of small production units.

Austria's plant constructing and exporting firms offer advanced plants and processes, many of them exemplifying non-waste-technology.

Nineteen approved and applied processes will be presented as well as three processes still in development.

(A) PRACTISED AND APPROVED PROCEDURES

1. The VEW-electro-slag-remelting Process

This process—the Hopkins process—has been known in principle since 1937. By remelting solid steel blocks (electrodes) in an electric conducting liquid slag, a raw ingot is produced by means of resistance heating which is adapted for further manufacturing.

The significance of this procedure was already early recognized and has been developed for application in large plants.

In the meantime this process has become known all over the world as "Böhler-ESU-Process" because of the following technical and economic advantages :

working with a short lifting ingot mould, the production of ingots of any length desired is made possible;

electrode-exchange method without interruption of the melting process, makes it possible to use fusible arc welding electrodes of any length without any refuse endings remaining;

energy economizing use of low-frequency alternating current, especially with ingot diameters of more than 800 mm, reduces electric losses with coincidental metallurgical advantages;

far-reaching automation of the melting process including the change of electrodes;

possibility of producing round, square, polygonal ingots, hollow blocks and shaped parts.

In comparison with usual procedures of producing ingots by casting ingot moulds, considerable technical and economic advantages are achieved by the "Böhler-ESU-Process" and the danger of environmental pollution is reduced.

The ESU ingot is built homogeneously and free from detrimental segregations, pores, flaws as well as from accumulations of non-metallic impurities. Thus the otherwise inevitable waste from the top and bottom of conventional ingots is avoided and an optimal yield of over 90 per cent (against 75 - 85 per cent, with forging often only 50-60 per cent) is achieved. The scrap percentage rate is reduced to a fraction of the risk normal until now, which also means that many power-consuming operations, such as hot forming and heat treatments, are saved.

This procedure gives rise to only insignificant quantities of waste products (scrap metal, slag, and exhaust-gas). In spite of the necessary expense of energy for the remelting process, production costs (for the production of high standard forging for example) can be charged lower with the ESU procedure than with conventional ingot production. That is especially true if waste products like run-down barrels are used as electrodes in the ESU-process without having been recast in an arc furnace. In this case one whole melting and casting process can be saved.

Further information may be obtained from :

Vereinigte Edelstahlwerke Aktiengesellschaft (VEW),
vormals Böhler-Schoeller-Bleckmann, Styria
A-1010 Wien, Elisabethstrasse 12,
Telephone : 0222/ 57 35 35
Telex: 1-1683

Supply of plants, licences for all steel grades, know-how and engineering.

2. *The VEW-BEST Process*

The production of heavy forgings from big cast ingots—needed above all to produce turbine shafts and alternator shafts as well as turbine pullies in electricity generation—has always been connected with an ingot waste of up to 50 per cent of the raw ingot weight and a high scrap percentage rate. That is because accumulations of non-metallic impurities, segregations, and flaws in the core of the ingot cannot be avoided when ingots solidify in a conventional way. Basic metallurgical analyses by VEW-research have now pointed the way to resolving these shortcomings at the crystallization of heavy forgings. A directed power (heat) feed during the solidification phase makes it possible to prevent the sub-cooling of the residual heat liquid in the inner part of the ingot. Thus the formation of crystallites which accumulate in the area of the ingot bottom and support the concentration of non-metallic contaminations is avoided. Moreover, an oriented crystallization of the whole volume is achieved (similar to the solidification of an ESU-ingot) which prevents the occurrence of segregations and flaws.

Technically this process, known as the "Böhler-Electro-Slag-Topping" (BEST) process, utilizes a water-cooled hob top on the forging ingot mould. It is, after fluid steel has been filled in, supplied with a necessary amount of heat by a metallurgical effective liquid slag, mostly by way of a fusible arc-welding electrode of the same or similar chemical structure as the material of the cast ingot. Thus contamination of the residual heat liquid is avoided which otherwise is brought about by exothermic blanket melts, the ascending non-metallic contaminations are taken up by the slag and the contraction in volume of the ingot cast is balanced by the addition of fluid steel during the solidification phase.

This procedure is patented in most industrialized countries and has by now proved its technical success on more than 1400 heavy forging ingots. In addition to its technical advantages against the conventional ingot casting procedure, its economic advantages have to be stressed, because of the avoidance of the otherwise necessarily produced dead head, which makes up to 25 per cent of the ingot weight with heavy forging ingots, and the avoidance of non-metallic contaminations of the bottom. The forging deposit is increased more than 90 per cent against about 50-60 per cent usual until now with big forgings. The risk of scrap material can, in spite of the intensive yield, be lowered to less than a third of the scrap percentage rate up to now.

Compared with the ESU-procedure the energy consumption is only 30 per cent while the costs for running the plant are significantly lower. The reduction of mill scrap on the one hand renders a more efficient exploitation of raw material and of expensive alloying elements possible and, on the other hand, environmental pollution is reduced.

The BEST process represents, according to today's state of technology, an ideal combination of a high degree of safety concerning the production of high-grade forging ingots, and of low production costs. Therefore, it seems possible to avoid the expensive development of other procedures as the ESU-process allows for ingot diameters of more than 1600 mm.

Further information may be obtained from :

Vereinigte Edelstahlwerke Aktiengesellschaft (VEW),
vormals Böhler-Schoeller-Bleckmann, Styria
A-1010 Wien, Elisabethstrasse 12,
Telephone: 0222/ 57 35 35
Telex : 1-1683

Supply of plants, of licences for all steel-grades, of know-how and engineering.

3. The Continuous One-step Production of Coppered Welding Wire

The automatic welding of low alloyed and unalloyed steels has gained great importance because of its economic advantages. Its application is still increasing. Thus the need of coppered wires, which are utilized in this process, is still increasing as well as the importance of their production methods.

Conventional procedures consist of the first drawing of the wire on high-speed drawing plants, the collecting of the wire to reels of 1 to 2 tons capacity, which is followed by a pickling process and the coppering. Finally the wires are finished in the usual way and reeled up.

Apart from disadvantages arising from the discontinuous manner of production, which make interbedding and high manipulation expenditure necessary, this procedure is connected with comparatively great environmental pollution. It results from the anneal heat treatment in gas-fired and oil-fired furnaces as well as from pickling in conventional souring baths.

To avoid these defects as much as possible, the following procedure has been developed. It is characterized by an operation where the wire continuously passes a wire drawing machine and annealing - pickling - coppering processes and a finish drawing plant till a large-scale coil is obtained ready for use. The continuous sequence of operation is made possible by the welding together of the wire reels treated.

The qualitative advantages resulting from this conception are to be seen in the extraordinary evenness of the finished welding wire, as all stages of the manufacturing take place under well-controlled conditions. Besides, the conditions of processing can be adjusted to particular requirements of the final product.

One of the most important advantages of this new procedure is found in its low polluting qualities. This is due to the electric resistance heating used for annealing the wire. Thus in contrast to conventional annealing in oil- or gas-fired annealing furnaces only small quantities of gas polluting the environment are produced. Furthermore, energy can be saved. A further favourable environmental effect is achieved by the electro-brightening process which is integrated in the whole procedure. This, in comparison with conventional souring baths, reduces the amount of polluting liquid waste matter to a great extent.

Further information may be obtained from :

Vereinigte Edelstahlwerke Aktiengesellschaft (VEW),
vormals Böhler-Schoeller-Bleckmann, Styria,
A-1010 Wien, Elisabethstrasse 12
Telephone: 0222/57 35 35
Telex: 1-1683

Supply of plants, of know-how and engineering.

4. *Powdered Metallurgical Manufacturing Process of Parts made of Sintered Iron and Sintered Steel*

In the powder metallurgical manufacturing process complicated machine parts made of sintered iron and sintered steel can economically be produced in large batch quantities. The process of production is divided into four operations :

1. production of the necessary iron powder;

2. pressing of the powder or a powder mixture into formed parts;

3. transformation into a metallic state by means of an adequate heat treatment, known as sintering;

4. heat treatment by means of calibrating and/or surface treatment of very dense parts with high strengths and close tolerances.

Powder metallurgy represents a process by which, due to direct shaping, scraps are avoided to a great degree. Furthermore, a reworking by dint of cutting processes is not necessary, thus, further waste matter is avoided, too.

Losses of metal powder during the manufacturing process amount to 0.5 per cent. The average amount of scrap material is about 2 to 3 per cent.

Further information may be obtained from :

Metallwerke Plansee Aktiengesellschaft
A-6600 Reutte/Tirol
Telephone: 05672/ 22 41
Telex: 5505 Plansee A

Sale of high-grade material, corresponding to the production programme, no granting of licenses, no supply of plants.

5. *Method of Operation to Produce Contact Material on Basis of Wolfram-copper Alloys and Wolfram-silver Alloys Respectively, and Heavy Metal Alloys on a Wolfram Base*

With the procedure used for producing semi-finished wolfram products high-grade wolfram is necessarily developed. The problem of the utilization of this expensive pure waste has been solved in the following way : after an adequate reworking this waste is used for the production of contact material on the basis of wolfram-copper and wolfram-silver respectively and heavy metal alloys on a wolfram base.

The wolfram waste in form of bar and sheet cuttings is crushed in mills and assorted in defined grain sizes. From this mechanically produced wolfram powder formed bodies are manufactured by admixing of copper and silver powder respectively and/or by subsequent infiltrating with molten copper or silver. These moulds are used for heavy-duty switch production.

The production of heavy metal alloys is one more possibility of re-employing the above-mentioned high-grade wolfram waste. For this purpose the waste is dressed to powder as described above. The powder is mechanically mixed with iron, copper, and nickel powder - depending on the alloy - to a total amount of 1 to 10 per cent.

The powder is pressed to shaped bodies which is followed by a heat treatment called sintering. By this process a metallic body of high strength is produced.

These heavy metal alloys are used when great densities are required such as for gyrating and balance masses. Besides, these alloys are widely utilized for screening radioactive radiation in technics and medicine.

For further information please contact :

Metallwerke Plansee Aktiengesellschaft
A-6600 Reutte/Tirol
Telephone : 05672/ 22 41
Telex: 5505 Plansee A

Sale of high-grade material, according to the production programme, no granting of licenses, no supply of plants.

6. Recovery of Hydrochloric Acid from the Pickling Process

Until 1950 sulphuric acid had been used all the world over for pickling the steel strip produced in steel mills. The spent pickling acid enriched with $FeSO_4$ had to be neutralized and the neutralization sludge dumped. In the new process hydrochloric acid is used instead of sulphuric acid. The iron chloride produced during pickling can be decomposed by means of heat to iron oxide and hydrogen chloride gas, thus permitting the "combined" as well as the "free" hydrochloric acid of the pickle solution to be fed back into the pickle process. This provides a closed cycle of the pickle acid, the only by-product of the decomposition by heat being iron oxide.

The process mainly consists of the spray roasting reactor, a preconcentration unit for the pickling acid, a hydrochloric acid absorption column, and an exhaust-gas ventilator. The pickling acid is preconcentrated in direct contact with the hot roasting gases and, subsequently, by spray nozzles atomized in the reactor. (The reactor is a directly heated empty tower where the concentrated pickling acid evaporates and the existing $FeCl_2$ is decomposed by heat to Fe_2O_3 and HCl-gas.) The Fe_2O_3 produced accumulates in the lower cone of the furnace and is removed from there. The hot exhaust vapours together with the entire HCl-gas are cooled in the preconcentration by means of direct contact with the spent pickling acid, washed, and enter the absorption column where the HCl-gas is washed out and is discharged as "regenerated hydrochloric acid" at the outlet of the column. The exhaust vapours freed from HCl are sucked out by the exhaust gas ventilator and discharged into the open air.

In 1968 more than 60 per cent of the entire world production of steel strip had already been pickled by means of hydrochloric acid. Newly built pickling plants are (almost) exclusively designed for operation with hydrochloric acid, and a great number of existing plants pickling with sulphuric acid have been converted to operation with hydrochloric acid. Since this trend is continuing, it may be expected by 1980 that about 90 per cent of the entire steel strip production will be pickled with hydrochloric acid and, thus, the pickling acid will be regenerated, too.

In closing, we would like to characterize the progress made by recovering hydrochloric acid out of the pickling process with the following figures.

By means of the closed acid circulation it was possible to decrease the consumption of technical hydrochloric acid (32 per cent concentration) to 2 - 2.5 kg per ton of steel strip. When pickling is done without regeneration about 18 kg of 32 per cent hydrochloric acid would be necessary for the production of 1 ton of steel strip, which has to be turned into 9 kg of $CaCl_2$, by means of lime discharged into the waste waters, thus resulting in approximately 20 kg of wet $Fe(OH)_3$ sludge which has to be dumped. The iron oxide produced in a regeneration plant, on the other hand, is partly sold at a good price to the electrical industry, but the major part is conveyed back into the blast furnace by the sintering belt. Thus, the iron removed during pickling is recovered, which means that a steel mill with a production of 800,000 tons of steel strip per year also produces 3000 to 4000 tons of iron per year.

With regard to the economy of the process itself it has proved over the years that the investment costs and operation expenses of a regeneration plant are amply compensated by the saving of acid and neutralization costs so that even for smaller units with a capacity of 700 litres per hour a regeneration plant of pickling acid may be running profitably.

Further information may be obtained from :

Ruthner Industrieanlagen Aktiengesellschaft
A-1121 Wien, Aichholzgasse 51-53
Telephone: 0222/83 95 01
Telex : 01-1273

Supply of plants.

7. Pickling of Special Steel Strips According to the Ruthner Neolyte Process

Plates and strips are repeatedly rolled and annealed. Thereby, a layer of scales is formed on the surface and has to be removed. This is normally done by means of pickling in acid.

The scales of special steel are extremely difficult to remove. Therefore, according to usual technology, particularly intense pickling processes are employed, e.g.

Descaling of special steel strip by chemical pickling.

The well-known dull-bright surface of special steel is usually obtained by pickling in a very aggressive mixture of nitric and hydrofluoric acid ("mixed acid"). However, this entails some disadvantages :

the aggressive acids represent a considerable accident risk;

poisonous nitrous gases, which cannot be completely removed from the waste gases, are produced during pickling;

an unnecessarily large portion of the basic material (special steel) is removed too—in a medium-size plant, approximately 100 tons per year;

great quantities of spent pickling acid are produced, the recovery of which still represents a problem.

Even conventional pickling in molten salt is a problem. The annealed strip is first treated at a temperature of 500°C in a bath mainly consisting of melted soda lye and additionally containing oxidizers or reducing agents. The scales in this bath should be altered in such a way as to reduce the subsequent treatment in acid. Among others, the following disadvantages result:

handling melted soda lye having a temperature of 500°C is very dangerous;

considerable drag-out losses entail heavy pollution of waste waters.

Electrolytic pickling according to the Ruthner Neolyte Process is advantageous in comparison with the two conventional processes described.

It is common knowledge that the pickling process is accelerated by electrolysis.

Sulphuric or nitric acid have always been used as pickling agents. However, the process of pickling in a neutral solution of non-poisonous salt developed by Ruthner Industrieanlagen A.H., Vienna, has been on the market for several years. This process offers three important advantages:

the scales are not removed in the pickle bath but form a sludge not disturbing the pickle process. The sludge may be removed from the pickle bath by means of a centrifuge. The pickle bath itself is not spent. Therefore, it need not be replaced at short intervals, as otherwise is usual, and does not pollute waste waters;

by the conditions chosen for this process only the scales are attacked and not the basic materials thus, on the one hand, saving material and, on the other hand, decreasing the sludge quantity;

In spite of higher investment costs this process is cheaper in operation than all others. Apart from this economic advantage and the decrease in environmental pollution an excellent quality of steel surface is obtained.

In the neutral pickling process, the strip is horizontally led through a pickle tank that is lined against corrosion with material which does not conduct electricity. In the tank a row of electrodes is installed above and below the strip. The electrodes are arranged in groups about 1.5 m in length and there is a distance of 0.5 - 1.0 m between each group. The groups are alternately connected with the positive and negative poles of a rectifier. The current flows from one group of electrodes through the electrolyse to the special steel strip, which to a great extent performs the transport of current towards the next group of electrodes which has the inverse role. Thereby, it takes in each case relative to the electrolyse the opposite polarity, i.e. it becomes alternately anode and cathode. In the anodic sections the scales are removed with a 90 per cent yield of current, whereas the basic material is rendered passive and scarcely attacked.

As occasion demands, descaling is connected with passivation in nitric acid or brightening of the surface in mixed acid.

For further information please contact:

Ruthner Industrieanlagen Aktiengesellschaft
A-1121 Wien, Aichheisgasse 51-53
Telephone: 0222/88 25 01
Telex: 01-1273

Supply of plants.

8. Procedure for Reworking Caustic Sludge, Obtained from Acid Polishing of Lead Glass, to Basic Lead Carbonate

Acid polishing is an important process in the section of the glass industry producing industry glass, lightening and decorative glass of high quality. The surface refinement of the work pieces and/or the production of the desired dimension within narrow tolerance limits is brought about by pickling with a mixture of hydrofluoric acid and sulphuric acid. This process is especially used for lead crystal glass (PbO content 24 per cent of weight at least) and special lead crystal glass (PbO content 30 per cent of weight at least). In the acid polishing process of lead crystal glass, caustic sludge waste is produced as a pollutant containing fluorine and lead compounds. This sludge has to be regularly precipitated from the caustic, for otherwise it can cover some areas of the glass and thus prevent the action of the acid. By such irregular pickling of glass, so-called polishing defects are produced.

The sludge is very soluble in water and can, therefore—even after neutralization—be dumped only in such a way that the lead contamination of underground water is excluded. This is a costly procedure.

The composition of the sludge depends on the composition of the lead glass. Chief constituents are lead sulphate and alkali fluoric silicate. This process is designed to obtain basic lead carbonate (with a fluorine content lower than 0.2 per cent of weight) from the sludge.

For reprocessing, the sludge, precipitated from the caustic acid, is first suspended in an alkali carbonate solution. When the reaction is finished the suspension is filtered off, the residue is dissolved in nitric acid and mixed with alkali ions. All fluoric and silicic acid still contained in the nitric solution is precipitated as barely soluble alkali fluoric silicate which is filtered off. From this filtrate basic lead carbonate of a high degree of purity is precipitated with soda.

For the application of this procedure only an acid-resistant sludge pump, a stirring vessel filtration unit as well as a drying plant equipped to dry the lead carbonate are necessary.

The cost of the used chemicals (natron, potash, nitre and silicic acid) is well covered by the value of the basic lead carbonate obtained. It can be used to special advantage in integrated lead glass production as a partial or full equivalent of minium.

Advantages:

1. There are no other known procedures for the utilization of caustic sludge. Alternative: The dumping of neutralized sludge as special waste.

2. The sludge contains the following chief components:

lead	30 to 40 per cent in weight
fluorine	about 30 per cent in weight

3. Chemicals necessary per ton of sludge:

soda	about 850 kg
nitric acid	about 350 kg
potash	about 75 kg

4. Polishing acid is a by-product of sludge production: from the mud settling pond is extracted: sludge: polishing acid = 1 : 1.75 kg. The polishing acid is conveyed back to glass production as a very precious material, thus it does not require a neutralization plant.

5. The finished product, basic lead carbonate, contains:

lead	about 80 per cent in weight
fluorine	less than 0.2 per cent in weight

6. Recoverable basic lead carbonate per ton of sludge: 250 to 300 kg.

7. Utilization of basic lead carbonate:

   substitute of minium at the lead crystal glass production;

   raw material for the production of other lead compounds

8. The waste water of the process is partly sour. It is, therefore, directed to the neutralization plant where fluoride is also precipitated as fluor spar.

Further information may be obtained from:

Fa. D. Swarowski & Co. Wattens
A-6112 Wattens/Tirol
Telephone: 05224/ 24 11

Know-how and engineering.

9. Production of Aluminium Fluoride from Waste Gas of Phosphoric Acid and Superphosphate Plants

In the decomposition of crude phosphates, fluorine containing waste gas is set free. Therefore, a procedure has been developed which makes not only the separation of this waste gas possible, but also its commercial utilization.

The fluorine containing waste gas is transformed into silica hydrofluoric acid by scrubbing with water.

The preheated silica hydrofluoric acid reacts on hydroxide of aluminium (Bayer Hydrate) to aluminium fluoride and silicic acid.

The crystalline silicic acid is centrifuged and, from the remaining AlF_3-solution, the aluminium fluoride trihydrate is recrystallized, which is separated in a centrifuge from the mother liquid.

The following drying process and calcination takes place continuously in a fluidized bed furnace. The process in question is an example of a perfect recycling of an atmospheric pollutant. The derived aluminium fluoride can be converted into cryolit with a sodium fluoride solution. Sodium fluoride is produced from silica hydrofluoric acid and soda in a reactor. The precipitated cryolit is separated from another liquid by drum filters and provides, after its calcination, a high-grade product to the aluminium industry. This process is very flexible as far as raw materials are concerned. Instead of silica hydrofluoric acid, sodium silico fluoride can be used and in place of soda, NaOH.

Further information may be obtained from:

Chemie Linz, Aktiengesellschaft
P.O.B. 296
A-4021 LINZ
Telephone: 07222/ 56 471

Licences, know-how and engineering.

10. Production of Ammonium Sulphate and Sulphuric Acid Respectively from Phosphoric Acid Waste

From the production of phosphoric acid significant quantities of waste gypsum are obtained. These have usually been poured into the sea or into rivers or have been dumped through lack of recycling possibilities. As dumping represents a danger to underground water it has been necessary to find new ways of processing and, in fact, two new procedures have been developed. In the first, waste gypsum is utilized for the production of the fertilizer ammonium sulphate, according to the equation $(NH_4)_2\ CO_3 + CaSO_4 = (NH_4)_2\ SO_4 + CaCO_3$. Natural gypsum or anhydrite is used, too. The ammonium carbonate is obtained from ammonia and CO_2-containing off gas or from waste lyes.

In the second process from gypsum as a sulphuric carrier, together with clay, sand, and coke, SO_2 and a high-quality portland cement, clinker are obtained in the rotary cement kiln according to the equation:

$$CaSO_4 + 2C \qquad CaS + 2CO_2$$

$$CaS + CaSO_4 \qquad 4CaO + 4SO_2.$$

The sulphur dioxide-containing off gases are catalytically oxidized to SO_3 after a dedusting process and from that a sulphuric acid of 96 per cent is obtained. As $CaSO_4$ can be utilized in the form of phosphoric acid waste gypsum, this procedure brings economic advantages in connection with plants for concentration of phosphoric acid and thus completes their sulphuric acid cycle.

Further information may be obtained from:

Chemie Linz Aktiengesellschaft
P.O.B. 296
A-4021 LINZ
Telephone: 07222/ 56 471

Licences, know-how and engineering.

11. Procedure for Reprocessing of a-tactical Polypropylen

A by-product which until now could not be avoided in the production of the high-quality thermoplastic synthetic material polypropylen is the so-called a-tactical polypropylen, a modification of the chief product differing from it in regard to the molecular configuration relating to space. Because of the difference in molecular structure the a-tactical polypropylen cannot be used as plastic and has to be separated from the main product as it influences its properties in a negative way.

This polymer amounts to about 10 per cent of the main product. Until now it has been burned without any retreatment because there have been no specific fields of use.

To solve the problem satisfactorily, that is to stop the production of a-tactical polypropylen, the development of new types of catalysers is above all necessary, a project which is under way but which until now has not had great success.

Therefore, forms of the waste product a-tactical polypropylen which makes its use possible have been sought. The most promising fields of use seem to be carpet-back coatings, isolating masses for buildings and cut-back bitumen.

A condition for the utilization of a-tactical polypropylen in these fields of use is an appropriate degree of purity, the absence of solvents and a form which can be manipulated. In the conventional process the a-tactical polypropylen—as distilling refuse discharged—contained a great amount of catalyst components from the polymerization process and a comparatively large proportion of solvents.

The first step, therefore, has been the development of a cleaning system which makes possible the removal of catalyst deposits. The a-tactical polypropylen produced in the polymerization process is dissolved in a diluent (n-heptane). The solution contains some percent of a-tactical polypropylen and elements of the polymerization catalyser aluminium diethyl monochloride and titanium trichloride. These contaminations are cleared by a multi-stage caustic scrubbing so thoroughly that the content of the remaining residue ash is only about 100 ppm and is neutralized too.

The petrol solution cleared in this way is pre-concentrated by flash vaporation. The pre-concentrate has a polymer content of about 60 per cent and is cleared of the solvent in a thin-layer evaporator. The a-tactical polypropylen is discharged as melt with a solvent content of less than 1 per cent. As the melt of the a-tactical polypropylen shows high viscosity with an adequate operating temperature of the thin-layer evaporator a special type of evaporator for highly viscous discharging products has to be developed.

The cleaned melt, free of solvents, is finally poured into moulds. After the solidification of the melt slabs are obtained in a form easy to manipulate.

Therewith all necessary conditions have been fulfilled to make a-tactical polypropylen marketable and utilizable in the above-mentioned field.

Obtainable selling price for purified a-tactical polypropylen is about 7.-- Schillings per kg.

Purification costs: about 1.50 Schillings per kg.

At the Petrochemie Schwechat about 7000 tons per annum of a-tactical polypropylen are produced (about 10 per cent of the produced polypropylen) and are offered for sale.

World discharge: about 300,000 tons a-tactical polypropylen per year.

For further requirements please contact:

Petrochemie Schwechat
A-2300 Schwechat
Telephone: 0222/ 77 66 01

Advice and know-how free of charge (no licence fees).

12. Process for Recycling of Stretched Thermoplastic Material in the Manufacturing Process

Until recently a scrap percentage rate of about 15 per cent in the manufacturing of thermoplastic granules into foil and woven silver respectively was not extraordinary.

A lot of time and money has been invested in finding a satisfactory solution. In the first phase attention was focused on working up the scrap material into so-called regranules which made a special plant necessary, as well as huge storage facilities to store the different kinds of scraps separated from each other. There was also the menace of contamination as the material had to be transported.

Direct utilization in the production was, therefore possible only with comparatively low-quality products. The greater part of the regranules were sold and the material was used in the die-casting industry.

In the second phase of effort the non-stretched edge strips were pulverized. As the bulk weight was similar to the granule, a mixture of these two components could have been moulded by extrusion. Such edge strip recyclings can be purchased everywhere and do not yet represent a development by Lenzing.

The third and important new phase: the problem of what to do with the stretched edge strips and the stretched folio waste was still an open question. If these stretched scraps are pulverized, the bulk weight is so small that a separation on the processing machine, the so-called extruders, takes place, and in this way obvious variations of quality occur which cannot be tolerated.

After numerous experiments a recycling unit was developed by the SFA department which allows a direct recycling of each waste, independent of the degree of stretching by a thermomechanical treatment without damaging the polymer.

Machines equipped with the Lenzing recycling unit produce with efficiencies of almost 100 per cent without appreciable waste.

Further information may be obtained from:

Chemiefaser Lenzing Aktiengesellschaft
A-4860 Lenzing
Telephone: 07672/ 25 11

Delivery of plants, engineering.

13. The Lenzing Magnesia Bisulphite Recovery Process in the Magnesia Bisulphite Wood Pulping Industry

The Lenzing magnesia bisulphite recovery process represents an approved process for recovering SO_2 and MgO in the magnesia bisulphate wood-pulping industry.

The substance and waste liquor deriving from the discontinuous pulping process are separated from each other in a multi-stage pressure filter washing, with an output of 1.6 tons of dry matter in the waste liquor covered per ton of wood pulp. The covered waste liquor (15 per cent dry matter) is concentrated in a multi-stage vacuum evaporating plant to a consistent lye with 55 per cent dry matter. This consistent lye is combusted in a furnace working counter-currently according to a principle developed by Lenzing in cooperation with the firm of Steinmüller. By this, heat is obtained (2.2 tons vapour/ton of wood pulp) and the chemicals magnesium oxide and sulphur oxide, bound in the lye, are thermically broken into a reactive form. By the flue-gas the regenerated chemicals are conveyed back to the recovery plant, where the desired magnesia bisulphite acid is produced by means of a wet filter and an absorption tower. On the following cleaning and concentration unit the regulation of the necessary acid quality for the pulping process takes place. The Lenzing magnesia bisulphite recovery process stands out because of the following features:

utilization of dirty magnesium oxides as, for example, Austrian calcined magnesites;

coverage rate of 99 per cent of waste liquors;

chemical recovery rates relative to covered rates liquids:

sulphur : 92 per cent,
MgO : 95 per cent;

only 0.02 per cent SO_2 concentration in the final flue-gas;

high-effective acid purification;

recovery system easy to manage.

Cost reduction per month (Austrian Schillings) achieved by the magnesia bisulphite recovery process for a plant of 225 tons per day (80 per cent beech, 20 per cent spruce).

(A) Costs of sulphur and limestone for calcium bisulphate

sulphur and handling	A.S. 1,100,000.--	
limestone and handling	A.S. 160,000.--	
total:		A.S. 1,260,000.--

(B) Costs of sulphur and MgO including costs for recovery plant

fixed costs, depreciation interest rates for recovery plant	A.S. 140,000.--	
general working expenditure for the recovery plant	A.S. 160,000.--	
sulphur and handling	A.S. 320,000.--	
MgO and handling	A.S. 110,000.--	
total:		A.S. 730,000.--
saving:		A.S. 530,000.--

(C) Reduction in price achieved by sulphur and MgO recovery including an increased yield of wood

increased yield of wood	A.S. 170,000.--	
saving by sulphur and MgO recovery	A.S. 530,000.--	
total:		A.S. 700,000.--

Further information may be obtained from:

Chemiefaser Lenzing Aktiengesellschaft
A-4860 Lenzing
Telephone: 07672/ 25 11

Licenses, know-how, engineering.

14. Production of Paper and Cardboard without Environmental Pollution

Pulp, loadings, process materials, water, energy in form of electric current and steam are necessary for the production of paper and cardboard.

250 tons of folding boxes are produced per day by one machine in an Austrian cardboard mill. Up to 85 per cent of raw materials are ungraded waste paper. A small quantity of pure cellulose and various process materials are added to this.

The waste paper is dissolved in water and conveyed into production. There the water is mainly used for the spray pipes. The waste waters produced during dewatering undergo a treatment. The daily quantity of water amounts to approximately 8000 m^3, which is with a 24 hours' operation 333 m^3/hour. The content of solid matters in the waste water averages 14,000 kg/day. Since this solid matter mainly consists of fibres and loadings it is obvious that they should be fed back into the production process.

The waste waters go first into a reactivator where the major portion of the fibre-containing material is separated and directly conveyed back into the production machine. An organic polyelectrolyte is added to the waste waters for better separation of the solid matter.

The waste waters leave the reactivator with an average content of solid matter of 30 mg/l. A filter station is installed for further improvement of the water quality. It consists of four closed filter boilers, each of them having a diameter of 3200 mm.

The filtering medium consists of quartz gravel and hydroanthracite in graded grain sizes. The filter does not only work as a surface filter, thus considerably increasing the retention time. Backwashing of the filters is done by means of compressed air and purified water. The water conditioned in that way leaves the filter station with a content of solid matter of less than 5 mg/l and may, therefore, be used again as perfect production water. The sludge separated during backwashing is conveyed back into the reactivator from where it is integrated into production.

Since slime and fungoid growth problems may result from the constant circulation of the water, bleach liquor is continuously dosed into the water system. In addition to the bleach liquor, sodium chloride is added at certain intervals. The costs for chemicals, including the organic coagulant, amount in total to Austrian Schillings 0.08 per cubic meter of water.

When evaluating the economy, the following results are obtained. The entire investment costs for the reactivator and the filter station amounted to 6 million Austrian Schillings, the price basis being 1974. The operating expenses, inclusive of energy and personnel, amounted in total to approximately A.S. 60,000.-- per month. On the other side, an amount of approximately A.S. 250,000.-- for raw materials was saved by recycling the sludge. Furthermore, by closing the circuit of waste waters, the temperature of the production water was increased to approximately 40°C. By this means, on the one hand, the output of the same cardboard machine was increased by approximately 15 per cent. On the other hand, energy in the form of steam could be saved for drying the cardboard. Taking these factors into account the plant pays itself off before long. In addition to an increase in productivity, environmental pollution could considerably be reduced, the quantity of waste waters decreased from originally 8000 m^3/day to less than 1000 m^3. These 1000 m^3 of residual water already contain unpolluted cooling and sealing waters.

Further information may be obtained from:

Fa. Ruthner Industrieanlagen A.G.
A-1121 Wien, Aichholzgasse 51 - 53
Telephone: 0222/ 83 95 01

Supply of plants.

15. Very Effective Purification of Radioactive Waste Water

For the cleaning of radioactive waste water from nuclear power stations, several methods are applied, the most effective one being the evaporation of the waste water. As water itself does not become radioactive, non-volatile radioactive substances can be concentrated by evaporation in the radioactive waste water, to be afterwards brought to a form insoluble in water, by enclosing in bitumen for final storage.

Unfortunately, in the evaporation of the waste water, radioactive substances are always carried along by the vapour, so that the distilled water that leaves the plant is not completely inactive as one would imagine, but still contains some residues of radioactive contamination. The proportion of the radioactivity of the liquid to evaporate to the radioactivity of the distillate, called decontamination factor, therefore, is one to one million at the utmost with conventional plants.

Based on intensive research and development during the last years, made by Gebr. Böhler in cooperation with the Austrian Association of Studies in Nuclear Energy Ges.m.b.H. in the research centre of Seibersdorf, the new two-stage "Böhler-Evaporator-Plant" for the elimination of liquid radio-active waste was developed, which guarantees a cleaning effect a thousand times higher than the conventional techniques had offered.

Heat consumption and the quantity of cooling water required of plants running according to the new procedure are not higher than with conventional plants. In regard to control adjustment and operating techniques the new plants do not pose any additional problems. The two-stage Böhler-Evaporator-System for liquid radioactive waste allows such a good cleaning of the waste water of nuclear power stations that it afterwards contains less non-volatile activity than normal spring water and again has the quality standards of potable water. In the utilization of the new process it is no longer necessary to dilute the waste water leaving the power station with great quantities of river or sea water in order to reach the prescribed standards before it is delivered. Power stations working according to the method do not release additional radioactivity attached to non-volatile carriers into the environment with the waste water and, therefore, do not need any water dilution, which is of special importance with nuclear power stations working with cooling towers, and especially with the utilization of dry cooling towers.

The first big plant delivered by VEW, working according to the new method, was opened in autumn 1975 in a nuclear power station (600 MWe) and has been running satisfactorily. Further plants are in process of construction.

Further information are to be obtained from:

Vereinigte Edelstahlwerke (VEW) AG
A-1011 Wien, Elisabethstrasse 12
Telephone: 0222/ 57 35 35.

Delivery of plants.

16. Heavy Degree Fermentors—an Integrated Component of Recycling and Environmental Protection Measures

The Vogelbusch-IZ-jet aeration system represents a new but already often approved procedure that makes the removal of waste products producing a negative effect on the environment and the recovery of reuseable final products possible. This system is marked by extremely high oxygen input rates up to 12 kg $0^2/m^3$/per hour, a low-power consumption outstanding homogeneity of the circulated aerated volume as well as by far-reaching independence of the reaction tank and low maintenance charges. The Vogelbusch-IZ-jet-system is already successfully utilized by the aeration of municipal or industrial waste water as well as for the aerobic fermentation in the biochemical and fermentation industry, as well as for the protein production.

An interesting example of the application of the Vogelbusch-IZ-jet-aeration-fermentor represents the yeasting of whey to obtain protein-enriched fodder —for foddering purposes and desired—lactose waste whey and skim milk. This product can either directly or, if transportation is necessary, in concentrated form to be used up as fodder for pigs. The high degree of economy of the whey-to-yeast production plant results in a better quality of meat of pigs that have been fed exclusively on yeasted whey, which is verified by numerous analyses. In Austria this meat is indeed put on the market with a specific market name and fetches up to 20 per cent higher prices than normal meat.

On the one hand, the liquid manure which arises from intensive animal keeping shows a high content of biologically decomposable substances and represents, on the other hand, a source of high odour inconveniences. While conventional aeration systems have until now not allowed biological removal of liquid manure without immense expense, the new aeration system opens up a promising field of use. These new aspects are also worth being applied to other fields, as for the treatment of human excrement.

Further information may be obtained from:

Vereinigte Edelstahlwerke Aktiengesellschaft (VEW)
A-1011 Wien, Elisabethstrasse 12
Telephone: 0222/ 57 35 35

Delivery of plants, licences, engineering.

17. Flare Gas Recovery Plant

In refineries and in the chemical industry burnable waste gas is constantly produced, even under normal operating conditions. These gases, obtained with a small over-atmospheric pressure, are mixed with steam and, for safety reasons, burnt on a flare stack into the atmosphere.

Flare gas is escaping from the safety valves in the various plants and is also issuing as process waste gas which cannot be directly used in the heating gas system. The gas is collected by a separate pipe system (blow-down system) and connected with the flare stack. By this method the loss of energy is considerable. Furthermore, each flare stack represents environmental problems from its light and gas emission.

A flare-gas recovery has been developed and is operated by the refinery Schwechat of the Österreichische Mineralölverwaltung-AG. This flare-gas recovery plant works without a separate gasometer and was, therefore, built at low cost. The recovered gas is then burned in the boilers of the power plant of the refinery.

The existing flare-gas system itself is not altered by the recovery plant, only the gas quantities are reduced which finally leave the flare stack, so that normally no combustion occurs.

The flare-gas recovery plant works in the following way:

The flare gas is sucked from the flare-gas system by the rotary compressor via the liquid separator. The compressed gas is then fed into the refinery gas system for further use. By a contact pressure gauge a quick-opening valve is opened at a certain adjusted minimum pressure in the flare-gas system, thus opening a by-pass over the compressor. The non-return valve closes now automatically and disconnects the refinery gas system from the flare-gas system.

The load of the rotary compressor is now small and the energy consumption of the driving motor is considerably reduced. When the pressure in the flare-gas system increases again and exceeds the adjusted pressure of the contact pressure gauge the by-pass and the chamber valves of the compressor are closed and flare gas is again fed into the refinery gas system.

To obtain a high degree of safety against an inadmissible reduction of the sucking pressure of the rotary compressor a second by-pass valve is provided which is directly controlled by the pressure in the flare gas system. If the sucking pressure becomes very low the driving motor of the rotary compressor is switched off by a pressure switch. By this safety system a vacuum in the flare gas system can never occur.

To allow for economic operation under various conditions the recovery plant can be equipped with several rotary compressors of different sizes working in parallel by manual selection.

The plant is open-air mounted, works fully automatically and has a low cost of maintenance.

Central control and far-reaching automation of the process, require no maintenance. Central control can be adjusted to specific requirements without impairment of functional safety.

Further information is to be obtained from:

Österreichische Mineralölverwaltung-AG
A-1090 Wien, Otto-Wagner-Platz 5
Telephone: 0222/ 42 36 41

Know-how, engineering.

18. Generation of Energy by Burning Wastes

Latest developments have proved that energy may continuously be generated from special wastes.

Economy depends on the particularities of the quality of each kind of waste. It may easily be achieved by installing after the post-combustion chamber a waste heat boiler adjusted to the special conditions. The following question is decisive: Is it possible or not to utilize the steam near the plant? In smaller plants it is opportune to produce hot water by means of the hot waste gases. However, in larger plants even the generation of electric energy is profitable.

With various types of special wastes particular precautions must be taken against dust and exhaust gases. Legal regulations may in any case be adhered to by installing appropriate electrofilters and exhaust gas scrubbers.

Furthermore, it is possible to run a plant profitably, even without a waste heat boiler, when wastes are against payment. With an incineration capacity of 4 - 5 t/h, amounting to 30,000 to 35,000 tons per year, and a calorific value of the wastes of 5000 to 6000 kcal/kg, the plant surely is profitable. This may be proved by a plant in operation for 4 years. Furthermore, neither nuisance nor impediment to neighbouring inhabitants or surrounding farms have resulted.

The number of persons to be employed per plant and shift is determined according to the quantity and the type of wastes supplied.

From 1 ton of waste with a calorific value of 6000 kcal/kg, 8 tons/h of steam with 20 atm. exc. pr. are obtained. With a price of A.S. 1.37/kg of fuel oil (heavy oil) A.S. 1.370.-- are saved when 10 tons of steam are generated per hour; transport charges for the wastes to the plant must be deducted.

The combustion and post-combustion chambers are fixed, only the water-cooled ash removal and charging machine enters the combustion chamber for a short time. Thus—unlike systems with moving parts of the furnace or movable furnaces—the furnace is only slightly mechanically stressed which entails the following two advantages: long working life and low costs for maintenance, and, furthermore, the opportunity of maintaining a relatively high temperature in the furnace.

Two important advantages, furthermore, result from this high temperature: practically no soot or smoke is produced and all combustible substances are completely burnt to CO_2, H_2O, and the corresponding oxides. Substances not combustible are baked thoroughly, glowed out, and oxidized.

The system developed together with Messrs. Polyma is particularly suitable for generation of energy because it is almost trouble-free and its failure quota is lower than that of any other system. It largely relieves dumping grounds for special waste and detoxication plants.

For further information please contact:

Fa. Ruthner Industrieanlagen AG.
A-1211 Wien, Aichholzgasse 51 - 53
Telephone: 0222/ 83 95 01.

Supply of plants.

19. Energy Recovery at Cracking Plants

The recovery process in question solved the problem of recovering energy from the hot flue gas originating from regeneration of the catalysator of catalytic cracking plants. The following demands were to be met:

1. to separate the highest possible dust content;

2. to expand the available enthalpy within the gas turbine;

3. to use the remaining carbon monoxide content in a steam boiler by either thermal or catalytic combustion.

These requirements led to specific solutions for the different parts of the plant. The cyclone separator protects turbine and boiler from the catalysator dust.

From the cyclone the almost dust-free flue gas enters into the turbine via four quick-shutting valves. In the case of the shutting of the quick-shutting valves, that stops the machine, the flue gas is fed in parallel to the turbine via the by-pass which is of great importance for maintaining the pressure difference, otherwise the process in the cracker might be disturbed.

The design as a four-stage reaction turbine ensures high efficiency as well as low gas velocities and large flow sections. In a four-stage turbine the flow sections are large compared to a single-stage turbine. Therefore, the large and sturdy blades are highly resistant to erosion. The influence of dust deposits on the blades is also much less than in the case of a single-stage turbine.

In designing the turbine, the high temperature and dust content of the flue gas had to be considered. Problems of cooling the material and protection against erosion had to be solved.

In order to make optimum use of the energy of the flue gas the carbon monoxide is burned after the turbine in a boiler. Depending on the amount of steam wanted, the CO is burned either catalytically or thermally. In case a separate steam power station is operated in the refinery, it is preferable to use catalytic combustion. The afterburners still required are necessary to raise the gas temperature to optimum combustion temperature. In case of high steam demand it is preferable to use thermal combustion. The afterburners have to be larger in order to reach thermal reaction temperature. Flue gas devoid of carbon is used in a heat exchanger.

Further information may be obtained from:

Österreichische Mineralölverwaltung-AG.
A-1090 Wien, Otto-Wagner-Platz 5
Telephone: 0222/ 42 36 21

Supply of plants, licences, engineering.

(B) PROCEDURES STILL IN DEVELOPMENT

1. *Environmentally Favourable Process of High Efficiency for Power Stations*

Based on thorough research and development work by Nuclear Energy Ges. m.b.H., Seibersdorf, in the field of high-temperature alkali metal technology, a new process for power stations can be proposed. Its application would decisively reduce negative effects on the environment caused by thermal power stations. In this triple-evaporation process, the utilization of three steam turbine processes making use of the operation media potassium, diphenyl and water, which are thermically connected in series, obtains a process efficiency of more than 60 per cent (Fig. 1). The efficiency of a corresponding gas- or oil-fired power station amounts to 56 per cent.

In comparison with steam power plants of modern design, power stations applying this new triple-evaporation process would need only about two-thirds of the fuel material needed by conventional plants, but with an invariable amount of output energy. This would also mean a reduction of the flue gas emission to two-thirds of the usual emission rate. The new process is even more efficient with the waste heat emission that is, for example, the amount of river water needed for the cooling of a power station, which could be reduced to less than half the quantity actually required.

To be able to utilize the triple-evaporation process commercially, a further development is still necessary. But the workability of the technicalities of this process can be regarded as already assured.

Further information may be obtained from:

Vereinigte Edelstahlwerke (VEW) AG
A-1011 Wien, Elisabethstrasse 12
Telephone: 0222/ 57 35 35.

Licences.

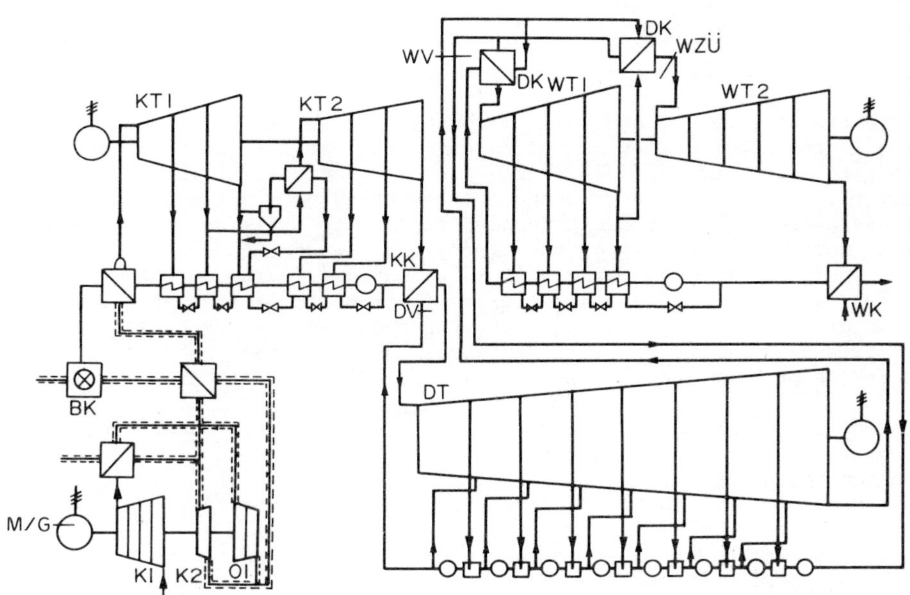

Fig. 1. Connection diagram of a gas-, oil- or coal-fired power station using the triple-evaporation process.

BK	Uncooled combustion chamber	KK	Potassium capacitor
DK	Diphenyl capacitor	KT	Potassium steam turbine
DT	Diphenyl steam turbine	M/G	Motor/Generator
DV	Diphenyl steam generating unit	WK	Water capacitor
G	Generator	WT	Steam turbine
GT	Flue gas turbine	WV	Steam generator
K	Compressor	WZÜ	Steam intermediate superheater

2. *Tyres made of Cast Plastic*

The Austrian company Polyair Maschinenbau offered the first compact plastic tyre, designated as LIM Liquid Injection Moulding. This tyre for passenger cars does not need the carcass usual with normal tyres. It is produced by a simple and cheap casting process. Several liquid raw materials are mechanically batched and mixed. The compound is injected into the moulds. After a certain curing time the LIM tyres are post-treated in a tempering cabinet. Then the finished tyre is obtained.

The production of the first practicable tyre had been made possible by the utilization of a new molecule which is a polyurethan-elastomer, obtained from petrol derivates, characterized by a high-dimensioned stability against tension.

In conventional tyres of any design, the elasticity of rubber has to be exploited in the construction of the carcass, but, at the same time over-stressing has to be prevented by inserting fibres. The utilization of polyurethan-elastomer—improving on rubber because of outstanding elastic and strain qualities—has brought about a decisive advantage: the omission of insertions. The only foreign substances of the LIM-tyre are insertions in the bead of the rim. A whole range of advantages includes savings in the weight of the tyre and, thus, will perhaps have a positive effect on the design of cars. The omission of the carcass, besides improving the road-holding qualities of the tyres, brings about a significant rationalization of the number of operations necessary to produce the tyre.

In connection with the comparatively simple technology of manufacturing polyurethan, it is possible—despite far-reaching automation—to keep the expense of the manufacturing machines, necessary for the series production, lower than is possible with the production of tyres made of rubber.

The production of LIM-tyres—from the conditioning of the raw material to the control and storing of the final product—takes place in a single fully automatic electronic operation.

A quality of the LIM-tyre specially worth mentioning and a significant advantage over rubber tyres is its high resistance to blow-outs. The side walls are so stiff that the tyre, without internal pressure does not stick to the clinch; there still remains some freedom of motion to do the necessary fulling. Even under hard wear the tyre does not risk damage from heating. Because of its rigid sides, the tyre has outstanding driving properties on bends. Accidents, caused by side slipping on account of a sudden escape of pressure at high speed, seem therewith to be reduced in comparison with conventional tyres.

By admixing of polyurethan-compatible chemicals it is possible to make the LIM-tyre on the one hand, hard to set on fire, and on the other to have an automatic extinguishing capacity. Thus, one more danger of conventional tyres can be reduced.

The LIM-tyre can be designated as a safety tyre. Regular standard rims are used along with standard assembly tools. There are tubeless LIM-tyres as well as tyres with tubes.

For the commercial utilization of this development, modern and environmentally favourable plants are at hand. The use of precision steels, the possibility of exact batching by electronic pre-selection, as well as a continuous manufacturing cycle, guarantee an optimum exploitation of raw material, which bring about production free from waste. The contact pressure moulding, used by Polyair, as well as the exothermic reaction of the finished product lead to a comparatively low energy consumption.

It is important to notice that there is a possibility of recycling. A run-down tyre can be converted to a granule which—mixed with a certain percentage of fresh raw materials—again represents a raw material system of high quality.

Further information to be obtained from:

Polyair Maschinenbau Ges.m.b.H.
A-2421 Kittsee, Burgenland
Telephone: 02143/245, 267

Licences.

3. Complete Recycling of Lees Production of Tartaric Acid

Even today natural tartaric acid can be technically produced only from residues originating in the making of wine. Because of a new wine technology the primary raw material cask-tartar is developed in ever-diminishing quantities. The yeast-containing sediment, the so-called lees formed after the fermenting process in the wine, can be regarded as secondary raw material. It contains about 20 per cent of salts of tartaric acid insoluble in water and about 30 - 33 per cent of organic dry matter in form of distiller's yeast.

The production of tartaric acid is generally based on the transformation of the tartrates being insoluble in water into tartaric acid, which is easily soluble, and the separation of tartrates from accompanying material. The raw tartaric acid is evaporated at 70 per cent and crystallized.

In the generally applied procedure, the lees process, the finely ground dry distiller's yeast is heated to 100-150°C in a rotary kiln furnace to make a separation possible. By this the accompanying organic material, especially yeast containing a high degree of protein raw material, is destroyed and enters into the waste water.

Only by using a new technology is a complete recycling practicable. By this the wine is resuspended in the water without drying and roasting, with a "specific cation-exchanger treatment" freed from metallic ions and by this lees as free tartaric acid. The developed thin tartaric acid solution is centrifuged from the yeast. In this process the separation of tartaric acid from the accompanying material is brought about without heating and so carefully that after the removal of the tartrates the protein-containing parts of the yeast of the raw material can be turned into a high quality feeding yeast in a conventional drying process. The potassium share contained in the lees is gained comparatively pure at the regeneration of the cation exchanger and turned into crystalline potassium sulphate.

The thin tartaric acid solution, obtained after the separation from the yeast-shares, is cleaned with an ion exchanger and concentrated to about 30 per cent. From this solution pure crystalline tartaric acid can be obtained by an evaporation at 70°C without recrystallization.

The chief advantages of this new process concerning non-waste technology are:

1. It is possible to utilize the secondary raw material from the tartaric acid production in such a way that the high quality and accompanying protein-containing material does not pollute the environment, but is used as fodder for the benefit of agriculture.

2. The raw material can be manufactured without drying and roasting.

3. The steam consumption of the production is significantly reduced.

4. Decrease of environmental pollution by strong residues ($CaSO_4$, chemicals).

5. The quantity of waste water is reduced and it is made biologically fully decomposable.

Based on the described process a complete recycling system has been developed and makes it possible to produce, with low thermal energy consumption, tartaric acid as well as feeding yeast and potassium sulphate from the waste product.

Further information to be obtained from:

ABO-CHEMIA
A-1140 Wien, Hütteldorferstrasse 216
Telephone: 0222/94 51 06

Some Aspects of Production Without Waste of Mineral Raw Materials in Poland

Stefan Gustkowicz

Committee of Science and Technology, Warsaw, Poland

1. INTRODUCTION

Mining is closely linked with the development of civilization. The more mature a civilization is, the higher is the demand for mineral raw materials. Dynamic development of various industrial branches requires more diversified chemical elements, rarer in their appearance with lower natural concentration and simultaneous mineral dispersion.

These factors lead to an increase in the total amount of minerals exploited and, consequently, to disproportionate increase in the amount of tailings after extraction of useful components. Let us take, for example, non-ferrous metal ores, radioactive elements or rare earth elements, the primary concentrations of which of about one-tenth per cent already are useful for exploitation and processing. In such a case the total of exploited rock consists of post-extractive wastes which consequently are more powdered and in every way more difficult to handle when stocked and processed.

Another factor influencing the amount of wastes is the developing mechanization of exploitation. Presently, in order to maximize the effectiveness of extraction from deposits, attention is paid less to the purity of the mineral than to the accuracy of exploitation. On this point, the example of mechanization of coal exploitation may be used, where an increased amount of ash and deads is observed. Similarities may also be noted in other industrial branches using mineral raw materials—in non-ferrous metallurgy, chemicals, sulphur and barite exploitation, building materials or phosphatic fertilizers.

The amount of waste substances from mineral utilization processes differs according to the type of mineral and to individual phases of the technological process. The percentage of those substances in the total amount of mass exploited by the Polish mineral industry in the years 1965-1970 reached 20 per cent and presently it is about 25 per cent.

Wastes call for expenditures for dumping on the one hand and, on the other, not only do they take space (which could be well used for agriculture) but also are the cause of air and water pollution. In addition each industrial branch using the mineral raw materials tends to fulfil its demand for them in the easiest way, i.e. directly exploiting the primary resources with regard only to the main element in the centre of interest from the economical point of view, the rest of minerals being dumped as mining or technological tailings. Such an organization of the mineral raw materials economy creates a certain feed-back: the increase of demand for mineral materials is accompanied by an increasing amount of useless waste commonly considered as not utilizable. It is judged that the accumulated amount of main raw materials as described above (with the exception of strip mine overburden) will reach in Poland up to 1990 about 7 x 10^9 tons, unless severe steps in order to economize this phenomenon are undertaken. These facts underline the necessity and importance of rational economy of those materials as a potential additional source and important factor in the field of natural environment protection.

It is a false opinion that both mining and technological wastes are entirely useless. When they are not consumed by the industrial branch which produces them, they may still be well used by other branches either directly or after more or less complicated processing.

This problem has been looked at with special attention in Poland. Relevant action has already been undertaken in the period 1960-1965, but it has become a fully planned action since 1970. Presently, this problem has acquired a fundamental character, both from the aspect of fulfilling the needs for mineral raw materials, and in the aspect of natural environment protection.

2. CLASSIFICATION OF EXPLOITED RAW MATERIALS AND WASTES

Complex utilization of mineral resources concerns a twofold activity: geological-mining and technological-processing.

Geological and mining activity in this sense covers, irrespective of attempts to maximize the extraction coefficient of a given material, exploitation of accompanying minerals or those appearing together. The main task then is to design and conduct the mining works in such a way that total exploitation of the main material would not make it impossible to exploit the accompanying material in the later stage. That there should be a tendency to simultaneous extraction of both main and accompanying minerals for the extraction of only one material at a time could lead to changes of geological and orogenetic conditions, which could make the exploitation of accompanying materials impossible.

The direction of the mineral resources economy as laid down in the law, requires anticipation already at the stage of geological exploration works. Consequently, all geological documentation should refer to all utilizable materials deposited, in the vertical as well as the technological directions of their utilization, together with an outline of the economy of an undertaking. Such a procedure has been followed by the Polish geological services since 1971.

Technological and processing activity concerns first of all the extracted material. The main task is aimed at designing and implementing a process or processes leading to an entire utilization of useful elements contained, and reduction of wastes in the form of either primary post-technological tailings, or secondary products of material processing.

The classification of mineral raw materials is determined by this twofold activity. The following division has been designed in Poland which takes into consideration the localization of materials appearing in the process of mineral resources utilization (see Fig. 1).

- principal and accompanying minerals obtained through the extraction process;
- fundamental concentrates, primary wastes and accompanying concentrates appearing during processing of a mineral;
- final products—secondary wastes and secondary or accompanying concentrates obtained in the result of further processing of fundamental concentrates.

The classification diagram presented in Fig. 1 depicts a fundamental cycle of

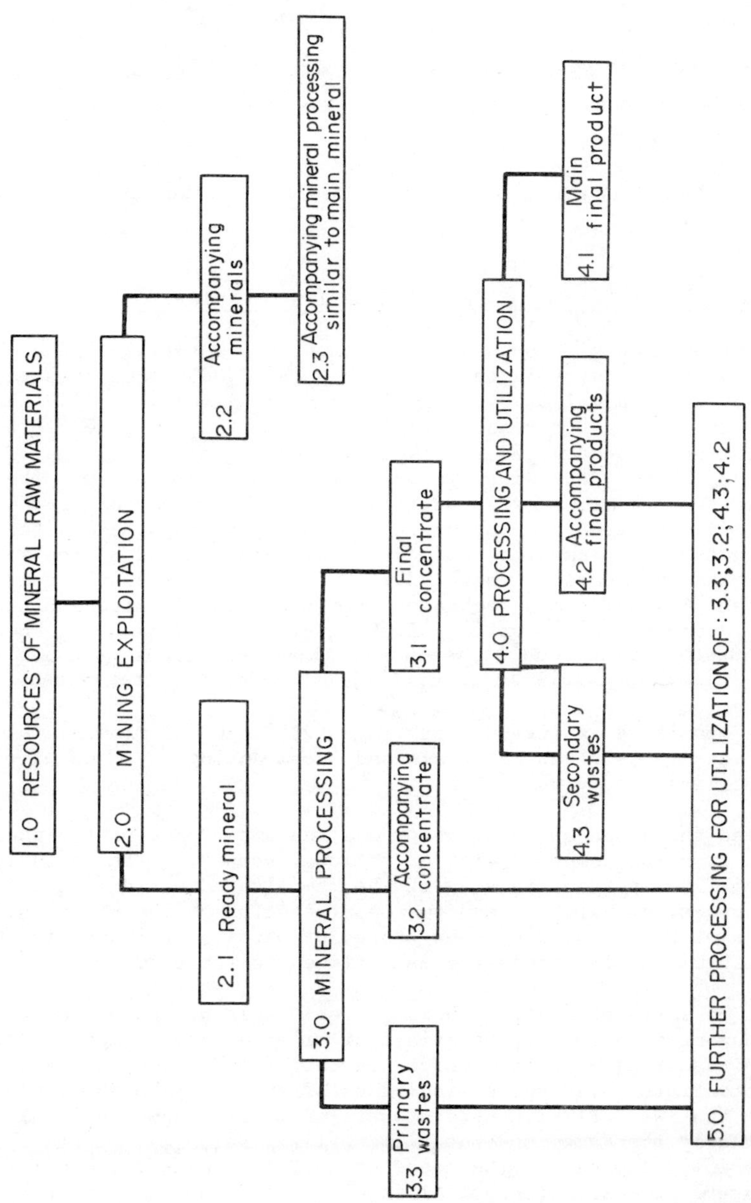

Figure 1. Scheme of mineral classification and products of their processing.

rotation of extracted mineral deposits. From the point of view of the production without waste of those raw materials the scope of intervention and stimulative action may be predicted. These are the following groups of raw materials:

primary and secondary wastes,

concentrates and accompanying products,

accompanying minerals,

as well as all products resulting from the processing of the above groups.

3. MEANS OF SOLVING THE PROBLEM AND ITS ORGANIZATION

The concept of production without waste of mineral raw materials is not new in Poland. Primarily the activity in this field concerned only some selected minerals which was determined mainly by economic factors. Already in 1959 the Polish - Hungarian Haldex Co. Ltd. was established, with the aim of exploiting the dumps of coal-schist extracted together with coal. As the result of introduced technology, the coal contained in primary wastes is recovered, whereas the secondary wastes are adapted for utilization in the cement and ceramic industries or as filling material. During the following years some new technologies were designed and implemented, but full-scale planned and harmonized actions were undertaken not earlier than 1970.[2] On the initiative of the Ministry of Science, Higher Education and Technology (formerly the Committee of Science and Technology) with the co-operation of the Ministry of Mining and Power Industry and the State Mining Council, some government decisions were issued with an aim of stimulating full-scale utilization of mineral resources.

The first stage consisted in introductory inventarization of mineral movements on the surface—mostly waste raw materials and accompanying ones—and of directions of their utilization.[3]

Analysis revealed the scope of the problem and the importance of the question of wastes, both from the point of view of rational economy of these additional resources, specially important in view of the complicated world-wide problems of natural resources, and with regard to the protection of the human environment. The amount of wastes produced and stored by particular industrial branches dealing with mining or processing is outlined in Table 1.

The compilation of waste production shown in Table 1 is not complete—it does not include a number of extracted materials of minor economic importance, and a number of other sources of wastes which together also present a certain quantitative potential. Nevertheless individual figures, although of vague character, show that we deal with huge amounts of mineral substance (already extracted, partially processed and ready to utilize) which may create an additional raw material resource for the needs of building industry, civil engineering, metallurgy and casting, chemical industry, agriculture, forestry and others. The results of analysis show that proper utilization of those resources should give the following advantages:[4]

increase of material resources actual disposable;

decrease, or even elimination, of importing of some materials;

TABLE 1

Present and expected amount of wastes connected with extraction and processing of main minerals in Poland.

Source of wastes	Amount of corresponding wastes in millions of tons					
	1970	1973	1975	1980	1990	1971-2000
Hard coal mining	28.0	43.5	46.6	66.0	73	1650
Brown coal mining (clays, sands, gravels)*	60.0	70.0	90.0	70	75	1800
Power industry	7.9	10.8	11.1	19	45	560
Chemical industry	8.6	18.1	33.0	35	36	500
Mining and metallurgy of ferrous ore	11.7	12.5	13.4	17	19	390
Mining and processing of non-ferrous metals ore	9.1	12.7	17.5	33	70	1140
Fire -resistant materials industry	3.0	3.7	4.0	5	6	120
Rock materials industry	15.0	18.99	33.0	35	36	600
Other sources	5.3	19.8	11.0	10	20	300
Total	148.0	200.0	260	290	380	7060

* With the exception of overburden.

decrease in cost of extraction of some minerals through the reduction of investments for building of new mines;

less danger to the natural environment as a consequence of liquidating old waste dumps and avoiding creation of new ones;

increase of economic affectiveness of extraction of principal materials;

enabling—in some cases—of enrichment of range of materials produced or even production of new materials;

reduction in transportation of some bulk freight, etc.

Joint utilization of wastes in 1990, according to preliminary calculations, should exceed 50 per cent (Table 2).

Table 2

Present and expected utilization of wastes and accompanying materials (without brown coal stripping mines overburden).

	1970	1973	1975	1980	1990
Total utilization of wastes in million t.	*ca*.20	60	90	148	200
Grade of utilization in per cent	23	30	36	51	53

With the aim of properly directing and coordinating the tasks, the Ministry of Science, Higher Education and Technology, aware of the size of the problem and prospective possibilities, decided in 1971 to include the whole range of activities in the framework of a centrally financed and co-ordinated project, previously scheduled for the years 1973-1975 and presently extended for 1976-1980. The project was given the title:

"COMPLEX UTILIZATION OF SOLID MINERALS, MINING WASTES AND MINERAL SECONDARY MATERIALS"

Accordingly, the Council of Ministers appointed on 26 May, 1970 the State Council of Material Economy with the following tasks:

design of combined annual and multi-annual plans of recovery and utilization of secondary materials, wastes and under qualified materials, as well as control of plan implementation;

stating the rules and organization for waste or secondary materials and the financial and economic conditions;

initiation and maintenance of wastes processing.

In accordance with the above and with the following directives of the Council of Ministers, the State Council of Material Economy designed a programme of complex utilization of mineral waste resources for the years 1973-1975 and then, in 1974, for the years 1976-1980 with perspectives until 1990. Besides, the Planning Commission of the Council of Ministers has laid down directives concerning the planning and designing of investments connected with the complex utilization of mineral waste resources.

Action has also been taken to prepare complex geological documentation of mineral raw materials resources.[5]

The organizational scheme of co-operation in the field of utilization of mineral materials is presented in Fig. 2.

4. DESIGN AND IMPLEMENTATION OF PRINCIPAL PROBLEM

Implementation of the main aim, i.e. the complex utilization of materials, comprises three phases of action:

- preparing of a catalogue of waste raw materials and analysis of possible ways of utilization and an introductory economical study;
- technological works;
- investment activity in the implementation phase.

With the general assumption of integration of intervention-stimulation actions, within all three phases, certain authorities are responsible for the results. The Ministry of Mining and Power Industry, through a subordinate "research-designing" unit, decided to implement the first two phases, whereas the guidance of technological or industrial implementation is maintained by the State Council of Material Economy (presently the Office of Material Economy). Both these organs of central state administration collaborate with all units which utilize mineral raw material in their activities. The scheme of such collaboration is presented in Fig. 2. The left part of the diagram concerns the first two phases of problem implementation, whereas the right part refers to inculcation activities.

The first and fundamental preparatory task in the field of utilization of accompanying materials and wastes was to prepare and maintain a catalogue of all raw materials and waste resources in Poland, both from qualitative and quantitative points of view, which would take into consideration their grade of utilization, i.e. "surface material movement balance".

On the other side of the balance, the quantitative demand for particular raw materials has been compiled, based on information given by industries which base their activity on extracted minerals. A comparison between what is at disposal in the shape of processing wastes of principal material, with actual needs permits an outline of the potential ways to put them to use, either in the form in which they exist or, after adaption to the receiver's needs, as a fully fledged substitute.

The next task was to provide an account of results of technological work so far in the field of waste adaptation to further utilization. Technological research in this area covered different stages of advancement; from laboratory to industrial; from works just touched upon to those entirely completed and included the design of processing projects. The first thing to do was to review the existing research results which had lain idle until the action began. A selection of technologies had been made from the point of view of their suitability for

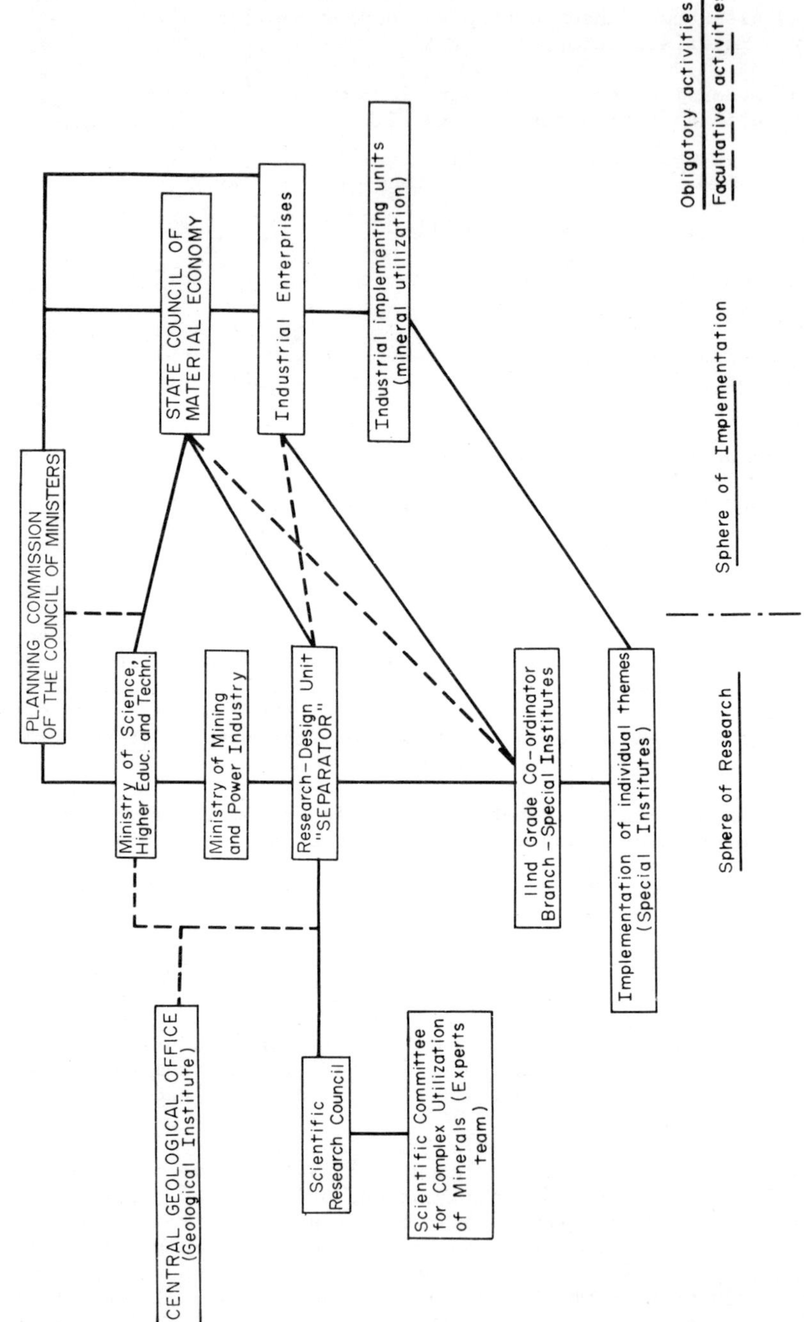

Fig. 2. Scheme of organization of co-ordinative activities in the field of production without waste of mineral raw materials (Data as at November 30, 1975)

waste utilization. In this regard, the accompanying materials and wastes can be divided into six groups:

for which technologies are mastered and industrially applied, and intervention is merely required in order to broaden their application;

for which technologies are mastered though not industrially implemented. In such case recapitulation of technologies, economical analysis of the undertaking and a statement of the reasons for non-implementation is required. Most commonly the reason is the lack of means to secure the necessary investments. There are also inherent obstacles, as , for example, fear of utilization of power industry ashes in concrete production for the building industry because of their negative influence on human health;

for which technologies are mastered but do not fulfil the requirements of implementation. Such being the case, the economic justification is revised, and in positive cases technologies are amended until they reach proper conditions of investment;

for which technologies are mastered but are not sufficiently economic. Analysis should state whether introducing modern processes or changing the direction of utilization will lead to improvement of economic effects;

for which technological research did not lead to any completion;

for which no research work was undertaken.

After having made the above classification, within each of these groups, different themes were listed with preference for those which were most effective and promising the quickest possibilities of implementation; it was assumed at the same time that at least 30 per cent of themes should remain in the "full cycle".

A full cycle is completed by setting down the output data or investment activities, i.e. a process design to be the base for technological and economic parameters.

In 1973 the Central Research-Design Institute of Minerals Utilization and Enrichment "SEPARATOR" developed the first plan of problem-coordination for the years 1973-1975 which covered twelve themes.

two themes concerned the utilization of the entire extracted material including the principal element; utilization of polyhalites and accompanying rock salt, as well as utilization of poor kaolin resources;

three themes dealing with accompanying materials: carboniferous montmorillonite clays accompanying hard coal, overburden clays and sands appearing with brown coal, and clay-like or sand materials in sulphur ore mines;

six themes connected with waste materials of processing technologies and power production;

one theme of an economic character in the framework of which the above mentioned theme analysis is conducted, and which maintains the "information bank" needed for central intervention and stimulation.

The work on the problem is conducted by a "Principal Problem Leader" having a permanent co-ordinating team at his disposal, there is also an advisory body consisting of experts acting within the Scientific-Technical Council of the co-ordinating unit. For instance, during implementation of the principal problem in the years 1973-1975, sixty reasearch units were employed, for sixty-one sub-themes, thirty-three were concluded with a "full cycle" scheme of works.

In the co-ordination project for the years 1976-1980 the attention was focused on waste materials of power production and of rock and fire-resistant wastes.

Each problem is centrally financed with the budgetary means at the disposal of the Ministry of Science, Higher Education and Technology through a centre responsible for the problem implementation, in a given case through the Ministry of Mining and Power Production.

Industrial implementation of technologies is financed from the investment funds of particular enterprises which is the most difficult element of the whole problem. In order to solve it it is necessary

- to state the needs of main potential consumers;
- to agree on the most attractive selling prices so that additional funds spent on adaptation to consumer needs would not exceed the price of the primary material which is substituted;
- to provide the means for necessary investments connected with adaptation and treatment of accompanying materials and wastes to utilization or transportation.

The whole of the implementation action requires a relevant, long-range agreement between suppliers of wastes and consumers. Presently this problem is being solved by individual agreements, though the set of final rules on prices and division of profits between suppliers and consumers has not yet been stated. The legislative activities in this field conducted by the State Council of Material Economy are continued.

Closely connected with the problem implementation an important question is the storage of wastes in a way which would allow future utilization. It is not a question of a technical or technological type which is easy to solve, but rather connected with the immobilization of investments over many years. This question requires a formal and legal type of solution.

The example of problem-solving connected with the complex utilization of raw material is the case of "polyhalites".[6] This example provides the means of co-ordination of activities.

"Polyhalites" is a mineral consisting of potassium and magnesium sulphates and of anhydrite.

Polish polyhalites ($K_2SO_4 \cdot MgSO_4 \cdot CaSO_4 \cdot 2H_2O$) appears together with rock salt and anhydrite; its texture is fine-grained and compact. The mineral is practically insoluble both in hot and cold water. Useless elements consist of anhydrite, a part of the polyhalite crystal lattice, and rock salt and anhydrite as accompanying deads. Together with polyhalite deposits a deposit of relatively pure rock salt is found.

The main task of exploitation is to extract polyhalites as the principal material and rock salt as the accompanying one. After the mining process there are available (according to nomenclature of amterial classification, Fig. 1):

2.1. polyhalite extraction,

2.2. rock salt.

Polyhalite processing leads to:

3.1. polyhalite,

3.2. rock salt or brine,

3.3. anhydrite.

Polyhalite processing in order to obtain potassium permits the production of:

4.1. potassium fertilizers of sulphate type,

4.2. potassium-magnesium fertilizers,

4.3. anhydrite and lyes.

Work on utilization of wastes 3.3 and 4.3 may be conducted in order to obtain evaporated salt, magnesium oxide for the needs of metallurgy, sodium and potassium lyes, magnesium sulphate and sulphuric acid, as well as grain-like anhydrite for the production of building binders and prefabricates with a resistance of 140 kg/cm^2. The part of the anhydrite not consumed by the building industry may be transported to the mine as filling material.

Polyhalite, although a very difficult material to process, is a very useful mineral with a wide demand for potassium fertilizer production. The same problems appear as far as utilization of post-productive wastes is concerned. The research workers have been given the following technical and technological conditions to fulfil:

- to maximize the extraction of K_2SO_4 and $MgSO_4$;
- to utilize fully liquid and solid wastes, without the need to store the solid waste and to transport liquid wastes (post processing lyes) to the Baltic Sea.

This difficult and economically complicated task was given to the Institute of Chemistry of Copernicus University at Torun, which acted as research co-ordinator. Practically all the technical universities in Poland have taken part. As the result, a number of technologies were obtained[6] which presently are being analysed with regard to their technological realism and economic effectiveness.

Preliminary analysis of the economic aspects of the undertaking, conducted before starting the laboratory research[7] had justified the technological research. The calculated time of amortization of investments with the gained profit taken into consideration, was about 1 to 3 years depending on the decision as far as different technological processes are concerned.

Our experience has supported the principle of concentrated research simultaneously executed by several units. Such an organizational scheme requires an energetic co-ordination of research projects by a technologically suitable unit, with gradual elimination of tasks with no prospect of success. In the case of polyhalites eight research teams were involved and fourteen technological solutions have been found. The teams have finally arrived at three possibilities of complex processes.

5. ECONOMIC ASPECTS OF COMPLEX UTILIZATION OF MINERAL RAW MATERIALS

Besides the unmeasurable benefits (environmental protection, increase of natural raw materials resources) complex utilization brings economic benefits both in terms of production by the plants utilizing the wastes and in those industrial units which produce the wastes.

Let us take, for example, the set of advantages resulting from coal-mining waste utilization.[8]

The main points of waste utilization are:

- levelling,
- filling,
- building ceramics,
- light aggregates,
- forestry and agriculture.

Levelling or sub level dumping, is the first phase of recultivation of ground destroyed by previous mining activities. The advantages of dumping in abandoned workings are in the lowering of costs of reserving ground which might otherwise be used for agriculture and at the same time minimizing the losses, since crops in such areas are not obtained. When wastes are stored in dumps 10 metre high, then to store 1 million cubic metres 1 hectare (ha.) of ground is necessary.

According to a Polish regulation introduced in 1971 the cost of expropriation varies according to the soil category, and the highest is 0.5 million zlotys for 1 ha. An equivalent is additionally paid for not using the ground for agricultural aims—the equivalent is 10 per cent of the land value and is paid for 20 years.

Filling of headings according to standards for obtaining a first-class filling material allows a 20 per cent addition of crushed wastes. In coal mining the wastes of the "Haldex" method are used, the cost of which is four times lower than filling sand. Besides, the costs of storage of stone in dumps are lowered by a value equal to the percentage of stone used for filling. In the case of the first-class filling material it reaches 20 per cent of storage costs.

Factories of building ceramics may process overburden clays from stripping mines as well as clay-like schists extracted with coal. Economic benefits result from:

utilization of clay directly in ceramics production without the necessity for intermediate storage;

cost difference between clay exploitation from stripping mines overburden and the costs of clay extraction (primary material) from deposits by ceramics factories;

elimination of costs resulting from expropriation of ground, thereby removed from agriculture which are paid both by the waste producer (coal mines) and by removal of the need to open new clay mines.

When in the case of open pit mines the profits of the above effects is equal to the value "A", then by utilization of extracted wastes of underground mines the effect increases and is about 1.7 of "A".

Light aggregate production on the base of clays and schists gives rise to benefits as a result of:

elimination of schist storage dumps (32 per cent);

difference of transportation costs of light and natural aggregates (45 per cent);

decrease of labour costs as a result of concrete production with utilization of light aggregates (2 per cent);

elimination of extraction of natural aggregate deposits (1 per cent).

Agriculture and forestry utilize for their purposes clay-like wastes which mainly serve to enrich sandy soils. The consumption of clay is about 40-60 tons per ha. which makes 25-30 m^3/ha. Frequency of this procedure in forests is every 50-100 years, whereas in the case of cultivated land it should be repeated every 7 years. Economical effects result from:

limitation of clay storage in dumps;

increase of agricultural crops and wood production.

The rate of effectiveness of utilization of 1 m^3 of wastes from coal mining[7] under Polish conditions is as follows:

levelling - 1,

filling - 12,

ceramics for building purposes - 93,

light aggregates - 233,

agriculture - 24,

forestry - 34.

Potential possibilities of utilization of produced wastes following from their utility for a given branch is listed below:

for levelling	97 per cent
for filling	85 per cent
for ceramics	40 per cent
for light aggregates	30 per cent
cement, ceramics	25 per cent
agriculture and forestry	5 per cent
roads and railways	10 per cent

Comparison of different variants of those data makes it possible to optimize activities in the field of waste economy. Solutions are different both in connection with individual regions, on the national scale, and overtime.

Other practical examples of the use of mine wastes besides those from coal mining are found in barite enrichment[9] and limestone rinsing.

The use of barite wastes for production of high resistance cements (400-600 kg/cm^2) makes it possible to simplify the technology of production by the elimination of mineralizers harmful to the environment (fluorosilicates), lowering of clinker saturation with calcium and aluminium, minimizing of necessary addition of potassium sulphate used as activator, optimizing of agglomerating capacity of material composition and of furnace brickwork, minimizing of fuel consumption per unit. General balance of those expenditures has shown that each utilized ton of barite wastes for cement production gives an economic benefit of about 10,000 zlotys, which corresponds to about 150 zlotys per ton of rapid hardening cement used in the building industry.

Effects of utilization of muds from limestone rinsing result first of all from limitation and elimination of necessity of storage, which would practically eliminate the use of agricultural land, and give additional raw material to the cement industry. Without going deeply into the details of accelerated sedimentation technology or the dehydration process, the financial advantage gained by elimination of: soil and crops losses, costs of sedimentation reservoirs, recovery of raw material and process water, and limitation of primary materials exploitation, allows for amortization of investments spent for research and implementation within 1.5 to 2 years.

It must be stated that both these wastes are already being industrially exploited on a large scale.

6. CONCLUSIONS

The set of problems connected with economizing and using mineral raw materials is very large and more complex than one could judge after the first research work undertaken within the past 5 years. Although some progress has been made, the most difficult element of the undertaking is to apply the results of research carried out in the normal course of production. A fundamental factor that has to be overcome is the unwillingness of industrial consumers to abandon primary materials extracted from their own sources in favour of the utilization of wastes produced by other industrial branches. It is frequently connected with heterogenity of waste materials, lack of transportation (mostly bulk or dust freight) or storage places.

Although the majority of solutions already being implemented permit financial profits at the level of individual undertakings in the form of investment amortization over a period of 2.5-5 years (a number of solutions were reached without any investment) for implementation of research results and technological utilization of the mass of wastes not utilized until now or utilized on a limited scale, the question of prices for post-production wastes is important. In many cases the prices are a negative element in the utilization process. On the one hand, a price which is too low does not stimulate the producer of wastes and on the other, a high price limits the consumer demand. Another factor is that prices for waste materials vary even when the latter are similar or the same. The factor of price is of special importance for the reliability of technological and economical analysis of future undertakings and on the whole of the implementation process. The role of waste producers should be stressed in this connection.

One possible solution developed by the State Council of Mining in order to stimulate the waste producers (so-called mineral secondary materials) to implement designed technologies, is to introduce taxation on wastes which are not utilized. Out of the funds created in this way, technological implementations would be financed and stimulative bonuses paid for supplying of waste materials in constant parameters. All profits gained by utilization of waste materials should be divided between producer and consumer. The latter should be selected directly after stating the direction of utilization and be obliged to implement a designated technology of waste material adaptation and to undertaking production of a final product based on that material.

Presently there are about 100 developed technologies of utilization of mineral waste materials and accompanying extracted minerals. Implementation of this task should be a dominant element in the years 1976-1980.

REFERENCES

1. B. Krupiński, W. Czachórski, T. Laskowski and Z. Nowak. The utilization of accompanying raw materials in Poland. VIth Int. Mining Congress, 1970, paper I-A.2.

2. W. Czachórski. Some remarks on fuller utilization of mineral raw materials. *Enrichment and Utilization of Minerals*, 1, 7-11 (1973).

3. B. Małecki. Pilot study of complex utilization of solid minerals and wastes and secondary materials of minerals. *Separator*, Katowice, Nov. 1971.

4. S. Gustkowicz and H. Walkowiak. Introductory analysis of advancement of works in the field of complex utilization of minerals, Committee of Science and Technology, Warsaw, July 1970.

5. S. Kozłowski. Principles of complex geological documentation, *Geological Survey*, 2 (XXIV) 61-66 (1976).

6. J. Tomaszewski. *Survey of Technologies of Chemical Polyhalite Processing Designed in Poland in years 1970-75*, Inst. of Chemistry, Copernicus Univ., Torun, Nov. 1975.

7. S. Gustkowicz. Information on potassium-magnesium salts resources, possibility of utilization, necessary investments and economy of undertaking, Committee of Science and Technology, Warsaw, Oct. 1971.

8. Z. Czerwenka. Forecasts on utilization of accompanying minerals, mine wastes and post-processing wastes in context of present utilization in mining and power industry. *Separator*, Katowice, Oct. 1973.

9. J. Peukert. Utilization of barite wastes for cement production, Industrial Institute of Binding Materials, Cracow, Nov. 1975.

Non-Waste Technology: United Kingdom Experience and Policy

R. Berry

Department of Industry, United Kingdom

Non-waste technology (NWT) in a literal sense is almost certainly unattainable, and this is recognized in the more limited definition* adopted by the ECE in January 1975. The United Kingdom certainly supports these principles of NWT, but at the same time considers that full and careful consideration must be given to the economic and social consequences of introducing such a technology in order to reduce waste.

Industries cannot be expected, for example, to introduce measures to reduce waste in their production processes if by so doing their commercial competitiveness is prejudiced. NWT must therefore be economically viable, and governments can perhaps best contribute to the efficiency of manufacturing industry by focusing attention on "best available means" that might be overlooked by industry often operating on a multinational basis.

It is important that we do not waste our "human" resources; and such measures as increasing the durability of products—desirable as they are—should be introduced gradually, at a pace which enables society, as well as the industrial and commercial sectors, to make the necessary adjustments without sudden disruptions or dislocations. A too-rapid move to longer product life would cause consequent reductions in manufacturing volume; this would militate against high employment, which is one of our socio-economic objectives. By contrast design with recycling or reuse in mind is an aspect which it seems could initially be encouraged, as it does not have the far-reaching effects of increased durability.

There are obstacles also to the introduction of substitute materials. Industry selects a material for particular use not on its abundance but on its price and cost of fabrication. Even where costs appear attractive the use of substitutes must be resisted until their reliability is beyond question, and this can take time when dealing with complex products.

Another point worth stressing at the outset is that full notice should be taken of the work already carried out by other international organizations such as EEC and OECD in the area of waste reclamation and recycling. This would avoid wasteful duplication of effort, of which fresh examples come to light almost every week, including overlapping studies and projects which

---

* The definition agreed by the Ad Hoc Meeting of Experts at Geneva in January 1975 (ENV/AC.4/2 of 15 January 1975) was: "Non-waste technology is the practical application of knowledge, methods and means so as to ensure the most rational use of natural resources and energy to provide for the needs of man and to protect the environment."

repeat work already done elsewhere.

GREEN PAPER: WAR ON WASTE

The United Kingdom took a decisive step towards the reduction of waste in society in 1974 when the Government decided that there should be a new national effort to conserve and reclaim scarce resources. A Green Paper entitled "War on Waste: A Policy for Reclamation" was presented to Parliament jointly by the Secretaries of State for the Environment and Industry in September 1974. The paper examined the whole subject of waste and in particular what the Government, local authorities, industry, and the public could do.

In order to co-ordinate the work and help develop new lines of policy, the Government established the Waste Management Advisory Council. The Council is jointly chaired by the Parliamentary Under Secretaries of State for the Departments of the Environment and of Industry.

The Council, whose twenty-nine members represent industry, local authorities, trade unions, academics and consumer and conservation interests, met for the first time on 18 December 1974. Its first task was to define priorities and establish a structure of subsidiary groups, the membership of which includes experts who are not members of the Council. Three Standing Committees were set up: General Policy and Coordination; Research and Development; and Information, Publicity and Education. A further tier of Working Groups was established for individual commodities and wider-based studies. These are Ferrous Metals, Non-Ferrous Metals, Packaging & Containers and Economic Studies. An Advisory Group on Waste Paper Recycling, which had been set up before the Council itself, was also brought under the aegis of the Council, in essentially the same role as the Working Groups.

First Report

A first report on the Council's activities was published on 12 January 1976. The report indicated that the Green Paper had made it clear that the Government's primary purpose in setting up the WMAC was to provide machinery for a detailed and wide-ranging examination of the possibilities of reducing waste and of increasing the amount of waste reclaimed for reuse and recycling.

Although the report concentrates mainly on where waste arose and what was being done to reclaim it, it also raised the question as to whether there was any need for the waste to have been produced in the first place. It suggested that a comprehensive waste-management policy should aim, within economic and technical constraints, to avoid creating waste and to make the most efficient use of materials at all stages of a product's life.

It stressed that design had a vital part to play, and explained that during manufacture the impact of design was two-fold: on the choice of materials and the production processes.

In view of market imperfections there should be, in some cases, a vigorous search for substitute materials. More abundant materials should be used in preference to the scarcer ones; the renewable instead of the non-renewable; and in general there should be a greater use of secondary materials.

Production processes needed constant and careful monitoring. Waste could be reduced by working to closer tolerances, by standardization and simplification. Specifications might be examined and could perhaps be slanted towards performance and conservation rather than particular materials and methods.

The report pointed out the important effect of design on product life. Many products could be made so as to be reusable. They could be made to be reliable and last longer, with emphasis being placed on ease of maintenance and on capability for component repair as opposed to exchange.

At the end of a product's useful life design would have an impact on the extent of recovery of materials from the product or its disposability. Ideally, recovery should be easy, economic and complete. Where composite materials are used they should be selected and put together with recycling in mind.

The report nevertheless recognized that the foregoing paragraphs described a situation which, however desirable, was unlikely to be fully achieved in practice. There were many constraints. Within industry there might be operating difficulties, limitations in capital investment programmes and special design standards requirements and restrictions. From the consumers' point of view initial costs might be higher, but there was also some premium on newness of certain products. Much was already being done to meet these problems, but more research and development could be devoted to solving technical difficulties in the field. The report concluded that a systems approach to the design of products might be developed that would aim at minimizing the total life costs of a product and not just its manufacturing cost.

The Council paid special attention to the prospects of better use of household waste, and underlined two points in this area:

(a) New ways of reclaiming household waste were being developed very rapidly, and traditional plant such as incinerators might quickly become obsolescent. Waste-disposal authorities should therefore try to exploit simple methods which required the minimum of new plant or equipment.

(b) Waste-disposal authorities would need to take account of all the options open to them, including some that were so far unproven. In doing so they should try to judge what the markets for reclaimed materials might be when the new methods of reclamation were in operation, and consider whether there were new markets to tap.

Other points made by the report were:

(a) More information was needed about the volume, nature and distribution of waste if further opportunities for reclamation were to be identified. The Control of Pollution Act, 1974 would require county councils in England and district councils in Scotland and Wales to survey all domestic, commercial and industrial waste arising in their area: these surveys should be invaluable as a basis for a strategy for reclamation.

(b) In considering possibilities for reclamation, account had to be taken of the costs and benefits to society as a whole, and of imperfections in the market which might distort costs and benefits.

(c) Assessment of the effect of reclamation on the balance of payments should take into account indirect as well as direct costs and benefits.

(d) Waste management in industry did not begin yesterday. Many sectors of industry already made substantial use of reclaimed and recycled materials. We must build on these foundations and encourage industry to make even greater use of secondary materials where this could be done economically. Industry had to bring sound economic sense to bear upon the War on Waste. Moreover, it made sense to reclaim materials only where there was a real saving in resources or other benefits compared with the alternative of disposing of the waste and using primary materials.

(e) After completing the review of Departmental programmes by looking at the Department of Industry's R&D programme, the following areas, *inter alia*, might be also looked at: utilization of organic wastes by biological processes, recycling of plastics, materials to be reclaimed from scrapped cars.

Future Programme

It is too early yet to indicate what results will accrue from the work of the Council. Waste management is a complex operation and a lot of hard work remains to be done. In most industrial sectors, for example, information on the use of secondary materials is inadequate and sometimes unreliable.

The Council is trying in the present economic situation to give preference to programmes which are likely to provide financial benefit or resource recovery within the next few years. Amongst the more important current studies are: segregation of waste at source, including ways of helping voluntary organizations to run waste-recovery schemes; mechanical sorting of refuse; short-term research into the use of refuse as fuel; use of waste paper in paper manufacture; reclamation of tin and steel scrap from cans, and improved recovery of non-ferrous metals.

To give the Council a keener cutting edge, certain changes have been made in its structure recently, the chief one being the appointment on 1 January 1976 of Dr. R.P.L. Berry (Chairman, Alcoa of Great Britain Limited) as the Director of the National Anti-waste Programme. His task is to develop new initiatives for waste management and to establish widespread contacts with industry, consumers, local authorities and others concerned with the use and conservation of materials. He is giving priority initially to reducing waste and encouraging greater use of reclaimed and recycled materials by industry. One of the Director's principal objectives is to create and encourage a positive attitude to using resources more effectively. He is currently involved in restructuring the WMAC and defining its strategy.

WASTE MATERIALS EXCHANGE

A Waste Materials Exchange was set up by the Department of Industry late in 1974, to be operated initially by the Department's Warren Spring Laboratory for a 2-year experimental period. It aims to reduce the cost of satisfactory waste disposal; encourage industrial thrift; and conserve national resources.

The exchange is a free and confidential service and operates by circulating to companies, by means of a quarterly bulletin, lists of items available and wanted. Since the first bulletin was issued in 1974, notifications (of wastes available or wanted) have grown to 2784, and 90 of these materials are known to have been taken up and disposed of through the Exchange. The "as new" value of the items transferred is well in excess of £1.5m. These results have been achieved with a total membership of less than 2600. This was considered to be much below the full potential and a further full-scale advertising campaign was launched at the end of January 1976. Already this has lifted membership to over 4000. It is confidently anticipated that there will be a related increase in the number of transactions and in the financial savings to the nation.

French Policies in Pollution-Free Technology

P. Chassande

Secrétariat d'Etat à l'Environnement, Ministère de la Qualité de la Vie, Paris, France

At the present time, in France as elsewhere in the world, it is no longer conceivable that the expansion, or even the simple maintenance, of present industrial activity should continue without taking into account the environment. It is now more a question that these very industrial activities be identified with its protection.

If the environment is in danger, there is no doubt that this is above all due to mass production, a production which, in every sector, requires concentration in limited areas and leads to ever greater squandering.

To protect the environment without detracting from the quality or the quantity of goods available to consumers, it will therefore be necessary to improve on production and to squander less.

The same fundamental principle is to be found as the basis of these two courses of action: recycling. In the case of the manufacturer, this consists of making available to the consumer market a greater proportion of the processed materials, and for the consumer, this means returning to industrial production a greater proportion of the non-consumed part of purchased goods.

In other terms, it would be preferable to have at every level recovery rather than the pure and simple dumping of wastes. The discharge of wastes inevitably adversely affects the integrity of resources taken as a whole.

In actual fact, pollution and nuisances generated by industrial activity do pose a threat to resources, and at times this is almost exclusively due to squandered material, be it raw materials, products in processing, or finished products. The efficiency of the system of production will thus be all the more lower as waste pollution is higher.

A reduction of pollution will therefore be a result of an improvement in efficiency, which can only be achieved by effort on the part of manufacturers.

All sectors of industry found themselves more or less suddenly confronted with their pollution problem, whether it was a matter of sewage, dusts, gas emissions or solid wastes.

To meet with the various situations, heads of industry have available two courses of action:

the first consists of an awareness that the creation of pollutants is a necessary consequence of industrial activity, and that this thus requires adding external facilities to production plants which will destroy such polluting material after it is brought into being;

the second course of action is radically different. It is internal. Ways are sought which will act on the very conditions leading to the formation of pollution so as to completely avoid its occurrence,or to replace it with a residue which is not harmful.

In this second scenario, the effluent (solid, liquid or gaseous) is no longer considered separately from the production facility. There is recourse in one of three approaches:

It is possible, first of all, to change the technical conditions leading to the formation of the effluent in order to replace it with a less noxious one, or even, to eliminate the effluent altogether;

This first approach consists of replacing the production process currently employed in the plant with a new, less polluting production method.

It is also possible to add on a secondary production activity involving the effluent, whereby the effluent will in turn be either less noxious or even become non-existent.

This second approach amounts to a recovery of effluents in order to turn them to a profitable use, whereby the polluting by-products of the main activity are upgraded to benefits quantifiable in money terms.

Finally, there, are, between these two extreme scenarios, intermediate approaches which can be considered, and which represent some combination of the preceding two: such is the case, for example, with recycling of liquid wastes of gas emissions.

It is these production methods perfected in a plant following an internal effort to reduce pollution which are commonly called integrated pollution-free technologies, or, as they are called in France, technologies propres.

Integrated pollution-free technology and external methods are therefore as different from one another as are prevention and cure.

Anyone who has at all taken part in the life at a factory knows the amount of effort needed to bring the various plant facilities up to an optimal program of production and then to keep them there.

The most noble course of action involving the production systems themselves will, therefore, without any doubt, entail some courageous resolve and well-considered efforts carried out with vigour.

The various presentations which will be held during these next few days by French industrialists in such highly varied sectors of activity as agriculture and food processing, paper-making, and the chemical and iron and steel industries are certainly the proof of the success and benefits founded on this policy, a policy which we have decided to pursue and to intensify.

Experience and Policy with Regard to Non-Waste Technology in Hungary

A. Takáts* and J. Francia**

**Hungarian National Council for Environment Protection Secretariat, Budapest, Hungary*

***State Committee for Technical Development, Budapest, Hungary*

The economic policy of the socialist countries is oriented to raising the material and cultural standard of living of the people by dynamic and proportional development of social production, by accelerating the rate of scientific - technical progress, increasing work productivity and improving quality in all branches of the national economy.

In the course of the solution of this fundamental task the attention is focused on the direct utilization of scientific and technical achievements in production in order to attain the advantages of quality improvement and increased efficiency. In this way the dynamic development of the productive forces may be organically interwoven with the other advantages of the socialist economic system. The development forecasts of the CMEA member states predict for the coming 15 to 20 years a multiple increase of the main industrial branches, energetics, chemical industry and metallurgy. The fulfilment of the national economic plans shows that these prognoses are completely realistic. It is a self-evident fact that the application of the present production technologies would lead to a proportionate increase in the amount of waste and to diminishing the non-renewable natural resources.

It is also obvious that the protection of the biosphere and the recovery of repeatedly utilizable raw materials could not be solved completely even by a rapid development of the necessary plant and equipment.

For the radical solution of the problems of environment protection the alternation of the conventional technological processes and methods leading to production with little or practically no waste and promoting the attainment of high technical-economic characteristics, as well as the most complete and rational utilization of the natural resources, are of a decisive importance. Therefore all the CMEA countries—with Hungary among them—are vitally interested in the development of non-waste technologies and in the utilization of the attained results.

According to the calculations of economists, mankind faces at present certain limits in the exploitation of the natural reserves of non-replaceable useful minerals and of the replaceable types of biological product. Large amounts of materials required for modern society are disposed after utilization, or during their processing into the environment. The material processes of production are not performed in a closed scheme, with repeated utilization in various spheres, but are based practically on the principle of single utilization, followed by the disposal phase into the environment, thus permanently damaging the various elements of the biosphere.

If due measures are not taken, every industrially developed country will reach a situation where the aggressive and toxic gases, effluents, mineral and organic compounds and solids disposed into the environment will stop the self-purification processes. Overburdening nature further may lead to the consequence that our descendants will be compelled to carry out extreme measures to eliminate the undesired harmful processes begun in our time.

The collective property of the production means in the socialist society offers a favourable possibility for the socialist state to promote—in the knowledge of the above consequences—the production based on non-waste technologies.

The aim of non-waste technology is to ensure, by creating a qualitatively new interaction between nature and society, the growth of production in accordance with the increasing demands of human society along with the rational utilization of natural resources and power. The principles of non-waste technology imply the prevention of waste occurrence, or transforming the production wastes into re-usable raw materials, simultaneously eliminating harm to the environment. In the wide-range realization of the economic laws of socialism the non-waste technology implies an ever closer interaction between the scientific-technical, political, economic, organizational and social components.

Regarding the process of intensification with all its requirements, we may state that the transition to non-waste technology is the repeated utilization, or the harmless disposal and recovery of wastes and the retrieval of natural materials. This is a multiple process aimed at attaining a closed "production-utilization" system, i.e. <u>non waste material processes.</u>

Thus non-waste technology must be regarded as an important component of the material-technical basis, ensuring simultaneously the solution of the problem of environment protection.

The evident harm to the human environment is overwhelmingly the consequence of pollution exceeding the tolerance limit of the natural processes and life phenomena, and endangering in this way also the health of man. Pollution is caused in the final resort always by the presence of some material, or energy, a waste derived from the production - consumption system, which is undesirable in the given connection. Another cause of the damage is due to utilization and consumption activities without due circumspection, scientific foresight and with disregard for the aesthetic aspects. In the course of such activities—:s—apart from other additional effects—certain land areas remain in an unusable form, i.e. waste areas are being created. On the basis of the above the conception of waste may be defined as follows: waste is the by-product and end product of any composition and state, appearing independently or as material combinations in carrier media, including waste energy and areas, created permanently during the life and the production - consumption activities of man, which cannot be utilized either in the production process, or in consumption in the present system of technical-economic conditions.

The above definition implies such aspects of the non-waste material processes, as

(a) its formulation is sufficiently general to be valid for every waste originating from any human activity and production - consumption system and permits the inclusion of very different things, as flue gases, gravel mine pits or drainage sludges in a unified system;

(b) it determines the conception of the waste on economic bases, emphasizing its relative nature, i.e. the fact that it is only valid for the given system of technical - economic conditions and consumption customs with the criteria of economy and efficiency interpreted in a manner which is only valid in the framework of the above system of conditions.

The definition does not state but implies logically that the non-utilized materials, energy and areas are excluded from the range of the human activities and must be replaced by winning new raw and basic materials and utilizing new areas.

The requirements of the reproduction and rational utilization of natural resources have already reached dimensions which go beyond the framework of individual national economies, facing us with great, new tasks in the field of international co-operation not only with the socialist countries, but also with other countries of the world.

In the long run, the increasing requirements of socialist society of the environment can be met on the present level of the productive forces, consumption and the production only, if a harmonious material and energy exchange is created between man and nature. It is here that one of the main lines of intensive development of the national economy must be established. Non-waste production processes in themselves can be realized within the production system only in very rare cases. A realistic aim in the individual production processes may be the creation of little waste, and the improvement of material economy. This aim may be reached by strict technological discipline and the maximal recirculation of the raw and additional materials involved in the production process.

The number of recirculation cycles is naturally limited by the fact that there exist no ideally pure materials and no reactions of 100 per cent yield, or ideal circulation processes to be realized. This means that the recirculation must be periodically tapped and in this case the formation of materials qualified as waste from the aspect of the given production process cannot be avoided.

In practice, non-waste production processes in the strict sense of the word may be realized in the production system only in the form of some vertical technology. The condition of this is to find production processes for utilizing the materials tapped from the material flow, which has been optimized already from the aspect of material and energy economy with or without some pre-treatment, and to transfer the same in a closed system to the site of their possible utilization.

Obviously the possibility of establishing non-waste production sub-systems must be studied in the context of a whole industrial branch. In the case of specialized national economies of small nations it may occur that the "non-waste closing" of certain technological lines is only realizable in the framework of international co-operation. Reducing, or possibly eliminating, the amount of waste originating from the production activity is partly a direct technical problem (elaboration of suitable production processes), partly an economic organization task (finding suitable partners), but it is one of the most important aims of environment protection.

The amount of waste arising during consumption is nearly identical with that originating from the production activity, even in moderately developed countries. Accordingly we cannot restrict ourselves to dealing with only how to make the production system free of waste. The amount of waste due to consumption may be reduced in two ways: by collecting as high a ratio as possible of the refuse materials from the dump, or before, in a utilizable form (glass, paper, metals, etc.). This method may be established in practice without any significant consequences, while leaving the present production and mainly the consumption system unchanged, by public enlightenment (we do not wish to treat here in detail the modes of realization, as, for example, waste selection, or selective collection, etc.).

The second method is producing consumption goods of essentially longer useful life (cars usable for 25-30 years, etc.). This method abolishes the application of the one mentioned before, but demands an essential change in the outlook of the producers and consumers, as well as important technical changes, whose impact cannot be foreseen as yet in many cases.

The aim of introducing non-waste technology is of a complex nature and may be realized only by state management, organization and control.

In the socialist countries the main tasks of the national economy are planned and their realization is controlled by the State.

Appropriate legal rules have been passed for the protection of the environment and the rational utilization of the natural sources and further legal regulations are being prepared for the improvement of the existing legal specifications, whose observation is strictly controlled by the respective state organs.

The CMEA member states carry on wide ranging technical and scientific work aimed at the conservation and improvement of the environment and at the rational utilization of natural resources. The economic need for the elaboration and introduction of non-waste technology and waste utilization methods follows to a great extent from the increase in the costs of production. The necessity of non-waste technology is supported by the following facts:

1. The costs of the non-reproducible natural resources, the raw materials and energy are increasing and we are forced ever more frequently to elaborate and learn the methods of exploiting geologically and hydrologically less advantageous deposits of raw materials.

2. The amounts of waste are increasing to an unprecedented extent, since on the one hand the production of goods is gradually increasing and, on the other hand, the significant part of the gaseous, liquid and solid materials involved in the production, as well as the energy, does not remain in the final product on the present level of the technologies but turns into waste emitted more or less into the natural environment. This process is enhanced by the rapid change of the consumption habits and the increasing ratio of short-life consumption goods.

As the introduction of non-waste technology is an economic problem too, it must be studied in two steps: On the microscale, i.e. from the aspect of the totality of the interests of every individual production enterprise and industrial branch, and on the macro-scale, i.e. on the level of the intersectoral approach, the economic life of the country and of internation relations, as not everything which is profitable for a branch is favourable for the national economy and vice versa.

The profitability of the introduction of non-waste technology and of the rate of introduction is different in the individual countries depending on such decisive factors as the different state of the raw material reserves, the structure of production, its different technical levels in the various industrial branches, the different labour situations, social establishments, etc.

An important element in evaluating the possibilities of introducing non-waste technology is the distribution of labour among the CMEA member states. The specialization of the individual countries in various fields of production implies the potentiality of introducing the most modern technological solution together with the possibility of introducing of non-waste technology.

Finally, in studying the economic problems also the fact that the economy of the individual countries and the whole cooperating socialist economy is part of the world economy must be taken into account, implying certain defined inter relations between the various political and economic systems, influencing the methods and rate of the introduction of non-waste technology. The economic calculations, on which the wide-range introduction of non-waste technology must be based, represent a complex and difficult task. At present the scientifically supported efficiency calculation methods and the experience of the wide-range introduction of the non-waste technological system are still missing; therefore economic problems must be studied on the level of the technical development research of non-waste technology, on different levels in the individual fields of technology both in the individual countries and in the form of common international research.

The economy of non-waste systems may be studied according to the following aspects:

1. By measuring the social losses due to non-observation of the principles of non-waste production and utilization.

2. By calculating the costs of the extension of non-waste production and utilization spheres and the efficiency of the results.

3. By studying the economic means (stimuli), by which the various economic units may be influenced in the interest of raising the level of non-waste production.

The phase of setting about the production of new goods requires a particularly wide intervention of the national economists, first of all in evaluating the efficiency of operations aimed at the extension of non-waste production and for the application of stimuli promoting the efforts in the interest of non-waste production in this phase of the process.

The rational distribution of the productive forces plays a great role in the modern socialist economy. Grouping the enterprises in territorial industrial complexes is emphasized. These enterprises are working in predetermined systematic interaction creating a unified production system, permitting the full application of little-waste, or non-waste technological processes. This particular role of non-waste technology permits the intensification of numerous economic effects, whose attainment under another arrangement would meet with difficulties.

The extensive economic development phase of establishing the fundamentals of socialism is practically closed in Hungary. The conditions of entering the period of intensive development of the economy have been fully created. In the years from 1976 to 1980, the period of the fifth 5 year plan, the main guideline of our economic policy is the vigorous raising of productivity. We wish to enhance at a fast rate the productivity of work, to improve the exploitation of the fixed assets, while simultaneously we will devote greater attention to the energy economy and to the reduction of raw and basic material consumption. According to the prescriptions of the plan we took an increase of 30 to 32 per cent of the national income during the plan period into account.

With regard to the fact that the given potentialities of the country are less favourable than the average, the realistic aim to be set can only be to prevent the worsening of the present situation, the increase of the damage and pollution levels appearing in the environment. It is to be noted that even this seemingly moderate aim can be attained with difficulty, if the pollution entering the country from abroad increases at the present rate. Therefore we wish to approach this aim gradually both in time (by eliminating first the most dangerous pollution sources and by concentrating on the modification of the most wasteful processes) and in space (by introducing our measures first in the most sensitive areas). The realization of these plans and conceptions on a state level in the near future is not alone promoted, but set as a task by the law No. II/1976. "On the protection of human environment", stipulating *inter alia* the following:

§ 10.(1). It is forbidden to expose the objects of the human environment declared as standing under protection against pollution, harm or other damaging effects, which might change their natural properties in a disadvantageous way, or worsen the conditions of human life.

§ 4.(3). The task of environment protection must be performed in coordination with the increase of production and the systematic rise of the living standard.

A necessary condition of meeting both requirements simultaneously is—according to our present knowledge—the gradual change-over to the non-waste economy.

In elaborating the concept of the environment protection activities, special attention must be devoted to laying the fundamentals of the technical - economical conditions that are important in respect of environment protection in the technical development of the producing and non-producing branches, to create the conditions of developing and employing production systems preserving the environment.

A research programme concerning non-waste economy—mainly in technical fields for the time being—has been started in Hungary including, as the most important items, the elaboration of new non-waste processes, to be applied in the chemical industry, by the University of the Chemical Industry in Veszprém, sponsored by the Hungarian Academy of Sciences and the Ministry of Heavy Industry, and a series of studies aimed at the re-utilization of industrial waste in various branches of industry, sponsored by the National Committee of Technical Development.

In several cases also enterprises, whose direct economic interests agree with the specifications of environment protection, introduce little-waste production methods or the utilization of their waste. Non-waste technology is the theoretical limit case of technology involving little waste. The expression "technology" includes the knowledge, the methods and means of creating material goods and of the processes, by which the raw material is transformed into final products. The component parts of the technology are also—doubtlessly—the buildings, machinery and equipment used in production, and a component of non-waste technology is the non-waste recovery of energy.

The introduction of non-waste technology is a wide - and long-range programme of primary importance, the realization of which will produce changes in the industrial and material spheres, in the interactions between utilizers and their living standards.

The problems of non-waste technology can be interpreted in the wide sense, extended to the spheres of production and utilization, energy and raw material reserves, the preservation and formation of the environment.

The creation of non-waste technology with the complex utilization of the raw material reserves, and its introduction into industrial production as soon as possible, is a highly important problem, since the violation of the ecological processes going on in the natural environment and its negative effects on human health may become critical and lead to limitation of industrial development itself.

Report from the Swedish Government

The Ministry of Agriculture, Sweden

INTRODUCTION

The world's total resources of raw materials and energy are limited. Their exploitation has a negative impact on the environment, both because the poorest countries are unable to make rational use of resources and because the rich countries waste resources. This fact has been established in many United Nations documents. The rich are wasteful with natural resources and are also confronted with environmental problems in the form of waste products and pollution. The poor countries lack the knowledge, technology and organization necessary for the effective handling of their scarce resources. The demand for international solidarity must lead to a redistribution of resources from the rich to the poor countries in accordance with the call for a new international economic order. To gain this end we must in the first place learn to use every resource more efficiently, at the same time as we safeguard the environment and conserve resources for coming generations.

The object of non-waste technology is to conserve raw materials and energy because:

- our total resources are limited and must be husbanded to supply the needs of coming generations;
- the use or handling of energy and raw materials can, as such, constitute a threat to our environment.

But a diminished wastefulness with raw materials and energy must never result in the increased exploitation of human beings. We cannot afford to save on materials and oil if this leads to more people starving—because they are out of work or are worn out by working too hard. The right to lead a decent life today and solidarity with coming generations are two demands which can clash with one another and face the politicians with very difficult decisions.

If we are to have a responsible management of resources which combines solidarity and thrift, both the economy and technical development must as far as possible be controlled by the community. The Swedish Government has often pointed out that the market forces do not always facilitate such management.

The Need for Technology Assessment

Under Swedish law, every major extension project in industry must be examined so that public authorities can make sure that national resources are used in the best way. We must safeguard both manpower and physical resources—land, water, energy and raw materials. Technical development in industry has such a great impact today on the world around it that every decision on technical development should be based on decisions that weigh up both the negative and positive effects for the community as a whole.

Sweden's economy is greatly dependent on its forests. Sweden has expanded the capacity of its forest industry to the limits permitted by timber resources. In order to secure raw material supplies in the future, production increases in the forest industry are from this year a matter of consideration by the Government. At the same time, public funds are being allocated to research on methods for increasing raw material supplies and for decreasing the loss of fibres at all stages of timber processing. Demands made by the community can also contribute towards the development of a new technology that conserves resources. The development of resource-conserving technology is one of the objectives of the National Board for Technical Development, which allocates a total of about 210 million crowns a year in government subsidies to technical research and development.

The Environment Protection Act, which was introduced in 1969, has also resulted in both the community and private enterprise making more use of raw materials which before were discarded and thrown out as waste or discharged as pollutants. This legislation requires that one of the criteria of consent for granting licenses for new constructions and extensions of industries and municipal installations is that the technology used for keeping discharges of wastes and pollutants at a minimum shall be the best available within reasonable economic limits. Discharges of mercury from the chloralkali industry into watercourses have declined from 15 to 20 tonnes a year in the middle of the 1960s to 2 tonnes in 1975. Production in the forest industry has doubled in the past 10 years, while discharges of biological oxygen-absorbent material have declined to barely one-half. Practically all built-up areas are today connected with some kind of purification plant; 60 per cent of these plants provide the highest degree of purification with a chemical or bio-chemical purification. A total of 2500 million crowns has been invested in purifying municipal waste water. Today only part of the sludge collected at the purification plants can be used as fertilizer, but research is now in progress to find ways of increasing the use of sludge as a fertilizer. Over a transitional period of 6 years, state subsidies amounting to 770 million crowns have been granted to existing industries for anti-pollution measures.

There are many examples of cases where the processing of a resource can lead to environmental damage. We need only mention carbon dioxide and sulphur dioxide emissions from oil combustion and the discharges of oil into the oceans during transportation from producer countries to consumer countries. There is also a risk that in the long run carbon dioxide emissions will bring about changes in the global climate. Sulphur dioxide emissions in Europe have caused the acidification of soil and water in Scandinavia due to the long-distance transport of these pollutants in the atmosphere. These emissions have very serious effects on the environment—the acidification of lakes and watercourses kills fish stocks and the acidification of soil jeopardizes timber growth. Other types of energy involve other environmental hazards.

Energy Management

Great efforts are being made in Sweden to conserve energy. The aim of Swedish energy policy is to keep the rise in energy consumption down to an average of 2 per cent a year up to 1990 and at a constant level thereafter. The heating of housing and other buildings today accounts for nearly half of Swedish energy consumption. So far the Riksdag has approved over 1000 million crowns in loans and subsidies for energy conservation in buildings. For the time being, nearly 200 million crowns has been set aside for subsidies to industry for the development of more low-energy installations. A further 360 million crowns will be channelled into energy research by 1978, and half of these funds will be used for energy-conservation research.

Thrifty Management Achieved by Organizational Changes and Improved Treatment of Waste

Non-waste technology need not only mean the development of new technology to reduce wastefulness with raw materials and energy. Sometimes, purely organizational changes can be just as effective in preventing the production of wastes, or in avoiding the necessity of using the resource at all. For example, an effective public transport system reduces the need for the use of private cars, which must increase automotive pollution and necessitate the expansion of the road network, etc. Another example is Swedish legislation on the handling of refuse, whereby the responsibility for collecting and treating refuse is concentrated to the local authorities in order to make it more possible to make use of waste as a resource. (More details on pp. 140 - 1).

Land Management

Even in a sparsely populated country like Sweden, land resources may be limited. In Sweden's national physical planning, the authorities have indicated which areas are to be reserved for nature conservation and open-air recreation in the future and which areas can be used by polluting industry. We also have restrictions on the use of cultivable land for building if other land is available. The most fertile agricultural land is also to be found in the most densely populated areas in south and central Sweden.

New Basis Data for Resources Policy

It cannot be asserted that we have a clearly defined policy for non-waste technology covering all areas, but the above account does show that attention has been paid to the problems in a number of sectors.

Swedish policy on raw materials and energy is divided into several sub-areas, e.g. agricultural, forestry, land, energy and environmental policy. The committees studying mineral policy, agricultural and forestry policy are three of the government committees now compiling new basic data for resource planning. The Secretariat for Future Studies is an official body which is studying ways of promoting technology assessment in the public sector. Its study "Resources and Raw Materials" looks into the linkages between several sectors and is to set out alternative avenues of long-range development to serve as basic material for a discussion of the kind of future we shall choose.

Swedish Environment Protection Legislation and Non-Waste Technology

Sweden has a number of laws covering various areas that are relatively well geared to the principles contained in the concept "non-waste technology", provided this concept is taken to mean that human activities are to be planned in such a way that losses of material and energy are kept to a minimum.

The legislation which makes the greatest impact on the problem under review is the Environment Protection Act and its related Environment Protection Ordinance. The Act is applicable to discharges of waste water, solid substances or gas from land, buildings or installations. Under this Act industries and local authorities who wish to build new installations or extend pollutive operations must apply for a licence to do so. One of the criteria on which such a licence is granted is that the best available technology is used as far as is economically reasonable in order to minimize the discharge of pollutants and waste products. This legislation also provides means for restricting discharges into water and air as well as controlling the handling of wastes, not only by means of external purification but also in respect of the selection of the processes used. It should be pointed out here that the Swedish authorities have pursued a policy of trying to tackle the problems as near the source as possible, i.e. at each separate production unit, instead of investing in large joint purification plants.

The Ordinance of Environmentally Hazardous Wastes regulates who may transport and finally take charge of this type of waste. Certain well-defined types of waste are classified as environmentally hazardous. This legislation makes it possible to co-ordinate waste handling so that certain constituents can be recycled.

One of the measures taken in Sweden to reduce the volume of certain types of wastes is a surcharge introduced on 1 March 1973 on some kinds of beverage containers. The purpose was to encourage people to buy beverages in returnable containers and thus reduce litter. Another advantage is a reduction of the energy used in the manufacture of containers and for the handling and treatment of waste. However, the transportation of returnable containers back to supplies reduces the gain in energy consumption. This particular surcharge was comparatively low—10 öre—and has had but little effect.

Sweden also has an Act on Products Hazardous to Health and to the Environment which regulates the manufacture, buying and selling and imports of products which may injure human beings or the environment. The Act prescribes that the contents of such products must be clearly marked on containers. Trade in and imports of a certain product can also be forbidden if this is considered necessary.

Within the framework of the municipal refuse collection monopoly, local authorities are required to collect waste paper for recycling. Waste paper is to be separated from other waste at source, e.g. by households, pending removal. Provisions on the separation of wastes and the extended public refuse-collection monopoly will gradually enter into force in local authority areas and be fully implemented before 1980.

Moreover, a municipal waste-collection monopoly for environmentally dangerous chemical waste will be set up in stages within 5 years. Industry is required every year to declare the nature, composition, volume and mode of handling of environmentally dangerous wastes. The final disposal of environmentally dangerous wastes is effected by recycling, neutralization or long-term storage. The responsibility for these operations will be taken over by a special, primarily public-owned, waste-treatment corporation.

Another measure is an authorization system for the scrapping of motor cars. Auto-scrappers can become authorized if they observe current legislation on human settlements, nature conservation and environment protection. An auto-scrapping premium of 300 crowns will be paid to car owners when they deliver their old car to an authorized auto-scrapper.

The Swedish Building Act prescribes that the establishment and location of industrial or similar activities, which are of substantial significance in the management of energy, timber or the country's aggregate land and water resources, shall be subject to a government decision. This gives the commmunity an instrument for minimizing energy consumption in industry and ensuring that Sweden's forests and mines are not exploited too quickly. The Nature Conservation Act regulates the extraction of sand, gravel and clay as well as operations that change the landscape, either by extracting natural resources or by the dumping of wastes.

The above account of Swedish legislation shows that in many respects it is possible to direct development along lines that keep energy and resource losses at a minimum.

There is, however, no legislation that can steer production in existing industries over from high-energy products to low-energy products. Nor can legislation force existing industries to make more efficient use of raw materials, even if the external environmental problems are solved by purification plants and efficient waste handling.

Some Areas where Legislation can Exercise Greater Restraint

It is thus clear that within the framework of current legislation there are areas where the principle of non-waste technology can be more stringently applied. One example is the treatment of municipal waste water. Were it not so polluted by heavy metals and chemicals, sludge could be made much greater use of as a fertilizer. The way to achieve this is to avoid unnecessary contamination of waste water coming into municipal purification plants or to introduce selective purification processes at the plants themselves.

The planning of industrial production can undoubtedly be improved to achieve a more optimal use of raw materials. In many industries, such as mining and chemicals, it is probable that only in exceptional cases will it be possible to achieve a 100 per cent utilization of raw materials. If wastes endanger the environment, the expense of treating waste may force industries to make fuller use of raw materials. However, as already intimated, laws empowering authorities to restrict trade in and imports of certain products could promote an advance along these lines.

Tightening up restrictions in current legislation on warm-water discharges would also reduce energy losses. This would be one way of enforcing the use of hot water for heating purposes. The discharge of hot gases from planned industries can, in principle, be influenced by provisions contained in building laws on energy conservation. In the case of existing industry, however, we have no laws that regulate energy losses due to the discharge of hot gases. But here, as in the heating of premises, price trends will probably have an automatic effect.

The term "non-waste technology" encompasses the whole field from raw material to waste products. The Environment Protection Act covers only part of this cycle. But in conclusion it can be said that this particular legislation has been an effective instrument in minimizing discharges of pollutants into air and water.

Energy Policy

The decision on energy policy passed by the Riksdag in 1975 covered guidelines for the 10-year period 1975-1985. Recommendations for the shaping of energy policy during the latter part of the 1980s will be put before the Riksdag in 1978.

The major features of energy policy are an integrated planning of energy policy for a longer span of time, and that goals are set up not only for energy supplies but also for energy consumption.

In the case of consumption, the energy policy decided on by the Riksdag aims to lower the rate of increase to an average of at most 2 per cent a year for the period 1973-1985. After that, growth should be further restrained so that if possible the country's total energy consumption can be stabilized at a constant level by 1990. The 2 per cent limit for the growth of energy consumption will bring total consumption up to about 540 tWh. The ceiling for the consumption of electric energy has been fixed at 160 tWh in 1985, an average rise of about 6 per cent a year from the base year 1973.

A number of immediate measures have been introduced to restrict the growth of energy consumption in accordance with the approved guidelines. The management programme includes both information and training activities and various kinds of financial assistance on the one hand, and on the other, planning, legislation and controls/restrictions. Committees now at work may well present other possible measures. The 3-year energy-research programme, which is part of the energy policy decided on in 1975, lays great emphasis on energy conservation.

The Energy Conservation Committee is largely responsible for supplying information on the desirability of conserving energy and how it can best be done.

The Committee's main terms of reference are:

- the continuous observation of and reporting on developments in the field of energy conservation;
- launching its own energy conservation campaigns, assistance to and co-ordination of the energy-conservation information supplied by other authorities;
- negotiations with local authorities, trade organizations and other interested bodies on conservation measures of various kinds;
- the compilation of basic data as regards restrictions aimed at conserving energy, e.g. bans of various kinds.

In the fiscal year 1975/76 the Committee has had a budget totalling about 6 million crowns. The greater part—5 million crowns—has been spent on outgoing information.

In addition to information, special training projects must be set on foot if all means of conservation are to be fully utilized. The energy-policy programme therefore includes intensified training courses and advisory activities in the energy field. The Riksdag has allocated a total of 5.75 million crowns, 3 million of which is for the fiscal year 1976/77, for these purposes.

In the Spring of 1974, the Riksdag for the first time allocated funds for financial assistance to energy-conservation measures. This assistance was primarily intended for measures in residential buildings, and in buildings used by local authorities, county administrations and business enterprises.

Since then, this financial assistance has been extended to cover industrial processes as well as certain prototypes and demonstration installations, etc.

Assistance so far totals 1430 million crowns, 406 million of which is for the fiscal year 1976/77.

The Riksdag has also allocated a total of 1060 million crowns, 300 million for the fiscal year 1976/77, for financial assistance to energy-conservation measures in housing.

New rules were introduced into the Building Statutes in 1975 for the purpose of reducing the consumption of energy in the heating of buildings.

Altogether, the new regulations are expected to result in cutting energy consumption in new residential building to about half of the average consumption in a dwelling built at the beginning of the 1970s. The saving of energy in a single-family house is estimated at an average of 13,000 kWh/year and for a dwelling in an apartment house at about 10,000 kWh/year. In terms of heating fuels, this is a saving of about 2 and about 1.5 cubic metres a year respectively. In the case of other types of buildings, the saving on energy has been estimated at the same at least as for residential buildings.

This proposal refers only to new construction. Obviously, there is a great potential for saving energy in the heating of existing buildings. The National Board of Urban Planning and the Building Research Council are now working on these problems as they relate to their respective areas. The National Board of Urban Planning is, for instance, drawing up regulations, advice and directions relating to energy conservation requirements in the reconstruction of buildings.

MANAGEMENT OF LAND AND WATER RESOURCES ENERGY AND CELLULOSE MATERIALS ACCORDING TO THE BUILDING ACT

Decisions concerning the management of land and water resources in Sweden were taken by the Swedish Riksdag in December 1972 and include guidelines for planning the management of natural resources in particular areas of the country, and guidelines indicating how particular activities should be treated in physical planning. Certain amendments to existing legislation were also made in order to secure the implementation of the guidelines.

The geographical guidelines adopted by the Riksdag refer solely to those parts of the country where competition for physical resources is, or can soon become, particularly acute and where supra-regional aspects are of special importance. This applies above all to the coasts but also to the mountain region and some river valleys.

The geographical guidelines are of two distinct kinds. Firstly, areas are demarcated whose scientific and recreational values seen in a national perspective are considered to be of such a kind that major environmental changes, e.g. through the siting of environmentally disruptive industry, should not be allowed other than on very special grounds. Secondly, certain guidelines are laid down for the location of the industry under consideration. Deliberations concerning the location of industry are dependent on factors which can change relatively quickly. Therefore, it must be possible to revise these guidelines relatively frequently in the light of technical developments, for instance.

Legislative changes included amendments to the existing building legislation in order to achieve a broader application at municipal level of the master plan procedure. It is mainly through municipal planning that the intentions of national physical planning can be implemented. The amendments to existing legislation also referred to changes in rules for compensation and compulsory purchase.

Further, a new rule in the Building Act provides that the location of industrial or similar activities which are of major importance to the management of the nation's land and water resources must first be submitted to the Government, which is to make a comprehensive assessment.

The implementation of the guidelines will be divided into two stages: a programming stage and a planning stage. The programming stage, which was terminated by 1 July 1974, consists of negotiations concerning the need for planning and other measures between the County Administrative Boards and the local authorities concerned. These negotiations are based on instructions from the Government and are to lead to programmes indicating how planning within the municipalities and counties should be accomplished and what other steps should be taken in order to safeguard the intentions of national physical planning. After a consultation procedure, the Government makes its decision concerning these proposed programmes.

One of the deficiencies of earlier legislation was that final decisions concerning the location of industrial activities, which were considered to be important for purposes of national physical planning, could not always be taken at a sufficiently early stage and on the basis of an all-round assessment. Often this was due to industry having chosen its location and gone very far in its projection and other preparations before state authorities were given an opportunity of making their own assessment. Different assessments were made on the basis of a variety of laws and thus lacked coherence.

To remedy these deficiencies, co-ordination has now been established regarding decisions affecting the location of industrial or similar activities which are of major importance to the management of the country's total land and water resources. A new rule in the Building Act provides that the question of establishment of certain industries is first to be considered by the Government. The Government shall hereby make a comprehensive assessment relating considerations of environment protection and planning to, for example, considerations of labour market policy, regional policy and industrial policy. In order to make the Government's decisions as broadly based as possible, opinions are normally requested from authorities, union bodies, industrial organizations, etc.

The rule in the Building Act which regulates the assessment procedure concerning certain industries was changed in 1975. Today it not only includes industrial activities which are of special interest in national physical planning, but also certain other industrial activities. Besides industrial activities which are of major importance to the management of the country's land and water resources the Government's permission is also now needed for industrial activities which are of major importance to the management of energy or cellulose materials for forest industries.

FUTURE STUDY OF RESOURCES AND RAW MATERIALS

In 1975 the Secretariat for Future Studies within the Ministry for Education initiated a study entitled "Resources and Raw Materials. Sweden's long-term supply in an international perspective." The study is now mid-way and is expected to produce a number of overviews of specific aspects of materials policy. The question of non-waste will be an important consideration in most of these surveys as well as in the final report which is expected towards the middle of 1977. There are, however, some fields of study which are particularly relevant to the subject of the ECE seminar.

Substitution

It is intended to produce a survey of mechanisms for the substitution of rare and of costly materials by more abundant ones. The survey will give historical examples of substitution, give an account of accepted economic theory for substitution, try to outline technological possibilities, discuss higher-level forms of substitution (functional substitution). Substitution is of obvious interest to Sweden, e.g. as regards raw materials for the petrochemical industry, where cellulose could presumably be substituted for petroleum.

New Technologies

The survey aims to present the prospects for more immediate break-throughs in several areas. Also here, some emphasis will probably be placed on the substitution of non-renewable by renewable materials.

Waste and Recycling

This over view will mainly reiterate findings of other studies on the scope of waste, its handling and recycling.

It is expected that part of the final report will be devoted to questions related to waste. Besides the obvious fact that a large share of the productive capacity is wasteful, i.e. sub-optimal in regard to known technologies, the study will discuss in general terms wasteful aspects of society. No specific research will be conducted beforehand. It is expected that proposals for such studies will be put forward.

*Production Sans Déchets en Belgique**

I. Van Vaerenbergh

Services du Premier Ministre, Programmation de la Politique Scientifique, Bruxelles, Belgique

PROGRAMME NATIONAL DE R & D SUR L'ECONOMIE DES DECHETS ET DES MATIERES PREMIERES SECONDAIRES

Le Gouvernement belge, conscient que le pays devra faire face, dans l'avenir à des problèmes dus au rencherissément des matières premières et des ressources énergétiques, a approuvé la proposition de la Commission Interministérielle de la Politique Scientifique de "donner une impulsion particulière à la recherche sur les divers aspects des problèmes concernant les matières premières, comme p. ex. la production sans déchets et le recyclage."

Le programme national de R & D en cette matière a démarré le 1er Octobre 1976 et se terminera en 1979.

1. OBJECTIFS DU PROGRAMME NATIONAL

1.1. Etablir une stratégie technico-économique visant à

réduire les quantités de déchets produits,
augmenter le taux de récupération et de recyclage,
traiter de façon appropriée les déchets non recyclables.

Cette politique présentera un intérêt du point de vue

environnement,
économies des matières.

Elle sera basée sur une étude économique

des procédés de fabrication permettant de réduire la production de déchets et de favoriser le recyclage et des méthodes de traitement des déchets non recyclables.

Enfin, elle devra être conçue de manière à convaincre

les producteurs et les consommateurs de son utilité.

1.2. Améliorer les processus de production en vue de réduire en amont la

*This document complements another report with the same title by A. Buekens on pp. 271-282.

production de déchets.

1.3. Récupération de matières utilisables provenant des déchets de production et des déchets de consommateurs: aspects technologiques.

1.4. Elimination de déchets non recyclables d'une manière acceptable tant du point de vue écologique que du point de vue économique.

1.5. Bilans technico-économique permettant d'orienter au niveau gouvernemental une politique cohérente et bien coordonnée.

2. PLAN GENERAL DU PROGRAMME NATIONAL

2.1. Analyse de l'inventaire des données belges concernant les déchets industriels et ménagers.

2.2. Développement de la technologie, approche économique et étude de comportement des producteurs et des consommateurs.

2.3. Examen des instruments de gestion et d'organisation et des directives à formuler.

3. ETUDE DE LA "PRODUCTION SANS DECHETS" DANS LE PROGRAMME NATIONAL

Les thèmes suivants, retenus dans le Programme National traitent de la production sans déchets:

3.1. Conception des produits: effets à attendre sur la réduction des quantités de déchets irrécupérables

Objectif: introduire au stade de la conception même des produits des criterès relatifs à la récupération et au recyclage après usage.

Dans une première phase exploratoire, on recherche des produits, qui grâce à une meilleure conception, ou à un emballage plus rationnel pourrait être récupérés dans de bonnes conditions de rentabilité. Cette étude aboutira à des projets de mesures gouvernementales d'une part encourageant par des aides financières les innovations conceptuelles facilitant la récupération et d'autre part définissant des conditions d'importation des produits étrangers.

Dans une deuxième phase on procédera à une étude technico-économique des produits retenus dans la phase exploratoire.

3.2. *Amélioration des processus de production en vue de réduire la quantité de déchets*

Objectifs: il s'agit de réaliser en amont des économies de matières premières en améliorant les processus de production, les modifiant ou en utilisant des matières plus appropriées. Le but de l'étude est d'économiser une partie des gros investissements en installations de récupération en aval de la production.

Dans une première phase l'Université Libre de Bruxelles, chargée de cette tâche, fera le point sur l'état de la question au niveau de la production.

Dans une deuxième phase on établira des solutions "préventives" d'amélioration des procédés, tenant compte de leur influence sur l'environnement.

Les thèmes mentionnés ci-après ne traitent pas de la production sans déchets au sens strict mais concernent des solutions curatives en vue de traiter et si possible de recycler les déchets "inévitables."

3.3. *Etude technico-économique d'une politique optimale de récupération en Belgique*

Objectif: une stratégie de récupération dépend en particulier d'un "marché de récupération" stable, afin que les produits de la récupération puissent entrer en compétition continue avec les matières primaires.

La première partie de l'étude sera consacrée à l'évaluation des procédés de triage et de récupération, de leur applicabilité et de l'équipement du secteur de récupération en Belgique.

Dans la seconde partie on étudiera le flux spatio-temporel des matières premières, l'évolution du marché de récupération en Belgique ainsi que les possibilités de substitution de matières primaires par des matières récupérées.

Enfin, on déterminera le coût des techniques de triage et de collecte.

En conclusion cette étude aboutira à la définition de directives pour l'organisation des activités de récupération. Toutes ces activités seront réalisées par les Facultés Universitaires Notre-Dame de Namur.

3.4. *Valorisation des boues d'épuration des papeteries*

Objectif: utilisation de boues d'épuration des papeteries comme substrat pour la sylviculture.

L'Université de l'Etat de Gand en collaboration avec l'Institut pour la culture des Peupliers à Grammont fera une recherche technique sur les différents paramètres qui conditionnent la qualité des boues d'épuration en tant que substrat pour la culture des différents types d'arbres.

Cette technique de valorisation sera évaluée économiquement et sera comparée à d'autres méthodes de traitement des boues.

Des papeteries intéressées contribueront à cette étude en fournissant les terrains et les boues, et en prenant soin du compostage des boues.

3.5. Traitement des boues d'épuration en combinaison avec des ordures ménagères

Objectif: il s'agit d'une étude technique tenant compte des facteurs économiques et écologiques. Elle sera réalisée par l'Institut d'Hygiène et d'Epidémiologie.

Cet Institut comparera les bilans énergétiques du traitement des ordures ménagères avec et sans addition de boues.

Ill déterminera aussi l'influence des boues d'épuration sur le processus de fermentation des ordures ménagères.

Le transfert eventuel d'éléments toxiques non dégradables des boues vers le sol (lors du déversement également du mélange ou des résidus après fermentation) sera examinée.

3.6. Boues industrielles à traiter par solvants

Objectif: traiter par solvants des boues industrielles en vue d'en extraire les éléments valorisables.

Le Centre de Recherche et de Contrôle Lainier et Chimique a établi le plan de travail suivant:

1. Etablissement de l'inventaire des boues rejetées

2. Etude technique des paramètres de l'extraction

3. Expérimentation sur des appareils pilotes

4. Examen des possibilités d'utilisation des matières extraites et du résidu.

3.7. Valorisation des déchets bio-industriels

Objectif: étude technico-économique du traitement de fumier avec et sans perte de valeur en tant qu'engrais.

1. Traitement du fumier en conservant sa valeur d'engrais

   filtration et tamissage (expériences en laboratoire et à l'échelle semi-industrielle);

   sédimentation et flottation (laboratoire);

   centrifugation et pressage.

2. Traitement du fumier avec perte de sa qualité d'engrais

épuration chimique;

épuration biologique aérobie.

3.8. *Utilisation de l'énergie des effluents thermiques pour la production de protéines par l'aquaculture basée sur le recyclage des déchets bio-industriels*

Objectifs: Utilisation de l'énergie d'effluents thermiques sur un cycle annuel complet de végétaux, de façon à obtenir un rendement d'élevage maximum à une température et avec une alimentation optimales pour chaque type d'organisme.

Détermination de chaînes alimentaires courtes pour

faire absorber des algues (cultées sur le déchet bio-industriel minéralisé) par des crustacés;

faire absorber les crustacés par des poissons.

3.9. *Aspects régionaux du traitement des déchets en Belgique*

Objectifs: 1. Limitation des déversements clandestins.

2. Etablissement d'une bourse de déchets.

3. Tri systématique des déchets.

4. Modélisation des systèmes de traitement et de récupération.

Recherche de mesures pour lutter contre les déversements clandestins, notamment par secteur de consommation

Evaluation des bourses existant en Belgique et à l'étranger. en collaboration avec les équipes responsables des autres aspects du programme national

Elaboration d'un projet global de bourse de déchets, pour la Belgique en tenant compte des mécanismes de marché prédominants

Mise au point de techniques de tri systématique des déchets ménagers, adaptée aux situations locales

Collecte des données économiques et techniques, nécessaires pour tester sur le plan régional, les modèles d'optimisation

état actuel de la récupération et du recyclage;

étude du flux de matières;

évaluation économique des solutions envisagées.

3.10. *Etude relative à la fermentation anaérobie des déchets organiques*

Objectifs: 1. Evaluation chimique des substances fermentescibles et des déchets obtenus après fermentation.

2. Etude technico-économique.

Analyses des matières premières et identification des substances inhibitrices pour la fermentation anaérobique.

Détermination des matières premières qui peuvent être mélangées en vue d'une fermentation stable et uniforme.

Analyses des substances résiduaires de la fermentation anaérobique de déchets agro-industriels.

Justification d'un éventuel traitement des dits résidus.

3.11. *Traitement des déchets par pyrolyse*

Objectifs: 1. Etude des rendements de pyrolyse de divers déchets.

2. Pyrolyse de déchets organiques.

3. Pyrolyse de déchets hydrocarburés.

4. Valorisation des gaz de pyrolyse.

Etude physico-chimique du traitement pyrolystique pour différents types de déchets sous différentes conditions; évaluation technico-économique

Etude technico-économique de la pyrolyse de déchets organiques

Etudes des caractéristiques générales et des propriétés chimiques et physiques des matières premières et de leur conditionnement pour la pyrolyse

Traitement des pyrolysats en vue de la récupération des fractions valorisables

Transposition des résultats au niveau d'un four prototype et, par après, au niveau industriel

Etude économique, tenant compte des impératifs liés à la protection de l'environnement.

Les autres thèmes du programme national, non mentionnés ci-dessus traitent de sujets qui ne concernent pas la production sans déchets proprement dite:

Récupération rationnelle des déchets: tri systématique.

Récupération de chaleur produite lors de la dégradation biologique aérobie de déchets.

Amélioration des propriétés physiques et mécaniques de la pâte à papier fabriquée à partir de papiers récupérés.

Etude technico-économique pour une politique optimale en matière de récupération et approche économique intégrée de la gestion des déchets en Belgique.

4. COUT TOTAL DU PROGRAMME NATIONAL

Le budget prévu pour ce programme national "Economie des déchets et des matières premières secondaires" est de l'ordre de 250 à 260 milliard de Francs belges.

Non-Waste Technology in Finland

Jali M. Ruuskanen* and Matti Vehkalahti**

**Civil Engineer, Finnish National Fund for Research and Development (SITRA), Helsinki 12, Finland*

***(Planner) Ministry of the Interior, Helsinki, Finland*

1. INTRODUCTION

It became quite evident during the preparation of this national report that an exhaustive analysis of the situation of non-waste technology in Finland could not be made without separate and extensive research work. This is owing to a number of reasons, the most important of which are described below.

The concept of NWT (non-waste technology) is a new and interdisciplinary concept and, in itself, is not a part of any particular branch of science, does not come under any particular professorial chair, and is not associated with any particular profession, and there is consequently no expertise available in Finland to cover the concept as a whole.

The concept of NWT is indefinite and undemarcated to such an extent that it merges into and overlaps concepts such as technology assessment, cost/benefit analysis, advantage comparison, the new technology, alternative technology, etc.

For these reasons, this report will content itself with providing some sort of description, with the aid of examples, of whether there are any operations in Finland that must definitely be regarded as being non-waste technology, and of what grade any such may be. The making of a more extensive analysis dealing with this question is being given consideration.

The outline in the ECE document ENV/SEM.6/PM/R.3 will be employed as a framework for the report.

2. NWT IN THE PRODUCTION OF RAW MATERIALS

(a) *Product planning*. Nothing particularly worth mentioning.

(b) *Process planning*. The employment of various biological processes in the production of the raw materials for the foodstuffs and fodder industry is being investigated. Some processes (protein production out of the wastes of the pulp industry) are already in the stage of production. The recovery and use of wood in its entirety is the subject of research and development. A more efficient recovery of wood, however, is likely to cause environmental damage, albeit producing savings in raw material.

(c) *Waste processes*. The processing of waste gypsum from the fertilizer industry into building and other panels is already in operation. The use of waste bark from the wood-processing industry, apart from its use in energy production, is the subject of research. The use of pine needles and of twigs is also the subject of continuous study. The exploitation of tree stumps is already at the manufacturing stage. One of the problems with this is the replacement of nutrient losses in the soil by means of artificial fertilizers.

3. NWT IN THE TRANSPORTATION OF RAW MATERIALS

(a) *Product planning.* Nothing worth mentioning.

(b) *Process planning.* Interior waterway transportation is being revived.

(c) *Waste processes.* Nothing worth mentioning.

4. NWT AT THE PRODUCTION STAGE

(a) *Product planning.* Although it is often admitted by now that product planning in accordance with the principles of NWT would frequently be called for, it has not really made a breakthrough. The importance of the selection of material is likewise recognized, as expressed in a lecture recently given by a Finnish expert in materials:

"The choice of material, product planning, production and regeneration should be combined into an optimum system when a new product is being planned, in which system the experts on materials who are familiar with the manufacturing process would be working from the start in close co-operation with the planners and the product engineers. Such co-operation is important not only to the quality of the product and to the economy of production but is of particular significance when the material operations are placed on a par with questions of energy and environment."

In the business of packaging, some "NWT achievements" in Finland might be mentioned. For instance, the returning of beer and soft drink bottles has become highly developed. The return percentage is close to 100. There is also standardization of bottles for alcoholic beverages other than beer, and their return is consequently high too.

In questions of transport, the aim is to stagger working hours and this has already been implemented to some extent, this matter too being indubitably involved with NWT.

On account of our northerly location, a great deal of energy in Finland is expended on heating. In order to minimize the thermal losses in heating, an extensive national clarification and investigation project is under way under the title of "Project on the Economy of the Heating of Buildings", and it has now become possible to make efficient use of its findings.

There is distinct interest in studying the heat pump and new energy forms. One way of utilizing solar energy involves greenhouses, i.e. cultivations under glass, the use of which is on the increase especially in the northern parts of the country, where the amount of solar energy is very great for 6 months of the year.

(b) *Process planning*. Naturally, there are plenty of examples from this field (for details, see the case studies presented by Finland). The advanced internal circulation of raw materials has been implemented in a great number of industrial branches, as the economic profitability of such operations can easily be demonstrated. The steel, the wood-processing and the glass industries may be mentioned as examples.

Research and development aiming at greater process yields has produced a great number of results. Among recent achievements, a mention may be made of the substantial increase in the yield of the sulphite pulp process.

The utilization of waste materials is familiar not only in the steel industry but also in the wood processing industry. The mills have recently tended to integrate, in the sense that the wood raw material brought into the industry is being entirely utilized irrespective of the species of wood or the quality of the raw material. Among the particularly pro-environmental and energy-saving industrial processes, a mention should be made of the copper flame-smelting method, which is based on a Finnish development. It makes effective use of the energy contained in enriched ores, thus requiring far less external energy than does the conventional method. On the other hand, the sulphur dioxide content of the combustion gases generated in this process is so high that the sulphur dioxide can easily be removed and then used for the production of sulphuric acid, thus reducing the harm to the environment.

One result of inter-Nordic product development is a new process for bleaching pulp which is called displacement bleaching, in which the space requirement is one-fourth of that needed in conventional technology, the amount of waste water one-tenth, and the heat and electricity consumption less than one-half.

Increasing attention has been paid to the saving of energy in industry ever since the onset of what came to be called "the energy crisis". An extensive investigation and product project is being carried out both in the process industry and in other industry, some of the results of which have already been applied in practice.

Aiming particularly at the saving of raw materials is a method called shell casting technology, which has already been in use in Finland for a long time.

(c) *The waste process*. The handling of the waste liquor of the pulp industry, aiming at the recovery of cooking chemicals and at energy production, is an old practice in Finland. There are examples of various internal waste raw-material cycling in many lines of industry. The exploitation of waste heat is a subject of special investigation in the energy-saving project mentioned above.

The repair and renewal of old dwellings should obviously be included among NWT activities. The Government has begun to support this activity rather vigorously in more recent times. The research and development work associated with this is also becoming livelier.

5. NWT IN THE DISTRIBUTION OF GOODS

(a) *Product planning*. The packing industry has carried out a lot of developmental work in order to develop various space-saving packages.

(b) *Process planning*. Obviously, a great deal remains to be done in this line in Finland, considering energy savings in transport.

(c) *Treatment of waste*. Nothing worth mentioning.

6. NWT AT THE REJECTION STAGE

(a) *Product planning*. The planning of products to such an effect that NWT principles would be implemented at the rejection stage is something new in Finland.

(b) *Process planning*. A good example of a waste treatment process in accordance with NWT principles is the "biofilter" recently adopted for industrial use and developed by a Finnish company. In this method, the condensates containing badly smelling and noxious compounds produced by evaporation plant and cooking plant at a pulp mill are treated by a natural, biological method. The discharge is conducted through a filter, the filling of which consists of the bark produced as a by-product of the wood processing industry and functioning as a substrate for a bacterial flora which lies at the heart of the process. The micro-organisms oxidize the sulphur compounds into a harmless form. Compared with the Steam Stripping Technique used for the same purpose, multiple savings are obtained in costs on energy, investment and operation.

(c) *Waste process*. Nothing worth mentioning.

7. NWT IN THE MACRO-SYSTEM

Research and development projects which may be regarded in a sense as being studies of NWT component systems are being carried out or are being implemented in Finland. These include:

energy project of the process industry (saving of energy);

energy project of other industry (saving of energy);

heat economy project for buildings (saving of energy);

study of materials policy (saving of raw materials);

project on the exploitation of industrial wastes, beginning with waste inventories which can be charted;

studies in the technology of recovery in the construction business (saving of energy and raw materials).

No overall studies such as energy analyses by product have been made. However, various studies and balances of the flow of materials have been carried out.

Some information on Finland

Area : 337,032 km^2

Population : 4.7 million

Population by industry (1970)	%
agriculture, forestry, fishing, etc.	20
manufacturing, etc.	26
construction	8
commerce, etc.	19
transport, communications etc.	7
services	20
	100

Industrial activity (1972)

establishments: 7071

wage earners: 404,000

power installed directly for running machines: 5.7 million kW

Domestic product by industries (1974): 72,200 million mark.

National income (1974): 63,200 million mark.

State of Non-Waste Technology: United States Experience and Policy

David Berg* and C. Lembit Kusik**

**Environmental Protection Agency, Washington, D.C. 20460, U.S.A.*
***Arthur D. Little, Inc., Cambridge, Massachusetts, U.S.A.*

OVERVIEW

The pace of the introduction of non-waste technologies (NWT) is slowly increasing in the United States. However, resource allocation decisions are still primarily based on market values that fail to take account of full social costs and benefits and tend to reflect relatively high discount rates. Progress in implementing NWT is reflected in the actions of all levels of government and by the private sector. The development of methodologies to evaluate the complex factors affecting policy decisions by government and the private sector has progressed rapidly, and considerations favorable to implementing NWT are, in general, receiving greater weight as a result. Major factors in the implementation of NWT in the United States include the following:

1. Increasing demand for raw materials and energy.
2. Increasing costs of extraction from dwindling domestic reserves.
3. The imposition of environmental cleanup costs as a result of declining environmental quality.
4. Increasing reliance on less dependable international materials and energy markets.

STATE OF NON-WASTE TECHNOLOGY IN THE UNITED STATES

The implementation of NWT technologies and the advent of governmental programs and policies has increased at a slow, but steadily accelerating, pace in the last 10 years. The acceleration in the implementation of NWT is attributable to a number of domestic and international factors. A key is the shift in the United States economy towards an increasing reliance on imports of basic raw materials and energy. Imports of all minerals rose from 10 per cent of domestic consumption in the early 1950s to over 15 per cent in 1970. Imports of some specific materials increased dramatically: iron ore from 5 to 30 per cent of domestic consumption and petroleum from 8 to 22 per cent (and by 1975 to about 40 per cent). A complementary factor is the increasing industrialization of the world economy, which is pushing upward the relative price of many basic supplies. In addition, key international markets have been influenced by conjunctive policies of producer nations.

Domestic factors have also affected the pace of NWT in the United States. In the post-war period the development of new industrial processes in the major materials-producing industries* in the United States slowed. Industrial investment in research and development settled at relatively low rates in the major materials-producing industries, and the result has been a slow rate of technology change in these industries. Sales in these major industries totaled $287.9 billion in 1975, while research and development expenditures were $2.7 billion, or less than 1 per cent of sales. If 1975 research and development investment is expressed as a percentage of sales, chemicals led with 2.6 per cent (or $1.3 billion); by contrast the steel, petroleum and coal, and paper and allied products industries together committed only 0.5 per cent of sales revenues to research and development.

Until the last several years, the substantial United States resource base—and consequently the low present value of materials and energy—provided a price structure which encouraged the use of existing technologies that are in many cases relatively material and energy intensive. Resource use and technology selection decisions are still made on primarily economic bases and at the plant or corporate levels; resources are, in general, priced in markets that may undervalue the long term benefits to society of the resources. Government policies have tended to be cautious about tinkering with the marketplace based on longer range or NWT considerations.

In the past several years underlying economic forces have shifted substantially, and government and industry are renewing their interest in NWT. Central to the shift has been the increasing importance of more costly foreign supplies of materials brought about by a growing economy and declining domestic sources. Another important factor has been declining environmental quality (only now beginning to improve) and strong public support for a government role in assuring cleanup. In 1975, of 256 air-quality-control regions (AQCR) in the United States, 112 were in violation of at least one annual primary air-quality standard (Total Suspended Particulates). Most AQCRs were in violation of the more stringent secondary air-quality standards. Despite this impetus to internalization, only some social costs are included in today's prices. Consequently, government forces are beginning to be exerted to aid the market in balancing competing factors in resource allocation decisions; environmental legislation (the Clean Air Act and the Federal Water Pollution Control Act) is forcing the use of emission and effluent controls.

NATIONAL LEGISLATION AND POLICIES

Government planners are taking a serious look at the potential for NWT methods for easing environmental problems, strengthening the economy, and alleviating resource shortages. This national interest is embodied in several laws, govern-

* Within the manufacturing sector five industrial segments account for about three-fourths of the energy use—or over 20 per cent of national energy consumption. These are (1) metals, (2) chemicals, (3) petroleum and coal products, (4) paper and allied products, and (5) stone, clay, and glass products. Of these five industries chemicals and paper and allied products have the strongest record for process development. The former seems to be on the verge of introducing important processes on a wide scale.

ment organizations, and governmental policies and programs. The two major emphases are environment and energy, and mechanisms used to implement NWT are geared to work through the market system. The environmental legislation internalizes some external or social costs of various economic activities by requiring polluters to add cleanup devices. The expenses incurred change the relative value-in-use of resources and permits market (and, hence, technological) reoptimization. Because not all costs are internalized and those that are reflect the expense of control, rather than the environmental damages themselves, the environmental legislation is somewhat indirect in its approach to NWT. The key energy provisions are less direct and weaker: support for research, development, and demonstration of energy-conserving technologies, and programs to promote a quicker marketplace response to changing conditions by providing information to consumers.

In the environmental arena laws have been passed which address all three media: air, water, and land (solid waste); and which single out other key problems: noise, radiation, toxic substances, drinking water. Additionally, the National Environmental Policy Act (NEPA) requires that for any major project involving federal action which may cause a significant environmental effect a statement evaluating environmental impacts must be filed that examines major alternatives to the planned project, including conservation NEPA establishes a process for the orderly consideration of most major projects in the United States and opens this process to the public and interested private and governmental bodies. Projects can be challenged if the environmental impact statement is shown to have omitted important alternatives or environmental impacts, or if environmental resource waste has not been identified.

There is another fundamental aspect to the environmental laws which encourages NWT. This is that no longer is the environment considered a "free" resource for depositing wastes. Under the laws, standards are established for environmental quality (by medium) and for setting forth emissions and effluent regulations for industries and/or specific facilities. Local or regional plans are developed which link individual polluters together in cleanup planning. Thus, costs of industrialization borne by society are being internalized to some extent in the prices of goods and services, and the economics of wasteful technologies are becoming less clearly favorable over NWT in several industries.

For example, 20 to 25 per cent would be added to the price of electricity generated at a single fossil fuel-fired power plant equipped with cooling towers to control waste heat discharge, flue gas-cleaning systems to remove sulfur oxides and electrostatic precipitators to control emissions of particulates. Another cost contributor is the siting of power plants away from the already dirty air in cities; this adds to the cost of transmission. The safe disposal of ash and flue gas-cleaning solids and liquid effluents from blowdown and ash ponds adds more to the internalized costs. In sum, the Environmental Protection Agency estimates that if all United States fossil-powered steam electric plants were equipped with all necessary pollution control equipment, approximately 10 per cent would be added to the cost of electricity from those plants. While this is a small increase in percentage terms, demand for electricity has been shown to be relatively elastic in the industrial and commercial sectors, and increasingly, though less, elastic in the residential sector. These influences on cost can thus be expected to encourage the use of conserving technologies.

The increasing cost of electricity is attributable in the main to the rising price of fossil fuels; a second important factor in recent years is the high cost of capital. The effects of environmental requirements are also beginning to be felt: perhaps 1 or 2 per cent was added by 1975 to the cost of United States electricity by environmental regulations. The effect of all of these factors is, however, the same: more efficient generation and use systems are encouraged. Several responses have been noted:

1. The growing use of life-cycle costing in selecting equipment and appliances.
2. The consideration of ways to make use of waste heat, such as by siting electricity-generating facilities near industrial complexes (as in Michigan).
3. The revision of electricity pricing structures to end promotional rates and encourage conservation by public utility commissions (state-level bodies).
4. The economics of prospective generating technologies, like combined cycle steam generation and potentially fluidized bed combustion, are becoming more favorable.

The Energy Policy and Conservation Act of 1975 (EPCA) strengthened the tools available to the Federal government to encourage conservation of energy. Yet, the measures remain weak and less direct in comparison to those of most European nations and are weaker than those addressing environmental problems. The EPCA provides authority for mandatory programs to label automobiles and major consumer appliances with an indication of their relative energy efficiency, and targets have been established for automobile mileage for the next several years. State programs to promote energy conservation are funded, and targets for (voluntary) energy conservation are to be developed for industrial groups. Further, EPCA provides $2 billion for loan guarantees for energy conservation and renewable energy resources projects. In short, fairly complete technology transfer and consumer education programs exist and are supported by several Federal agencies, including the Energy Research and Development Administration (ERDA), the Department of Commerce (DOC), the Department of Housing and Urban Development (HUD), the Department of the Interior (DOI), the Department of Transportation (DOT), the Environmental Protection Agency (EPA), the National Bureau of Standards (NBS), the National Science Foundation (NSF), the Federal Power Commission (FPC), and the General Services Administration (GSA, the Federal procurement agency).

Federal policy calls for a sizeable reduction of energy use by government agencies. Virtually every government agency, bureau, and department has its own energy-conservation program designed to reduce energy use within its activities. In Fiscal Year 1974, the most recent for which data was available, government use of energy was reduced by over 20 per cent, an amount equal to one-half of one per cent of national energy use.

Policies involving direct economic incentives or outright proscription of excessive energy use have generally not been applied as tools to implement NWT. One exception is the investment tax credit which encourages capital investment by directly reducing income taxes by varying percentages of the investment. This mechanism applies, however, to all capital investment and is not limited to NWT. The new tax law adds an alternative mechanism: firms may accelerate the depreciation (over 5 years) of pollution-control investments and take a 5 per cent tax credit. Another tool is the tax-exempt pollution-control bond, which permits lower cost financing of the portion of major private-sector capital expenditures devoted to pollution-control equipment.

Of increasing importance is the experimentation now underway by some states to encourage NWT. An excellent example of this is in Oregon and Vermont where deposit charges are mandatory for beverage containers. These experiments have to date resulted in reduced roadside litter, a shift towards refillable containers and no detectable adverse economic consequences. A recent FEA study indicates that consumers could save over $2 billion annually and that energy requirements could be reduced by over 70,000 barrels of oil equivalent, if deposits were required nationally. The study suggests that long-lasting, reuseable containers would be used in place of disposables to a significant degree, thus reducing materials and energy consumption.

RESEARCH AND DEVELOPMENT

Another major approach to encouraging NWT in the United States is government support for research and development. NWT-related research and development falls in three major areas: energy, environment, and NWT methodologies. Several agencies are involved.

In Fiscal Year 1977 an expenditure in excess of $100 million is planned for energy conservation-related research and development by several agencies of the Federal government. State and local governments and private sources will add many more millions of dollars of support. Much of this work will affect the efficiency of use of other resources, as well. The principal topics addressed are: industrial processes and process modifications, buildings design, buildings equipment, conversion of wastes to energy, efficient transmission of electricity, and transportation improvements, particularly automobiles. Related areas of concentration include efficient agricultural practices (e.g. no-till farming), materials research, and product design.

Environmental research and development will total over $400 million in Fiscal Year 1977. The work involves several areas which have importance for the development and implementation of NWT. These are:

1. Study of the fate and effects of pollutants—work necessary for understanding the causes and extent of social costs associated with the various activities of industrialized systems.
2. Development of control technologies and of environmental standards—this work translates progress in understanding effects into requirements and techniques to minimize the most serious environmental problems.
3. Assessment of emerging energy supply technologies and consuming technologies from an environmental perspective—work seeking to anticipate external costs and, thus, lay the basis for systems designed to avoid them.

Research and development into NWT methodologies is sponsored at every level of the United States government and by private parties, as well. Among the government agencies involved in sponsoring or performing this R&D are the National Science Foundation, the National Academy of Sciences, the Council for Environmental Quality, the Environmental Protection Agency, and the Congressional Office of Technology Assessment. Several million dollars is dedicated to this research and in excess of $25 million will be spent to apply these methodologies.

METHODOLOGIES

One major area of emphasis in the United States has been the development of methodologies for making decisions involving the complex of technical, economic and social issues which must be addressed in the gradual shift towards NWT. Four methodologies that have been pioneered by United States researchers are cost/benefit analysis, input/output analysis, technology assessment and resource analysis. The application of these tools has contributed to decisions affecting every important natural resource.

Cost/benefit (C/B) analysis was applied early on in the United States to public works projects. Proponents used C/B techniques to define water resource, energy, recreational and other benefits in common units. The same techniques were used to develop successful opposing arguments to several projects. More recently C/B techniques have been applied to wide-ranging economic and social decisions. Notably, this tool has aided government decision-making in the difficult field of electricity rates.

Input/output analysis is a United States development pioneered by V. Leontief of Harvard. This tool tracks resource flows and can be used to identify key points of waste, opportunities to close loops, and other opportunities for NWT. Another methodology is resource analysis, which defines complete materials balances and overlays them with an evaluation of energy inputs. A final step of resource analysis is the application of economic tools for decision making to the arrayed data.

Technology assessments complete the range of leading methodological development in the United States. Technology assessment is a policy analysis tool which is interdisciplinary, systems-oriented, iterative and cumulative. In predicting the ramifications of technology alternatives it combines quantitative, qualitative, and judgmental aspects; as such, it is the least precise, most "artistic" of the methodologies of NWT. Leading assessments have been performed on the development of offshore oil (Atlantic coast), the alternatives for more efficient automobiles, and geothermal energy.

A leading methodological issue in the United States is the degree of detail that is necessary in depicting resource flows for the conduct of policy options assessment. The continuing development of information processing technology on the one hand and data on the other facilitates the consideration of enormous quantities of information in these studies. At this time it is unclear where researchers will settle.

PROSPECTS FOR NWT IN THE UNITED STATES

As the study of NWT has progressed it has become increasingly clear that tradeoffs must often be made between conservation of one resource and another. Other key factors, such as the reliability problems of new technologies, complicate major resource use decisions. An excellent example of a highly complex situation of this type is in the iron and steel industry.

One process being actively considered by the iron and steel industry is direct reduction with electric arc furnaces. The direct reduction technology would offset the conventional-blast-furnace-coke-oven-basic oxygen-furnace technique now the domestic state of the art. A report prepared for the EPA suggests that a few plants may be built in the United States by the end of the century applying direct reduction, depending on the results of delicate and complex calculations. Advantages are the potential for using steam coal rather than metallurgical coal, potential pollution-control advantages (in eliminating the need for coke ovens and blast furnaces) and potential cost advantages. On the other hand, direct reduction requires about 40 per cent more energy per ton of steel than the conventional route.

Ultimately, the decision with respect to any single facility may revolve around secondary issues like transportation cost and other site-specific economic conditions. Further, in a mixed economy like that in the United States, most of the responsibility for resource decisions lies with the private sector. Thus, the national movement towards non-waste technologies is likely to continue to be largely dependent on individual judgements by economic players. Government decisions will play a growing role to the extent that public support grows for the internalization of external costs. The strains of growing populations and of continuing industrialization would seem to be a sufficient force to encourage that role in the United States. These considerations appear to apply to all countries, although the usefulness of specific NWTs will vary depending local considerations.

Experience and Policies in the Field of Non-Waste Technology in the Federal Republic of Germany

J. Orlich

In a highly industrialized country such as the Federal Republic of Germany, which is dependent on importation of raw materials, further development of low raw material, low waste technologies is a permanent responsibility of Science, Industry and the State. The latter establishes the corresponding framework structure by passing new environmental conservation acts, provides orientation aid in the form of programmes and grants financial support for to encourage research and innovation. Furthermore, it makes use of its influence on education, this applying particularly in the field of the universities. The main responsibility and also the greatest interest in the development and application of non-waste technologies, however, lies with industry itself. The industrial sector has, to this end, set up its own institutions such as Rationalization Boards, Waste-exchange Markets and Expert Committees. It invests large sums of money both in research and in training staff. New impetus for implementation of non-waste technologies arises first and foremost as a consequence of private sector initiatives arising from industry.

The Federal Government passed its first environmental protection programme in 1971. Even this first programme laid down waste prevention and reduction as important points. On the basis of this environment programme both the Federal Minister of the Interior and the Federal Minister of Research and Technology have established and implemented comprehensive research programmes for development of non-pollution technologies, waste prevention and waste utilization.

The research programmes are statistically documented each year. The amount of money made available by these two ministries alone, amounting to approximately DM 20 million yearly, is used up totally by industry and scientific institutes.

In 1975 the Federal Government for the first time presented a special waste-management programme. This programme was drawn up in co-operation with experts from industry and the relevant associations as well as experts and scientists working in administration and research institutions. On the basis of a systematic overall analysis, industry and consumers are provided with target guides and orientation aids aimed at bringing about behaviour which is sensible from the point of view of waste management.

The two main aims of this programme are reduction of waste on the production and consumer levels and raising of the level of beneficial utilization of waste.

Reduction in waste quantities is to be achieved by which will:

decrease production waste;

encourage application of environmentally favourable production processes;

supervise material usage with respect to the end-purpose of products;

increase the durability and ease of repair of products;

increase re-usability of products.

It is intended to increase profitable utilization of waste materials by:

using them as raw materials in production processes;

exploiting their energy content;

reintroducing them into the biological cycle.

A systematic data survey has been carried out for most types of waste. To implement the aims quoted above a comprehensive catalogue of measures has been compiled.

The waste management programme attaches special significance to specifically-directed information campaigns amongst the public as run by consumer associations, civic initiative organizations and environmental protection associations. By this means non-pollutant designs can be furthered for specific products under their influence. However, this waste-management programme not only attaches importance to providing consumer information. In the industrial sector, too, information must be provided on an in-depth level on the problems of waste management. Primarily medium-size and small companies need to catch up on objective information about waste-management problems. In this respect the decision taken recently by the legislative authorities to introduce the institution of a works representative for waste assumes particular importance. This waste officer, who in future will have to be employed in the majority of industrial concerns, must amongst other duties ensure that low-waste technologies are applied in his company in order to reduce generation of waste and also to guarantee that production-specific waste is if possible put to some beneficial use.

A similar role is played by the institution of the air pollution inspector, an office which was introduced in the Federal Anti-pollution Act of 1974. Not only must he ensure that certain limiting effluent emission values are not exceeded in his company but also that as a basic principle low-emission production processes are employed.

In its 1976 Environment Report (2nd Environment Programme) the Federal Republic has again explained the order or priority of waste-management aims.

The most important aim according to this report is likewise the reduction of waste arising both on the production and the consumer sides. This applies particularly to medium- and long-term aims. Direct prevention of waste as a rule is the most efficient means of relieving the environment of pollution and bringing about a saving in the consumption of raw materials and energy. Second in order of importance is the beneficial utilization of those types of waste whose arisal cannot for technological or economical reasons be avoided. By such measures, efforts to reduce waste are not replaced but rather a supplementary approach is provided which is required for reasons involving the environment and raw material policies. The 1976 Environment Report of the Federal Republic also stresses that the problems of reducing waste as well as the development and, above all, implementation of low-waste technologies must initially be tackled and put into practice by the companies concerned themselves.

Because the increasing shortage of raw-material reserves and in view of the heavy dependence of the Federal Republic of Germany on raw material imports, the Federal Government has this year (1976) presented a comprehensive Raw Materials Safeguarding Programme. This programme pursues three aims: expansion of the raw materials base, saving of raw materials and recycling of raw materials.

To save raw materials it is intended to introduce methods which will:

replace raw materials and primary materials in short supply by high-availability raw materials;

make use of accompanying substances and by-products;

lower the specific material application;

reduce production losses;

improve material yields;

increase the flexibility of raw material input and of the range of alternative processes available;

raise the product quality or product life-expectancy, e.g. by decreasing susceptibility to wear-and-tear and corrosion;

encourage manufacture of re-usable products.

To increase recycling of raw materials it is intended to introduce methods which will

utilize valuable materials regained from spent products;

make use of production waste;

favour product designs suitable for recycling.

A large number of individual research projects have been implemented in connection with each of these major aims.

In the Federal Republic of Germany there are two acts directly controlling the utilization of residual substances. These are the Waste Oil Act and the Animal Carcase Disposal Act. Other anti-pollution laws do not contain provisions which demand direct application of low-waste technologies or waste utilization. They do, however, operate indirectly in this direction.

A law was passed in 1972 governing the disposal of waste materials (Waste Disposal Act). The strict provisions of this Act and the sometimes drastic disposal fees arising out of the application of the instigator principle, particularly in the case of dangerous waste or special-category waste, have already done much to encourage industry for cost reasons to develop low-waste technologies and waste utilization processes. Thus, for example, two-thirds of the 320 tempering shops concerned in the Federal Republic have converted their salt bath nitration plants to some other process in which no cyanide-containing spent salts are produced.

The legislative powers in passing the Waste Disposal Act have included provisions for prohibiting certain types of packaging materials. The share of packaging waste in domestic refuse is at present more than 50 per cent by volume. The resulting efforts by the packaging industry to prevent packaging waste from continuing to rise unchecked are clearly evident. A particularly vivid example is the glass bottle. Whereas in 1970 a 0.33-litre single-trip glass bottle weighed 235 g its weight has now been reduced to 185 g.

In the meantime financial backing by the Federal Government has enabled research to develop a light glass bottle which weighs only 70 g (0.3 l). Such bottles are now about to be brought on to the market. Beyond these efforts the glass industry together with the waste material trading industry have started to organize separate collection of bottles so that more production plants can be switched to operation on melted-down spent glass.

Similar trends can be observed in the tin-plate industry. In this sphere the weight of a beverage can, for instance, has been reduced by 25 per cent in the last 5 years. The coating of tin has been decreased by 80 per cent. Efforts are aimed at developing a completely tin-free thin-walled can, recycling of which will no longer cause any problems for the steel industry. The efforts of the tin-plate industry have likewise been given new impetus towards greater recycling as a result of the Waste Disposal Act.

Laws governing anti-pollution in the case of water have had similar repercussions to the Waste Disposal Act. In this connection, mention must be made in the case of the Federal Republic of Germany of the Water Conservation Act and the Waste Water Levies Act. After a transition period of several years the Waste Water Levies Act will come into force in 1981. Various sectors of industry have already started to adapt their processes to these requirements. The paper industry and in particular the cardboard industry in their production plants have been making efforts to switch to closed water cycles so that their processes will then be practically free of waste water. Production processes producing either no waste water at all or only small quantities thereof are also undergoing trials in the cellulose-production sector.

In the sphere of anti-pollution relating to the atmosphere the Federal Anti-pollution Act, with its various provisions, has had noticeable repercussions on the application of low-effluent production processes. These provisions at present quote detailed limiting values for fifty different types of particle-containing effluents and approximately 120 gaseous substances. Industry is endeavouring not only to remain within these limits but as far as possible to drop below the required levels by installation of air-scrubbing plants. To an increasing extent these concerned are endeavouring to fulfil the imposed anti-pollution requirements by converting plants to low-effluent and thus low-waste production processes.

The Anti-pollution Act demands issue of permits for establishment and operation of approximately fifty different types of plant. These plants not only have to adhere to the maximum permissible effluent-emission limits but also have to be designed in accordance with the latest developments of technology. This guarantees that newly developed low-pollution production processes are actually put into practice.

Further development of low-pollution technologies will likewise continue to be supported by State Research and Innovation Promotion Programmes. As an example of no-effluent technology, mention might be made of no-solvent paint application, a technique which has just recently been perfected.

Apart from Industry there are a large number of state and non-state institutions dealing with development of non-waste technologies. The Federal Republic, in establishing the Federal Office of the Environment, has created an authoritative institution which will support these other institutions in the implementation of research programmes and by provision of advice in the field of anti-pollution measures, waste management and drafting of legal and administrative laws.

While providing other facilities, the Federal Office of the Environment has built up a comprehensive information system for environmental planning (UMPLIS). Within this information system, data files are being prepared in the field of waste management which will, for example, supply the following details:

Quantities of products causing waste as well as waste quantities and their spacial distribution (waste register).

Description of waste according to composition, dangerous properties, possible means of disposal and destruction (waste type data file).

Information on utilization processes and users of waste with details of secondary products, their fields of application and possible marketing opportunities, quality demands and costs (utilization data file).

This information system will help to implement specifically directed measures to reduce generation of waste and to encourage profitable utilization of waste. Corresponding measures are being compiled in the field of research and development in each case in terms of programmes, much of this working being supported with considerable monetary backing. This information system not only serves as advice material for state institutions but is also at the disposal of interested parties and the public as well as industry in general.

The number of non-state institutions and research institutes dealing with research into low-waste technologies in a wide variety of fields is very great. As an example of interdisciplinary methods which frequently have to be applied in this connection, mention might be made of the Institute of System Technology and Innovation Research at Karlsruhe.

Technical universities and technical colleges have long been taking into account production processes aimed at saving raw materials in educational courses for future engineers and constructors. Corresponding approaches are becoming increasingly important for raw material and anti-pollution reasons. Aspects such as easy-to-repair design of consumer goods have become central educational subjects at design schools with the aim of reducing waste quantities.

To a large extent industry has recognized the need to intensify the development of raw-material saving and low-waste technologies.

As in the past, industry has proved that it either already possesses or is quick to acquire the necessary technical know-how in this field.

Industry is anticipating imposition of further expensive anti-pollution stipulations in the field of air and water conservation as well as steep rises in waste-disposal charges. Moreover, as a result of rising raw-material and energy costs, the industrial sector will be forced to increase its efforts to develop low-waste technologies. German industry, which in many sectors is predominantly dependent on importation of raw materials, will presumably follow the trend towards manufacture of products which are technically of higher quality and which will thus as a rule have longer life-expectancy.

In view of complex economical interlinkage with the international environment a study restricted to one individual country cannot lead to satisfactory results.

First and foremost it is important to make the relationships more transparent for politicians, producers and consumers. From the orientation thus arising new impetus can be expected.

Expériences et Politique de la Yougoslavie.

Contribution of the Government of Yugoslavia

I. CERTAINS ELEMENTS DE LA POLITIQUE NATIONALE EN MATIERE DE LA TECHNOLOGIE NON-POLLUANTE

Ayant conscience de l'importance de l'environnement et de sa qualité et tenant compte de ses possibilités concrèts du développement économique, la Yougoslavie, pays socialiste et pays en voie de développement avait connu une évolution au rythme accélérée.

Dans ce contexte, il conviendrait de souligner ce qui suit:

(a) la Yougoslavie appartient, dans son ensemble, aux pays ayant un environnement préservé, à l'exception de quelques grandes villes ou centres industriels et énergétiques;

(b) le développement technologique de la Yougoslavie se fonde, pour une large part, sur les technologies importées; par conséquent, les problèmes qui surgissent dans ce domaine sont des problèmes propres aux technologies respectives.

En conclusion, l'image de l'environnement en Yougoslavie, vu dans son ensemble, porte des caractéristiques des pays en voie de développement, alors que la situation dans les centres industriels et les grandes villes est indentique à celle des pays développés.

Plus le pays se développait sur le plan socio-économique, plus le danger de voir compromettre l'environnement était prêter une prise de conscience accrue de la nécessité de préserver permettant l'application d'une politique de préservation de l'environnement ont été améliorées.

Cette politique repose sur le principe qui pourrait être défini, en résumé, comme un *développement conforme aux conditions de l'environnement.* Ce principe est confirmé aussi bien dans la Constitution yougoslave* que dans les chapitres du Plan de développement.

Un développement conforme aux conditions de l'environnement devrait, avant tout, empêcher l'introduction de nouvelles sources de pollution et de dégradation dans tous les milieux et permettre, à la fois, de combattre les sources déjà existantes et d'en éliminer les conséquences fâcheuses là où la mise

---

*"Quiconque veut utiliser le sol, les eaux ou d'autres biens naturels est tenu de le faire de manière à assurer les conditions de vie et d'activité de l'homme dans un milieu sain." (Constitution, article 193).

en danger de l'environnement est déjà manifeste. L'orientation stratégique consisterait, donc, dans la recherche des voies et des méthodes de développement qui ne mettent pas en danger l'environnement.

Cette politique a été formulée à partir de l'attitude selon laquelle le futur développement de la société devrait se fonder sur une introduction progressive des technologies non polluantes aussi bien dans le cadre de la réalisation des nouveaux projets d'investissements que sur le plan de la reconstruction des installations en place.* En procédant de cette manière on permettra le dépérissement naturel-physique et techno-économique-des sources de pollution, toutes les forces et les moyens disponibles pouvant être mis, à partir de ce moment-là, au service de la promotion de l'environnement.

Le projet de loi sur l'environnement (en cours de discussion à l'Assemblée de la RSF de Yougoslavie) pourvoit les bases concrète d'application de la politique de l'environnement. L'article 16 de la loi en question stipule, entre autres, que l'amélioration de la production doit être concue et réalisée sur la base de l'introduction des procédés technologiques propres à permettre la préservation de l'environnement (technologiques non-polluantes et sans déchets, recyclage, valorisation des déchets, etc.). Les réglements d'application de cette disposition légale sont définis, de manière coordonnée, par les républiques et les provinces autonomes.

Il est notoire que la technologie non-polluante, considerée sous un aspect à court terme, est plus onéreuse que celle qui ne tient pas compte de l'environnement. Malgré ce fait, de nombreux exemples ont démontré, dans notre pays, qu'il est possible de trouver un intérêt économique à appliquer la technologie non-polluante à condition que le problème soit considéré à travers le prisme de la rentabilité sociale à long terme qui englobe, entre autres, l'élément "prix" de l'environnement.

II. CERTAINS RESULTATS ET EXPERIENCES ACQUISES SUR LE PLAN DE L'INTRODUCTION DE LA TECHNOLOGIE SANS DECHTS

Les examples cités ci-dessous devraient illustrer, en partie, les efforts que nous déployons en vue de promouvoir l'application des technologies sans déchts:

A Belgrade (Mali Makiš) un projet relatif à la construction d'une cimenterie selon le principe de circuit fermé de production, a été établi; un autre projet, portant sur la construction d'une fonderie à base de technologie non-polluante, est en cours d'élaboration.

Des etudes approfondies ont été enterprises aux fins d'introduction des circuits fermés de production dans d'autres domaines, à savoir: la production du fer à partir de la boue rouge obtenue dans la production de

*"Les plans de développement et de production assureront, entre autres, une utilisation plus rationelle des ressources naturelles, de l'énergie et des matières premières secondaires, aussi bien que la diminution des déchets et la substitution de la technologie polluante par la technologie non-polluante." *(Plan de développement social dans la période 1976-1980*, chapitre II, par.12.)

l'allumine, la production de l'acier à partir des cendres de pirite pelotonnées, après l'extraction du cuivre (200.000 t. environ, par an). Les études portent également sur les possibilités d'utilisation de la chaleur dégagée dans les acieries pour le chauffage.

Dans le cadre de trois usines faisant partie du bassin minief et métallurgique de Bor, on a introduit le procédé de désulfuration des gaz déchappement produits lors du grillage du minérai de cuivre, permettant d'obtenir l'acide sulfurique utilisé aux fins différente, et tout particulièrement pour la production d'engrais chimiques.

Un nouvel atelier de lavage et de criblage de charbon a été mis en place dans le cadre du bassin énergétique de Tuzla.

Des efforts ont été déployés afin que les gros producteurs d'engrais chimiques à base d'azote deviennent aussi producteurs d'oxygène qui pourrait être utilisé dans les procédés de combustion pour améliorer l'oxydation et pour réduire les quantités des gaz d'échappement; certaines usines (Brod, Beograd) produisent déjà des chaudrons de capacités différents et destinés à l'utilisation pour différents combustibles.

Dans le cadre du bassin Brod-Prahovo, ainsi que dans l'entreprise "Zorka" (Sabac) des efforts sont faits pour améliorer l'absorbtion des gaz issus de la production des superphosphates et pour utiliser l'acide silico-fluor-hydrique aux fins de production du criolite synthétique, de AlF^3 et du fréone.

A Voivodina, un "Programme de développement d'équipements destinés à la protection et la promotion de l'environnement" est en cours d'élaboration.

Il est envisagé, dans plusieurs combinats ou fermes agricoles, d'introudire le procédé de filtration du fumier, permettant d'utiliser les matières sèches pour l'enrichissement du sol, l'eau obtenue par ce procédé étant, à la fin, complètement épuré (Sabac—projet élaboré par l'Institut pour la mecanisation dans l'agriculture de Belgrade).

Sur le plan de l'introduction des procédés propres à diminuer la pollution de l'atmosphère

des carburateurs permettant d'atteindre un niveau de combustion plus élvé ont étéréalisés par l'usine "precizna mehanika" et sont utilisés dans les machines automobiles "Renault" montées en Yougoslavie;

les nouveaux projets de centrales thermo-électriques prévoient la préservation complète de l'environnement, la protection contre les gaz, la suie et les cendres, la chaleur et assurent la protection presque complète des eaux (Obrenovac);

il est envisagé d'assurer la gazification des grandes villes et des industries (Sarajevo);

une nouvelle méthode a été élaborée un vue de controler la diffusion des polluants de l'air dans les milieux urbains et industriels;

un nouveau type de source d'énergie chimique a été réalisé au niveau de laboratoire, pouvant être utilisé, en perspective, pour la traction d'automobile.

Il existe, par ailleurs, des conditions favorables à l'utilisation de nouvelles sources d'énergie (solaire, éolienne, etc.), mais les recherches concretès dans ce domaine n'ont pas encore été entreprises.

Un encouragement est apporté aux efforts visant à permettre l'utilisation de nouvelles matières premières et matériaux qui devraient remplacer ceux dont la production va à l'encontre de la politique d'une utilisation rationelle des ressources naturelles et de la préservation de l'environnement.

Conformément à la politique de l'environnement et de l'introduction des technologies non-polluantes, ainsi qu'à l'esprit de l'Acte final d'Helsinki, on est en train, en Yougoslavie, d'examiner la possaibilité de prendre une initiative visant l'établissement d'un accord sur l'homologation des technologies non-polluantes. Cet accord constituerait une contribution à la réalisation de la conception européenne d'un développement harmonisé et conforme aux conditions de l'environnement. Par la même, on aiderait les pays en voie de développement à éviter de se trouver en face des problèmes et des dépenses rencontrés actuellement par le monde développé.

Industrial Experience. Introductory Report

D. Moyen (Rapporteur)

Assistant Director, Institut National de la Recherche sur la Securité, Paris, France

First of all I should like to thank the authors of the various papers on which I have been asked to report for the quality of their work which has greatly simplified my task. The six documents presented, i.e.:

Recovery of proteins from potato vegetation waters, by Mr. Huchette, from Etablissements Roquette Frères (France),

Profitable industrial uses for wheys, by Mr. Bertrand, Ministry of Agriculture (France),

Dyeing in a solvent medium, by Mr. Laurent, GIE STX (France),

How and why we have chosen integral recycling, by Mr. B. Maréchal, Société ISOREL (France),

Recovery of the iron contained in pickling solutions and residual ore etching waters as magnetite, by Mr. Lefort, Société Pont à Mousson (France),

Waste exchanges: Better management for another growth, by Mr. Deloy, editor of the magazine *Nuisances et Environnement*,

relate industrial experiments in various fields. Analysis of these documents will demonstrate that they all have points in common. In each case, the effluent has no longer been considered separate from the production shops, and an attempt has been made to act upon the pollution-generating conditions to completely avoid pollution or to replace it by a harmless residue.

From among the ways chosen, the first consists in "grafting" upon the effluent a secondary productive activity, the effluents of which are less harmful or even non-existent.

It is in this way that the recovery of proteins from potato vegetation waters, the profitable industrial uses for whey, the recovery of the iron contained in pickling solutions and residual ore etching waters as magnetite, enables the recovery of the pollutant by-products from the major activity in order to put them to profitable use.

The pollution problem posed by potato starch-works is inherent to the very composition of the raw material used. Schematically, it can be said that the tubers are starch reservoirs which soak in the vegetation water since, quantitatively, for 100 kg of potatoes, the balance shows:

20 kg of starch,
2.5 kg of pulp,
77.5 kg of vegetation water.

The art of potato starch manufacturers consists in separating these different phases. But, if the starch and the pulp are marketed, the more or less diluted vegetation waters are discharged. When we consider that their COD is on the order of 50,000 mg/l and that a plant processing 4000 tonnes of potatoes per day discharges approximately 3000 m^3 of potatoes per day, we realize the importance of the problems to be solved.

This vegetation water, or "red water", contains about 4.5 per cent of soluble material, in the form of:

nitrogenous compounds (approximately 2 per cent), mineral elements (1 per cent), sugars, organic acids, lipids.

While awaiting pursuance of the research to enable total recovery, the recovery of only part of the proteins constitutes a first step toward reducing the discharges.

The first operation in a starch-works consists in grinding the potatoes in graters to free the starch. It is right from this first production stage that thought must be given to the recovery of protein for:

first of all, the quality, by using reducing agents to protect the vegetation water against any oxidation;

next, the quantity, by covering the maximum amount of vegetation water under the least expensive conditions.

After centrifugation to separate 70 to 90 per cent of the vegetation water depending on the means used, the used is acidified to a pH of approximately 5.0—close to the isoelectric point of the proteins—and then brought to a temperature of 100 to 105°C in tubes.

Approximately 55 per cent of the proteins are flocculation and coagulated; this flocculate is then separated by centrifugation, concentrated by evaporation and then dehydrated by spraying or on a roller. The final product obtained is a creamy powder containing 8 per cent water, 78 per cent proteins, 2 per cent fat and 2 per cent mineral matter.

One-third of the organic matter responsible for the COD has been removed; this treatment thus ensures the elimination of 10 kg of COD per tonne of processed potato.

This achievement emphasizes:

First of all, the importance of research to develop protein-recovery techniques. In fact, this research has led to a patent filed by Société Roquette Frères.

Then, the fact that research work is never ended. Thus, for example, tests are under way to improve the efficiency of the operation, to find profitable uses for the overflow remaining after protein extraction and to find better market openings and the most profitable uses for the proteins obtained. These tests are particularly important. To give but one example of this, intensive tests have been carried out in the field of nursing foods for calves through partial substitution with milk powder, in comparison with other animal or vegetable replacement proteins. In this particular case, Société Roquette Frères erected a stable with 350 stalls to conduct this study which proved the excellent quality of the products obtained and the possibility of incorporating them in the diets at high levels, up to 8 per cent of this figure representing a 30 per cent milk protein substitution rate, without affecting the performances before weaning.

It should thus be pointed out that, although it is meritorious to develop techniques making it possible to recover pollutant by-products, the problem of finding profitable uses for them is of prime importance to the companies. There are two ways for the companies to solve this problem:

use the recovered by-products to manufacture or produce other products which are more and more diversified in order to enlarge their markets;

make use of waste exchanges.

This first approach is the one discussed in Mr. Bertrand's paper on the profitable uses of whey. Indeed, conventional technology for clean disposal of this residue consists in drying it to obtain powder. But the pressure placed by the French Administration on industrials to keep them from discharging whey into rivers led to an ever-increasingly abundant production of whey powder resulting in glutting of the market, whence the tendency of the major drying plants to limit production so as not to induce a slump in prices. Obviously, this limitation induces more and more numerous direct illicit discharges of whey. To remedy this situation, other profitable uses had to be found.

Tables 1 and 2 at the end of Mr. Bertrand's paper (pp. 218, 219) show different types of possible profitable uses for the whey.

The whole whey can be treated with the industrial effluents and the cost would be 16 ± 4 francs/m^3 of whey treated.

As we said, the conventional profitable use consists in producing powder for cattle feed. This solution leaves hopes for 9.7 ± 27 F/m^3 of treated whey.

The increase in the cost of energy and the fact that passage through the spraying tower requires an exceptional amount of energy have led us to ask ourselves whether or not it would be possible to use the concentrated whey in the form of molasses in the animal feed industry, instead of using the product in the powder form. Research led to a positive answer and the mean profit thus obtained is 18.5 ± 19 francs, i.e. a much surer profit than in the case of powder.

Other profitable uses consist in separating the proteins from the lactose.

Examination of the whey composition (93 per cent water, 5 per cent lactose, 1 per cent protein and 1 per cent mineral salts) shows that this product is disbalanced for nutritional uses, the lactose content (sugar which is assimilable by young mammals but much less so by adults) being too high. Other types of treatments thus consist in isolating the proteins and the lactose to find separate profitable uses for each.

Let us come back to the table. It can be seen that we obtain much greater revenues for the protein fraction, it being possible to use the latter in human nutrition, either in dietetics or in recycling in cheese. In this particular case, separation of the proteins through ultracentrifugation and their incorporation in cheese gives a mean profit of 101.6 ± 45 F/m^3 of whey.

Table 2 shows the possible combinations for profitable uses of the lactose and the protein fraction.

The complexity of the operations to be carried out or the small amount of wastes to be treated might incite some industrialists not to find valuable uses for the latter themselves, but place them in the commercial circuit as such. The appearance of waste exchanges favouring re-use through knowledge of the available or sought-after wastes and providing contacts between industrialists has been a considerable aid to rehabilitation of the recovered products which have become "secondary raw materials" which is the expression used in the documents issued by the European Economic Commission in Brussels.

Historically, the waste exchanges created as from 1972 in Western Europe resulted from the conjunction of several factors of unequal importance, but all of which contributed to the achievement of the same goal:

the concern for more economic management of the natural resources which appear to be nearing depletion;

the development of new re-use techniques;

the desire of some industrialists to diversify their procurement sources by making use of national potentialities:

the awareness of the commercial value of some wastes and of a supply and demand market in this sector.

1. *Savings of raw material and energy*. In most industrial sectors, France, and quite often all the European countries, are heavily dependent upon extra-European procurement sources as concerns both energy products and basic raw materials. Wiser use of some wastes or by-products connected with improved recovery should enable an appreciable savings in national currency through a reduction in imports. In the case of France, such a policy could lead to a savings estimated at 5000 million francs as of 1975, between 1980 and 1985.

2. *Enlarged range of reutilization possibilities*. Quantitative and qualitative knowledge of numerous, more or less complex wastes has induced multiple research work and development of new technologies.

A vast field of innovativeness has been opened, coinciding with a certain degree of rehabilitation of the recovered products.

Nevertheless, a major obstacle too often exists: the non-incentive nature for re-use which the specifications for State and municipal contracts can present. Thus products containing a certain percentage of recoverable materials do not find applicables due to specifications from a past era. In this case, the Administration is greatly responsible.

3. *Search for procurement sources within the national framework*. Before waste exchanges were created, the only examples of the utilization of wastes owed much to the empiricism and to the chance meetings among industrialists and, sometimes, to the intermediary role of a regularly supplied professional "recovery agent".

Almost all industrialists long preferred the security of fresh raw materials to which was added the low level and the surprising stability of the prices on the stock market.

In the history of economics, the beginning of the 1970s will always be considered the period of instability and, in general, that of increased raw material and energy prices. And this led a growing number of industrialists to question a system which gave too great a preference to fresh raw materials over re-usable wastes or by-products, without any real economic bases.

4. *Establishment of a dialogue between waste producers and re-users.* A structure for informing and placing the industrialists in contact with each other proved to be necessary.

Indeed, for the waste producer, when the wastes were disposed of under conditions which were harmless to the environment represented an additional cost, an expense. As for the waste re-user, in the case of complex products he found himself without any monetary reference since there were no recovery prices quoted on the market. This situation merely consisted of a "confrontation" between two partners.

It is tempting and often interesting to bring competition into play, be one in the position of the buyer or the seller. The industrialists were aware of the existence of a supply and demand market for all types of wastes and by-products. But it needed to be arranged by giving it a necessarily light structure.

Thus, in most Western European countries, waste exchanges were born as of 1972. The waste and by-product exchange for France was created by the magazine *Nuisances et Environnement* on the initiative of Mr. Deloy, at the end of 1974. Then, in June of 1975, the Chemical Waste Exchange appeared, using the columns on this same magazine and those of *Chimie Actualité* to place its advertisements.

The basic advantage of having chosen *Nuisances et Environnement* as a support is that this magazine is read in all branches of pollutant activities, since it is a multi-industrial, technical-economic magazine and not a trade magazine. Indeed, those products and wastes which cannot be re-used within the same industrial branch may be so in other sectors of economic activity.

In one year of publication, 186 advertisements were published: 152 offers of wastes (82 per cent) and 34 demands (18 per cent). Even though there was an average of three replies per advertisement, this figure proves to be somewhat misleading since, although there were no replies to some of the advertisements (especially those concerning chemical wastes of a highly complex composition), others received abundant mail, sometimes reaching ten or twelve letters.

Moreover, the waste exchanges do not pretend to find an immediate buyer or seller for all types of waste. On the other hand, they play the role of an indicator which can create a certain degree of dynamics and open a market for wastes the re-use of which had not yet been envisaged. On the community level, the re-use of the greatest possible amount of by-products provides a large savings on raw materials which are often imported, and therefore a lesser degree of dependency upon foreign markets and the market prices for the latter.

This problem of finding profitable uses for the by-products in the effluents is also posed in the pickling industries and the ore-etching industries. In these industries, acids are used to pickle ferrous metals or to etch ores having greater or lesser iron contents. The problem studied and solved by Mr. Lefort of the research centre of Société Pont-à-Mousson was to recover and put the products found in the industrial effluents to profitable use.

Indeed, in most cases these effluents are neutralized and oxidized, and the result is very important quantities of wastes which are bulky to a greater or lesser extent and thus difficult to handle.

The development of a waste-free technology consists in modifying this treatment in order to recover and recycle by-products.

When metals are pickled with sulphuric acid, the spent baths usually contain Fe^{++} ions and excess SO_4^{--} ions. At the end of the conventional lime neutralization operation, there is a mixture of SO_4Ca, $2H_2O$ and $Fe(OH)_3$, nH_2O which cannot be recovered. Thus, a process was developed to transform all the iron contained in the water into magnetite: this enabled its recovery (as an oxide) and its recycling in the iron and steel industry for example.

One of the ways of precipitating magnetite consists in proceeding as follows:

the solution to be treated in divided (batch and continuous operation) into two-thirds and one-third portions, the two-thirds of the volume are neutralized and oxidized so as to transform all the Fe^{+2} into Fe^{+3} in the form of ferric hydroxide;

both portions or fractions (two-thirds and one-third) are then mixed and neutralized and, after a while, a black magnetite precipitate begins to form.

Other, more sophisticated processes offer the advantage of enabling the production of magnetite in a single tank.

Tests are currently being carried out to develop the special magnetic separators required by the extreme fineness of the magnetite produced. These tests should lead to the production of instruments making it possible to recover the magnetite for less than 5 dollars/tonne of Fe_3O_4, including the amortization and the operating costs.

The basic use for the magnetite produced in this manner should be its recycling in the iron and steel industry, either in agglomeration, or in Kaldo type steel mills, or following oxidation the liquid baths are cooled through the addition of rich ores or scrap; these two categories of materials could be advantageously replaced by magnetite.

Such a market opening would be necessary to make profitable use of the hundreds of tonnes of magnetite which can be recovered per day through neutralization of the industrial effluents resulting from sulphuric acid etching of ores containing titanium.

Thus the profitable use of the by-products contained in the effluents of a given industry and their taking into account as raw materials by other branches of the industrial activity can be a way which leads to the creation of a waste-free industry. A particular spectacular case of this concerns the recycling of pollutants within the very manufacturing system which created them in the first place.

This is the solution which has been adopted in the ISOREL Plant in Casteljaloux. This plant produces hardwood fibreboard and insulating board (total of 60,000 tonnes per year), each type of board counting for one-half of the total production.

These boards are manufactured from waste wood through the wet process and in accordance with the following brief outline:

the waste wood is reduced to chips;

the chips are slightly cooked and then ground;

the resultant fibres are placed in suspension in water to form the pulp;

this pulp is refined;

then it is machine drip-dried to form the "mats" which, after complete drying and polymerization, become the boards.

The pollution consisted of the fibres lost in the water used for manufacture as well as the soluble wood components, such as the resins and sugars in particular which ran off with the water at the time of machine drip-drying of the mats. It was on the order of 110 kg of suspended matter and oxidizable matter per tonne of finished product, such that, if nothing had been done, the pollution from the plant would have been 22 tonnes per day, i.e. 145,000 inhabitant equivalents.

Ever since 1953 a certain number of treatment methods had been tried in the plant:

decantation,

filtration,

flocculation,

centrifugation,

electrostatic precipitation,

biological and physical/chemical purification.

These last methods gave less discouraging results, but reached prohibitive operating costs; thus ISOREL Company decided to imagine what would happen if, instead of placing the fibres in suspension in water to form the pulp, they were placed in suspension in another liquid which would be conserved. This led to the idea of recycling, and then to complete closing of the valves.

Now the fibres are thus transferred by a fluid which is quite different from what it used to be, since it is a water charged with 90 to 100 g of dry extract per litre.

All of the above took place little by little since it was necessary to transform practically the entire plant, all the habits and all the routines of the men, be they labourers, foremen or engineers.

Needless to say, the production tool was not designed for such a change, and it was therefore necessary to make often minute changes which required much time, analysis, trial and error, tinkering, etc. In particular, the board refining, drip-drying and pressing processes had to be modified.

The major constraint of recycling is obvious, but the respect of this constraint required an in-depth analysis of the role of the water in the production process and that of the seemingly most insignificant elements of the production installations. The goal to be reached was that of making sure that the quantity of water which entered the circuit was exactly equal to *that which disappeared* through evaporation during manufacture.

This goal was reached at the end of 1972; the pollution level which was 110 kg per tonne of finished board was reduced to 0.8 kg per tonne, which represents a 99.3 per cent efficiency. Since then, this level has been brought down to 0 per cent.

This new process has definite advantages as concerns:

the capital outlay: relatively low in comparison to that for a conventional purification station—2,500,000 FF against 6,000,000 at the least;

the efficiency: particularly high: 100 per cent and no mud to be eliminated as is the case with a purification station;

the operating expenses: negligible—the equivalent of 25 to 30 kW/h per tonne of finished board;

the savings of raw materials: since this process does away with all material losses and leakage, practically 1 tonne of dry board is produced per tonne of dry wood;

the technical flexibility and the savings: this process makes it possible to incorporate all the necessary additives into the pulp, regardless of their toxicity with respect to the natural surroundings, since there is no loss; for example, to produce 12 per cent asphalt insulating board, the exact equivalent of the 12 per cent of asphalt retained are used.

Problems have also been solved as far as the commercial quality and the technical qualities of the boards are concerned.

Finally, the last paper concerns the STX Process.

Here again, it was a matter of developing a new production technique for the dyeing industry to solve the ever-increasing problems of procurement and pollution of the waters.

Even though, in fact, the treatments for 1 tonne of textile requires an average of 400 to 600 tonnes of water, they also produce a great deal of pollution which, unfortunately, ranks the textile industry among those which are the most pollutant.

In order to solve these two problems, numerous research workers tried to replace the water by organic solvents.

Attempting to replace an element as fundamental as water in the dyeing processes is no easy matter, not only for reasons of strict routine, but especially since all the know-how and improvements accumulated to date are based on this unalterable parameter. The following points must be taken into consideration in the search for a solvent-medium process:

1. It must really solve the procurement and water-pollution problems. Moreover, it must be capable of being implemented in accordance with the work safety and hygiene standards.

2. Its principle must be as simple as possible, in particular making allowance for the use of a homogeneous dye bath.

3. The quality of the dyed items must at least equal that obtained with water.

4. It must, at least at first, be capable of being implemented with the fibres and dyes currently available on the market.

5. Finally, its operation cost must at least be of the same order of magnitude as that of conventional water techniques. In particular, this presupposes simple and economic recycling of the dye baths and maximum recovery of the solvents used.

The research undertaken by Société STX gave the desired results with polyamide fibres. The principle behind the process is that of replacing the water by two solvents:

methanol, which acts as an ionizing medium, expands the fibres and renders the dyes soluble;

perchlorethylene which plays the role of a diluent having very interesting physical and chemical properties (inflammability, low toxicity, low specific heat and latent vaporization heat values).

The dyes are soluble in the perchlorethylene/methanol solution, but insoluble in perchlorethylene alone. During the dyeing process, when the reactional balance is reached, the methanol is eliminated from the system. In this way, the dyes become insoluble in the medium and stick to the fibres, which enables re-use of the solvents. Depending on the allowable rate of impurities in the dyeing medium, a purge value is set (approximately 15 per cent). The solvents are then regenerated through simple whitening and aeolotropic distillation. The concentrated wastes obtained (oils and oligomers) are post-treated by steam stripping to reduce the residual perchloerethylene content and facilitate their destruction by mere combustion. Studies are under way to treat the solid wastes (filtration additives) in order to re-use them.

The thermal and chemical pollutants of the water are almost entirely done away with: for a shop with a daily treatment capacity of 6 tonnes of carpet: DOC $<$ 10 equiv./h/day with the STX Process against 10,000 to 15,000 equiv./h/day with conventional dyeing.

The perchlorethylene losses vary from 2 to 5 per cent in relation to the treated textile, depending on the operating conditions (equipment maintenance, care in handling the solvents, etc.), which emphasizes the need to demand that the staff acquire not only new technical skills, but also a new way of working.

Even though the process gives satisfaction from the technical standpoint, there nevertheless remains an important point for its industrial application and that is its economic aspect and its profitability.

As far as the capital outlay is concerned, the cost of the equipment assembled and ready to operate is slightly above that of conventional installations.

However, the operating costs (Table 1) are definitely below those of the conventional processes, especially as concerns three items:

the consumption of auxiliary products which is nil,

the lack of the need to treat the water,

the energy production factor

TABLE 1 *Comparative dyeing costs*

	Barques (FF/kg)	STX process (FF/kg)
Dyes	0.3067	0.3067
ADP*	0.2628	-
Steam	0.57	0.21
Electricity	0.2410	0.17
Water	0.225	-
Solvent	-	0.051
Total	1.6055	0.7377

*Auxiliary dyeing products

Energy savings per tonne of treated carpet

	STX		Barque	
	Measured values	%	Measured values	%
Consumption of electricity (in kWh)				
Materials handling	70	12		
Pre-drying, drying	350	50	780	78
Dyeing	250	36	220	22
Regeneration	15	2		
	685	100	1000	100
Consumption of steam (tonnes)				
Materials handling	0.9	13.2		
Pre-drying, drying	0.3	4.4	4	21
Dyeing	2.8	41.2	15	79
Regeneration	2.8	41.2		
	6.8	100	19	100

Before giving the floor to the authors of the various papers so they can answer the questions you would like to ask, I would like to draw the following conclusions:

First of all, the implementation of clean techniques, even though it sometimes makes use of simple solutions resulting from good common sense, most often requires in-depth research. All the industrial examples cited here were only possible after intensive laboratory, pilot-installation and factory tests. ISOREL spent 4 years of testing in order to convert its Casteljaloux plant to the closed-circuit process, and at least as much time was spent in developing the STX process or that of recovering the starch proteins. This demonstrates the importance of the technological factor in changing the behaviour of the industry with respect to the resources and the natural surroundings.

The introduction of constraints due to the environment into industry can and must therefore have an effect which does not weaken the environment but rather induces harmonious development. The ills created by current technology can only be done away with through the progress of this very same technology.

Next, the implementation of a waste-free technology can only be achieved with the active participation of the entire staff. For we must realize that the development of a waste-free production technology does not suffice to ensure its success. Once the decision has been made to change the production process, everyone in the plant must be involved in its application. Indeed, it results in changes in the production mechanisms and thus in the habits of the staff members.

Finally, the particular cases which have been discussed demonstrate that the waste-free production technologies have sound technical and economic arguments.

First of all, from the technical point of view, since the pollution-control system being integrated into the manufacturing process, it automatically absorbs the variations, be they temporary or lengthy changes in the production capacity of the plant. Thus reliability and consistency. This fact is brought out in the papers received from Mr. Maurent (STX process) and Mr. Maréchal (integral recycling in a board manufacturing plant).

From the economic point of view, the consequences of the clean technologies are usually interesting for the company. If we do not consider for a moment the reduction or elimination of the pollution, the implementation of a clean process is often profitable in itself due to the gain in production capacity it brings about. Moreover, if we add to this the savings made in the payments of rights, taxes, penalties and other expenses which "polluting industrialists" in the various countries are required to pay because of their pollution, very few economic and financial balances for clean technologies are negative.

And, most of all, since pollution must, in any case, be fought, it is just to compare the cost of pollution prevention to that of the capital layout and the operating costs which would be incurred by an external installation of identical efficiency. You can therefore realize more clearly for yourselves that a clean technology is of a beneficial nature.

Introductory Report

László Markó (Rapporteur)

University of Chemical Engineering, Veszprém, Hungary

This introductory report summarizes the following seven papers submitted to the seminar:

L. Markó: "Synthesis with high stereo-selectivity as the basis of low-waste organic technologies",

D. Kalló: "The use of natural zeolites in the chemical industry",

V. Cziglina, L.L. Dzsida, Z. Meleg and A. Takáts: "The utilization of brown coals outside of energy fields",

S. Märkki: "Outokumpu flash smelting method",

B. Westerholm: "Methods of conserving raw material and energy and protecting the environment in chemical and electro-chemical plating plants",

J.T. Ling: "Developing conservation-oriented technology for industrial pollution control",

K.E. Kulander, L.-G. Lindfors, E. Lohrdén: "The Nordic organization for waste exchange".

There are practically no common principal features in the papers under consideration except the obvious intention of presenting some methods and examples for reducing environmental pollution. It has therefore been regarded as most appropriate to summarize their content briefly, emphasizing non-waste cases instead those of waste treatment.

Three of the papers deal with more general aspects of the problem: Kulander, Lindfors and Lohrdén describe the Nordic Organization for Waste Exchange (NOWE) which has functioned since November 1973 and aims to furnish information about the supply of, and demand for, waste or residual products, but is not concerned with well-known wastes like paper, scrap metal, etc. In the first two years 270 items from different companies were advertised and this resulted in 147 mediated contacts, 27 per cent of which were positive. The best results were registered with inorganic chemicals (45 per cent positive) but no results were achieved with acids.

Ling stresses the importance of developing methods (technologies) that prevent the formation of pollutants or utilize them, instead of removing already existing pollutants. The equation

Pollutant + Know-how = Resource

symbolizes the ideal solution of pollution problems and this is the goal of the so-called 3P-programme of the company (Pollution prevention pays) which has already resulted in significant achievements in diminishing air pollution and waste water (a few details are given).

I have stressed the importance of organic synthetic routes with high selectivity in the diminishing of low-value by-products causing disposal and waste problems. The use of asymmetric homogeneous catalysts enables the solution of the as yet unsolved optical selectivity problem. Using such catalysts, optically active organic compounds (many biologically useful compounds like drugs belong to this class) may be synthetized with selectivities of above 90 per cent instead of below 50 per cent reached until now.

Three other papers deal with more or less special cases and describe technologies producing little waste.

Cziglina, Dzsida, Meleg and Takáts treat the multi-purpose utilization of brown coals instead of their simple use for energy production. The method proposed consists of extraction with alkaline hydroxide solutions which furnish a leached coal containing less sulphur and having a higher calorific value (thus giving rise to less pollution) and an alkaline solution of humic acids which may be used as such or from which the humic acid may be recovered by acidification. Both the solution and the free acid are useful in agriculture either as fertilizing or stimulating agents. Furthermore, the by-products of this extraction process may be used together with almost worthless coal dust for recultivation of land spoiled by mining operations.

Märkki deals with a flash smelting process of copper ore concentrates that has two important characteristics, a low need of external energy (only 20-25 per cent of the total energy demand has to be covered from outside, the rest is furnished by the oxidation of the iron and sulphur content of the raw material) and a high SO_2 content of the off-gases which can thus be processed with standard methods and units to sulphuric acid or elementary sulphur. The process has been in operation since 1949.

Westerholm reports on problems of saving raw materials and energy and, at the same time, reducing pollution in chemical and electro-chemical plating plants. Three important measures have been taken to achieve the above-mentioned goals: organic solvent vapours are recovered by active carbon filtering (efficiency about 95 per cent), process water is saved by using multi-stage rinsing and counter-flow (saving is 90-99.8 per cent) and waste rinsing water is treated with ion exchangers to remove and recuperate detrimental or valuable cations (and anions). The efficiency of the last operation highly depends on the concentration of waste waters which in turn is determined by the rinsing technology.

The last paper in this group reviews the properties and uses of natural and synthetic zeolites. Kalló emphasizes in this connection those applications of natural zeolites in the chemical industry which in one way or the other have some effect on environmental problems. Three aspects are mentioned as most important in this connection: use of zeolites as molecular sieves in the production of linear bio-degradable detergents, selective adsorption of SO_2 or nitrogen oxides from flue gases and the application of zeolites as ion exchangers to remove monovalent cations, for example radioactive ^{137}Cs from waste waters.

Introductory Report

M.F. Torocheshnikov (Rapporteur)

Mendeleev Institute of Chemical Technology, Moscow, USSR

The steady growth of the world's population has been accompanied by increasing industrialization in most countries and, inevitably, by a sharp increase in the consumption of the world's mineral resources, namely, those found in the lithosphere (minerals), the hydrosphere (water and the salt it contains), and the atmosphere (hydrogen, oxygen and rare gases). At the present time over 100 billion tons of various materials are extracted each year from the earth. Industry uses not only natural raw materials but also a large amount of secondary materials, such as industrial wastes and, as they are called in the United States, "urban minerals" (urban waste).

Virtually all the elements listed in Mendeleev's periodic table are used in industry at the present time, as may be seen from the figures given in Table 1.

TABLE 1

Industrial use of elements

Year	1869	1906	1917	1937	1970
Known elements	62	84	85	89	104
Elements used	35	52	64	73	87

Experts have on many occasions expressed concern at the use of mineral resources on a massive scale which might well lead to the rapid depletion of sources of certain raw materials, such as phosphorites and apatites, mercury and lead ores, etc. The enormous expansion of industrial production, which contributes to the development of agriculture, transport and the urban economy, has also considerably increased the volume of emissions—very often harmful—into the biosphere. These emissions affect the natural environment as well as mankind. The pollution of the biosphere has become a global problem calling for immediate solution, particularly by the development of non-waste technology that does away with emissions of harmful substances into the environment. The development of non-waste emission-free production methods is implicit in the rational utilization of raw materials and energy in the light of their cycles in nature.[1]

This new technology must be developed in the context of a thorough understanding of natural laws and the relationships that exist between industrial cycles and natural processes. Attention was drawn to the need for this kind of approach to industrial activity by V.I. Lenin, who stated that: "The laws governing the external world of nature, which may be sub-divided into mechanical and chemical (this distinction is extremely important), constitute the basis for man's technological activity. In his practical work, man faces an objective world on which he depends and which determines the nature of his activities."[2]

The Final Act of the Conference on Security and Co-operation in Europe, dated 1 August 1975, drew attention to the relationship between the productive activities of mankind and environmental protection: ". . . the success of any environmental policy presupposes that economic development and technological progress must be compatible with the protection of the environment and the preservation of historical and cultural values; that damage to the environment is best avoided by preventive measures; and that the ecological balance must be preserved in the exploitation and management of natural resources."

Considerable attention is paid to environmental protection matters in the USSR. Quite recently the Supreme Soviet of the USSR adopted important decisions on this problem, namely, "Measures for further improving the protection of nature and for the rational use of natural resources" (1972) and "Measures for the greater conservation of mineral resources and the better utilization of useful minerals" (1975).

In accordance with current USSR legislation on the protection of nature, various branches of industry, and particularly the chemical-based industries, have taken far-reaching measures to improve production methods and to develop processes in which non-waste and emission-free technology is used. Generally speaking, these measures in the case of the chemical and iron and steel industries are aimed at the development of:

(a) integrated production flow diagrams in which all the various components of raw materials are fully utilized; state plans for the development of the national economy now contain targets for the improved utilization of useful minerals;

(b) production flow diagrams with complete water recycling in order to reduce industrial demand for fresh water (chemical and metallurgical plants use 30 per cent of all water supplied to industry);

(c) energy/technological flow diagrams in which process heat is used, so that production of certain substances such as ammonia becomes energy-producing instead of energy-consuming.

Specific measures adopted in industry in the USSR to develop low-waste emission free technological processes are described below.

BASIC CHEMICAL INDUSTRY

Steps have been taken at a number of plants producing acids to recover substances from waste-gases, such as sulphur dioxide, nitrate oxides and fluoride compounds. Sulfite-bisulfite ammonia solutions are used to absorb sulphur dioxide. A method for the recovery of this gas using adsorbents is being developed.[3] Selenium, tellurium and rhenium are being recovered at a number of plants from solid wastes in the course of the sulphuric acid production process. Plants producing nitric acid are taking steps to eliminate nitric oxide emissions by the use of the catalytic method. Research on the catalytic oxidation of nitric oxide is under way. The introduction of this process will virtually put an end to nitric oxide emissions and increase nitric acid production.[4] A considerable amount of work is being done to improve methods of using fluorine compounds.

The trend in the production of ammonia, which is the most important chemical product of all, is towards the construction of very large units with a daily output of 1360 tons and more. The operation of large units facilitates the solution of a number of important problems, such as heat utilization, and considerably reduces energy consumption (from 1250 to 1600 kWh to 100-50 kWh per ton of NH_3) as well as consumption of process water (from 500 to 100-50 m^3), this result being achieved by the use of air cooling.[5] The history of the development of ammonia production confirms the dialectical law concerning the transition from quantity to quality, in that gradual quantitative changes in ammonia production (greater daily output) have brought about considerable qualitative changes in ammonia production technology.

Of considerable importance in the creation of large power and technological units is the approach adopted to their design, which makes it possible to plan units that operate under optimal conditions. Some interesting work on the development of a methodology governing such design work has been done by the Mendeleev Chemical and Technological Institute in Moscow on the basis of the exergetic principle of thermodynamic analysis (the efficiency of a system is evaluated by means of a thermodynamic efficiency factor that depends on intensity coefficients and exergetic losses by elements of the system). Technico-economic analysis, as well as thermodynamic analysis methods, are used in the solution of problems relating to the structure of the flow diagram.[6] In 1972 various units with an output of 450,000 tons of ammonium nitrate per year were commissioned in the USSR. Thanks to the new technology used, these units produce no effluents polluted by ammonium nitrate, and their emissions of ammonium nitrate dust have been reduced by the installation of devices for cleaning the air from the granulating towers and evaporators.[5] The chemical fertilizer industry has also adopted a number of measures to improve production processes. The Almalyk chemical plant, for example, which produces ammophos, has introduced a closed-circuit recycling system for the re-use of effluents which have been rendered harmless and treated and contain a maximum of 20 mg/l of suspended matter.

Research is being carried out with a view to eliminating the compound fertilizer drying process (which generates a great deal of dust) and obtaining the finished product directly at the granulation stage.[5] Comprehensive research is being done on the complete utilization of all the components of the apatite-nephelite ores of the Kola peninsula, phosphogypsum and ash.[5] Methods have been developed to reduce emissions during the production of chromic salts[5] and barium salts.[7]

THE ORGANIC CHEMICAL INDUSTRY

A number of branches of the organic chemical industry, and specifically the varnish and paint industry, have taken steps to reduce the amount of effluents connected with the production of many resins and pigments, to develop a range of water-soluble and water-emulsion paints not requiring the addition of solvents, to find substitutes for raw materials as, for example, in the production of bakelite varnishes (urotropin for formaldehyde) so as to eliminate wastes, etc. Considerable headway in eliminating harmful emissions has been made in the chemical fibre industry.[5]

One of the important achievements of the organic chemical industry is the design and construction of a large chemical combine incorporating an integrated water supply and drainage system which is connected to the water system of an adjacent town ([5] and [8]). Effluents which cannot be piped into the local process water recycling system are discharged, depending on their content, into the following drainage systems for effluents:

(a) polluted by organic compounds,

(b) containing up to 3 g/l mineral compounds, and

(c) containing up to 5 g/l mineral compounds.

These effluents are piped to the appropriate mechanical treatment plant, to a biochemical treatment plant, to the second-stage treatment plant where activated microporous anthracite and ion-exchange resins are used, and then to the deminbalization plant. Wastes from the treatment plants are made into fertilizer containing ammonium nitrate (in the ion-exchange resin regeneration process), organic mineral fertilizers, etc. Approximately 96 per cent of the water can be recycled when this system is used. Fresh water consumption at the combine is about seven times less than that at plants with a similar output using the old technology.

THE COKE-OVEN INDUSTRY

One of the unpleasant phases of coke production is connected with the disposal of effluents containing phenols. Approximately 0.35 m^3 of such effluent is produced per ton of coke. In a number of cases the effluent is used to quench the coke and it is at this time that harmful contaminants are released into atmosphere. Systems in which phenols can be removed biochemically in a mixture with process water in the cooling cycle have been operated successfully during the past 4 years at four coke-oven plants in the USSR.[9] Use of this method has prevented scale formation in the heat-exchanger, reduced emissions of harmful substances into the atmosphere, put an end to the discharge of theoretically pure water beyond the limits of the plant, and reduced the consumption of scarce process water, particularly where the dry quenching method is used.

THE IRON AND STEEL INDUSTRY

The iron and steel industry has reorganized its water supply and discharge system, reduced the volume of effluents, and taken steps to use slag for the production of certain building materials.

Advanced research is being conducted on the elimination and utilization of harmful components, such as carbon monoxide and sulphur dioxide, in the waste gases of iron and steel plants, and particularly sintering plants. A process has been developed for the elimination of carbon monoxide from the waste gases of sintering plants by means of a catalyser device without preliminary scrubbing and heating.[10]

Experimental tests to eliminate sulphur dioxide from the waste gases of sintering plants in Venturi tubes with a daily capacity of 160,000 m^3 by the use of a limestone slurry have yielded encouraging results.[11]

A large plant for the direct reduction of ferric oxides by the use of reducing gases is to be built in the USSR. This plant will produce virtually no emissions.

NON-FERROUS METALLURGY

During the past few years non-ferrous metallurgical plants have taken steps to modify the technology they use so as to reduce emissions.[12] As a result, they now produce sixty metals and their compounds in addition to their basic metal production. This has been achieved in many cases by the use of sorption and extraction processes for the recovery of compounds of associated metals from the liquid slurry of flotation plants.

Hydrometallurgical processes are being used on a larger scale at certain plants, particularly for the leaching of copper ores. A considerable proportion of the wastes of concentration plants is used as building material (gravel, rubble and sand) or as stowing material in mining. The non-ferrous metallurgical industry uses large amounts of water, particularly for the processing of low-concentration ores. As a result of the measures taken to reorganize their water supply and discharge systems, 120 non-ferrous metals plants at present recycle their process water, i.e. 97 per cent of their total water requirements, and sixty plants have completely done away with effluent discharges into natural and artificial bodies of water.

The efficiency of dust-removal devices has been improved and steps taken to process dust in order to reduce the volume of emissions produced by non-ferrous metallurgical plants. In order to prevent the creation of dust by tailings heaps their surfaces are consolidated with bituminous-schist emulsions or with vegetative cover consisting of perennial grasses and bushes.

THE POWER INDUSTRY

Power stations, like chemical and iron and steel plants, are major consumers of mineral resources, i.e. fuel and water. The amount of water used by electric power stations is roughly equivalent to one-half of the total amount of water used by industry. Moreover, owing to their enormous fuel consumption, electric power plants pose a very large number of environmental problems. Suffice it to say that the bulk of the sulphur dioxide released into the atmosphere is attributable to power-station operation.[13]

A considerable amount of research has been done in the USSR on the removal of sulphur dioxide from power-station waste gases by means of solid and liquid absorbants, and the methods that have been developed are to be introduced in practice.

Our Seminar, attended by experts from a large number of countries, will certainly contribute a great deal to the solution of this noble task which is of importance to all mankind, namely, the design of industrial plants producing only useful products. It would be well, in our view, to consider the following problems during the course of the discussion:

development of the basic thermodynamic principles of processes for the recovery of impurities (trace) from multi-component systems, as such principles are necessary for the elaboration of non-waste technological processes;

development of highly efficient equipment for separation processes (gaseous, liquid and solid wastes);

solution of the problem of using soda industry wastes;

the complete utilization of emissions of acid gases (sulphur dioxide, nitric oxide, fluorine gases, etc.), alkaline (ammonia) and neutral gases (nitrogen emissions from oxygen-producing plants);

utilization of the solid wastes of the basic chemical industry, such as phosphogypsum, ash, halite, etc;

development of methods for the recovery of mercury, cadmium, lead and other heavy metals;

development of processes for the production of biodegradable detergents;

development of measures to prevent phosphorous and nitrogen salts from entering bodies of water (prevention of eutrophication);

solution of the problem of compounding pesticides which decompose easily under natural conditions (photochemical and biological);

organization of advanced technical and economic research work on the problem of using substances present in waste gases and low-concentration effluents;

development of processes for the utilization of emissions of organic substances, solvents, etc;

development of basic thermodynamic principles of processes for the utilization of the waste heat of industrial enterprises, and particularly low-potential heat.

REFERENCES

1. *Biosfera* (Biosphere), publ. "Mir", Moscow, 1972.
2. V.I. Lenin, *Complete Works*, Vol. 29, p. 169.
3. S.A. Anurov, V.I. Smola, V.A. Zinkovsky, L.G. Guseletov, N.V. Keltsev and N.S. Torocheshnikov, *Khim. Prom.* (Chemical Industry) No. 6, 445 (1974);

   S.A. Anurov, N.V. Keltsev, V.I. Smola and N.S. Torocheshnikov, *J. Phys. Chem.* 48, No. 8, 2124 (1974).
4. V.T. Leonov, V.F. Marchenkov and N.C. Torocheshnikov, *J. Phys. Chem.*, 48, No. 1, 239 (1974); *Khim. Prom.* (Chemical Industry) No. 2, 41 (1974).
5. V.L. Bakula, Major activities of Ministry of Chemical Industry towards orientation of non-waste technological processes.*

6. V.V. Kafarov and V.L. Perov. Energetic and technological aspects of the optimum designing of chemical processes.*

7. P.G. Akhmetov, *Khimiya i Tekhnologiya Bariya* (Barium chemistry and technology), publ. Khimiya, Moscow 1974.

8. V. N. Yevstratov and M. I. Kievsky, this volume, p. 301.

9. V.Y. Privalov, G.I. Pankov, B.P. Sukhomlinov and V.S. Vinarsky. Creation of non-effluent water supplying systems for coke and chemical plants.*

10. A.I. Reznik, Author's certificate No. 434970.

11. B.G. Kazarov, V.A. Pinaev, L.A. Dekhtyareva and V.I. Budanov, *Khim. Prom.* (Chemical Industry) No. 10, 763 (1971).

12. Y.N. Sriadoch. Rational water use and creation of non-waste technology of non-ferrous ore concentration.*

13. *Cleaning Our Environment. The Chemical Basis for Action.* Report Amer. Chem. Soc., W. 1969.

14. J.A. Cook, *Survey of Modern Industrial Chemistry.* Ann Arbor Science Publ. Inc., Mich., 1975.

*Only available in Russian. The text of the paper may be obtained from the ECE Secretariat, Palais de Nations, C. 1211, Geneva 10, Switzerland.

Protein Recovery from Liquid Potato Wastes

M. Huchette

Roquette Frères Company, Lestrem, France

As with a great many industries of the same type, potato starch plants have from the very beginning been involved with problems of liquid-waste disposal. As early as the 1950s, several solutions were already under investigation, ranging from treatment with chlorine to bacterial beds hastily built with blast furnace slag heaps, and which were clogged up 48 hours after going into operation. Then there was the more or less rewarding attempt to spread wastes in fields in a region with clayey soil, which quickly became saturated, or attempts at evaporation, creating ponds, and even flocculation. All this out of concern to avoid any inconvenience to such downstream water users as laundries and breweries faced with using water already polluted by an upstream factory.

How many samples were taken, and how many analyses were carried out along the whole length of this river Lys! And all the worries from fishing firms after us from sunrise to sundown, betrayed as we were by a tenacious, dogged foam which would have been the pride of the best detergents of today.

Alas! Before cleaning up these wastes, it was going to be necessary to clean out the pocket book. And with a last-ditch effort, with a genuine concern to resolve one day our famous problem, the exodus began. Management made the decision to relocate the plant to a region with better prospects for natural liquid-waste treatment. Geological studies of the region undertaken with the University at Lille together with agreements with farmers for field spreading led to a site selection. Pipelines by the tens of kilometers were laid in order to spread these liquid wastes which did not want to lend itself to any physical or chemical treatment.

For the firm of Roquette Frères, under duress and coercion, this was then the founding of an enterprise called "Grandes Féculeries de Picardie", located at Vecquemont, near Amiens, on the Somme river, a river which was one of the most abundant in fish. What daring and foolhardiness !

At the time, no one thought of anything except field spreading—which had been the most rational solution to the problem of liquid wastes, due to their fertilizing qualities—and also because there had been so much effort and know how applied to field studies and to the distribution of quantities over each land parcel. No one thought that this solution could be anything but final.

It became necessary, however, to cope with ever-increasing ecological requirements and successive increases in plant capacity.

To reduce the pollution, it therefore became necessary to find other approaches which would be compatible with the survival of the industry.

The only other approach was to be found in not destroying the effluents, but rather in extracting out of the effluents everything they would be able to give up.

In recent years, all research efforts have converged in working along this line.

First of all, let us review from top to bottom traditional systems used in potato starch plants for the purpose of eliminating to the extent possible any addition of unnecessary water which would be likely to increase the flow of liquid wastes, and thereby make treatment more difficult.

In order to make any progress, it was necessary to give evidence of daring, and above all, to commit enormous amounts of investment.

It is then necessary to examine in what manner protein could be isolated, protein being the main constituent which pollutes in the "red waters", as they are called in the trade.

This protein is inevitably present in a natural state within the treated raw material, and it is therefore necessary to stop and examine it in order to gain an understanding of the problem which liquid-waste disposal represents for a potato starch plant.

There is no doubt that this raw material is well understood, and professionals in the field up to now have done their upmost to extract it under the best conditions, this being the main and the highest in quality of the constituents, the starch which in France is called fecula.

To extract this starch, all of the steps employed are physical ones, with the result that the potato starch plant can only collect that which it is mechanically capable of retaining, restoring to nature that which nature would agree laboriously, or even impossibly, to take back.

In this recovery of waste products, those in the field have felt, and still do so feel, that it is not an equal contest.

When considering plants with a daily capacity of 3000 to 4000 tons of potatoes, this is then several thousand cubic metres of liquid wastes which must be treated in order to solve the pollution problem, which is, as can be well understood, an expensive operation, and especially so if it is completely non-productive.

The recovery of only a part of the proteins would represent a first step in the reduction of liquid wastes, but it is essential that research presently underway in numerous fields of technology involving manufacturing, characteristics and the applications for products make it possible to contemplate total recovery in the future.

Let us return once again to the raw material.

Potato starch, when cleaned, essentially consists of 77 per cent water, and therefore 23 per cent dry matter composed firstly of insoluble matter in the form of starch (16-17 per cent) and cellulose (1.5 per cent). There is as well soluble matter (about 4.5 per cent) in the form of the following:

- nitrogen compounds (approx. 2 per cent);
- mineral elements, as a residue from calcination (1 per cent);
- sugars (saccharose, glucose, fructose);

organic acids (citric acid, oxalic acid);

lipides (and especially linoleic acid);

and some other minor constituents such as phenol compounds, tyrosine, hihydroxyphenylalanine, caffeic acid, chlorogenic acid, which, under the action of enzymes (tyrosinase, phenol oxidase or catalase), redden upon contact with air.

To be sure, industrial yields do not vary greatly as a function of seasons, regions, or varieties, and they stay rather close to the following percentage figures:

20 per cent potato starch—hydrated to 20 per cent—more than 90 per cent of the starch is therefore recovered in the marketable form;

2.5 per cent pulp, consisting of 30 per cent starch, 30 per cent hemicellulose, 15 per cent non-digestible cellulose, 2.5 per cent inorganic matter, 5 per cent nitrogen containing substances, and 12 per cent water.

Schematically, it could be said that the tubers are reservoir of starch steeped in water destined to become liquid waste, since quantitatively, the results show, for 100 kg of potato:

20 kg of potato starch;

2.5 kg of pulp;

77.5 kg of "red water"

It is obviously this 77.5 kg of "red water", as it is called in the trade, which up to now has made up the liquid waste more or less extensively diluted.

This liquid waste entails a BOD of some 30,000 and a COD of close to 50,000, and therefore the COD:BOD ratio is about 1.7.

After the failure of industrial attempts at flocculation a few years ago, due in part at the time to inadequate or non-existent means for the separation of sludge, attempts at heat coagulation were given one last chance.

Potato protein, also called albumin by those in the field, contains according to the literature 60 - 70 per cent globulin and 20 - 40 per cent glutelin, which is entirely soluble and which in fact partly coagulates with heat.

It is this costly, but also productive, approach which was selected. The firm of Roquette Frères registered a patent on 4 January 1974 under patent application number 7 400 310 which describes the process in detail.

After cleaning of all that which could be carried in from the ground, such as pebbles, earth, straw, and residues, and after abundant washing, the tubers are ground in graters using traditional methods, which results in as fine a paste as possible, without, however, the cellulose being micronized in the process.

It is necessary to think protein right at this grating stage:

First of all, quality, by protecting the waste water against any oxidation through the use of reducing agents such as SO_2 or the bisulfite.

Then quantity, by recovering the maximum amount of liquid waste at the least cost.

To this end, separation by centrifuge is undertaken in the first stage (see protein facility flow sheet below) which recovers 70 to 90 per cent of the liquid waste according to the stage and the resources employed, while the fecula and the pulp are elsewhere subjected to various refining operations, whereby they carry off 10 to 30 per cent of this liquid waste water.

This water—still protected—and which is in the form of a beautifully yellow-colored liquor, especially when compared to raw liquid waste, has approximately the following composition:

A dry extract: 50-60 g/liter, composed of:

Proteins (N 6.25) : 48 to 52 per cent;

Inorganic matter : 15 per cent.

For a residue on calcination of about 20

Organic acids : 22 per cent;

Sugars : 13 - 14 per cent.

It is then acidified to a pH of about 5, close to the protein isoelectric point, and then brought up to 100-105°C in the pipes.

A flocculation and coagulation are thus obtained whose quality depends on a multitude of factors, which has a bearing, of course, on the subsequent separation of the flocculate, whose density closely approaches that of the medium.

To separate this flocculate, enormous power, very high speeds, and huge machines will be necessary which will recover on the one hand a sludge which can be dried pneumatically or on rollers, then ground, mixed, and packaged, which represents mostly 55 per cent of the total protein, but only a third of the soluble dry extract; and on the other hand, a liquid, an overflow, which has the following composition:

Dry extract : 36 g/liter

composed of :

Proteins (N 6.25) : about 35 per cent;

Inorganic matter : 20 per cent

for a residue on calcination of 25;

Organic acids : 25 per cent;

Sugars : 18 per cent.

and which can be concentrated to 500 g/liter, then dried in turn, either by spraying or on rollers. (Fig. 1).

In its final state, the dehydrated flocculate is in the form of a cream powder which is very fine (all particles less than 60 microns), giving off the characteristic smell of potato puree, with a neutral taste, easy to put and keep in suspension in water, and which contains the constituents in the following analysis:

8 per cent water;

78 per cent protein N 6.25 (85 per cent of dry);

2.5 per cent fats, essentially consisting of:

60 per cent linoleic acid;

27 per cent palmitic acid;

7 per cent steatic acid;

2.5 per cent of inorganic matter, essentially consisting of:

phosphorus : 1.3 per cent as P_2O_5;

potassium : 1 per cent as K_2O;

calcium : 0.35 per cent as CaO;

magnesium : 0.08 per cent as MgO.

The amines shown in Table 1.

TABLE 1
(grams of amino acids per 16 g of total nitrogen)

Aspartic acid	13.4	Methionine	1.83
Threonine	5.4	Isoleucine	5.47
Serine	4.79	Leucine	9.53
Glutamic acid	12.03	Tyrosine	5.10
Proline	4.82	Phenylalanine	6.06
Glycine	4.60	Lysine	7.37
Alanine	4.54	Histidine	2.13
Valine	6.70	Arginine	4.60
Cystine	1.54		

Of particular note is the high lysine content which is comparable to that found in animal proteins, whereby the level of available lysine is greater than 95 per cent.

As regards the soluble matter in the concentrated state, they are in the form of a very viscous syrup, colored, and with organoleptic properties approaching those of the hydrolysates.

In the dry form, the powder remains very hygroscopic, the flake somewhat less so when the product has been dried on rollers.

These organoleptic properties, and this hygroscopicity, are to be explained by the analysis seen in Table 2.

TABLE 2

Average analysis of potato solubles

Total nitrogen	5.5 per cent
Amine nitrogen	1.5 per cent
Reducing power (expressed as glucose)	11 per cent
Saccharose	7 per cent
Residue on calcination	26 per cent
Phosphorus	1 per cent
Sodium	0.6 per cent
Potassium	10.5 per cent
Magnesium	0.6 per cent
Calcium	750 mg/kg
Iron	180 mg/kg
Copper	22 mg/kg
Zinc	70 mg/kg

This is marked by a high content of inorganic matter—mostly potassium in the form of phosphates, chlorides, and sulfates, and by sugars, saccharose, glucose, and fructose, and, of course, proteins (32 per cent).

This is a protein which is completely imbalanced since more than 50 per cent of the nitrogen enters into the composition of the aspartic acid and glutamic acid, as noted in Table 3.

TABLE 3

Amine composition of potato solubles
(grams of amino acids per 16 g of nitrogen)

Aspartic acid	33.48	Methionine	0.81
Threonine	1.70	Isoleucine	1.44
Serine	1.73	Leucine	0.97
Glutamic acid	22.28	Thyrosine	1.07
Proline	1.44	Phenylalanine	1.44
Glycine	1.27	Lysine	2.30
Alanine	1.51	Histidine	1.22
Valine	2.48	Arginine	5.19
Cystine	0.60		

+ y-amino butyric acid.

The examination of all these tables and all of these results will no doubt lead to an initial conclusion about the treatment yield.

Protein has been recovered in at least 50 per cent of its coagulated form, which represents a third of the organic matter responsible for the COB, figures that are to be regularly found in the overflow after coagulation.

In other words, this treatment ensures the elimination of at least 10 kg of COD per ton of potato.

In order to achieve perfect results, it would of course be necessary to subject solubles escaping flocculation to other treatments leading to beneficial re-uses.

A concentration from 3.75 to 500 g/liter is costly, and the resulting syrups have not found any other outlets than in a mixture with the pulp.

Dehydration is even more costly, and it is above all difficult.

The field spreading currently practised runs the risk of being brought into question. Investigations must therefore be pursued. To do this, the most modern methods must be employed, making use of ultrafiltration, reverse osmosis, ion exchange resins, or various microbiological processes.

It is to be recalled that these waters are essentially formed by:

amino acids, and especially aspartic and glutamic acid;

organic acids, and especially citric acid;

minerals, and especially potassium;

and sugars.

It is therefore possible to contemplate the beneficial re-use of some of these fractions.

Studies in progress at the present time have, unfortunately, not led to any concrete results. It is nevertheless certain that results can only be obtained with a major investment.

Further efforts are also still needed in the area of product quality and product applications (Figs. 2 and 3).

In the field of animal feed, however, the quality of the product is beginning to be well recognized, and it is valued by both manufacturers and animal raisers.

Numerous, very intensive tests have been carried out, and in particular in the field of feed for the nursing of calves as a partial substitute for milk powder. Comparison tests have also been run on substitutions for plant or animal proteins.

For more than a year now, 350 cattle boxes which the Roquette Company has in its experimental cattle sheds have been exclusively devoted to this study.

Other testing programs conducted by official organizations such as INRA ("Institut National de la Recherche Agronomique", Centre de Recherches, located at Rennes) or ITEB ("Institut Technique de l'Elevage Bovine", CANA Station Experimentale, located at Ancenis), or by manufacturers and cattle raisers have, in the case of our "Protein 74", demonstrated a digestibility factor approaching 90.

In experiment 3375, which is in part shown in the following graphic illustration, values for rate of substitution of 13.5 per cent and 20 per cent milk protein have been, respectively, 89.2 and 86.3.

In this same experiment, performances, carcass yields, mean daily weight gain, consumption index, and meat quality were comparable in both the experimental and control groups.

Several testing programs carried out by ITEB on calf raising have demonstrated that it was possible to use "Lysamine" (patent registered by the firm of Roquette Frères under number 186 124) in large amounts in feed, up to 8 per cent, whereby that figure represents a 30 per cent substitution rate for milk protein. This had no effect on performance prior to weaning.

To be sure, other testing programs are in progress. We have already been able to determine, however, that potato protein can be substituted for all of the replacement proteins currently in use.

This is the same case in other fields, such as in pig, cattle, and poultry breeding.

In the case of pig-breeding, and in particular in the case of pigs for fattening, tests have been carried out with substitution rates close to 50 per cent.

With respect to cattle, all of the oil cake was replaced.

With poultry, identical, if not superior performances were achieved through a complete replacement of all of the fish protein.

It is unfortunate in these fields that the potato protein can not be sufficiently priced, since it comes into competition with other, less costly proteins, and with soya in particular.

Other outlets will therefore have to be sought and outside of the industrial fields where substantial quantities have already been used in, for example:

- the glue and adhesive industry;
- building material industry (manufacture of plyboard, conglomerates, chip board with or without the use of resins of the urea-formol, melamine-formol, and phenol-formol types);
- floor preparations.

TABLE 4

ITEB EXPERIMENTS (calf rearing)
Characteristics of feed for nursing

	Control Feed				Test Feed			
	0 375 a 0 375 b		b 0 475 c 0 675		0 375 a 0 375 b		b 0 475 c 0 675	
Milk powder	60				39		43.5	
Lipides	21				21		21	
Lactoserum	12				16		17.5	
Lysamine	0				8		6.5	
Protamyl	0				11		8	
Pregel starch	2				2		2	
Natural starch	2				2		0	
Dextrose	2				0		0	
$CaPO_4H$	0				0		0.5	
C.m.v.	1				1		1	
Dig. prot. mat.	21.5				21.5			
Lipides	21.0				21.0			
Digest energy	4849				4834			
Dig. en. prot. Mat./dig.en	25.0				25.0			
<u>Performances</u>								
Total number of calves	32	32	24	32	32	32	24	32
Total number of calves weaned	32	32	23	31	32	30	24	31
Daily mean gain	830	772	698	740	815	786	705	721
Mean	764				760			

We would of course think of the field of human nutrition. This is, without doubt, a prime field for potato protein. The product will have to merit entrance into this field, and it will only merit it to the extent that it is qualitatively worthy of a role in human nutrition. Potato protein is intrinsically capable of it if, for this purpose, reference is made to its composition, to testing programs, and to official opinions such as the one expressed by investigators at the "Max Planck Institut für Ernährungsphysiologie" at Dortmund (West Germany), who provided unquestionable proof to the effect that potato protein has as great a biological value as egg protein. For this, it will be necessary to improve the quality with further research employing new means.

Present research is devoted to perfecting such improvements, with the conviction that there can only be progress in the area of pollution when there are at the same time profitable uses for recovered products.

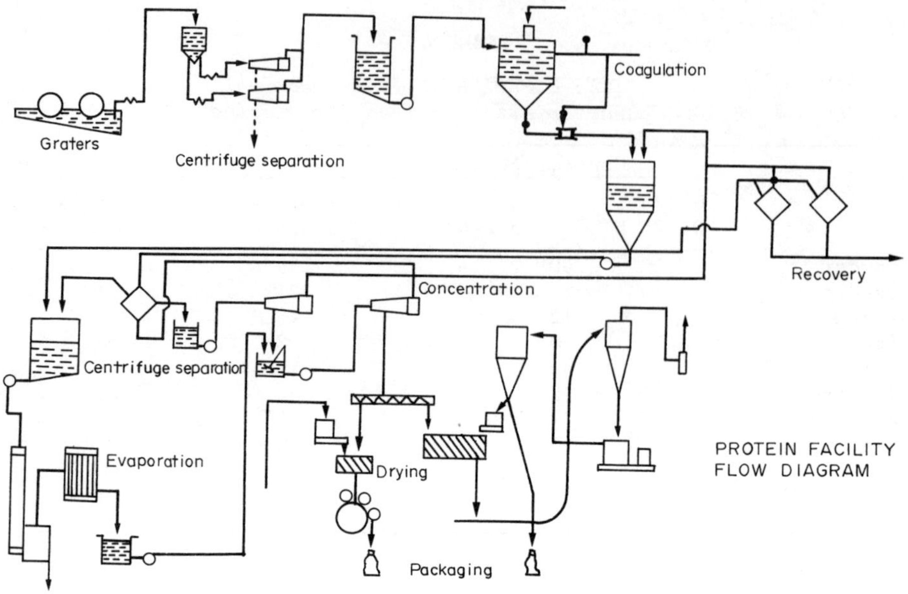

Coagulation
Graters
Centrifuge separation
Recovery
Concentration
Centrifuge separation
Evaporation
Drying
PROTEIN FACILITY
FLOW DIAGRAM
Packaging

Fig. 1

EXPERIMENT 3375
Chromium oxide balance
Protein digestibility, %
0 5
0
0 5
T 1 2
Milk powder 71 61 57
Milk protein 24 20.8 19.2
Potato protein 0 3.2 4.8
Feed digestibility 94.2 93.5 92.6
Protein digestibility 89.2 86.3
2 4 6 8 10 12
Mean weight gain, weeks

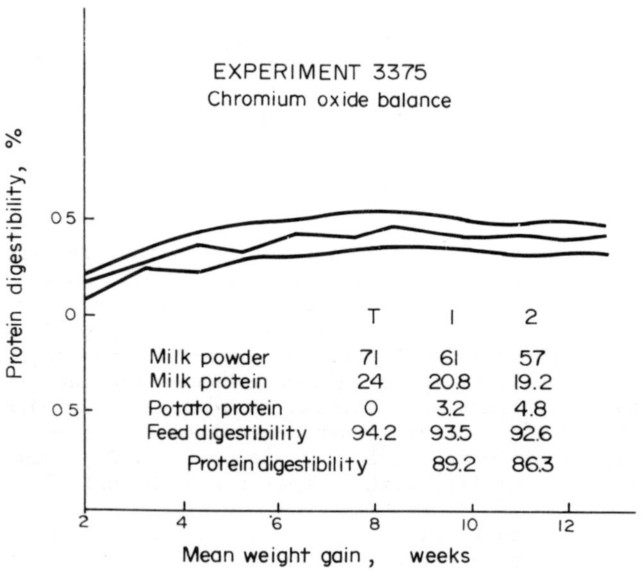

Fig. 2

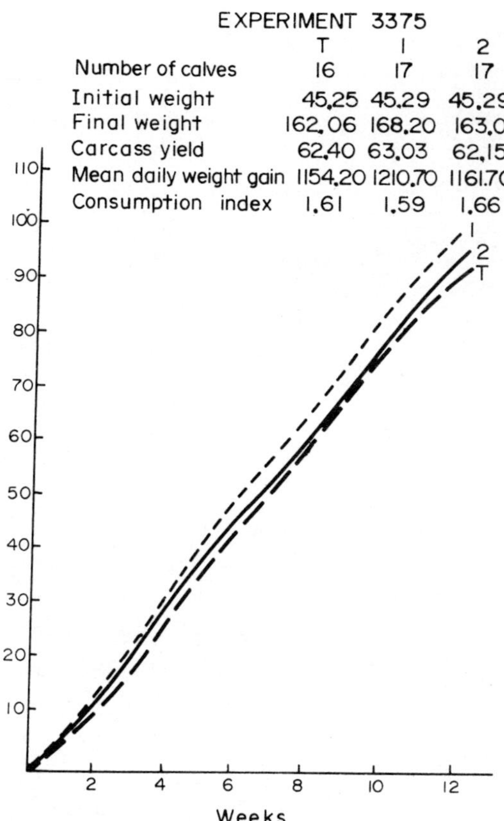
EXPERIMENT 3375
T
1
2
Number of calves 16 17 17
Initial weight 45.25 45.29 45.29
Final weight 162.06 168.20 163.00
Carcass yield 62.40 63.03 62.15
Mean daily weight gain 1154.20 1210.70 1161.70
Consumption index 1.61 1.59 1.66
110
100
90
80
70
60
50
40
30
20
10
2
4
6
8
10
12
Weeks

Fig. 3

Profitable Industrial Uses for Whey

F. Bertrand

Engineer, Centre Technique du Génie Rural des Eaux et des Forêts, Ministère de l'Agriculture, Antony, France

Whey is a by-product of the cheese industry. In France it can be estimated that for approximately every million tons of cheese produced each year, some 7 million cubic meters per year result as a by-product.

Current interest in this by-product goes hand in hand with concern for pollution matters, savings in power consumption, and the slump in whey powder prices in 1975.

It would therefore be of interest to review the transformation techniques presently employed or which could possibly be developed, and also to evaluate the commercial benefits of the various possibilities for profitable uses which have been studied by estimating the probable margins entailed in the various products to be obtained, together with the scope of their market potential.

I. SOURCE AND COMPOSITION OF WHEY—ITS POLLUTING POWER

1.1. There are already two types of whey.

(a) "Traditional" whey. Arising from the draining of cheese produced by classical, mechanized or unmechanized methods, this by-product can be described as a bacterial culture medium which for the most part consists of the following:

93 per cent water,
5 per cent lactose,
1 per cent protein,
1 per cent inorganic salts and other substances.

(b) The new whey. Arising from the production of cheese based on milk treated by the MMV process, it results from the ultrafiltration of milk through a semi-permeable membrane. It essentially differs from the preceding one in that it is almost sterile, it has the composition of a simple "lactose juice", and does not contain any proteins. This "new" whey only contains the following:

94 per cent water,
5 per cent lactose,
1 per cent inorganic salts and other substances.

1.2. Polluting Power of Whey

Pollution associated with whey is approximately equal to 30 to 50 g of BOD_5

per liter, this being 0.7 to 1 population equivalent depending upon the origin. This capacity to pollute is essentially due to the lactose, which brings about 80 per cent of the pollution as against barely 20 per cent for the protein fraction.

The discharge into the environment of whey produced per day in France corresponds to a pollution equivalent to that of one-half of its population.

II. METHODS FOR PROFITABLE INDUSTRIAL USES

2.1. *General Principles*

An examination of the compositions mentioned above is in itself sufficient to show that methods for the profitable use of these by-products by transforming them into better balanced foodstuffs in human nutrition and animal feed will lead to the following:

the removal of the limiting factor of water, which restricts intake by cattle due to its bulk, and makes transportation to special treatment plants very costly;

the removal of the limiting factor of lactose, which is a sugar which can easily be assimilated by young mammals, but a great deal less so by full-grown animals.

In the case of traditional whey *to be employed as best as possible for the protein fraction,* this fraction consists of high-quality, soluble proteins which are exceptionally easy to digest.

2.2. *Treatment of Whey by the Simple Elimination of Water*

(a) Concentrated Whey. Two concentration methods are complementing one another at the present time:

Concentration when cold by reverse osmosis: this recent, membrane method provides for the pre-concentration of serums with about 6 to 20 per cent of total dry matter. The increase in viscosity and osmotic pressure of the concentrates limits the level of concentration to a factor of between 3 and 4. In the case of pre-concentration facilities with a capacity of some 50 m^3 of whey per day, this process is more economical when referred to the ton of eliminated water than traditional concentration when hot. It is therefore very suitable for the purpose of reducing transportation costs of the by-product between cheese manufacturers and whey treatment plants.

Concentration when hot allows the production of concentrates with 45/55 per cent dry matter destined for drying. For the manufacture of powder, the maximum amount of water is to be removed on concentration, power consumption for 1 kg of eliminated water being 5 to 10 times higher during drying than at concentration. Provided that there are specially designed "finishers" in operation, "molasses" containing 60 per cent or more of DM can also be produced from whey. This DM (dry matter) is sufficiently dehydrated to be used directly as a constituent of cattle feed.

(b) Whey Powder. Spray drying is presently the most widespread and it takes up more than two-thirds of the French whey production. It gives by far the best results leading to the *prior crystallization of lactose* in pre-concentrated whey.

Drying on rollers (Hatmaker) provides an inadequate yield and has been all but abandoned. By contrast, it is possible that "ball" dryers, less bulky and consuming only half as much power, will in the future take the place of the classical Spray towers.

2.3 Individual Separation and Profitable Uses of Proteins and Lactose

2.3.1. Extraction and use of proteins. This series of treatments, of course, only involves traditional whey, since, in the case of lactose juices from the application of the MMV process, the proteins completely remain in the cheese.

The main uses for proteins extracted from whey can be, whatever be the extraction process employed:

recycling in cheese production;

use in human nutrition (baby food and dietetic foods);

use in the food, cosmetic, and chemical industries.

(a) Traditional Processes. Separation of proteins when hot by coagulation in an acid medium:

Centri-Whey Process (Alfa-Laval)

The protein complex from whey to be reincorporated in processing milk can also be used for other purposes.

For every 1000 liters of treated whey, 50 to 60 kg of protein milk is recovered containing about 10 per cent protein for 16 per cent dry matter.

Bel Industries Process (protein fraction)

The entire process produces on a continuous basis yeasts and proteins. The extraction of proteins by heat and in an acid medium making up the first phase is more efficient than in Centri-Whey. For each cubic meter of whey, it extracts about 4 kg of dry product containing 72 per cent proteins.

(b) Separation of Proteins by Ultrafiltration of Whey. This involves filtration at a molecular scale, the installation for which is rather similar to that used in reverse osmosis. The larger membranopores selectively retain proteins while letting the water, the lactose, and the salts pass through. To reach high protein concentrations, it is essential to recycle the concentrate and to "wash" it to facilitate elution of the lactose and salts.

(c) Other Processes. Numerous procedures for protein extraction have been investigated, but very few have reached industrial application in France:

precipitation by the formation of a complex with an inorganic salt;

precipitation by alcohol;

other physical means: dialysis, gel-filtration;

and finally, the extraction of freed lactose, with, as a result, the mother liquids low in lactose and rich in proteins. When dried, they give powders with high protein contents.

2.3.2 Treatment of lactose juices

(a) Decomposition by microorganisms. This approach allows the application of a broad range of technology and leads to a great product variety:

proteins from the cellular compounds of microorganisms;

products from the decomposition of lactose by microorganisms, and by-products of their metabolism:

ethyl alcohol;

lactic acid and vinegar;

vitamins, antibiotics, enzymes, fats, and the like.

Production of lactic yeast by the Bel process. The whey, free of proteins and enriched with salts, is seeded with yeast. Yeast creams are extracted by centrifugation from the drawn-off unfermented portion. Their filtration yields a plasmolysed "paste" under heat. It is then dried on rollers. In this manner, 16 - 20 kg of dried yeast extract containing 45 - 50 per cent protein is obtained from a cubic meter of whey.

Production of protein whey using the Devos process. This process is suitable for the valorization of acid whey since the yeasts which are used to consume lactic acid.

The yeast containing whey is pasteurized, concentrated, and dried in a tower, which yields a powder containing 15 - 18 per cent proteins (this being 1.5 - 2 times more than an ordinary whey powder).

Alcohol production. Even though these are long-standing methods, they have not been given any industrial application due to their high cost. This notwithstanding, it is possible to produce 30 liters of alcohol from 1 cubic meter of whey.

(b) Lactose extraction—increase in sweetening power: manufacture and purification of lactose. Concentrated whey is heated to bring the lactose into perfect solution. The crystallization of lactose in the concentrate is then carried out in tanks in which the controlled temperature is slowly lowered in stages.

Centrifugation separates the crystals from the mother liquid. Lactose for industrial or nutritional purposes is then dried, graded, and packaged.

In the event that traditional whey is treated, the mother liquids can be dried so as to yield about 40 kg of powder with 25 per cent protein.

To obtain lactose conforming to Pharmacopeia standards, additional refining will be necessary. Approximately 30 kg of lactose are recovered from 1 cubic meter of whey.

Enzymatic hydrolysis of lactose. Even though this process finds no industrial application in France, this method can be developed to supply an industrial substitute for sugar. Even if lactose "sweetens" about 6 times less than saccharose, the glucose - galactose mixture obtained at the end of enzymatic hydrolysis of a concentrated solution gives a "milk honey" whose sweetening power reaches 70 per cent of that of saccharose, and which, by not crystallizing, could become of particular interest in the canning and pastry industries.

III. ESTIMATES OF POSSIBLE PROFITABLE USES

The estimated profitable uses represent the possibility for firms which transform this by-product to pay the cheese manufacturers who produce it. This purchase cost which is possible for the raw material are to be estimated on an "all expenses paid" basis for the various channels, i.e. after a firm, whatever its status, have paid for the costs of transportation, transformation, and all of the overhead expenses, such as waste-water treatment, management, and marketing in particular, whereby the return on invested capital has already been charged to the various cost items.

These profitable uses are summarized for each type of product in Table 1. They have been established on "fragile" assumptions with respect to the price of the finished product which often are highly variable. Costs estimates of transformation made by the "model" method are, by contrast, more reliable and the spread between low and high costs are essentially due to the economy of scale.

The order of magnitude for the additional, possible profitable uses for both the lactose and the protein fractions are presented in Table 2.

Even though these figures were established by computing specific, exact assumptions, which of course nevertheless will vary with location and over time, the relative values still are of significance.

The following, practical conclusions are most often to be retained from an examination of the figures:

Separate profitable uses for the proteins and the lactose fraction of whey give better economic results than an overall use through drying.

The profitable use of protein from whey often finds its best outlet in the field of human nutrition: dietetic products and baby foods do not appear to give really better economical results than the recycling of proteins in cheese manufacturing, for which the market is at least ten times larger.

The profitable use of lactose juices can attain high levels for products of a pharmaceutical type with narrow markets, where, however, high prices are uncertain (lactose, ethyl alcohol). In the case of this fraction of whey, yeast extracts would appear to be a good channel capable of taking large tonnages for cattle fodder, especially if the transformation facilities are large enough to provide a proper economy of scale.

TABLE 1 *Profitable uses by product*

			Approx. yield per m³ of treated whole whey	Total sales assumption in FF/kg per liter of product		Total cost estimates F/m³ of whey		Mean valorization F/m³ of whey
						Min.	Max.	
	Lost (in Sewage Treatment)		-	-	-	12.0	20.0	+ 16 ± 4
Whole whey	Dehydrate	Molasses	100 kg of conc. at 65% dry m.	0.40	0.60	22.4	40.6	18.5 ± 19
		Spray powder	60 kg of powd.	0.80	1.30	41.6	64.9	9.7 ± 27
Separation of proteins and lactose	Protein fraction	Lost (in ST)	-	-	-	2.4	4.0	- 3.2 ± 0.8
		Lactose-free spray powder	35 kg of powd.	0.90	1.50	20.3	33.0	+ 15.3 ± 17
		Dry protein extract (Iel)	5 kg	5	8	14.2	24.9	+ 13 ± 13
		Protein concen. ultrafiltration	5 kg	20	30	14.2	24.9	+ 105.5 ± 30
		Proteins centri-whey	50 kg	2	3	33.1	44.7	+ 86.1 ± 31
		Proteins ultrafiltration	10% inc. in cheese yield	(1.5)	(12.8)	14.1	25.7	+ 95.1 ± 41
	Lactose fraction	Lost (in ST)	-	-	-	9.6	16.0	- 12.8 ± 3.2
		Lactose	30 kg	1.9	3.5	45.5	63.5	+ 26.5 ± 33
		Lactose hydrolysate	30 kg milk honey	0.6	0.8	40.6	48.4	+ 115 ± 11.9
		Yeasts	20 kg dried yeast	2.5	3	38.4	58.1	+ 6.8 ± 15
		Alcohol	28 liters	1	4.5	19.0	30.9	+ 52 ± 55

TABLE 2 *Valorization of lactose fraction*

Additional profitable uses for proteins and lactose from whey (average assumptions)			Lost in ST	Lactose	Lactose hydrolysis	Yeasts	Alcohol
	Lost (in Sewage Treatment (ST))		-16	+ 23.3	+ 8.3	+ 3.6	+ 48.8
Valorization of the protein fraction	Cattle Fodder	Lactose free spray powder	+ 2.5	+ 41.8	+ 26.8	+ 22.1	+ 67.8
		Dry, denatured powder extract	+ 0.2	+ 39.5	+ 24.5	+ 19.8	+ 65
	Dietetic and human foodstuffs	Dry protein extract	+92.7	+132	+117	+112.3	+157.5
		Protein concentrates from ultrafiltration	+73.3	+112.6	+ 97.6	+ 92.9	+138.1
	Recycling in cheese	Meat coagulated and centrifuged proteins	+82.3	+121.6	+106.6	+101.9	+147.1
		Proteins from ultrafiltration	+88.8	+115.3	+113.1	+108.4	+153.6

Dyeing in a Solvent Medium: STX Process

M. Laurent
France

INTRODUCTION

Since it was founded, the only goal of the firm of STX has been that of aiding the textile industry to benefit from the use of organic solvents to resolve the ever-increasing problems in the area of water supply and water pollution.

Inasmuch as the treatment of 1 ton of textile requires 400 to 600 tons of water on the average, this represents a major use of water and brings about, as well, an extensive amount of pollution. This places the textile industry unfortunately among the major polluters.

This state of affairs is so preoccupying that we have, for several years, been witnessing the appearance of new techniques and processes which attempt to reduce this in scope, if not to resolve it completely.

To be sure, all of these efforts have already begun to bear fruit, but they have still not totally eliminated polluted wastes or that other problem so closely connected with it, that of the water supply.

It is precisely in an attempt to solve both of these problems at the same time that numerous research teams have tried to replace water with organic solvents.

From the very start, this original approach was greeted with a great deal of optimism and the entire textile industry became enthusiastic about the promising future which was held forth.

At the time, this was to perhaps underestimate the magnitude of the task at hand, and the difficulties to be overcome.

To attempt to replace such a basic constituent as water in the dyeing process is not at all easy. This is so not only because of strictly commonplace reasons, but especially so because all of the knowledge and improvements to date are based on this immutable factor.

When synthetic fibers first made their appearance, only the strong growth of these fibers, their advantages, and their future provided solutions to the major problems presented by dyeing them. This treatment still remains rather problematical as concerns, for example, polyester and acrylic fibers.

However, this phenomenon of the mobilization of massive amounts of resources has not as yet reached dyeing in a solvent medium. This is, perhaps, one of the factors which has delayed the materialization of some projects.

If, in any event, it is true that we are beginning to see some progress in the area of perfecting new dyes adapted to a solvent medium, these studies (and especially their outcome) are long and laborious, and so much so that numerous investigators have attempted to carry out a transition phase using water containing emulsions so as to profit from the acquired knowledge in aqueous dyeing.

But this course is not easy, and we are well aware of the problems associated with it.

How should the use of solvents in the dyeing field be approached ?

From the outset, the investigation of a solvent medium process should be guided by a consideration of the following points:

1. It is necessary that it really solves the problems of water supply and water pollution. Moreover, it must enable operations which meet with standards of safety and industrial health.
2. It is necessary that its principle be as simple as possible, by providing, in particular, for a uniform dyeing bath.
3. It is necessary that the quality of the dye be at least equivalent to that which is achieved in water.
4. At least initially, it must be able to operate with fibers and dyes already available on the market.
5. Finally, its operating costs must be in the same order of magnitude as those of traditional aqueous techniques. In particular, this assumes a simple and economic recycling of dyeing baths, and the maximum recovery of the solvents which are employed.

It was in bringing together all of these limiting factors that we were able to establish a definition for the STX process, to study its actual design, and to ensure its application, i.e. to bring it out of the laboratory.

To bring this program to a successful completion, we were of the opinion that we had to proceed step by step, and to undertake the analysis fiber by fiber and application by application.

To this end, the polyamide fiber very quickly came to the fore by virtue of its behavior, and due to the nature of the dyes required in its dyeing. In addition, polyamide is used in the raw state and in heavy tonnage, which is the case with the rugs with which we began. We then turned our attention to the field of ladies' tights. Nevertheless, whatever the article to be dyed, the processing procedure remains basically identical in all respects.

Throughout the remainder of this paper,we shall consider the case of dyeing rugs made of polyamide.

SOME CONSIDERATION ON DYEING IN AN AQUEOUS MEDIUM

As with any chemical reaction, the process of dyeing requires three factors to be present:

the reagents: dyes;

the substrate: the polyamide fiber;

the reaction medium: water, serving as the ionizing medium, as solvent for the dye, as the fiber-swelling agent, and as the medium for the transfer of heat and matter.

The shift of the equilibrium of the reaction in favor of the fibre is carried out by raising the temperature of the medium and through the addition of the so-called "auxiliary products" in dyeing, such as acids, inorganic salts, solvents and the like.

The yield of the reaction not being 100 per cent, it is necessary, after dyeing, to carry out a rinsing of the material, which takes place with a minimum consumption of demineralized water of 90 to 100 m^3 per ton of rug.

THE STX PROCESS

Considering the preceding comments, it is then certain that the choice of a processing solvent would be guided by the search for a certain similarity with some of the aspects of dyeing in water. A dye bath composition would then be sought which enables the dissolution of solvents as well as the swelling of the fiber to be dyed.

After a systematic examination of the available solvents, we then reached the following conclusions:

That the use of a single solvent just as it is did not have in itself properties sufficient to meet with all of the conditions previously mentioned, and that it would be necessary to add auxiliary products to it.

That the necessary auxiliary product or products could very well be simply a second solvent, whose properties would be added to those of the first one.

That the methyl alcohol/perchlorethylene combination would be perfectly well suited to the dyeing of polyamide, providing in addition the possibility of visualizing the dyeing of other fibers. In actual fact, a bath consisting of 90 per cent perchlorethylene and only 10 per cent methanol can keep a sufficient quantity of dyes in solution. Thanks to this pair of solvents, we were able to bring together all of the chemical characteristics of water in conventional dyeing:

methanol plays the role of dye solvent, fiber-swelling agent, and ionizing medium,

perchlorethylene, by virtue of its physicochemical characteristics (non-inflammability, and very low specific heat and latent heat of vaporization), is the ideal heat-transfer medium.

	Perchlorethylene	Water
Specific heat : Cp kcal/kg/0°C	0.25	1
Latent heat of vaporization: Lv kcal/kg	56	539

At this stage, we noted that a change in the composition of the bath through a removal of the methanol, i.e. a removal of the dye solvent, brought about a disequilibrium leading to a total loss of capacity. This feature, together with the absence of using any auxiliary dyeing products, creates both technical and economic advantages:

both solvent can be re-used immediately without being subjected to long and costly distillations (about 15 per cent of the bath);

it is not necessary to rinse the textile prior to drying. Drying takes place right in the machine, thus ensuring an effective control of solvent losses.

PROBLEMS AND DISADVANTAGES

Presented in this light, the STX process appears to be some sort of a panacea. You might well surmise that this program did not see the light of day without presenting a few problems, these being the same problems to be found in any system operating in a closed circuit, and which are due to the contamination of the bath. The impurities are introduced into the system by the textile. We might look more closely at this matter here.

Water

The introduction of water into the methanol/perchlorethylene mixture leads to a lack of a homogeneous mixture which makes it impossible to achieve a uniform dyeing. The initial 4 per cent water content of the polyamide rug is reduced to 1 per cent by pre-drying of the material right within the dyeing.

Residual fiber water is extracted by the methanol and carried by it out of the flash. In order to avoid any accumulation of water in the system, the methanol/perchlorethylene mixture is dehydrated in a distillation column, the water remaining at the bottom of the tower.

Oils

Textiles contain 0.5 - 1 per cent textile oil. As they are soluble in perchlorethylene, they contaminate the bath. There are two aspects to this phenomenon:

it serves to clean the textile, which is somewhat positive since this reduces the likelihood of stains;

it burdens the bath with foreign matter, which, if it does not at all affect the dyeing process itself, does become bothersome in high concentrations (spots).

The maximum oil content in the bath thus being established experimentally, a certain run-off value is derived from it. The extracted perchlorethylene is cleaned in an ancillary facility and the oil residues are easily destroyed by combustion.

Oligomers

Oligomers are the product of incomplete polymerization. With a very low molecular weight, they readily go into solution in hot dye baths, to appear once again in a solid form should, for example, there be a fall in temperature. The oligomers of polyamides are extracted from the fibers during treatment and go into solution in the bath which is thus contaminated. If the oligomers are soluble in methanol, then they are not soluble in perchlorethylene, and when the alcohol is removed during dyeing, the oligomers then precipitate. For this reason, we have developed a filtration system which is specially adapted to this exact case. The filtration/oligomers adjuvant mixture is then treated to remove the perchlorethylene by steam stripping which gives an exceedingly low, 1 per cent residual content of sludge in the solvent. Studies are now in progress to find a solution to recycling these sludges.

POLLUTION PROBLEMS

1. Dyeing in an Aqueous Medium

It is obvious that all of these impurities just mentioned with respect to dyeing in a solvent medium are also to be found in an aqueous medium. Because they are so well known, we shall not dwell here on water-pollution problems in conventional dyeing. Nevertheless, it should be stated that to dye 1 ton of rugs, it is necessary to make use of 90 to 100 m^3 of previously demineralized water. This water, which is discharged at an average temperature of 60^oC, essentially contains the following:

unspent dyes:
  with acid dyes, 5 - 10 per cent of the quantity initially used;

auxilliary dyeing products:
  wetting agents, regulators, swelling agents, acids, inorganic salts and the like;

products in solution or emulsion:
  oligomers and oils.

All of this obviously requires post-treatment in a water purification plant, a post-treatment which, even though employing known techniques, is generally complex and costly to operate.

We would like to note here in review that, independent of heat pollution, chemical pollution from a conventional dyeing plant with a capacity of 6 tons of rugs per day is from 10,000 to 15,000 equiv/h/d.

2. STX Process

2.1. Water Pollution. In the STX process, water is used as processing water; i.e. it is essentially used as a vehicle for heat exchange. We need to have available, to dye 1 ton of rugs, about 350 m^3 of water.

This water, which does not at any time come into contact with the solvents, is therefore not chemically polluted and can be directly re-used in other procedures where a temperature below 30-35°C is not required, or used once again as cooling water after passing through air-cooling units, for example.

2.2. Solvent Losses. Were we to analyze the different sources of solvent losses, we would find that there are three of them:

absorbed solvent;

solvent in solution;

solvent carried off in the air.

Absorbed Solvent. Polyamide, after drying in hot air, still holds 3 - 5 per cent of its weight in perchlorethylene. This content is reduced to 0.3 - 0.5 per cent by carrying out a post-spraying of the rug. As mentioned earlier, there is also residual perchlorethylene in the filtration adjuvants and in the oils. This represents at most 1 kg of solvent, this being 1 per cent in relation to the treated textile.

Solvent in Solution. The slight solubility of perchlorethylene in water at 30°C explains the negligible loss of solvent by this route. Were we to consider that in the entire rug-dyeing process, there is approximately 500 kg of water in the presence of the solvent (fiber water, condensate of steam from the sprayer...), then there is a perchlorethylene loss in the order of 100 g per ton of treated rug (solubility of perchlorethylene in water: 100 ppm). This water is recovered and re-used with the processing water. If it were discharged, it would cause a pollution (OCD) of less than 10 equiv/h/d.

Solvent Carried Off in the Air. The drying of the textile takes place right within the dyeing. At the end of the operation it is essential, in order to avoid bad odors, to carry out a deodorization of the drying loop.

The consumption of perchlorethylene can be reduced to a negligible amount with the addition of an adsorption facility using active carbon.

Another source of loss is in the various movements of liquids in the vats. Its magnitude will depend on numerous factors, such as, among others, atmospheric conditions.

PROCESS PROFITABILITY

Even though the process provides satisfaction from the technical viewpoint, there still remains an important factor to be verified with respect to its industrial application. This concerns the economical and profitability aspect.

TABLE 1

STX Processes. Solvent losses compared to treated textiles

	Perchlorethylene (%)		Methanol (%)
	Without A.C.	With A.C.	
Solvents in solution in 500 liters of water	traces	traces	
Absorbed solvents			
textile	0.3 - 0.5	0.3 - 0.5	
filtration adjuvants	0.05	0.05	
oils	0.05	0.05	
sub-total	0.4 - 0.6	0.4 - 0.6	
Solvents in air			
drying	0.8	traces	1
vat vents	0.5	0.5	1
sub-total	1.3	0.5	2
total	1.7 - 1.9	0.9 - 1.1	2

Were we to base an analysis on results obtained in our experimental plant, we would then have the principal characteristics as follows:

Rug length: 4.06 m

Capacity per run: 800 - 1200 kg (2000 m^2)

Treatment time (dry to dry)(at present): 3½-4 hours

Annual production: 1500 to 1900 tons

At equivalent capacities, the cost of equipment assembled and ready to operate is slightly above that for an assembled and complete installation of 4 boats together with their drying oars (see Table 2).

Sources of savings are essentially on three items:

- the consumption of auxilliary products, which is nil;
- there is no need for water treatment;
- the power cost item (see Table 2).

TABLE 2

Comparative Costs of Buying

	Boats (FF/kg)	STX Process (FF/kg)
Dyes	0.3067	0.3067
	0.2628	—
Steam	0.57	0.21
Electricity	0.2410	0.17
Water	0.225	—
Solvent	—	0.051
	1.6055	0.7377

Savings in Power per Ton of Treated Rug

	STX		BOST	
	Absolute Value	%	Absolute Value	%
Power consumption (kW/h)				
Handling	70	12		
Pre-drying, drying	350	50	780	78
Dyeing	250	36	220	22
Regeneration	15	2		
	685	100	1000	100
Steam Consumption (tons)				
Handling	0.9	13.2		
Pre-drying, drying	0.3	4.4	4	21
Dyeing	2.8	41.2	15	79
Regeneration	2.8	41.2		
	6.8	100	19	100

CONCLUSIONS

During the course of this presentation, we have seen that the solvent can provide:

a decided response to pollution problems;

a considerable reduction in power consumption;

and, especially low dyeing costs.

Nevertheless, a great deal of work still remains to be done, because the dyeing potential of the process is not as vast as that of known process using water, and, even though polyamide dyeing is now perfectly possible, there are still other fibers.

Moreover, further efforts are needed to expand this process into other dimensions, even though the process does make it possible to industrially dye manufactured articles.

As we have seen, it is difficult to rely on the bases of classical dyeing using water. The research and development of a dyeing process in a solvent medium is a long and difficult task, as it is a complex multi-discipline effort bringing together specialists from a variety of fields in chemistry, technology, design and analysis, dyes and fibers.

Such a team effort is the main factor in our success in the research and development of this process.

In the name of all the engineers and technical support personnel who have made this presentation possible, may I thank you for your courtesy and attention.

How and Why we Choose Integral Recycling

B. Maréchal

ISOREL, Tour Roussel Nobel, Puteaux, France

HISTORICAL ACCOUNT

The Isorel plant in Casteljaloux, France, first began producing fibreboard in 1946. Two different types of boards are produced:

hard board: thicknesses of 2, 3 and 4 mm; density: approximately 1.00;

insulating board: thicknesses of 10, 16 and 25 mm; density: approximately 0.250.

The two production lines have a current total capacity of around 65,000 tonnes per year, or 200 tonnes per day, broken down approximately as follows:

50 per cent insulating board,
50 per cent hard board.

These boards are produced by the wet process from waste wood as outlined below:

The waste wood is reduced to chips.
The chips are slightly cooked and ground (thermo-mechanical pulp).
The resultant fibres are suspended in water to form the pulp stock.
This stock is then refined.
It is then machine-drained to form the "mats" which become the boards:

(a) the hard boards are pressed and polymerized in a hydraulic press;
(b) the insulating boards are simply dried and polymerized in a tunnel drier; the boards are then cooled, refrigerated and cut to standard commercial sizes prior to storage and shipping.

POLLUTION (era of the past)

The pollution stemmed from the fibres lost in the industrial water and from the soluble wood components, such as resins and sugars in particular, which ran off with the water when the mats were machine dried.

The Adour-Garonne Basin Water Technical Commission—a public organisation which prepared the creation of the "Basin Agencies" in 1967—was assigned to measure the quantity of pollutants contained in our effluents as a representative index of that discharged by the companies in the same field of activity. These analyses were to serve as a basis for preparing the schedule for calculating the fixed fees imposed as of 1969. The results of these examinations showed that, at the time, we discharged 100 kg of suspended matter and oxidizable matter per tonne of finished product, which meant that our pollution level equalled that of a town with *110,000 inhabitants,* i.e. 16,500 kg/day.

We were not unaware of the gravity of the situation. Even if we had wanted to, the increasing severity of the complaints from fishermen and residents along the Avance River downstream from our plant would not have allowed us to plead ignorance.

As early as 1958, that is to say long before the pollution-control campaign was conducted throughout the general public, Isorel Company started looking for a solution to the problem. A great number of *physical treatment* methods were thus tried:

settling,
filtering,
flocculation,
centrifuging,
electrostatic precipitation, etc.

Then came a series of experiments with chemical and biological methods.

We shall not dwell on each of these experiments and their respective results. To be sure, the results improved with time, at least on the experimental scale, but we ran into the following difficulties:

- the very high investment costs required to treat all of our effluents;
- a prohibitive running cost for our products;
- the imperfection of each purification process which would have left a non-negligible degree of pollution as well as risks of technical hitches and accidents;
- the impossibility of compensating for these costs and difficulties through the development of a profitable use for the purification residue, even though we did search for and try ways of transforming this residue into antibiotics, yeasts, fertilizers, cattle feed, etc.

Thus, from 1953 to 1968, we conducted *research work* on our own, both technically and financially, and we kept seeing the victims of the pollution as well as the administrative bodies to which they addressed themselves grow more and more impatient (see Fig. 1).

SOLUTION

Since that time, the Administration has realized that there are pollution problems to be solved. The Basin Committees and Basin Financial Agencies were set up. The immediate result of this was that a certain number of industrialists became aware of the fight against pollution at which time different companies made offers for more efficient equipment. This meant that we were no longer the only company to tackle this problem. We were led to try a certain number of items of equipment manufactured abroad, but the results were still not satisfactory. This is why, one day, after having met with defeat using equipment on which we put trust, we tried to imagine what would happen if, instead of suspending the fibres in water to produce the pulp stock, we suspended them in another liquid which we would then conserve. This led to the idea of recycling, and then of closing the valves.

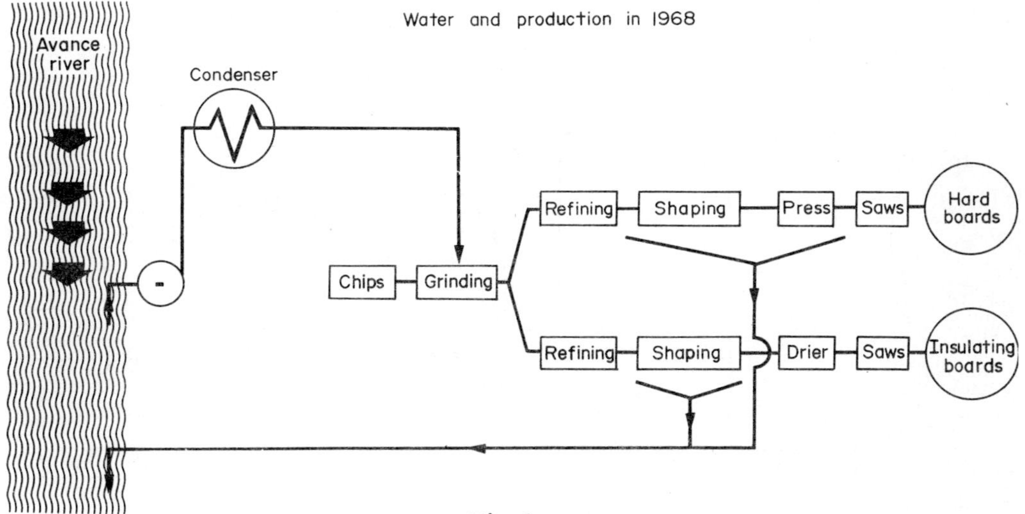

Fig.1

Since then we have closed all the valves and we now actually have a fibre-conveying fluid which is quite different from the one we had before, since it now consists of water charged with 90 to 100 grams of dry extract per litre. All this happened little by little since we practically had to convert the entire plant, change all our habits and the entire routine of the staff members, be they workers, foremen or engineers.

It goes without saying that the production tool was not designed for this and we therefore had to make new arrangements, sometimes quite minute, which required much time, analysis, trial and error, tinkering, etc. In particular, we had to change the refining, draining and board-pressing processes.

The major constraint with recycling stems from good common sense, but its respect required in-depth analysis of the role played by the water in the production process as well as of the items in the production installations which even appeared to be the most significant. The problem was to reach a point where *exactly the same quantity of water* entered the closed circuit as *that which was lost* through evaporation during manufacture.

I shall not enter into the details concerning all the adaptations required, since the sum of knowledge acquired during this period constitutes know-how which we are now negotiating in different countries. What I can say, however, is that we had to make modifications in parts of the plant which were quite remote from the actual board-production equipment and which were truly above suspicion in the beginning.

We consider we reached our goal at the end of 1972, since the pollution level which was formally *110 kg/tonne of board produced* was brought down to *0.8 kg/tonne*, which corresponds to a 99.3 per cent efficiency. I should like to point out that these figures are those found during the various measurement campaigns carried out by the Basin Financial Agency throughout our period of adaptation. This level has now been reduced to zero.

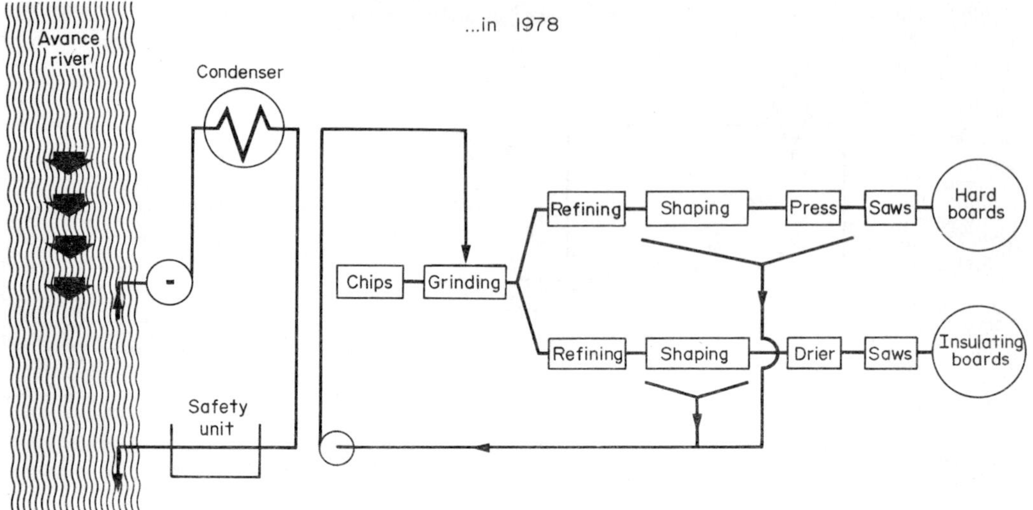

Fig. 2

SAFETY

The closed circuit is without a doubt the most ideal set-up, but we know there is always the possibility of an accident (see Fig. 2). Thus, a break in a pipe would not only pollute the river to an even greater extent than usual since the receiving environment has lost the habit of reacting to the pollution, but it would also greatly trouble board production.

This is why we installed a *buffer tank* between the plant and the river. In normal conditions, this tank collects the water from the cooling circuit and rain water, but it has a sufficient capacity to retain the full contents of the industrial water circuit in the event of an accident.

A *luminous panel* in the plant permanently monitors all sensitive points in the water circuits and can give warning in the case of abnormal functioning. In particular, there is a device which constantly samples the water in the buffer tank, controls its quality and determines whether or not discharge into the river should be stopped.

If, for example, one of the pipes break, the industrial water (thus pollutant) would be channelled from the buffer tank to two *aerated lagoons,* as soon as a warning is given.

The lagoons would retain the effluent in safe conditions, owing to aeration of the effluent by means of turbines. They would also be used to collect the contents of the industrial water circuit when the plant is shut down for paid holidays, for example. In both cases, the fluid is returned to the plant when it is ready to restart operations, or it is gradually returned as make-up water as the need arises.

Apart from the installation of the devices, or rather in order to complete them and guarantee their efficient operation, a real water-leakage hunt had to be conducted in the plant to ensure tightness of the closed circuit. Here again, the work (sometimes of minor importance) concerned parts of the plant which, *a priori*, it would have been difficult to suspect as having anything to do with the pollution problem.

RESULTS

The results were not checked but followed through measurements taken by the personnel of the Adour-Garonne Basin Agency with each work progress phase.

To obtain a valid comparison, we retained as a common basis the quantity of pollution per tonne of board produced. It can be seen that the level of 110 kg per tonne in 1968 dropped to 0.8 kg/tonne in January 1973, or, in other words, there has been a reduction of more than 99 per cent. What is more, the few 100 kg of effluent discharged each day was the result of the washing of roofs and floors, mainly by rain water: this equals the pollution level of a community with 700 inhabitants. Today, nothing more is discharged from the plant.

At the same time, the Basin Agency measured the improvement in the quality of the river water based on the percentage of *dissolved oxygen*. The point of measurement located closest to the plant (4 km downstream) indicated a rate of 120 per cent in relation to the oxygen-saturation rate in Jauary 1973 as against 65 per cent in 1968 (see Fig. 3).

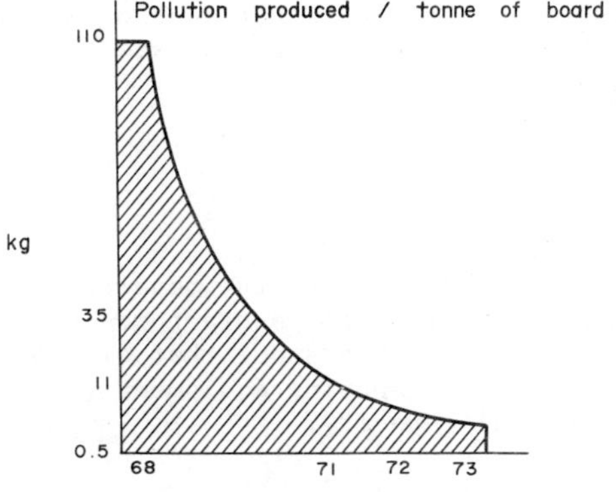

Fig. 3

I personally think it is interesting to point out that these results were achieved in a plant using its initial equipment-old Asplund 600 mm diameter grinders—operating with a steam pressure of 14 bars and at maximum capacity—therefore, under the least favourable conditions. It should be stated that the Isorel Company did not intend to spend the least amount of money to renew the wet process equipment as long as the pollution problem still existed.

Ever since a solution was found, we have set up a renovation programme for this plant, and especially for the pulp department which is now equipped with modern fibre-production machines operating at a lower steam pressure, thus greatly reducing the amount of soluble matter in the pulp—this affords greater facilities in relation to the past situation.

ADVANTAGES OF THE PROCESS

Below is an interesting list of the main advantages of the process used:

Relatively low *investments* in comparison with those of a conventional purification station—2,500,000 francs against at least 6,000,000 francs.

Particularly high *efficiency* (greater than 99 per cent) and no sludge to be removed, whereas a conventional station would have difficulties in achieving 80 per cent efficiency and would produce quantities of sludge which would raise problems of removal, since they are extremely difficult to treat or process.

Negligible *operating expenses*—equal to 25 to 30 kWh per tonne of board produced.

Raw material savings. Since this process does away with all material loss and leakage, practically 1 tonne of dry board is produced per tonne of dry wood (7150 tonnes/year including 40 - 50 per cent of recovered fibre).

Technical flexibility and savings. This process makes it possible to incorporate all the necessary additives in the pulp, regardless of their toxicity with respect to the natural environment, since there is no loss; by way of example, in order to produce 12 per cent asphalt insulating board, we use exactly the same equivalent of the 12 per cent of asphalt retained. The same holds true for the phenol adhesives and was for the hard board when they are incorporated.

EFFECT ON QUALITY

Below we have given our remarks concerning the development in the commercial and technical properties of the products manufactured during this process finalization period, it being well understood that all the products were obtained with grinding, refining and shaping equipment installed in 1945 and operated at greater than normal capacity.

Commercial Quality

How has the process affected the commercial quality of our boards? Before the process was used, we obtained 97 per cent of first class commercial board and 3 per cent second-choice board.

When recycling was first used, the percentage of first-class board dropped off to an average of 80 per cent, with minimums of 60 per cent. Faced with these highly disappointing results, we centred our efforts around the systematic, in-depth statistical analysis of the origin of the defects in order to master and subsequently eliminate them. Among the major defects we noted:

sticking,
burns,
veining,
multiple spots, etc.

The results of our studies allowed us to determine the fact that these defects stemmed from the gradual increase in the concentration of matter in solution and in suspension in the industrial water, due to integral recycling.

After several years of perseverance, we have been able to bring the percentage of first-choice board back up to the initial level, despite measures to increase the production capacity which were implemented in parallel with the fight against pollution - it should be pointed out that in the Casteljaloux plant we have now reached a rate of ten press-loads per hour with the 3 mm thick board.

Technical Quality

How has the process affected the technical quality of our boards? We can answer this question by means of Figs. 4 and 5 for the following properties:

density,
bending,
absorption.

It should also be pointed out that, in all cases, the boards on which these graphs are based were manufactured without any bonding or waterproofing additives, but simply with a product to adjust the pH.

Naturally, these mechanical and physical properties can be improved through the addition of resins and waxes to the pulp, respectively.

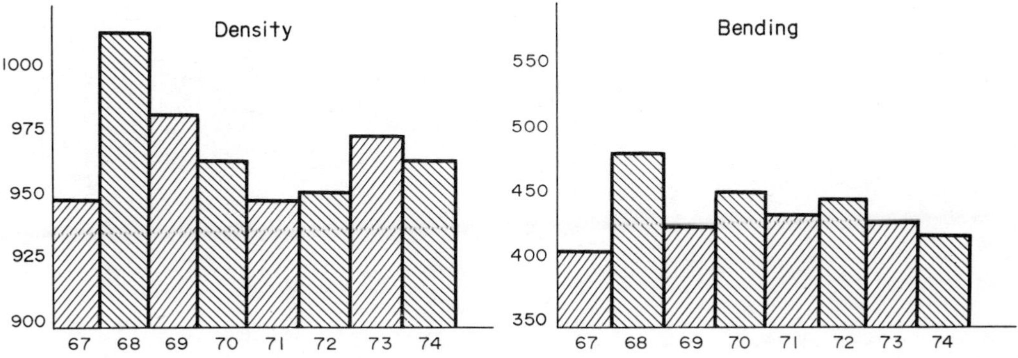

Fig. 4

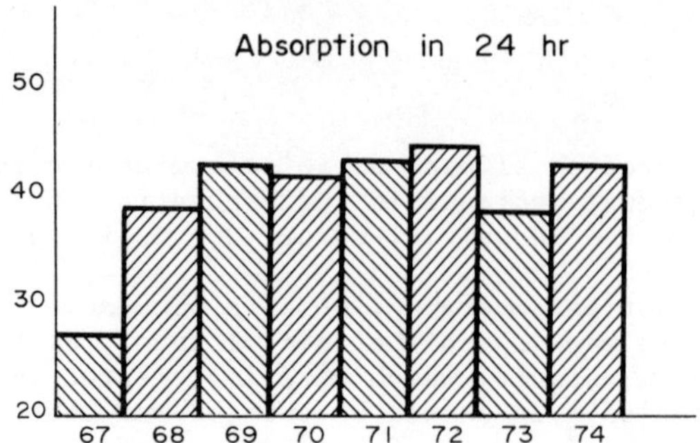

Fig. 5

It can be seen from Figs. 4 and 5 that the property which seems to be reduced by the recycling of water is the absorption of water (in 24 hours), but this defect can be overcome through the incorporation of a very small amount of wax—tests have shown that with only 0.5 per cent of dry wax in relation to the pulp, this property can be brought back to normal levels.

CONCLUSIONS

We have filed a certain number of patents in France and abroad. We have set up a company to market this process and rights have already been granted to severa of our colleagues in various countries.

Recovery of the Iron Contained in Pickling Solutions and Waste Ore Etching Solutions, in the Form of Magnetite

D. Lefort

Centre de Recherches de Pont-à-Mousson, 54700 Pont-à-Mousson, France

I. INTRODUCTION

In numerous industries in which acids are used to pickle ferrous metals or to etch ore with greater or lesser iron contents, there are waste solutions left containing products which it would be desirable to recover and utilize.

In the most favourable cases, the iron and other salts are recovered through crystallization; these salts are then roasted and raw materials such as sulphur, iron oxides, etc., are obtained; but in most cases the waste acid solutions are neutralized and oxidized thus producing large quantities of more or less gelatinous residues which are bulky and difficult to handle. What is more, we do not know how to dispose of them. Fortunately, cases such as the Mediterranean red mud incident, where industralists dump their wastes without any prior treatment, are becoming more and more rare.

The account that follows describes a process which offers the following advantages:

- it is practically just as easy to implement as the conventional neutralizing and oxidizing installations,
- for equivalent amounts of effluent, its operating expenses are lower;
- most of all, in some cases it allows recovery and recycling of the by-products.

We will discuss its use in two types of common industrial activities:

1. Installations for pickling ferrous metals to prepare them for rolling.
2. Installations for processing titanium ores to produce titanium dioxide (TiO_2).

II. FERROUS METAL ETCHING INSTALLATIONS

Before rolling, drawing or working ferrous metals in any way whatsoever, the oxides which have started to form on the surface must be removed. These oxides are usually removed in two operations: mechanical removal followed by acid etching (HCl, H_2SO_4, etc.).

If, for example, H_2SO_4 is used to etch the metal, we have etching solutions which basically contain Fe^{++} ions and excess SO_4^{--} ions. Most often these solutions are neutralized (using lime, for example) and oxidized in the air through agitation; at the end of the operation this produces a mixture of $CaSO_4.2H_2O$ and gelatinous Fe $(OH)3.nH_2O$ suspended in water. This non-recoverable mud is difficult to dehydrate and, in any case, leads to a residue which can only be dumped.

At first we looked into the possibility of obtaining a crystalline product instead of the ferric hydroxide; by two-stage neutralizing and oxidizing it was possible to obtain a mixture of a goethite (FeO·OH) or lepidocrocite and ferric hydroxide; the only advantage of this mixture was that it was easier to filter. Then, and this is the process which will be further described, we studied the way of converting all the iron contained in the water into magnetite: it could then be recovered (as an oxide) and recycled in ironworks for example.

1. *Description of the Process*

Etching solutions have approximate Fe^{++} and SO_4^{--} contents of 100 and 200 g/1 respectively. At such concentrations it is difficult to neutralize and oxidize these solutions; as soon as they are neutralized a thick mud is formed which it becomes impossible to stir in order for the iron to oxidize. On the other hand, as is usually the case, part washing and rinsing waters are mixed with the etching solutions; this increases the overall quantities of SO_4^{--} and Fe^{++} ions to be handled, but it also dilutes the solutions around 10 to 20 times.

When such is the case, it becomes impossible to implement the process; this same condition is also necessary to produce ferric hydroxide, geothite or lepidocrocite.

One way of then precipitating the magnetite is to proceed as follows:

the solution to be treated is divided (continuous or batch) into two-thirds and one-third fractions;

two-thirds of the volume are neutralized and oxidized to convert all the Fe^{++} into Fe^{+++} in the form of ferric hydroxide;

both fractions (two-thirds and one-third) are then mixed and neutralized and, after a certain length of time, a black magnetite precipitate forms. The speed at which this magnetic precipitate forms depends on the purity of the products (presence of corrosion, and foam inhibitors, etc.); it varies from 20 to 60 minutes. In order for the magnetite to form, the mixture must not be aerated; otherwise, the Fe^{++} would continue to oxidize into Fe^{+++} (see Fig. 1).

The two-thirds-one-third ratios result from the fact that Fe_3O_4 can also be written $Fe_2O_3 \cdot FeO$ from which it can be seen that two-fifths of the iron is in the form of Fe^{+++} and one-third in the form of Fe^{++}.

2. *Another Process for the Production of Magnetite*

It can be thought that the Fe^{++} oxidizes into magnetite as follows:

(a) Oxidation of part of the Fe^{++} into Fe^{+++}

$$Fe^{++} + \tfrac{1}{4}O_2 + \tfrac{1}{2}H_2O + 2OH^- \longrightarrow Fe^{+++} + 3OH^- \qquad (1)$$

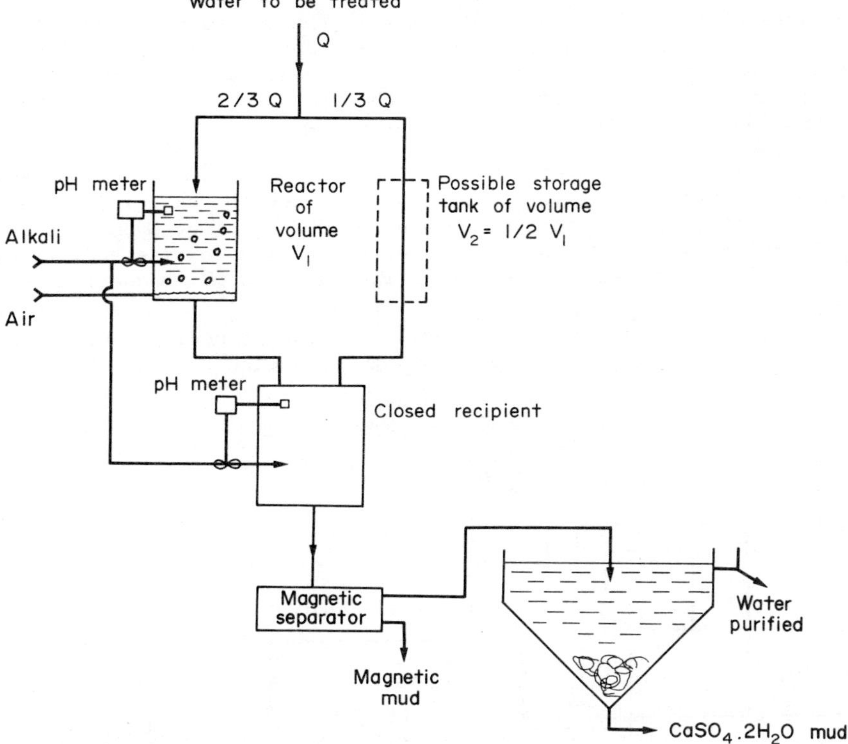

Fig. 1 The 2/3 - 1/3 ratios result from the fact that the Fe Fe_3O_4 Can also be written Fe_2O_3 .FeO from which it can be seen that 2/5 of the iron is in the form of Fe^{+++} and 1/3 in the normal form of Fe^{++}.

This is a quick reaction which is limited by the speed with which oxygen is introduced into the solution.

(b) Forming of an Fe^{++}/Fe^{+++} complex

$$2Fe^{+++} + Fe^{++} + 8OH^- \underset{1}{\overset{2}{\rightleftarrows}} Fe^{II}.Fe^{III}(OH)_3\ (OH)_2 \quad (2)$$

This is a balanced reaction.

(c) Conversion of the above complex into magnetite

$$Fe^{II}.2Fe^{III}(OH)_3\ \ (OH)_2 \longrightarrow FeO{\cdot}Fe_2O_3 + 4H_2O \quad (3)$$

This last, slow reaction is the one which requires rather strict pH control and allows or inhibits the productions of magnetite.

If there is no oxygen present or if large quantities of oxygen are introduced into the medium, there is no reason for reaction (2) to result in decomposition of the complex. In this case, magnetite is formed.

Thus, the process described in the preceding section consists in the oxidation of two-thirds of the Fe^{++} into Fe^{+++} which is then mixed with the one-third of Fe^{++} which is not oxidized. Magnetite is also obtained if, in a single reactor, oxidation of the Fe^{++} can be controlled and stopped when two-thirds of the latter have been oxidized: this is achieved by controlling the Redox (oxidation-reduction) potential.

A 5-g/1 solution of $FeSO_4$ neutralized to pH 9-9.5 corresponds to a Redox of less than -600 mV. If this solution is aerated and the pH held constant, the Redox remains at values of this magnitude for a long time; but, as soon as the Fe^{+++}/Fe^{++} ratio reaches 2, the Redox suddenly begins to rise and, if the aeration is stopped as soon as the potential reaches -520 mV. magnetite is obtained.

This process, which is more sophisticated than the preceding one, offers the advantage of being able to produce magnetite in a single tank (Fig. 2).

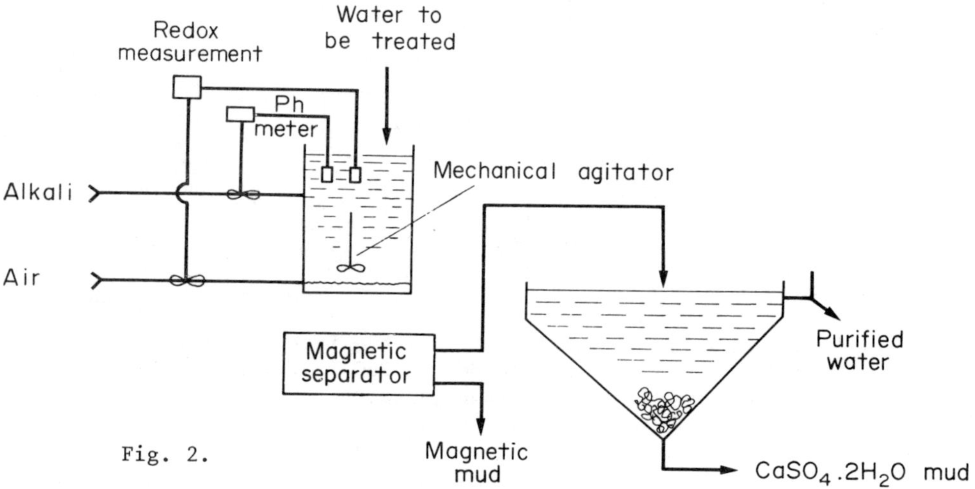

Fig. 2.

3. *Properties of the Mud Obtained*

The magnetite obtained using these processes is extremely fine and exists in the form of particles of less than 1 micron. These particles are completely distinct from the rod-shaped $CaSO_4 \cdot 2H_2O$ crystals measuring 1 micron in diameter and 10 to 20 microns in length.

Even if no attempt is made to recover the iron, the process leading to the formation of magnetite already has the following advantages:

- reduced energy consumption since only two-thirds of the iron needs to be oxidized;
- but especially, a much improved settling capacity of the mud obtained; with the usual mixtures of etching solutions and washing solutions, and with lime (or caustic soda) neutralization, 2.5 to 3 times less mud is actually produced than when 100% of the waste solutions are neutralized and oxidized.

Magnetite has yet another advantage over ferric hydroxide and geothite—its filterability. Magnetite actually filters 2 times easier than ferric hydroxide. Unfortunately, this advantage cannot be put to use since, from a practical standpoint, when they are pure, the two precipitates are regarded as being non-filterable. The filterabilities of $Fe(OH)_3$ and Fe_3O_4 are around 10^{10} s^2/g at a pressure of 0.5 kg/cm^2, whereas in order to be filterable at this pressure, this figure should be around 10^7 to 10^8 s^2/g.

On the other hand, with $CaSO \cdot 2H_2O$, which has a filterability of 0.7×10^7 s^2/g, filterability values of around 1.4×10^8 s^2/g are obtained for a mixture of $CaSO_4 \cdot 2H_2O$ and magnetite; the same proportions of $CaSO_4 \cdot 2H_2O$ and $Fe(OH)_3$ give filterabilities of around 2.3×10^8 s^2/g. As far as the filtration times are concerned, whereas it would take 1 hour to filter a cake containing ferric hydroxide, it would only take 40 minutes to filter a cake containing magnetite.

4. *Recovery of Magnetite*

The magnetite produced is so fine that it is difficult to recover it economically using the separators currently available on the market. They are intended to recover rather large magnetic particles (10 to 100 microns) and are therefore designed for large flow-rates. Moreover, the dimensions of their gaps are such that the fine particles in the middle of the liquid jets do not have enough time to be attracted by the magnet poles; instead, the currents entrain them outside the field and several passes must be made before large quantities are fixed.

Tests are currently being conducted to modify the existing separators, and they should lead to the development of devices making it possible to recover magnetite for less than 5 dollars/tonne of Fe_3O_4, including amortization of the installations and operating costs.

The magnetite recovered in this way exists as mud with a 70 - 75 per cent moisture content; depending on the operating conditions, a product is obtained which contains 0.5 - 5 per cent of impurities (in this case, $CaSO_4 \cdot 2H_2O$).

5. *Use of the Magnetite*

The magnetite recovered using the process described above is too fine to be used in the preparation of dense liquid (ore separation) or ferrites (electronics industries).

It can, however, be used as a paint pigment; depending on the drying technique, a black or reddish-brown pigment is obtained.

Considering the tonnages which can be recovered, the major use for the magnetite will be its recycling in ironworks, as we will see below.

There are two major possible ways of achieving this:

(a) Through agglomeration in which the ore and coke fines are moistened with water; in this case the magnetite can be suspended in water.

(b) In Kaldo-type steel mills in which the liquid solutions are cooled after oxidation by adding rich or scrap ores (these two types of ore could be advantageously replaced by magnetite when it has a sufficiently low sulphur content).

III. TITANIUM ORE PROCESSING INSTALLATIONS

One of the most common processes for obtaining titanium dioxide (TiO_2) consists in etching ores or titanium-bearing residues with sulphuric acid. In addition to titanium, these solutions contain numerous cations which, for a large part, are iron.

Let us take Canadian slag etching as an example. Following hydrolysis and recovery of the TiO_2, waste solutions are left which, in addition to the residual TiO_2 (5 to 8 g/l), can contain:

20 - 25 g/l of iron,

5 - 7 g/l of magnesium,

3 - 5 g/l of aluminium,

plus small amounts of numerous other elements (V, Mn, Cr, etc.).

These highly acid solutions contain 200 - 250 g/l of SO_4^{--} ions.

A large number of industries producing TiO_2 are located on the sea and, without any prior treatment, daily dump thousands of cubic metres of solutions having compositions close to that indicated above, or simply diluted with other washing solutions.

Among the treatment processes sometimes used, there is the neutralization - oxidation process, as in the case of pickling solutions; but the current implementation of this process does not make it possible to profitably utilize or recover the by-products. This is why the magnetite process has been developed.

Due to the large amounts of ions to be precipitated (SO_4^{--}, Fe^{++}, Ti^{++++}, Al^{+++}, Mg^{++}, etc.), the process must be carried out in two stages:

(a) Neutralization at pH 4 - 4.5 without oxidation. In this stage, limestone or lime is used to precipitate large amounts of the following ions:

SO_4^{--} in the form of $CaSO_4 \cdot 2H_2O$

Ti^{++++} in the form of $Ti(OH)_4$

Al^{+++} in the form of $Al(OH)_3$

These white precipitates are easy to filter and can be used as raw materials for the production of second-quality plasters (poorer mechanical properties than pure plaster).

(b) Production of magnetite in the residual solution. Once the above ions have been removed, the residual solution basically contains iron, magnesium and sulphate ions; application of the process described for pickling solutions leads to the production of a precipitate consisting of a mixture basically containing $CaSO_4 \cdot 2H_2O$ and magnetite (Fe_3O_4). With pH values of around 9 - 9.5, we are actually at the very limit at which $Mg(OH)_2$ starts to precipitate.

Therefore, the waste solution will still have a few ions (magnesium for the most part).

Contrary to the tonnages produced with pickling solutions, in the TiO_2 production industries there are enormous quantities which daily amount to hundreds of tonnes of by-products to be recycled or recovered.

The problems raised with magnetite are restricted to the conditions for handling such a fine product.

On the other hand, for the other by-products which, in addition, are not pure, the tonnages recovered are much higher than the local demands for them.

IV. CONCLUSIONS

This process, which consists in oxidizing the iron contained in various waste solutions into magnetite, already marks a great step forward in relation to the conventional neutralization - oxidation processes; it supplies by-products, one of which alone (magnetite) should make it possible to amortize the operating costs and the costs for the reagents required in the treatment installations.

Moreover, even though they still cannot be completely put to profitable use, the other by-products have physical and chemical properties which make them much easier to handle (non-gelatinous crystalline products which are much easier to settle and filter, etc., and they have reduced volumes and weights.

The last stage in the work described above would therefore consist in finding other outlets for these calcium hydrosulphate by-products.

Waste Exchanges: Improved Management for a New Type of Growth

J.C. Deloy

Editor in Chief, Nuisances et Environnement, 40 rue du Colisée, 75008 Paris, France

INTRODUCTION

The production of waste, both during and on completion of an industrial process, is always the result of a technological failure. The reduction, or even the total elimination, of waste is therefore an objective to be reached, but within the framework of maintenance or productivity and competitiveness of the firm.

This International Seminar should allow presentation of the results obtained in certain industrial branches by the implementation of production methods with no waste or with a minimum of waste.

These achievements must be developed and encouraged. However, if we are realistic, we must believe that for a long time yet, most industrial activities will engender—in the production stage—large quantities of waste and by-products.

These must be considered as a resource which has not found a new use or valorization. From that time on, any pure and simple destruction can be closely assimilated with wastage. Only three means are acceptable:

- recycling, i.e. the reincorporation of the waste into the production process giving rise to the same product;
- recovery of calories in the case of waste with a high calorific value, which cannot be recycled and which at present cannot be re-used;
- re-use, i.e. the incorporation of the waste into another production process in order to obtain a new product.

If we can consider that the first of these means has been studied and applied for some time by certain branches of industry and that the second, currently being developed, is only a last resort, waste exchanges assist re-use through the knowledge of available or desired waste matters and the contact between industrialists which they provide.

THE REASONS BEHIND WASTE EXCHANGES

Historically, the waste exchanges created in Western Europe from 1972 onwards result from the combination of several factors, which do not have the same degree of importance, but which all have the same aim:

- preoccupation with more economic management of natural resources which appear to be on the road to exhaustion;

development of new re-use technologies;

the desire on the part of certain industrialists to diversify their sources of supply, by calling upon national potentials;

the awareness of the commercial value of some types of waste and the existence of a market in this field (offer and demand).

1. Making raw materials and energy savings. France, and very often all the European states, are heavily dependent in most industrial sectors on sources of supply which lie outside Europe, for both energy products and basic raw materials.

More judicious use or certain waste materials or by-products, allied with improved recovery should allow for appreciable savings of national currencies by reducing imports. For France, this type of policy could mean savings, between 1980 and 1985, of 5000 million francs (1975 value).

2. Widening the range of the possibilities for re-utilization. Quantitative and qualitative information on many types of more or less complex waste materials has given rise to a great deal of research and development of new technologies.

A vast field of innovation has been opened up, coinciding with a certain rehabilitation of the recovery product which has become a "secondary raw material", according to the expression found in documents issued by the European Economic Commission in Brussels.

A major obstacle still exists far too often, however: the lack of incentive for re-utilization found in specifications for government and community contracts.

Thus products containing a certain percentage of recovery products do not find an application in the specifications of another age. This largely involves the responsibility of the Administration.

3. Looking for sources of supply on the national level. Prior to the creation of waste exchanges, the only examples of re-utilization of waste owed much to empiricism and to the chance meeting between industrialists and sometimes to the "middle-man" situation of a regularly supplied professional waste recoverer. For a long time almost all industrialists preferred the security of being supplied with virgin raw materials, especially since costs were remarkably stable.

The beginning of the seventies will remain, in economic history, the period of instability and in general the period of rises in the price of raw materials and energy. This has led an increasing number of industrialists to question a system granting too much favour, with no real economic basis, to virgin raw materials with respect to re-usable waste or by-products.

4. Establishing a dialogue between waste producers and waste re-users. An information structure and means for creating contacts between industrialists has become necessary.

In fact, for the producer of waste products, their elimination under conditions which did not adversely affect the environment represented a large item of expenditure.

The re-user of waste products found himself, in the case of complex products, without a monetary reference because there was no recovery price, but only an agreement between two partners. It is tempting and sometimes even worthwhile to bring competition into play, whether one is selling or buying. Industrialists were aware of the existence of a market involving supply and demand for all types of waste and by-products. But this market had to be developed and given a light structure.

HOW WASTE EXCHANGES OPERATE

The waste exchanges came into being in most of the Western European countries, beginning in January 1972. The following countries have these exchanges: West Germany, Switzerland, Austria, Belgium, the Netherlands, the Scandinavian countries (Denmark, Sweden, Norway and Finland), Italy, the United Kingdom and France.

1. *Initiative and date of creation.* The first observation to be made is this: most of the European waste exchanges were created on the initiative of chemical industry employer associations, following requests by their members. There are, however, a few exceptions: West Germany, Austria and Switzerland also have exchanges for all types of waste, organized by industrial or commercial associations. The French waste and by-products exchange was created by a magazine specializing in problems of the environment. The British waste exchange was set up by the Ministry of Industry. The first European waste exchange was set up in the Netherlands in 1972. The same year saw the creation of the Ecochem waste exchange by the Belgian Federation of Chemical Industries. But it was in 1973 that the largest number of waste exchanges were created, including the first West German exchange in January, the Austrian exchanges in February, the Italian and Swiss in March and the Scandinavian exchange in November. In July 1974 the West German Republic set up a second exchange for all types of waste, created by the Chambers of Commerce and Industry.

France was therefore something of a late arrival in the field and, at our instigation, the magazine *Nuisances et Environnement* created the Exchange for Waste Products and By-products in December 1974.

In June 1975 the Chemical Residue Exchange was created; this exchange uses space in our magazine and in *Chimie Actualités* to place its announcements.

2. *Types of waste and distribution of announcements.* Influenced by their creators, most exchanges are concerned with waste products of chemical origin and waste from other sectors which can be re-used by the chemical industry.

But the West German Industrial Exchange, the Scandinavian Exchange and the French Exchange for Waste Products and By-products publish information on waste from all industrial sectors, because what cannot be re-used within one industrial branch may possibly be used in other economic sectors.

The truth of this is shown by the Belgian Chemical Exchange which decided a few months ago to handle offers of waste from all sectors of industry, in order to improve its efficiency.

Apart from the "recoverability" of waste products, the widest possible distribution of announcements is an important factor for success. And most of the chemical exchanges have used and are still using a system of distribution of supply and demand by both sending circulars to members of the chemical employers' association and publishing announcements in the association's bulletin or periodical. By creating our Exchange for Waste Products and By-products, we wished to offer advertisers a maximum chance of finding a buyer or seller, depending on their wishes. We do this through the magazine *Nuisances et Environnement*, which is a multi-industry economic and technical periodical having the advantage of being read, in all branches of the polluting industries, by both commercial and technical management responsible for the elimination of waste, within the firm's pollution-control programme. The excellent results recorded are doubtless related to this fact.

3. *Type of participation.* All waste exchanges publish announcements offering and demanding waste, and some of them also accept offers of services from firms specialized in the detoxication and destruction of waste which is not yet reusable. Industrialists send the text of the announcement they wish to publish to the secretariat of the waste exchange.

In order to simplify the procedure, the forms to be filled in by the advertiser urge him to give a fair amount of basic data: description and detailed composition of the waste product, the quantities offered or demanded and their frequency, the geographical area concerned (province or department). The text may be followed, at the advertiser's request, either by the name and the full address of the firm, or by a code number for complete anonymity.

The announcements are distributed in accordance with the frequency of the publication (weekly, monthly, two-monthly and even sometimes half-yearly).

Replies go directly to the advertiser if he has given his name and address, or to the exchange secretariat which forwards mail to those using a code number.

This is very important since the secretariat does not know what is contained in the replies which are almost always forwarded in sealed envelopes, and it cannot influence in any way the subsequent negotiation which may or may not take place between industrialists.

4. *The first results.* However, when everything is weighed up, this formula, which is essential for satisfactory operation of waste exchanges, shows its limits.

It is easy to give the number of advertisements published and the number of replies received, but it appears somewhat more difficult to add together quantities of very varied types of waste products in order to determine the tonnages which the environment has been spared up to now. It is almost impossible to obtain information from advertisers on the results of their negotiations. Many arguments are put forward, basically psychological, commercial or industrial, which are, however, a well-known handicap for preparing statistics.

Nevertheless, if we must make use of statistical data, we shall confine ourselves to a preliminary balance sheet for one year of the Exchange for Waste Products and By-products, as shown in the publication *Nuisances et Environnement*. At the time of writing, we had published 186 advertisements: 152 offers of waste products (82 per cent) and 34 demands (18 per cent). These included advertisements for the Chemical Residue Exchange (around 50 per cent of the total), with which we signed an agreement in November 1975.

We have recorded 502 replies for 162 coded advertisements (the only ones which we are able to count) giving an average of three replies for each advertisement. This average may be misleading since if some advertisements (especially chemical waste with a highly complex composition) receive no replies, others can receive up to ten or twelve letters.

We have just mentioned the problem of highly complex chemical waste products, most of which are at present unable to be recycled or re-used because there does not yet exist a process for separating the component products, which are often valuable products. A very wide field of research has now been opened up in order to valorize waste products and to provide savings of raw materials.

CONCLUSION

Waste product exchanges do not pretend to find an immediate buyer or seller for all types of waste. Rather, they act as "discoverers" which can create a dynamic service and show the existence of a market for waste products whose re-utilization has not yet been considered. This is the industrial point of view. At the community level, the re-use of the majority of waste products and by-products makes considerable savings of raw materials, which are often imported, and therefore reduces dependence on external markets and on the prices of these materials.

Should we not therefore consider it a priority measure to reduce imports of costly virgin raw materials by encouraging re-use, in the same way that export is considered a priority measure for earning money? A formula for incentive should be studied by public bodies in order to recompense firms which incorporate into their production a percentage of recovered materials, thus making an important contribution to our trade balance.

This would mean the re-establishment of a certain balance within waste product exchanges, where at present supplies of waste products largely outweigh demands.

Metals in the Organic Chemical Industry: Problems and Aids for Non-Waste Technologies

László Markó

Veszprém University of Chemical Engineering, Veszprém, Hungary

This report is intended to focus attention on the special and yet rather universal role of metals in the organic chemical industry and its environmental pollution problems. Such a point of view is thought to be a somewhat less usual one, since the wastes of the organic chemical industry are always regarded as organic compounds (and from the side of quantity this opinion is of course justified) which cause problems mainly because of their high concentrations in gaseous and liquid effluents.

It should be noted, however, that the problem of organic wastes is almost in all cases a problem of quantity and not that of quality, since practically all organic compounds are biologically degradable (although at rather different rates, of course), whereas many of the metals contained in different industrial wastes are principally poisonous for the biosphere, and are therefore not eliminated—or eliminated in the wrong way, e.g. mercury—by biological processes. Elimination—or more precisely transformation to biologically inert forms—of metals dissolved in sewage or dispersed in the soil or in the air has thus to await for much more slow geochemical processes which means that there is an acute danger of accumulation of even seemingly minute amounts of poisonous metals.

Metals and metal compounds are used in the organic chemical industry mainly for three purposes: as catalysts, as inhibitors and as chelating agents. It is obviously impossible to go into a more detailed analysis of all these aspects of the problem, so I should like to restrict the treatment to the role of metal-containing catalysts. Even this is a rather diversified subject, since—as we shall see later on—metal-containing catalysts play both a positive and negative role from the point of view of pollution.

To mention negative effects first, the processing and utilization of spent heterogeneous metal-containing catalysts is solved mostly only in the case of large chemical enterprises or plants because of the relatively small amounts which make most working-up procedures not profitable. A significant exemption of this rule are the catalysts containing the rather costly metals like platinum (from the reforming processes of the oil industry) or palladium (widely used as a hydrogenation catalyst, for example, in the pharmaceutical industry), but already such metals as silver (catalyst for the oxidation of methanol to formaldehyde) may cause problems in an economy with a relatively small chemical industry. The metal causing most nuisance is probably nickel, however, which is perhaps the most widely used catalyst component in the organic chemical industry both because of its effectiveness and cheapness, the latter making regeneration a non-desirable burden.

The regeneration of nickel from spent catalysts may seem at first to be a very attractive task since, if regarded as ores, these materials are very rich in nickel. The main problem is, however, that they are contaminated with large amounts of organic material making the usual ore-processing technologies impracticable. Experiments have shown that one way to utilize these wastes is to burn their organic content, reduce the nickel oxide thus formed to metallic nickel and use the latter for the production of nickel carbonyl.

It should be mentioned at this point that nickel catalysts play a significant role in methane conversion, a basic process of both the petrochemical and the fertilizer industry, too, and due to the large volumes processed here such plants are also producing significant amounts of spent nickel caralysts. These do not contain organic material but are contaminated by sulphur originating from the methane and originally causing the deterioration of the catalyst. The above-mentioned working-up procedure may be used here too.

Just to mention briefly some other metals contained in spent catalysts of the organic industry: chromium, copper and zinc in different hydrogenation catalysts of the oxide type, vanadium in vanadium pentoxide-based oxidation catalysts, cobalt and molybdenum in desulphurization catalysts of oil refining, etc. All of these need special attention and processing.

One should bear in mind, however, that the main role which these metals play is the selective control of organic chemical reaction problems so their effect is clearly highly positive. It may sound trivial, but one of the fundamental requirements of non-waste technologies is the availability of highly selective processes producing minimum amounts of by-products and in this respect catalytic processes rank high in the list and in many cases could also serve as models for other organic reactions. Biochemical reactions are good examples for highly selective processes and it is well known that practically all reactions within a living organism are catalysed by the most efficient and selective catalysts we know, the enzymes.

Thus the metal containing catalysts of the organic chemical industry are a highly important tool in reducing waste products by increasing the selectivity of chemical processes. Many of the catalysts used approach the ultimate goal of 100 per cent selectivity for the reaction under consideration already rather well, so this may be regarded as a problem already largely solved. What remains to be done is the—obviously rather laborious—task to develop as many such catalytic processes as possible. I should like to mention, however, that there is one principal problem of catalytic reaction selectivity which has been solved only in the last few years and which therefore is a much less well known problem: the enantiomeric or optical selectivity.

It is well known that the most biologically important organic compounds (amino acids, carbohydrates, alkaloids, etc.) are optically active substances which exhibit their characteristic biological properties only in one of their enantiomeric forms. Thus, for example, all amino acids are members of the L-series and all living organisms use and synthesize only these enantiomers. On the other hand, if these compounds are synthesized in the laboratory by the methods generally used, both enentiomers (i.e. in the case of the above-mentioned amino acids the L and D-forms) will be produced in equal amounts. This obviously means 50 per cent waste in the latter case.

Research performed on homogeneous organometallic catalysts has provided now a principally new solution to this problem in form of the so-called asymmetric homogeneous catalytic reactions. The basic idea is the application of complexes containing chiral ligands as catalysts, for example asymmetric phosphine complexes of the platinum metals. These catalysts induce their asymmetry on the reaction products and thus imitate the function of enzymes, which have the same property. Up till now the catalysts for asymmetric hydrogenation are the best developed and the highest "optical yields" achieved with these are already above 95 per cent in the case of amino acid synthesis.

Much remains to be done in this new field of catalysis but the efforts seem to be well justified since this is one of the ways by which nature has solved its own perfect non-waste technology.

The Use of Natural Zeolites in the Chemical Industry

Dénes Kalló

Head of Department for Hydrocarbon Catalysis, Central Research Institute for Chemistry, Academy of Sciences, 1025 Budapest, Hungary

1. STRUCTURE OF ZEOLITES

Clarification of the applicability of zeolites in the chemical industry requires a review of their structure in order to present the properties of these materials.

Crystal Lattice

Zeolites are crystalline silica-aluminas. The three-dimensional lattice is built up of

$$\begin{matrix}-O\\-O\end{matrix}\!\!>\!Si\!<\!\!\begin{matrix}O-\\O-\end{matrix} \quad \text{and} \quad \left[\begin{matrix}-O\\-O\end{matrix}\!\!>\!Al\!<\!\!\begin{matrix}O-\\O-\end{matrix}\right]$$

tetrahedra bound by oxygen bridges. Different membered rings are formed in this way which are further connected with each other forming a sterically regular channel system in the crystal. The pore sizes are of molecular dimensions.

SiO_4 tetrahedra are isomorphally substituted by AlO_4^- tetrahedra. The substitution is random only the neighbouring location of AlO_4^- tetrahedra is not allowed. [1]

The negative charge of AlO_4^- is compensated by different cations. These formations either in the pores or on the outer crystal surface are the active sites responsible for the catalytic and adsorption properties.

Derivatives

Electrostatically bound mono-, bi- or trivalent cations are exchangeable.

When the compensating cations are protons Brönsted acidic sites are formed. H-forms can hardly be prepared by direct ion-exchange i.e. by a simple treatment with mineral acids, for more or less dissolution of aluminium from the lattice will inevitably take place, [2] leading to a damage in crystallinity. The lower is the Si/Al ratio and the looser the crystal structure the greater is the destruction of the lattice. H-forms of high catalytic activity and strong base adsorptivity are therefore prepared by thermal deammoniation of ammonium forms obtained by ion exchange with an ammonium salt. [3]

The reduction of transition metal ions on the cationic positions results in

H-forms and metal agglomerates.[4]

Acidity of the Brönsted sites may cause dissolution of the aluminium.[5,6] This self-destruction of the lattice can be observed if hydrated protons are present in the zeolite crystal of lower stability with a low Si/Al ratio, as, for example, molecular sieves A, X, Y, chabasite, stilbite, etc.

By heating an H-zeolite, water is evolved from two protons and a lattice oxygen. Brönsted sites are dehydrated in this way to Lewis sites above 400^0C. The latter formations have a rather narrow range of stability.[7] They transfer quickly to amorphous microdomains.[8] Therefore Lewis sites cannot be rehydrated.[5,7]

If the negative charge of the zeolite lattice is compensated by metal ions, the water attached to them is polarized or dissociated depending on the strength of the electrostatic field of the cation.[9] This effect of bi- and trivalent cations is very pronounced. A similar acidic behaviour of the cationic sites is observable as in the case of H-forms.

Cations modify the properties of zeolites, e.g. pore size, temperature range of crystal stability, catalytic and adsorption behaviour.

In zeolite lattices there are only few kinds of positions where cations may reside. Each type represents a homogeneous distribution of the sites. This is why zeolites are selective catalysts and adsorbents.

Catalytic properties can be modified by introducing different cations into a given zeolite. It has been proved, however, that the same cationic form of different zeolites may show different catalytic behaviour. This is the reason why natural zeolites can be used as catalysts in different forms, namely there are some zeolites that cannot be synthetized yet.

2. NATURAL ZEOLITES IN HUNGARY

The low costs and accessibility of natural zeolites support their use for adsorption purposes, where larger amounts are required.

Two sedimentary zeolites occurring in Hungary have been investigated in detail:[5,7,8,10,11,12] clinoptilolite and mordenite. Both are mined in Tokaj-Hills.

Clinoptilolite occurs in a purity of about 60 wt. per cent and the mordenite of above 80 wt. per cent. The accompanying minerals are quartz, feldspar and volcanic glass, the catalytic and adsorption effects of which are negligible.

Both zeolite-containing materials are mechanically strong enough and can be used after being simply smashed and sieved. Their thermal and chemical stability is satisfactory owing to the high Si/Al ratio. The crystal lattice of clinoptilolite collapses at 700^oC that of mordenite above 800^oC. Mineral acids do not destroy the crystal structure by dissolution of aluminium. H-forms of both zeolites are stable after water adsorption.

3. APPLICABILITY OF ZEOLITES IN NON-WASTE TECHNOLOGIES

On the basis of their most important properties zeolites can be utilized as (1) molecular sieves, (2) selective adsorbants, (3) ion exchangers, and (4) anionic matrices of catalysts.

Zeolites as Molecular Sieves

The pore sizes of zeolites are in molecular order of magnitude. The pore diameters of the actually most important natural zeolites, of clinoptilolite and mordenite enables the penetration of only straight-chain aliphatic hydrocarbons whereas the branched chain aliphatics or aromatics cannot be absorbed. This separation may be useful in the manufacture of detergents not containing tertiary carbon atoms, impeding the biological decomposition.

Stereospecific selectivity in the cracking of paraffins on clinoptilolite can be attributed to molecular sieve effect. [12]

Selective Adsorption on Zeolites

In contrast to conventional adsorbants, e.g. charcoal or silica gel, the boiling point or critical temperature of the adsorbates is not decisive in adsorption on zeolites. The adsorption ability of unsaturated compounds is higher than that of saturated ones.

The *adsorption of sulphur oxides* from industrial pollutions is of predominant importance. The adsorption of sulphur dioxide was measured from N_2 - SO_2 mixtures. [13] It has been found that the adsorption rates and capacities for zeolites are higher than for charcoal. The SO_2 adsorption capacity of natural mordenite at room temperature is nearly 150 mg/g which is slightly affected by the cations present in the zeolites; the differential heat of adsorption varies between 4500 and 6000 cal/mole. [14] At 150°C the adsorption is 32 mg SO_2/g-mordenite. [15] Adsorption isotherms of SO_3 on H-mordenite have been determined between 25° and 400°C. [16] SO_2 can be adsorbed from air on clinoptilolite- and mordenite- containing stones up to 4-5 wt. per cent at 120°C and desorbed at 350°C. [17] Synthetic molecular sieves 3A, 4A, 5A, AW 500, 13X selectively adsorb sulphur dioxide from dried industrial gases; after desorption the adsorption capacities decreased. [18] In the presence of water vapour zeolites of high Si/Al ratio are more stable in SO_2 adsorption. [19,20] Therefore, natural zeolites seem to be applicable for these purposes.

The *adsorption of nitrogen oxides* from effluent gases on different synthetic zeolites was investigated. [21,22] H-erionite and dealuminated mordenite were found to be the most effective and stable. Acid resistance is of decisive importance in these cases, too.

The selective *adsorption of hydrogen sulphide and mercaptanes* from hydrocarbon gases can be easily achieved on synthetic molecular sieves A and X. [23] The zeolites are regenerated either by desorption lowering the pressure or by partial oxidation. [24]

There are suggested methods for *separation of* saturated and unsaturated *hydrocarbon gases* on molecular sieve A,[25] X[26] as well as for separation of aromatics and saturated aliphatics on molecular sieve X[27]. The latter can be carried out on natural zeolites pretreated with sodium hydroxide.[28]

Zeolites as Ion Exchangers

The ion exchange selectivity of zeolites depends on the crystalline structure or more precisely on the cationic positions in the lattice. Clinoptilolite is, for example, very selective toward monovalent cations which explains its applicability in purifying waste waters.

Ammonium ions can be removed even in the presence of earth alkaline cations.[18,19] Regeneration is carried out by treatment with $Ca(OH)_2$ [30] or NaCl [31] solution. Nitrifying bacteria on the surface of clinoptilolite crystals oxidize ammonium ions to nitrate ions.[32]

Clinoptilolite has a very high selectivity for caesium. Thus the removal of ^{137}Cs from radioactive waste water is very effective [33] even in the presence of calcium or strontium.[34]

Zeolite Catalysts

The structure of zeolites allow the preparation of a wide range of catalysts. They have been investigated predominantly in hydrocarbon transformations. It seems likely that there are further possibilities for their application in catalysis. The use of zeolites in this respect is so diverse that it cannot be reviewed in this brief compilation.

General Remarks

Natural zeolites as easily accessible materials may have many kinds of application in non-waste technologies. However, the possibilities have not been clarified so far. Further investigations of given type of zeolites are required to solve practical problems of environment protection.

REFERENCES

1. J.V. Smith, In *Advances in Chemistry Series*, Am. Chem. Soc., Washington, 1971, Vol. 101, p. 171.

2. R.M. Barrer and M.B. Makki, *Can. J. Chem.* 42, 1481 (1964).

3. J.B. Uytterhoeven, L.G. Christner and W.K. Hall, *J. Phys. Chem.* 69, 2117 (1965).

4. J.A. Rabó, C.L. Angell, P.H. Kasai and V. Shomaker, *Disc. Faraday Soc.*, 41, 328 (1966).

5. H. Beyer, J. Papp and D. Kalló, *Acta Chim. Budapest*, 84, 7 (1975).

6. F. Koubowetz, J. Papp and H. Beyer, *Zeitschr. f. anorg. u. allgem. Chem.* (in the press).

7. E. Detreköy, P.A. Jacobs, D. Kalló and J.B. Uytterhoeven, *J. Catalysis*, 32, 442 (1974).

8. H. Beyer, E. Detreköy, D. Kalló and J. Papp, *Acta Geol. Hung.* (in the press).

9. J.W. Ward, *J. Catalysis*, 22, 237 (1971); *ibid.* 14, 365 (1969).

10. J. Papp, D. Kalló and G. Schay, *J. Catalysis*, 23, 168 (1971); *Vses. Conf. Het. Cat.*, Moscow, 1974, No. 46.

11. H. Beyer, D. Kalló and G. Schay, *Rev. Rom. Chim.*, 17, No. 1, 2, 29 (1972).

12. H. Beyer, *Acta Chim. Budapest*, 84, 25 (1975).

13. S. Kasaoka and B. Sasaoke, *Kogyo Kagaku Zacchi*, 74, 2213 (1971).

14. J. Valyon, J. Papp and Czárán Lné., *Magyar Kêmiai Folyóirat* (in the press).

15. S.A. Anurov, *et al.*, *Trans. Mosc. Khim.-Tekhnol. Inst. 1974*, 79 (3-4), 79 (5-7).

16. Sh.O. Minasyan, *et al.*, *Zh. Fiz. Khim.* 44, 3136 (1970).

17. H. Ishikawa, *et al.*, *Waseda Daigaku Rikogaku Kenkyusho Hokoku*, 51, 46 (1971).

18. D.A. Martin and F.E. Brantley, *Bur. Mines Rept. Invest. No. 6321*, p. 15, (1963).

19. J.K. Tamboli and L.B. Sand, *Proc. 2nd Int. Clean Air Congr. 1970*, p. 861.

20. L.I. Vinnikov, *et al.*, *Khim. Prom.* 49 (2), 125 (1973).

21. E.B. Krasnyi, *et al.*, *Khim. Prom.* 42 (7), 523 (1966).

22. I.L. Varshavskii, *et al.*, *Sb. Tr. Vses. Zoach. Politekh. Inst.*, 80, 133 (1973).

23. U.S. Patent 3 078 634, 3 078 640, 3 078 641.

24. Ger. Offen. 2 315 113 (Grace W.R. and Co.).

25. U.S. Patent 3 078 636.

26. U.S. Patent 3 078 644.

27. U.S. Patent 3 078 643.

28. Y. Miyata, *Kogyo Kagaku Zacchi*, 73, 2090 (1970).

29. Battelle-Northwast, *Water Pollut. Contr. Res. Ser. 1971,* No. 17010 ECZ 02/71.

30. Jap. Patent 74 64, 253.

31. M. Funaki, *et al., Kogyo Yosui,* 188, 63 (1974).

32. R.C. Sims, *et al., Environ. Letters,* 4 (1), 27 (1973).

33. L.L. Ames, U.S. At. Energy Comm. HW-62607 (1959).

34. G. Lenzi, *Conv. Probl. Acque Ital., Atii Conv.,* Milan, 1965, p. 451.

The Utilization of Brown Coals Other Than for Energy Production

V. Cziglina*, L. Dzsida and Z. Meleg*****
**Collieries of Tatabánya, Tatabánya, Hungary*
***"VIDUS" Enterprise of the Collieries of Tatabánya, Tatabánya, Hungary*
****Hungarian National Council for Environment Protection, Secretariat, Budapest, Hungary*

The trend of utilization of various energy resources is capital not only for branches of industry, but also for the environment in respect of the quantity and quality of gaseous and solid combustion products. For this reason it is certainly worth while to consider a multi-purpose processing and utilization of one of the most important energy carriers: old and young brown coal (lignite) and at the same time to insist upon novel, economic processes, which protect man and his environment.

The utilization of brown coals in addition to energy production came into prominence for practical purposes only of late. This new aspect has drawn the attention of experts to solvent processes by which, first of all, bioactive components of the coal are extracted.

ORGANIC COMPONENTS OF BROWN COALS

By means of coal petrography examinations, hundreds of rock-forming organic minerals can be found in brown coals, but modern systematization puts them into three main groups, namely into those having bituminitic, huminitic or oxynitic main components.

Bituminites are the least-oxidized substances, their calorific values are high, their flames are long as a result of their plasticity and they improve the briquetting characteristics of brown coals.

Humic and *oxynitic* parts of the mixture are predominantly mixtures of organic acids in a solid state and in elementary C:H:O composition contain 10 - 40 per cent oxygen, 5 - 6 per cent hydrogen and only 50 - 80 per cent carbon. Nitrogen and sulphur are permanent accompanying substances or parts of these compounds. As a result of their composition their calorific values are comparatively low. Being brittle they deteriorate briquette quality. At the same time their biologic effect on living organisms is very conspicuous.

In the case of plants, humic compounds favourably influence the intake of nutrients through leaves and roots, oxygen supply during critical periods and the thermodynamic processes of plant life. They also have an effect on living organisms of higher and lower classes than plants. As accompanying matter of humic substances extracted from brown coals, organic compounds may also be found, which themselves are stimulants and promote utilization of humic substances. Huminoid brown coals are carriers of biologic energy.

TYPES OF BIOACTIVE PRODUCTS AND THEIR MAJOR EFFECTS

From raw humus-carriers and their extracts, fertilizing, stimulating and disinfecting agents are produced. *Fertilizing agents* are introduced into the soil in quantities of 1 t/ha at least. They improve the physical structure and water economy of soils and the culture of microorganisms; they contain or take over and conserve—in an ideal way for plants—nutrients and damp acid and alkali impacts. Of *stimulating* agents only a few kg/ha have to be used; their aqueous solutions are suitable for irrigation, for drenching propagation matter and, first of all, for sprinkling leaf surfaces. They accelerate plant ripening and coloration and render it uniform; they increase (green weight) quantities and internal content values of crops, stimulate radication development and improve callus formation of grafts. The improved proportion of biologically important protein components is also noteworthy, considering the internal content of plants. *Disinfecting* with the above can be primarily effected by seed steeping.

BASES OF SOLVENT COAL PROCESSING

The method of solvent coal processing developed in Hungary, but basically relying on early knowledge, is a technology on the border line of coal dressing and coal chemistry. Considering further utilization of coal it also may be regarded as a coal-refining process, which does not interrupt procedures but only connects them into series.

In case of alkaline digestion (Fig.1) humic compounds are dissolved by the monovalent alkali and immediately form a monovalent humic metallic salt (humate and humate solution, respectively) with the cations of the alkali. When treating the liquid obtained in this way with a strong mineral acid, the humic acid will appear as a loose gel containing a comparatively small quantity of dry substance only. The gel is concentrated and dried, respectively, first by mechanical and then by thermal treatment. Drying or concentrating of the humate solution can be carried out most favourably in case of power plant application of this technology.

It is probable that until now a large-scale spread of the agricultural use of humic substances obtained by solvent extraction has been very much limited by the circumstance, that the solvent, its concentration and the temperature of gel drying were not always well selected.

The yield and prime cost of bioactive products may be modified even in the ratio 1:101 by the quality of the raw humic carrier and by the different ways of carrying out dissolving operations. Experiments in Hungary have proven that older brown coals, e.g. those of the Eocene epoch, are very good basic materials for a quantitative and qualitative recovery of bioactive products.

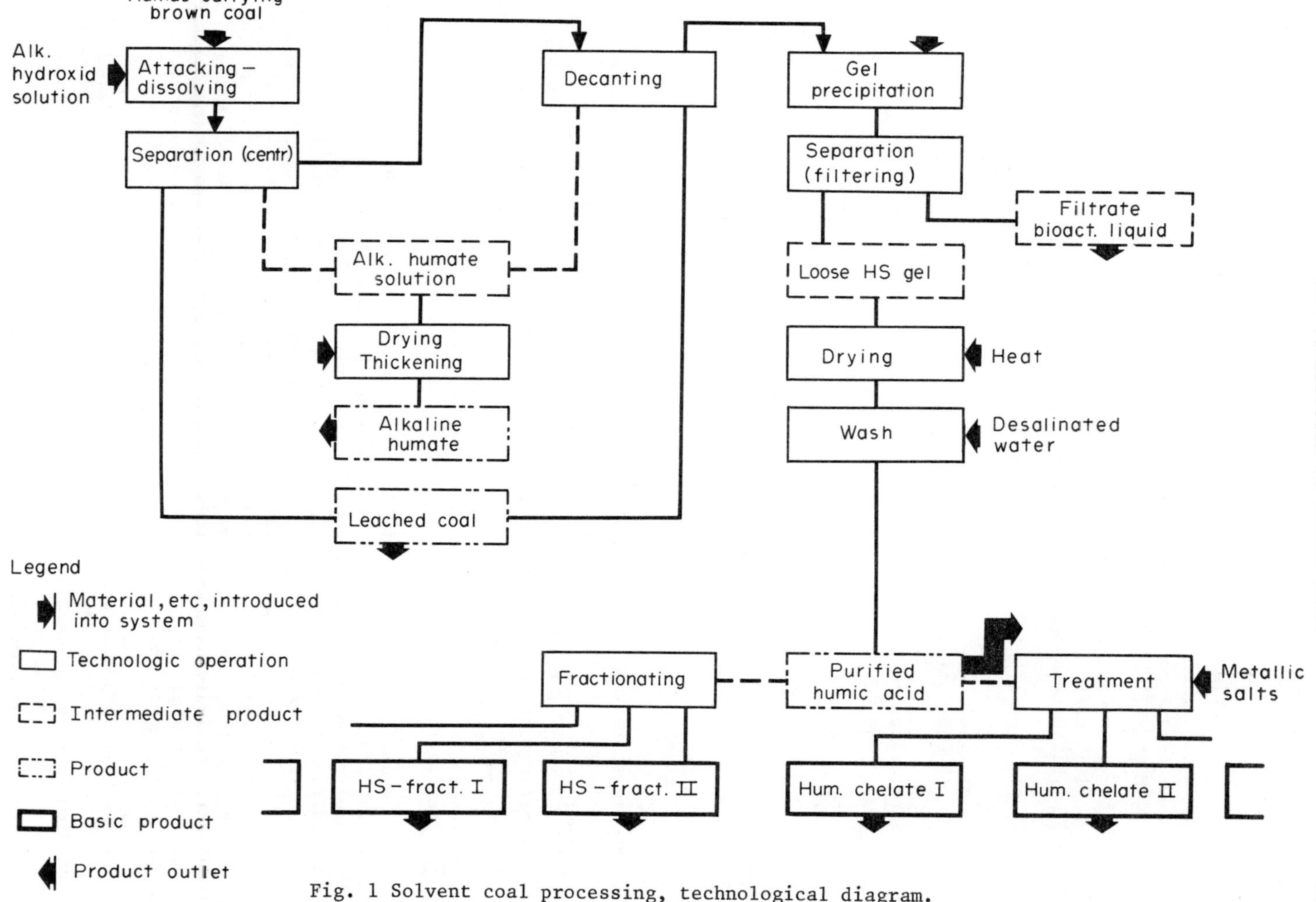

Fig. 1 Solvent coal processing, technological diagram.

TABLE 1 *Test results of brown coal types*

Age and denomination of coal	Ash_V %	$Moisture_V$ %	HS_V %	HS_O %
Pleistocene bituminous peat	43	1	34	60
Pliocene lignite	17	19	27	42
Miocene I brown coal	11	16	20	27
Miocene II brown coal	44	5	0	0
Eocene I brown coal	*11*	*21*	*64*	*95*
Eocene II brown coal	11	16	37	50
Lias hard coal	20	3	1	1

The ash, moisture and humic acid (HS_V) content values of the material examined may be misleading sometimes. Namely, if the moisture of the sample is essentially lower than that of the commercial product (e.g. in case of peat in Table 1) the index HS_V will show a higher value than the real one. Table 1, however, draws a picture of the values of fossil biological energy carriers. Among the Hungarian brown coal types, Eocene brown coal has been enhanced by underlining as being, to our knowledge, the richest in humic compounds. In this case the sample tested is a commercial product, the humic acid content of which exceeds 60 per cent in raw state and 90 per cent related to ash-free dry substance.

In Fig. 2 shown from the point of view of structure examinations the molecular weight distribution of four humic carriers may be seen. Curve *a* shows an extreme case: the molecular distribution of a humic acid (fulvic acid) prepared from a salicylate medicine; curves *b*, *c*, and *d* reflect the molecular weight course of extracts obtained by the same technology from a garden mould (compost), a fen peat of the Hévizfürdő region and the Eocene brown coal mentioned above. Molecules of different sizes—or more precisely molecular groups (micellae)—are fractionated in the most uniform manner in the humic acid of brown coal. The individual fractions contain the functional groups responsible for bioactive properties, chemosorptive capacities and redox characteristics in the given proportions.

One of the basic products of the solvent process developed in Hungary, the alkali humate (condensate, dried product), is readily soluble in water. Humates of bivalent metals are hardly soluble in water, but disperse readily and contain part of the metal in chelate bonding. Biological effects of Cu, Zn, Co and Mn humates are also promising for plant cultivation, and in particular for

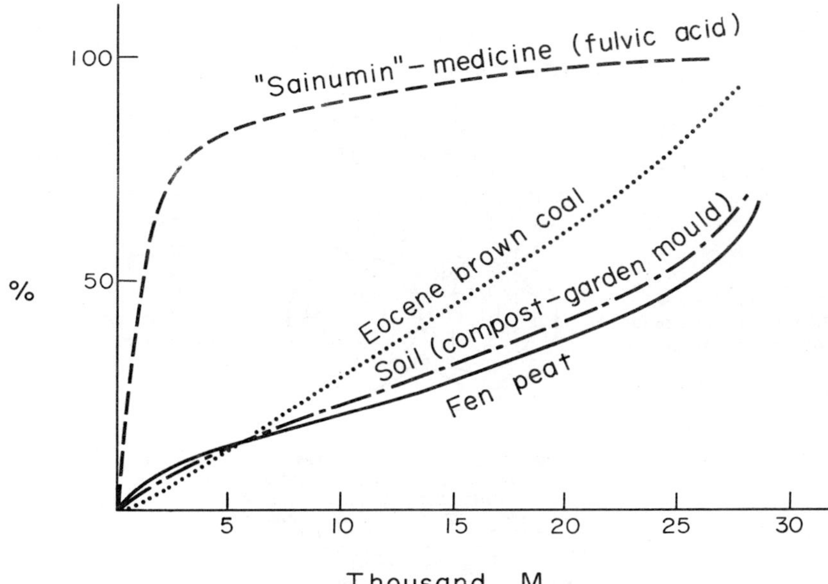

Fig. 2 Molecular weight (particle weight) distribution curves of humic acid types.(From Jate, 1974, modified by Dzsida, 1976.)

animal husbandry, admixed to the mineral premix of fodder, replacing conventional anorganic metallic salts.

It should be mentioned here that products of medium molecular weights obtainable by additional solvent fractionating of humic acids (fractions of himatomelanic acid) are showing bactericidal effects and in Hungary proved to be advantageous, e.g. for the medical treatment of gangrenous wounds. Prior to this the English scientist Lord Energlyn (W.D. Evans, Nottingham) has found a similar substance introduced by him under the name vitricite effective against tuberculum $H_{37}R_V$.

"REFINING" OF LEACHED COAL

It has been mentioned that from among the main components of coal, huminites and oxynites deteriorate briquetting characteristics and are combined with air-polluting sulphur as well. Both substances are mostly carried over into the extract in the course of processing. The carbon content of the organic components of the leached product is enriched after the extraction and its calorific value becomes higher.

The study of the forming of anorganic components and structural water content of coal and of humic acid, respectively, leads to a surprising perception. The ash content of humic acid is—usually—lower than that of raw coal, but the ash content of humic acid of younger (Pliocene and Miocene) "earthy brown coal"

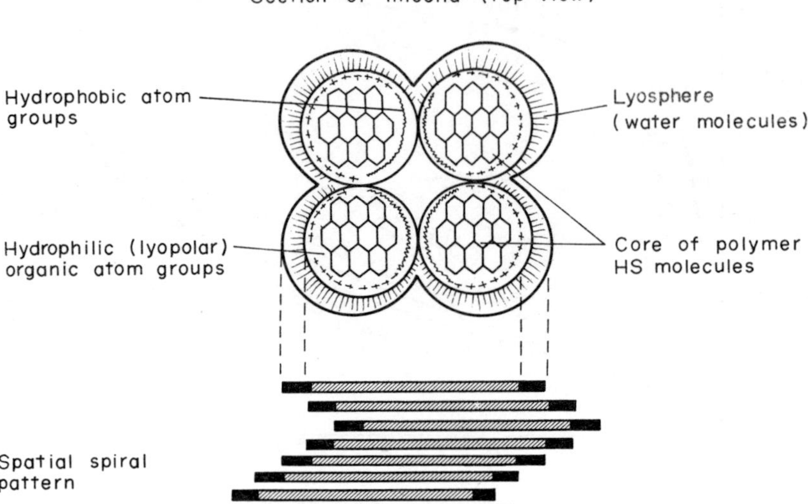

Fig. 3 Diagram of hydrophilic humic acid molecule groups. (From Adge *et al*, modified by Dzsida, 1976.)

types may be much higher. From certain lignite types—as a matter of fact, not humic acids, but bioactive humic acid—clay mineral complexes can be dissolved, sometimes with an uncombustible content exceeding 30 per cent. Humic acids (including the humic components of complexes) have micella structures (Fig 3) and lyogel characteristics. They are enclosed by electrically arranged water hulls, the internal water molecules, which carry positive charges, are connected to the OH-radicals of the carboxy groups of the humic molecules by Van der Waals forces.

The combustion heat of some raw coal types and their humic acids ($E_{sz,o}$ and E_{HS}) and the difference between them are given in kcal/kg in Table 2. The trend visible from the table is reflected by the changes in the C_o and H_o contents of coals and humic acids as well: in the humic acids examined the former diminished by 7 to 10 per cent and the latter by 0 to 0.9 per cent.

On the basis of our knowledge and considerations until now we think it possible to develop a large-scale industrial technology, the primary aim of which would be the refinement of raw coal by means of the solvent method, the costs being covered by the counter-value of the bioactive substance obtained as a by-product. The quantity of the latter may amount to 5 to 35 per cent of the basic material. Waste heat, which would otherwise only degrade the environment, can also become utilizable.

The advantages of a fuel with a lower ash, moisture and nitrogen content and higher calorific value are obvious, regarding clean air and numerous other as-

TABLE 2

Age and type of coal	$E_{sz,o}$	E_{HS}	$E_{sz,o}$	E_{HS}
Eocene bituminous brown coal I	7137	5283	1854	
Eocene bituminous brown coal II	6896	5216	1080	
Pliocene earthy brown coal (lignite) fly dust	6187	5114	1073	

pects. (It is worth considering that a big part of the world's brown coal reserves consist of earthy brown coal!) Processing of the bituminitic product poor in huminites, but rich in tar, wax and resin, may lead to considerable results.

UTILIZATION OF BIOACTIVE SUBSTANCES ON COAL BASIS FOR ENVIRONMENT PROTECTION

In Hungary, it is primarily the collieries of Tatabánya which are dealing with the utilization of brown coal in addition to energy production. As a matter of fact manufacturing fertilizer type products has been started for the purpose of repairing damage caused by mining operations, i.e. for biological recultivation. Into this product type the waste of the extraction process, raw humus-carriers (e.g. in the shape of unmarketable coal dust polluting the environment) can be introduced and their natural content of nutrients can also be enriched with macro-, meso-, and microelements.

According to the Hungarian mine law the ruins of undermined structures have to be pulled down and removed, the pits of open-cast minings backfilled, dumps and their banks, etc., levelled and the restored terrain has to be bound by vegetation and returned to sylviculture and agriculture for tillage.

A special task in the course of coal winning and crushing is prevention of the formation of flue dust and of the environmental pollution caused by it. Because of its explosiveness, flue dust may be very dangerous as an air pollutant (e.g. in an above-ground crushing plant). It is an important Hungarian conclusion that flue dust intensively inclined to oxidation and chemosorption can—following a simple treatment—fix cations and by means of them organic anions and in respect of biological effects can approximate the products of extraction. They can be rendered suitable for disinfecting industrial waste (e.g. that of glue factories) and agricultural waste (e.g. that of animal keeping) and to convert these wastes into a product free from foul smell, from breeding grounds for flies, etc.

The surface of pits and dumps forming in the course of mining operations consists of sterile "wild earth" or flue ash, slag, etc. An ancient—and at the same time very costly—method of restoration consisted in scraping off the arable humus soil, depositing it and, after having levelled the dumps, replacing it onto the surface. The microflora of the arable land, however, will be transformed within the deposit, its useful (aerobic) microorganisms will perish, its nutrient reserves may be exhausted and thus the replaced earth will not possess the same biological value as before.

Advanced biological recultivation following technical landscape correction will reject the above method. Instead, a readily and quickly humifiable and/or humified fertilizing agent with an adequate nutrient content will be worked into the upper layer of the "wild soil". This agent will at the same time be favourable to the establishing of a microflora. It is primarily this purpose that the carbomineral mixtures mentioned before are used for, but they also are suitable for the amelioration of barren soils. Worked into culture-soils, as tested on many plant types, they have given surplus yields as high as 20 to 30 per cent as compared to patches treated with fertilizers only.

Rocks coming to the surface in the course of mining are highly different from each other and their types will also determine the length of restoration.

If sandy, clayey media come out of the overburden of opencast mines, a few years will be sufficient for biological recultivation. In case of dumps of harder (shaly, etc.) rocks, fly ash and similar materials, the situation will be more difficult, although atmospheric condensations will also participate in their reutilization. Experience has shown, however, that sowing them with grass and planting with trees at a satisfactory tempo seems feasible, in particular if water supply can be provided for separately. By rational use of coal-based stimulating agents and of carbomineral agents (raw humus-carriers enriched with fertilizers from time to time) it was possible to improve root-forming of saplings and render root-taking surer when planting with trees. By means of systematic biological renaturalization such areas can be rendered suitable for agricultural and even for horticultural tillage.

Agricultural land is being reduced by industrialization all over the world, but the increase of pollution makes rapid restoration of worked-over areas indispensable and likewise the amelioration of barren territories and a steady increase of harvest results. This work can to a high degree be promoted by the utilization of bioactive substances as extensively as possible.

*Non-Waste Technology in Belgium**

A.G. Buekens
Professor, Vrije Universiteit, Brussels, Belgium

INTRODUCTION

Belgium is a small (30,500 km^2), but densely populated (9 million people) and industrialized country. Waste generation rates can be estimated at[1]

3 million ton/year household refuse,

7.5 million ton/year industrial waste,

0.05 million ton/year problem waste.

An increasing part of the household refuse is disposed of by incineration, pulverizing or composting. Most industrial wastes are tipped, but it is increasingly difficult to find suitable tipping sites and to obtain a tipping license.

Industry is becoming conscious of this problem, and also of the high intrinsic value of the incurred losses of materials and energy. The tightening regularity requirements in the field of pollution abatement are an important stimulus for developing non-waste technology. A considerable part of the industrial wastes is water borne, and while the intrinsic value of the materials lost in the wastewaters is as a rule not excessive, the cost of their biological oxidation or other treatment rapidly becomes prohibitive. For this reason recycling of process water after separation of waste components became an economic necessity in many industries.

Specific waste problems can be tackled by single industrial enterprises, or by groups of related industries. In Belgium many associations of industries have their own government-supported research centre, called "De Groote centres".

IRSIA founded a special committee, named Conacord, to help industry in solving its problems of waste recycling and disposal. IRSIA is an Institute, sponsored by the Ministry of Economic Affairs, which encourages scientific research in industry and agriculture by supporting part of the cost of the research programme. At present Conacord is already supporting a number of waste recycling projects and is investigating others.

The "Prototypes" department of the Ministry of Economic Affairs gives loans, bearing no interest, to industries, which are developing new equipment or processes. In the field of waste technology loans were given to Ets. George, for the development of cryogenic shredding of baled cars, electric motors, rubber tyres, etc., to S.A. Denaeyer, for the development of a rotary kiln furnace

* This document complements the report by I. Van Vaerenbergh, pp.147-153.

for incinerating industrial wastes, and to an undisclosed company, for the development of methods of re-using animal manure.

The Services of Scientific Programmation also decided to launch a research programme centred on waste recycling and disposal. This programme is discussed in a separate paper presented by Mr. Van Vaerenbergh.

The Belgian Federation of Chemical Industries (Féchimie) was among the first to launch a Waste-exchange programme.

The waste generated by one company may prove to be useful to another. Due to the short distances and the excellent means of transportation this programme can be relatively successful in a small country such as Belgium.

Obviously non-waste technology consists of solutions to specific problems and varies with the type of industry considered. A number of case studies, which are related to Belgian Industrial conditions, will now be considered.

COLLIERY SPOILS

The coal-mining activity in Belgium is steadily declining. Many coal veins are narrow and inclined. Mechanical extraction of coal has a mean volumetric yield of 79 per cent, and a mean gravimetric yield of 69 per cent[2], so that important quantities of shale are brought to the surface. This material often can only be used for landfilling purposes. Coal washing, moreover, yields sludge, which is settled in basins or lagoons.

In the southern part of the country a number of large heaps of colliery spoils still date to the nineteenth century and sometimes contain appreciable amounts of coal. Removal of these heaps is often beneficial, if only from an urbanistic point of view. These spoils can be dry sorted into material which is acceptable in power plants and in rejects. In the manufacturing of cement both coal and shale can be valuable. At Roselies, a heap of colliery spoils is transformed into light aggregates by a high-temperature treatment in a rotary kiln.

Unfortunately, the recovery of old colliery spoils is not always feasible. The composition of the spoils is quite heterogeneous and largely unknown. Space should be available to install sorting equipment and, after sorting, a new heap of spoils is formed, which is not always much smaller than the original one.

Table 1 gives some data on the use of old spoils and sludges.

In the coal mines mine gas is captured, both for safety and economic reasons.

THE IRON AND STEEL INDUSTRY

In the blast furnace iron ore is reduced to "hot metal" (pig iron). The slag collects undesirable impurities and is granulated by quenching with a vigorous water spray. The granulated slag is a desirable material, e.g. in the cement industry. Inadequate quenching can lead to the formation of large blocks of slag, which generally are useless and form a waste product.

TABLE 1

Use of old colliery spoils and sludge (1000 tons, 1975)

	Spoil heaps			Sludge basins
	Mixed	Sludge	Total	Sludge
Combustible shales				
ash content < 40%	319	-	319	65
40-60%	191	68	260	310
> 60%	852	342	1193	100
Total	1362	410	1772	475
Supplied to				
power plants	425	41	466	180
cement factories	757	471	1228	85
France	103	30	132	-
Germany	33	-	33	-
Total	1318	542	1859	265
Available	110	16	125	318

The blast furnace gas is used to heat the Cowpers, i.e. a number of regenerative air preheaters. The larger part of the gas is available for in-plant or nearby industrial uses. The dust removed from the blast furnace gas returns to the sintering plant, together with the dust of the steel-making and the coking plant, and the iron oxide of the steel rolling plant. These materials are sintered on a continuous grate to a synthetic iron ore, which is recycled into the blast furnace.

After desulpherization by means of calcium carbide and removal of the slag the hot metal is charged into the steel-making furnace, together with a carefully controlled amount of ferrous scrap. In the basic oxygen furnace, the undesirable elements, such as carbon and silicon, are oxidized with the aid of water-cooled oxygen lances. Meanwhile, lime, flux and ore are added. The hot metal is transformed into steel and poured into ingot moulds. The slag is solidified and broken; the contained iron particles are recovered magnetically, but no market exists yet for the broken slag. The heating value of the generated gases is also wasted. Recovery of this gas would allow CO-emissions to be reduced.

In terms of tonnages ferrous scrap is the most important secondary material. Scrap can be recycled into the steel mills, the blast furnace, the foundries of cast iron, steel or ferro alloys, or to be used in cementation (copper precipitation from its solutions) or in the chemical industry.[3]

According to its origin, ferrous scrap can be subdivided in:

home scrap, arising in the steel plant. Home scrap has a known origin and a high purity and is recycled immediately, as a matter of good housekeeping;

prompt industrial scrap, generated by the manufacturing industries. This scrap is also recovered almost automatically, and sold to the steel mills or to a scrap merchant;

quality can be high, e.g. the steel plate from the car manufacturers, or low, e.g. borings and turnings, contaminated with oil and dust;

old or obsolete scrap, consisting of iron and steel discarded by all kinds of users. Important quantities of scrap arise in car shredding and ship-wrecking. The quality of old scrap is highly variable. Poor quality scrap cannot be recycled in the steel-making process, because of excessive contamination by foreign metals or materials.

Scrap requirements vary with the kind of steel-making process. In the basic oxygen furnace the required melting heat is provided by oxidation of the impurities in the hot pig iron. About 25-40 per cent of the charge consists of scrap, which prevents overheating of the charge.

Scrap can be melted directly in an electric furnace, to yield less demanding products, such as reinforcing bars. Foundries also accept large quantities of scrap. Cans, borings and turnings can be charged into the blast furnace.

The availability and competitive position of ferrous scrap is influenced by:[3]

new scrap processing and upgrading techniques, e.g. cryogenic car shredding, removal of non-ferrous contaminants and detinning;

the introduction of continuous casting, which reduces scrap generation at the casting stage from 20 per cent to 6 per cent, and eliminates the need for ingot moulds and stools. During conventional casting losses arise when shifting the pour from one mould to another, when ingots stick to the mould, etc.;

the replacement of Siemens-Martin or of Bessemer furnaces by the basic oxygen process or by electric furnaces. This shift increases scrap consumption in Belgium, but reduces its use in most other countries;[1]

the nature of the steel products. Sheet production generates much more scrap than does the production of rails or beams;

the possible breakthrough of direct reduction by natural gas or desulphurized coke oven gas;

the possible implementation of material conservation strategies, involving an increase of the lifetime of products, or the re-use of components without remelting.

The US Bureau of Mines studied the possibility of increasing the amount of scrap in the charge, by pre-heating the charge with the hot off-gases.

The recycling potential of scrap is highly affected by the presence of foreign elements, such as copper, tin and lead. These components are not removed during steel-making, whereas phosphorus and manganese report only partially to the slag. Automotive scrap contains copper from electrical wiring and brass. Cans contain up to 0.5 per cent of tin. Solder contains lead and tin.

The adverse effect of impurities is important especially when making steel sheet. Alloy elements, such as nickel and molybdenum, moreover have an unpredictable influence upon steel upon heat treatment. Cast iron and reinforcing bars are less demanding applications.

THE NON-FERROUS METALLURGICAL INDUSTRY

Most ores contain a number of valuable metals, which can only be recovered by means of sophisticated refining processes. Thus non-ferrous metallurgy yields many successful applications of non-waste technology. Belgium is a major refiner of non-ferrous ores and metals and has a vast amount of know-how in this field.

Pyrometallurgical refining methods yield a slag, which still contains valuable metals. All metal-bearing slag is recycled into the blast furnace, the slag of which has a low metal content and is discarded. Most of the blast furnace slag is exported to the Netherlands, to build dikes and reclaim polders.

The electrolytic winning of copper yields a sludge by-product, from which the noble metals are extracted in a cupola furnace. The resulting slag also returns to the blast furnace.

The electrolytic winning of zinc proceeds over:

- the roasting of the sulphide ore;
- the dissolution of ZnO in sulphuric acid;
- the purification of the $ZnSo_4$ solution;
- the electrolytic deposition of the metal.

The dissolution of ZnO leaves a non-soluble fraction, which is further separated into soluble Zn, Cu, Cd and Fe salts, and a Pb-containing residue. Iron is precipitated as a goethite or jarosite residue, which normally forms a waste product.

Ironically, the waste iron oxide has a markedly higher iron content (35 - 50 per cent) than the small quantity (116,000 tons/year) of iron ore produced in Belgium (34 per cent Fe). The oxide waste is improper for steel production because of the presence of excessive amounts of valuable, non-ferrous metals.

S.A. Vieille Montagne developed methods to produce the goethite as an iron oxide pigment, which can be used to colour concrete and asphalt products. When bound with 10-15 per cent of lime the waste forms a suitable ballast material.

OTHER MINERAL WASTES

The most important waste product from the Belgian chemical industry is phosphogypsum. No market exists for this product, natural gypsum being plentiful in the north of France. Phosphogypsum is disposed of by tipping. The chemical industry tries to convince the Ministry of Public Works to use gypsum in the foundations of smaller roads, as a substitute of sand.

S.A. Solvay is investigating the possibilities of incorporating the sludge, arising in the production of soda ash, in building materials.

S.A. Cemstobel developed a proprietary method for transforming sludge and solid wastes into a solid material, to the leaching losses of which are particularly low. It is believed that this method yields better results than competing and more expensive processes.

In a new research programme, S.A. Cemstobel hopes to produce solid silica adsorbents from waste acids and blast furnace slag. Both the Cemstobel and the Solvay research programmes are supported by IRSIA's Conacord-committee.

THE GLASS MANUFACTURERS

A large quantity of waste glass arises during the production of glass bottles: during quality control about 10 per cent of the bottles are rejected, broken and eventually returned to the furnace.

Two Belgian companies are active in recycling cullet to the bottle manufacturers.[4] Most of the recycled cullet (*ca.* 75,000 tons/year) is of industrial origin and arises in the bottling division of breweries, dairies, etc. The amount of post-consumer cullet is rapidly increasing, due to the organization of selective collection of glass bottles and to the presence of glass receiving containers at supermarkets. Post-consumer cullet currently represents some 10,000 tons/year.

THE PULP AND PAPER INDUSTRY

Like the steel and the glass manufacturers the paper industry is capable of absorbing its own waste, which consists of broke (during a breakdown of the paper sheet), of trimmings, etc. The better qualities of wastepaper are also actively recycled, but mixed wastepaper in Belgium has only a limited market (fluting of corrugated paper, felt and roofing paper), due to the absence of de-inking facilities. According to a recent report of Cobelpa, some progress is to be expected in this field.[5] Shortening of the fibres, and stringent mechanical and hygienic specifications preclude total recycling.

S.A. PRB (Balen), is studying the chemical modification of cellulose fibres by means of monomers, which are later polymerised. The resulting material is then hot-pressed, and the mechanical properties of the product are evaluated as such, and at various temperatures and moisture contents. The product is also analysed to determine up to which extent the monomer added to the cellulose, or formed a homopolymer. The research is supported by IRSIA.

S.A. Denaeyer (Willebroek) was successful in developing markets for bark-derived compost and concentrated or dried residual lye, arising in pulp marking by the calcium bisulphite process.

THE CHEMICAL AND PETROCHEMICAL INDUSTRY

Many companies in the chemical and petrochemical industries are investigating the possibilities of re-using by-products. The main difficulty of recycling waste fractions is the complexity of by-product compositions on the one hand and the difficulty of meeting the often stringent product specifications on the other.

THE PLASTICS INDUSTRY

Industrial wastes arising in the plastics industry are either regranulated locally or sold to specialized firms (S.A. Ravago, S.A. Deltaplast, Papeteries de Genval). The problem has been reviewed in recent literature.[6,7] Problems arise when the plastic waste consists of mixed, composite or thermosetting plastics.

CHEMICAL RECOVERY OF POLYURETHANE FOAM

Large amounts of wastes are generated, during the transformation of foams, an appreciable part of the material being cut away to arrive at the desired form. Waste generation in flexible foams (mattresses, cushions, sponges) represents 20 per cent of the raw material, which corresponds to about 14,000 tons/year or about $14,000,000 for the sole PRB company. Only a small part of these wastes can be recovered, e.g. as a carpet under-layer, a foam agglomerate, etc. Also the head and the sides of the polyurethane casting constitute a source of waste material, which at the time being can only be incinerated.

The polyurethane waste could be hydrolysed by superheated steam, at about 300°C and atmospheric pressure. Hydrolysis yields the original polyol, and the amine corresponding to the original isocyanate (generally toluene diamine).

S.A. PRB studies the optimal conditions of hydrolysis and the possible markets for the recovered products. The toluene diamine can be re-used in the manufacture of diisocyanate. The polyols consist of a variable mixture of unknown polyol compounds, depending on the type of foam hydrolysed. Their re-use possibilities are also being investigated, with the aid of IRSIA's Conacord.

At present, the problem of hard forms is less urgent, because both consumption and specific rate of waste generation are smaller. The method of chemical recovery, moreover, is more complicated.

RECOVERY OF THE UNDERLAYER OF PHOTOGRAPHIC FILM WASTES

Large amounts of waste are generated in the manufacture of photographic film. Depending on the production stage at which the waste arises, the film underlayer is covered with one or more layers of foreign material and possibly with dirt.

The polyethylene terephtalate (PETP) film underlayer can be recovered after removal of the soil and the other layers of the film structure. The recovery has been demonstrated in batch equipment, but the large amounts of film scrap can be treated more economically in continuous equipment.

A pilot-plant, consisting of washing and rinsing equipment, and of filtering centrifuges, is being operated with the aid of an IRSIA's Conacord grant. The problem is not straightforward, since a finite contacting time with the cleaning liquor is necessary, to allow for the dissolution of the film layers to be removed. A complete elimination of the cleaning liquor is required, to obtain a complete removal of the undesirable components.

RECYCLING OF PETP-FILM WASTES

From an economical point of view PETP-film scrap should be recycled as directly as possible. For this reason a laboratory research program was initiated, in which the film scrap was recycled in the PETP-granulation step, and the properties of the resulting material were studied.

Preliminary results showed that the scrap-containing PETP-granules have the same viscosity and diethyleneglycol and carboxyl content as the virgin material. The presence of film scrap, however, has a negative influence upon the colour of the resulting film, and the presence of "burned" cuttings and of appreciable amounts of cellulose triacetate cannot be tolerated. Part of the foreign material can be removed by filtration of the PETP-solution, prior to trans-esterification.

This research is also supported by an IRSIA grant.

RECYCLING OF MIXED PLASTIC WASTES

Recycling of mixed plastic wastes usually yield materials with inferior mechanical properties, most families of commercial plastics being "incompatible", i.e. forming discrete phases with little mutual adherance.

Since several years the "Centre de Recherches Scientifiques et Techniques de l'Industrie des Fabrications Métalliques" (CRIF) studied the mechanical and rheological properties of "alloys" of polyethylene, polyvinylchloride, polystyrene, and of other commercial plastics.

The influence of mixing and moulding on the molecular weight, the molecular-weight distribution and the constitution of the polymeric material is studied. The phase distribution can further be influenced by addition of various fillers, or by especially synthetized sequential or graft copolymers.

This research study is also financed by IRSIA and aims at:

- the recovery of the mixed plastic wastes, contained in household refuse, and of unrecyclable composite industrial plastic wastes;
- a better understanding of the structure and properties of mixed plastics.

The UCB-Sidac division also studied the solubility diagrams of ternary polymer alloys and holds two patents on the co-extrusion of non-compatible polymers.

AGRICULTURE

Traditional agriculture forms a good example of non-waste technology: each waste is either re-used, recycled or burned. The introduction of industrialized techniques in many cultures and in cattle and poultry breeding disrupted this equilibrium, and so did the use of chemicals, pharmaceuticals and biocides. IRSIA's agricultural division founded a committee to study the alteration of the agricultural environment, and another to study the chemical fertility of the soil. These and other committees sponsor the study of a number of topics, related to wastes and their re-use:

1. Study of the re-use of animal manure, sewage sludge and industrial wastes. Chemical analysis (nitrogen and trace elements) of these compounds and determination of their influence upon soil fertility and water and vegetation (Prof. Cottenie, Ghent).
2. Study of the odour nuisance caused by the re-use of animal manure. Analysis of the stable atmosphere (Prof. Schamp, Ghent; Prof. Petit, Gembloux).
3. Study of the survival of pathogens in animal wastes (Prof. Van de Voorde, Louvain).
4. Study of the influence of manure on the productivity of meadows and fields, the mineral composition and the condition of the soil (Agronomical Service of Belgium).
5. Study of the optimal conditions for composting household refuse, bark, mixtures of bark with sludge from the wastewater treatment of paper mills, and mixtures of bark with manure, which served for growing mushrooms (Prof. De Boodt, Ghent).
6. Study of the fertilizing effect of industrial wastes upon forests (Centre d'étude des sols forestiers de la Haute-Belgique).

DAIRIES

In new dairies production losses are minimized by automatic control of the various operations (rinsing, cleaning, sterilizing, milk transportation, etc.).

The continuous production of butter eliminated the butter losses in the wash water, which occurred during batch churning.

Originally, cheese whey was fed to hogs. Later, this practice was no longer possible, because of the distance between the cheese factories and the hog breeders. Recently, techniques (drying, ion exchange, ultrafiltration) have been introduced which allow this waste stream to be recovered.

VEGETABLE CANNING INDUSTRY

The vegetable canning industry also recycles important quantities of process water. The number of possible cycles is limited by the need of safeguarding the bacteriological quality of the water.

A large canning factory contractually returns all its wastes to its suppliers of vegetables. The wastes are used for hog feeding, or spread in meadows.

THE SUGAR INDUSTRY

The sugar industry is a remarkable example of successful non-waste technology. The required amount of process water and the resulting quantity and pollution load of the waste-waters have been drastically reduced.[8] The waste products can be turned into a long series of useful products.[9]

The sugar beet is beheaded and the leafs removed in the fields. At the factory the beet is washed and cut into slices. The sugar is extracted with warm water in diffusers, yielding a liquid extract and a solid pulp. The pulp is pressed and the free water is recirculated to the diffusers.

The crude juice is purified by addition of lime. Excess lime is precipitated with CO_2 and removed in a filter press.

The purified juice is concentrated in multiple effect evaporators and crystallized. The sugar crystals are recovered in a centrifuge; the remaining mother liquor, the molasses, still contain an appreciable amount of sugars.

About 0.5 - 1 m^3 water is needed for washing and transporting 100 kg of beets. The transport water is recirculated in a closed circuit after settling of the earth (10 - 50 per cent of the gross weight), but it soon attains a high BOD-load by dissolution of organic substances from beets which were damaged during unloading or transport. The high BOD-load soon leads to odour and foaming problems in the circulating water.

For this reason an internal cleaning step is necessary. Coarse material, such as leaf rests, is retained on a screen. At Tienen 96 - 97 per cent of the high BOD-load (5-6 g O_2/l) is removed after an aerobic treatment of 3-4 hours in a RT Lefrancois fermentator vessel.[8] The vessel was developed with the aid of the "Prototypes" Division of the Dept. of Economic Affairs.

The solid wastes are re-used as far as possible:

- the leafs are used as a fodder;
- the pulp is dried up to a moisture content of 10 per cent and used as a fodder. The pulp production amounts to 52 kg dry pulp/ton of beets;
- the filter cake, containing 75 per cent of $CaCO_3$ and 25 per cent of organic material is sold as a soil conditioner to the beet planters. The filter cake represents 5 kg dry cake/ton of beets;
- the molasses form a raw material for the production of alcohol, citric acid, fodder and numerous chemical compounds. Molasses have a moisture content of 20 per cent and represent 40 kg (dry)/ton of beets.

Numerous processes have been developed for the valorization of molasses, e.g. the fermentation to alcohol by yeasts, the bacterial decomposition to acetone, butanol, lactic and other acids, the decomposition by moulds to various acids, the catalytic hydrogenation to sorbitol and mannitol, the extraction of amino acids, betain and vitamins, and the pyrolysis to various compounds.

The fermentation of molasses to citric acid yields wastewaters with a high protein content. After partial evaporation this wastewater is used as a fodder complement, which is rich in natural proteins.

TANNERIES

The Centre for Leather Research (a De Groote centre) is conducting a research programme, aiming at

1. the reduction of water consumption from a former 175 m^3/ton of crude skin to values of 55 - 60 m^3/ton in the future. At present the consumption level is at about 100 - 110 m^3/ton;
2. the development of technical methods of recycling the sodium sulphide, used during the elimination of hair and upper skin with lime;
3. the recovery of chromium oxide from the sludge and of proteins from the wastewaters;
4. the elimination of fat from the wastewaters, arising in the manufacture of wash leather.

CONCLUSIONS

Non-waste technology can be defined as a sum of techniques, which allow the rates of waste generation to be reduced, individual components to be separated and recovered from the wastes, and new uses to be developed for by-products and former waste products. In this way non-waste technology leads to materials, energy and environmental savings, and eliminates or eases waste-disposal problems.

Private industry is actively helped in developing non-waste technology, by financial support of IRSIA's Conacord committee and by the Prototypes Service, both of the Ministry of Economic Affairs.

REFERENCES

1. A. Buekens, *Vaste Afval. Ontstaan, Verwerking en Beheer,* Monografieën Leefmilieu Nu - De Nederlandse Boekhandel, 1975.
2. Ministry of Economic Affairs, Administration of Mines, Service of Statistics, Montoyerstreet 3 - 1040 Brussels; Statistics 1975—houille—cokes—agglomérés métallurgie—carrières.
3. M.B. Bever, "Signification of Scrap in Western Economies," "Recycling of ferrous metals," presented at the University of Nottingham, 1975.
4. Sweeping, de Broqueville Avenue 12, B-1150 Brussels.
5. Cobelpa, de Craeyerstreet 14, B-1050 Brussels.
6. A. Buekens, *Facts and Fiction in the Recycling of Plastics,* KVIV, 10 March

1975.

7. ASTM Special Technical Publication 533 (1973), Report on the ASTM - Symposium of 25 October 1973.

8. R. Pieck, *Le Problème de la Pollution des Eaux de Sucreries et les Solutions Adoptées Aujourd'hui*, Colloque sur le traitement des eaux résiduaires industrielles, Mars 1974, SRBII.

9. H. Olbrich, *Die Branntweinwirtschaft*, January 1971, pp. 22-28.

Outokumpu Flash Smelting Method

Seppo Härkki
Outokumpu Oy, Helsinki, Finland

INTRODUCTION

The energy crisis in Finland after World War II together with a sharp rise in the price of energy led Outokumpu Oy to consider replacing the electric furnace smelting of copper concentrates by some other method. The flash smelting method developed in these circumstances offered a considerable saving of fuel through the utilization of the reaction heat of sulphides and yielded a high recovery of sulphur in the form of acid or elemental sulphur. The pilot plant testing of the new process was carried out in the years 1947-1948, and the process became operational on a commercial scale in 1949.

For 27 years the flash smelting method has been developed further both metallurgically and technically so that it is still one of the most advanced processes in its field both from the economic and the ecological points of view.

Apart from the utilization of the latent reaction heat of the concentrates, important features of the process as regards energy consumption are the far-developed waste-heat recovery and saving of energy. The oxygen enrichment of the process air introduced a few years ago has considerably increased the specific capacity of the smelting unit and made the process almost independent of additional fuel. The steam recovered from the waste heat can be used as operating power in the oxygen plant and for generating electric energy, which makes a process like this a considerable advantage today when the prices of fuel and electric energy are on the increase.

PROCESS DESCRIPTION

The processing of copper concentrates at the Harjavalta Works of Outokumpu Oy using the flash smelting method is shown as an example in the flowsheet in Figs. 1 and 2.

From the concentrate storage the feed materials, concentrates and silica sand are fed automatically in the right proportions onto a belt, which conveys the charge to the dryer. In the dryer, which at Harjavalta is a directly oil-fired kiln, the concentrate is thoroughly dried. The finest particles leaving the kiln as flue dust are collected in an electrostatic precipitator and returned to the main concentrate flow. The dried charge is transported by a pneumatic conveying system to the feed bin of the flash smelting furnace.

The flash smelting furnace consists of three sections: a reaction shaft, a settler and an uptake. The oxidation reactions and smelting are effected in the vertical reaction shaft. For this purpose, the dried, fine-grained

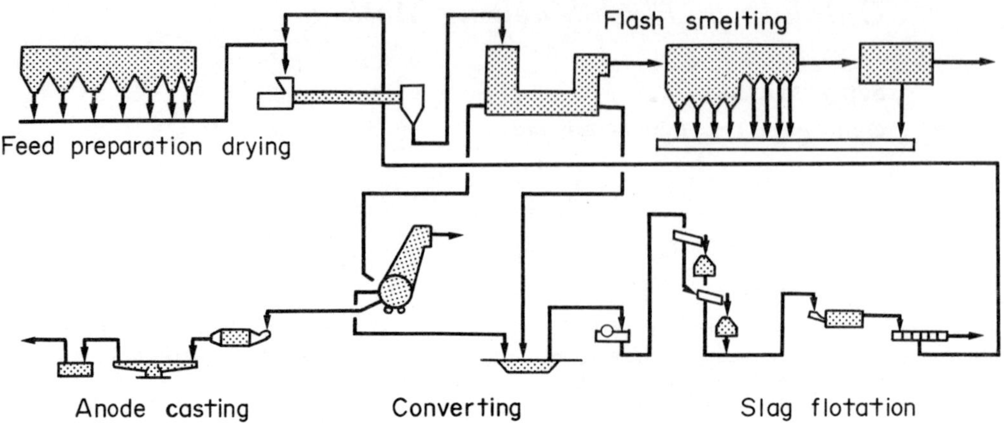

Fig. 1. Flowsheet of the Harjavalta copper smelter

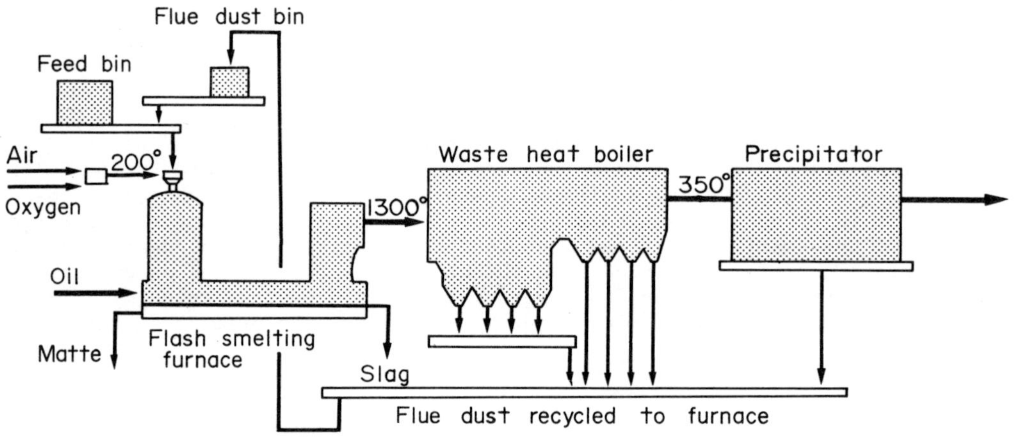

Fig. 2. The flash smelting furnace arrangement (oxygen-enriched air)

concentrate, flux, and usually the flue dust are led into one or several concentrate burners on top of the reaction shaft. The preheated process air is also led into the concentrate burner. The air and concentrate feed mix and form a suspension which is blown into the reaction shaft.

The heat requirements are supplied almost completely by the oxidation of iron and sulphur. To balance the heat requirements, additional fuel, preheated air and/or oxygen can be used. To ensure favourable operating conditions for the furnace, special attention is paid to obtaining a uniform feed and an even concentrate suspension.

Molten particles are separated from the gas stream in the settler part of the furnace. Matte drops pass through the slag layer to the bottom of the settler. Iron oxide and other slag-forming compounds collect in the slag layer, where the main part of the slag reactions occur.

The matte is usually composed mainly of Cu_2S and unoxidized FeS, the used oxygen-concentrate ratio determining to what extent the FeS is oxidized and, consequently, the matte grade.

Under normal flash smelting conditions, the copper flash furnace operates with 65 per cent copper matte. However, tests of long duration carried out on a full production scale in recent years at Harjavalta have shown that copper matte containing about 80 per cent copper can be produced continuously by flash smelting without essentially increasing the copper content of the waste slag.

Totally automatic control of the flash smelting process is possible. By installing a continuous or semi-continuous concentrate feed analyzer, automatic process control can be effected by an on-line computer to optimize the smelting conditions.

The exhaust gases from the flash smelting furnace contain some molten or semi-molten particles. Volatile compounds tend to be concentrated in the dust. To recover heat and dust, the smelter gases pass through the uptake section of the furnace to a specially designed waste-heat boiler before entering the electrostatic precipitators. The steam from the waste-heat boiler can be utilized for generating electric power, preheating process air or producing oxygen. The flue dust from the boiler and electrostatic precipitators is usually recycled in a closed conveying system and fed back to the furnace.

From the electrostatic precipitator the gases are led to the sulphuric acid plant by exhaust gas fans with an automatic draft control.

From the settler section of the flash smelting furnace the molten copper matte or metal and slag are tapped. The matte is transported in ladles to the converters. The high matte grade makes the blowing time of the converter short, which means that the size or number of the converters can be reduced. Blister copper is transferred to an anode furnace, refined and cast into anodes.

The cleaning of the flash smelter and converter slags can be carried out in an electric furnace or by slag flotation. In electric furnace cleaning, the importance of oxygen potential for the recovery of valuable metals is a well-known parameter. The decrease of sulphur content in slags has also been found to improve the results. A high recovery of zinc and lead in fumes can be attained by selecting a suitable reduction temperature. In slag flotation, the right cooling rate and a sufficient sulphur content are the main requirements for the optimum recovery of metals.

Typical assays in the Harjavalta smelter are given in Table 1 as an example of copper flash smelter operations.

TABLE 1. *Typical assays in the Harjavalta copper smelter*

	Cu %	Ni %	Fe %	S %	SiO_2 %
Concentrate feed	21.9	0.11	30.3	32.0	8.6
Typical flash furnace matte	64.1	0.85	10.6	21.5	
Flash furnace high-grade matte	78.5	0.75	0.9	19.0	
Typical flash furnace slag	1.5	0.05	44.4	1.6	26.6
Concentrate from slag flotation	20.8	0.54	29.9	10.8	15.6
Tailing from slag flotation	0.3	0.10	44.0	0.4	

SULPHUR DISTRIBUTION

The SO_2 content of the flash smelting off-gases is in general very high, and as the gas volume is very stable, the gases are most suitable for producing sulphuric acid. High sulphur recovery is one of the advantages of flash smelting, especially now that air-pollution problems are becoming more and more evident all over the world.

The distribution of sulphur in the flash smelting system differs from the conventional smelting methods in that the sulphur is confined essentially to the gas lines of the furnace and converter as a rich SO_2-bearing gas, the former carrying the main load. This is illustrated in Table 2 for a concentrate containing 25 per cent Cu, 29 per cent Fe, and 32 per cent S, smelted to four different matte grades.

TABLE 2. *Sulphur distribution (%)*

	Matte grade (% Cu)			
	45	55	65	75
Feed	100	100	100	100
Furnace gas	55	65	72	76
Converter gas	45	35	28	24

The combination of these gases from the Outokumpu flash smelter and the converter gases enables a high recovery of sulphur with standard operating units. The sulphur losses in the smelter are outlined in Table 3.

TABLE 3. *Sulphur losses (%)*

	%
Drying	0.2
Smelting	0.2
Converting	0.5
Anode furnace	0.1
Slag	1.2
Total loss	2.2
Atmospheric emission	1.0
Fixed sulphur (= slag losses)	1.2

The total recovery of sulphur at the smelter is 97.8 per cent.

The average recovery of sulphuric acid from the gases at Harjavalta is 99.7 per cent giving an overall recovery of sulphur as acid of 97.5 per cent. The total emission to the atmosphere is 1.3 per cent of the total feed.

The sulphur discarded with the slag is not dangerous to the surroundings, because it does not oxidize into gaseous products nor dissolve in water.

ENERGY CONSUMPTION

In recent years, many papers have been published on the energy consumption and the economy of various copper production methods. The growing interest in the conservation of energy is due to the fact that the mining of lower-grade ores has been on the increase and at the same time the prices of energy and fuel have risen.

The flash smelting process is a practical example of the conservation of energy: firstly, because the heat from the exothermic oxidation reactions is used for smelting the charge; and secondly, because the heat from the furnace off-gases is efficiently recovered as high-pressure steam in a waste-heat boiler.

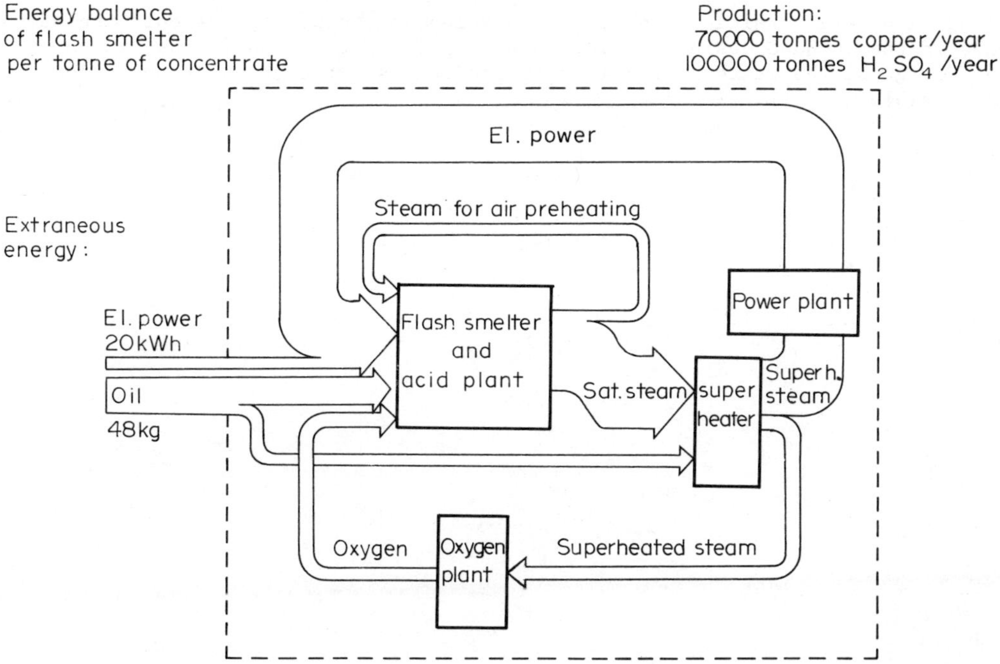

Energy balance
of flash smelter
per tonne of concentrate

Production:
70000 tonnes copper/year
100000 tonnes H_2SO_4 /year

Extraneous
energy:
El. power
20kWh
Oil
48kg
El. power
Steam for air preheating
Flash smelter
and
acid plant
Power plant
Sat. steam
super
heater
Superh.
steam
Oxygen
Oxygen
plant
Superheated steam

Fig. 3.

Figure 3 shows a schematic energy balance for a copper smelter using oxygen-enriched process air, with a smelting capacity of 1000 tons of concentrate a day and a production rate of 70,000 tons of copper and 100,000 tons of sulphuric acid a year. The energy balance in question covers the whole smelter and takes into account all the energy requirements for the entire production process, from wet (8 per cent moisture) concentrate up to anode copper, and on the waste gas side, up to sulphuric acid production. The scheme shows how electric energy and oxygen can be generated from steam recovered from waste heat by using waste-heat boilers.

Because of the good energy conservation of the flash smelting process, the demand for extraneous electric energy and fuel oil is very small. Only 20 kWh of electric energy and 48 kg of fuel oil per ton of concentrate are needed to maintain production.

TRENDS IN THE DEVELOPMENT OF THE FLASH SMELTING METHOD

Metallurgical research on the flash smelting method is now concentrated on the use of oxygen or high oxygen enrichment of the process air. The aims of the research and development work are, thus, to make the flash smelting method completely self-sufficient and independent of extraneous fuel, and to improve the conditions for producing high-grade matte or blister copper directly with sufficient slag losses and for treating different kinds of complex and impure copper concentrates as economically as possible.

Along with metallurgical research and development work, improvements are being made in the furnace design to allow smelting in extreme conditions, and also in the design of the waste-heat boiler to produce as efficient waste-heat recovery as possible.

Methods of Conserving Raw Material and Energy and Protecting the Environment in Chemical and Electro-Chemical Plating Plants

Bengt Westerholm

Metal Division, UPO Osakeyhtiö, Lahti, Finland

The sensitivity of most metals to corrosion always causes considerable losses. For this reason, the working life of those corrosion-prone raw materials is extended by either careful alloying or by protecting the surface of the basic material with a layer of some inert material.

Apart from organic surface treatments, metallic plating is also used very effectively to protect corrosion-prone substrates. In most cases this type of plating is much more economical than alloying because even a layer only a few microns thick may extend the life expectancy of the part by 400 or 500 per cent.

Some metals can be rendered inert chemically.

The inorganic plating plant in this picture contains a great number of operations in which it is possible to save both raw materials and energy.

RECOVERY OF ORGANIC SOLVENTS IN PRE-TREATMENT

A vital pre-treatment before the plating process begins is the thorough cleaning of the work piece of all grease, wax, oil, etc. Cleaning takes place in a solvent bath or a solvent spray. Such solvents as trichlorethylene and perchlorethylene are quick acting and efficient. Chlorinated hydrocarbons have the serious disadvantage of being harmful to the environment. The disadvantages of solvent bathing can be overcome with effective air-conditioning and by filtering the gases.

Active carbon has a well-known strong ability to bind other molecules. When air containing solvent gases is led through a layer of active carbon the gas molecules are adsorbed because of the force between the carbon molecules and the surface of the carbon.

A filtering unit is usually made up of two tanks each filled with active carbon. The solvent gases are forced through one of the filtering tanks until the active carbon is saturated. Then the other tank is brought into operation and the saturated carbon is desorbed with low-pressure steam. The mixture thus obtained is then cooled and the solvents recovered, regenerated and recycled. This system is economical since the recovery process is highly efficient—approximately 95 per cent—the process itself does not require any moving mechanical parts, energy costs and raw material costs are low and the whole operation can be automated.

In addition to environmental protection, a strong trump in favour of active

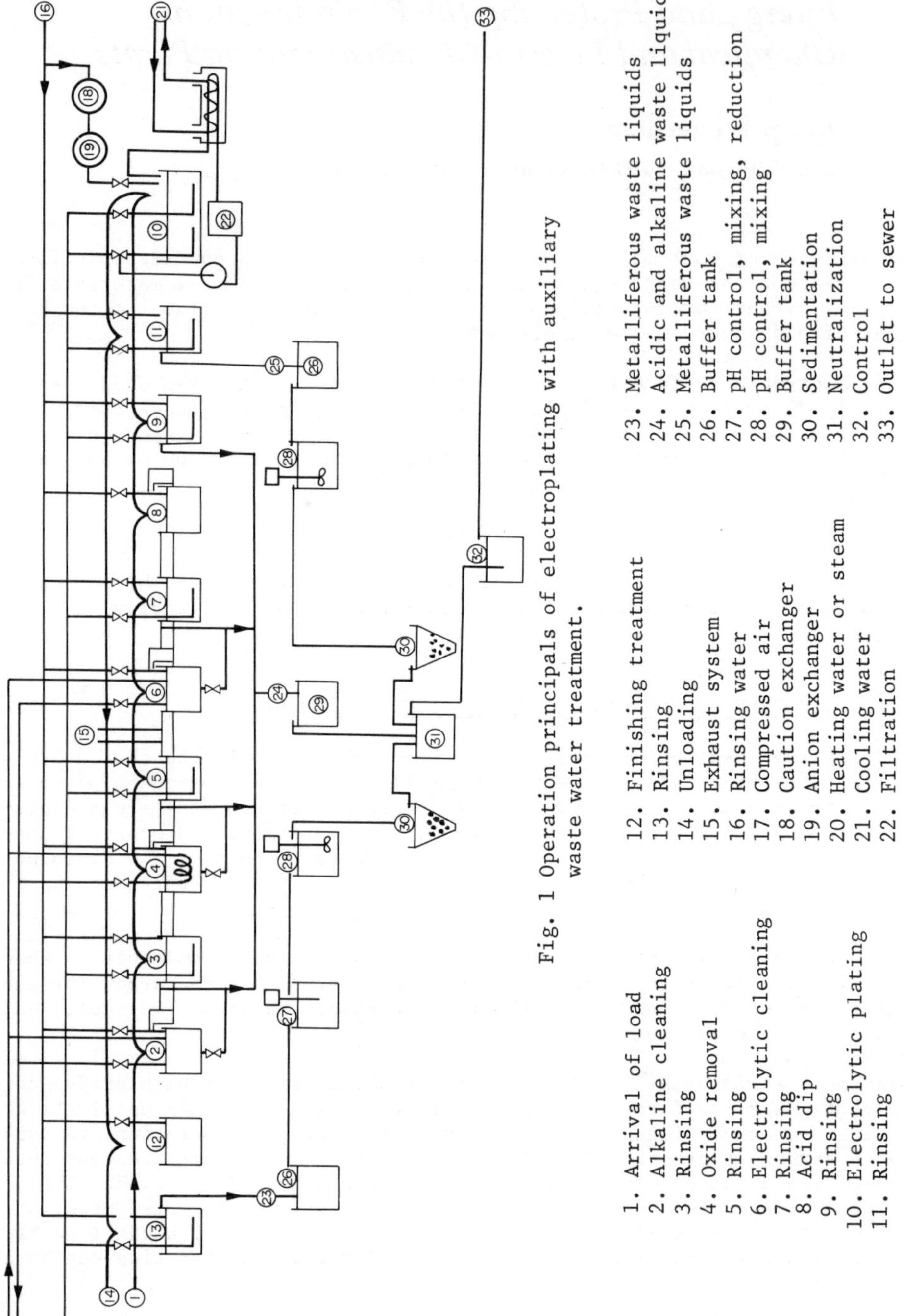

Fig. 1 Operation principals of electroplating with auxiliary waste water treatment.

1. Arrival of load
2. Alkaline cleaning
3. Rinsing
4. Oxide removal
5. Rinsing
6. Electrolytic cleaning
7. Rinsing
8. Acid dip
9. Rinsing
10. Electrolytic plating
11. Rinsing
12. Finishing treatment
13. Rinsing
14. Unloading
15. Exhaust system
16. Rinsing water
17. Compressed air
18. Caution exchanger
19. Anion exchanger
20. Heating water or steam
21. Cooling water
22. Filtration
23. Metalliferous waste liquids
24. Acidic and alkaline waste liquids
25. Metalliferous waste liquids
26. Buffer tank
27. pH control, mixing, reduction
28. pH control, mixing
29. Buffer tank
30. Sedimentation
31. Neutralization
32. Control
33. Outlet to sewer

carbon filtering is its economy. This process, which earlier contained no provision for solvent recovery, would repay its own investment costs in a few years even in unfavourable conditions.

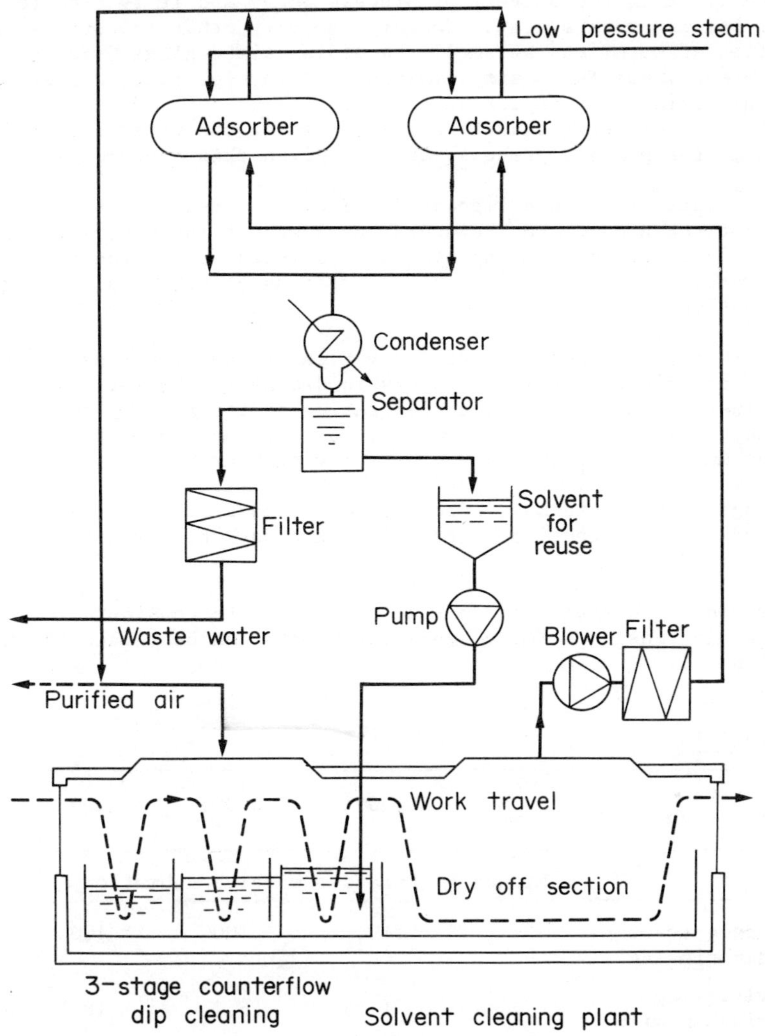

Fig. 2. Recovery unit for solvent cleaning plant

In Calculation 1 you can see investment costs and operating costs for one unit.

The equipment is adaptable for use in the rubber, viscose, plastics, paint and paper industries. These industries use hundreds of kilos of solvents per hour per unit. The raw material saving here must be quite obvious.

CONSERVATION OF PROCESS WATER

In terms of sheer quantity, water is the most significant raw material in the plating plant. It is the solvent of process baths and it is used to rinse the work pieces between work stages. Compared to many other industries, perhaps, the quantities of water may be small. A medium sized plant uses 20 or 30 m^3/ of water per hour, but the costs involved in using the water are significant. On top of the basic price of the water there is the cost of dealing with the waste (contaminated) water and the cost of heating the water for hot rinsing. In addition to the plating process, UPO has given this question serious study.

Diluted waste water is a sure sign of ineffective rinsing. It increases the capital investment needed and the operating costs of the purifying plant. The economic objective of the rinsing stages is a reduction in the actual quantity of rinsing water used and an increase—as great as possible—in the degree of its contamination.

Since the work piece cannot be rinsed any cleaner than the rinsing water, in single-stage dip rinsing it is necessary to discharge the water into the drainage system in the same degree of concentration as the rinsing requirement for the work piece. It is possible, however, to make much more efficient use of the water by multi-stage rinsing and by spray rinsing.

In multi-stage rinsing the most effective is cascade rinsing in which the water flows counter to the direction of travel of the work pieces, from the cleanest bath to the dirtiest one.

The water savings achieved with this method are quite considerable compared to single-stage dip rinsing. The cleaner the pieces can be rinsed the greater the saving (Table 1).

TABLE 1.

Rinsing criterion	$\frac{C_n}{C_o}$	0.0001	0.001	0.01
	Stages	Relative water consumption, %		
Single-stage dip rinsing	1	100	100	100
Multi-stage rinsing with parallel water flow	2	1.95	6.2	18.7
	3	0.65	2.7	10.8
Multi-stage rinsing with counter flow of water	2	1.0	3.1	9.6
	3	0.2	0.96	4.25

Spray rinsing is economical both in terms of water consumption and space saving. The application of the method is, however, limited because unless the piece is quite flat there are shadow areas which the spray cannot reach. In an enclosed galvanizing drum spraying is not possible without special arrangements.

UPO, however, has developed and patented an interior drum spray rinsing method which consumes only 20 or 30 per cent of the amount of water used in a single-stage dip rinsing system.

The electrical conductivity of water is generally a useful property for determining the quality of the rinsing water since the salts, acids and alkalis increase its conductivity and conductivity is almost directly proportional to its concentration of salts. With automatic-control equipment water flow is regulated according to load fluctuations.

The sensible use of water includes detailed knowledge of the water consumption of the entire factory, the object being to re-use the water as many times as possible. Cold, it is suitable for the cooling rectifiers and process baths after which the water is led to the rinsing baths. In rinsing it may be used several times and, finally, in rather alkaline waste water, the water may be decontaminated in the purification plant.

The thermal energy in the waste water is not often considered to be worth recovering in this type of industry. However, it seems very likely that with the rising cost of oil, a heat exchanger would be an excellent investment and pay for itself in its very first year of operation.

RINSING WATER TREATMENT AND RECOVERY OF CHEMICALS

The chemical waste in plating plants causes problems of many kinds. There is no standard solution to these problems. Each plant and installation requires individual consideration and detailed instructions on what to do with the contaminants—cyanide, 6^{+}- chromium, heavy metals, etc.—in their waste water. Parallel to conventional purifying systems, the new ion-exchange technology together with improved rinsing techniques has revealed new possibilities in the treatment of industrial waste water. Development is towards closed systems more and more, the object being an installation that is totally enclosed, where all rinsing water is purified and re-used and all useful chemicals are recovered and recycled.

The selective filtering properties of ion exchangers are based on their use of large molecule compounds with active molecule groups.

The number of groups of these molecules determines the ion exchange capacity and their chemical characteristics depend upon whether it is a cation exchanger or an anion exchanger. Cation exchangers have the ability to bind positively charged ions and release hydrogen ions (H^{+}) while anion exchangers bind negatively charged ions and release hydroxide ions (OH^{-}).

It is possible to regenerate the resin in the ion exchanger, in other words to return to its original active state. In a cation exchanger this regeneration takes place with the aid of acid and in an anion exchanger with sodium hydroxide.

The ions collected by the resin in the exchanger are discharged as acid or alkali eluate into a collector tank prior to further treatment. The unusable

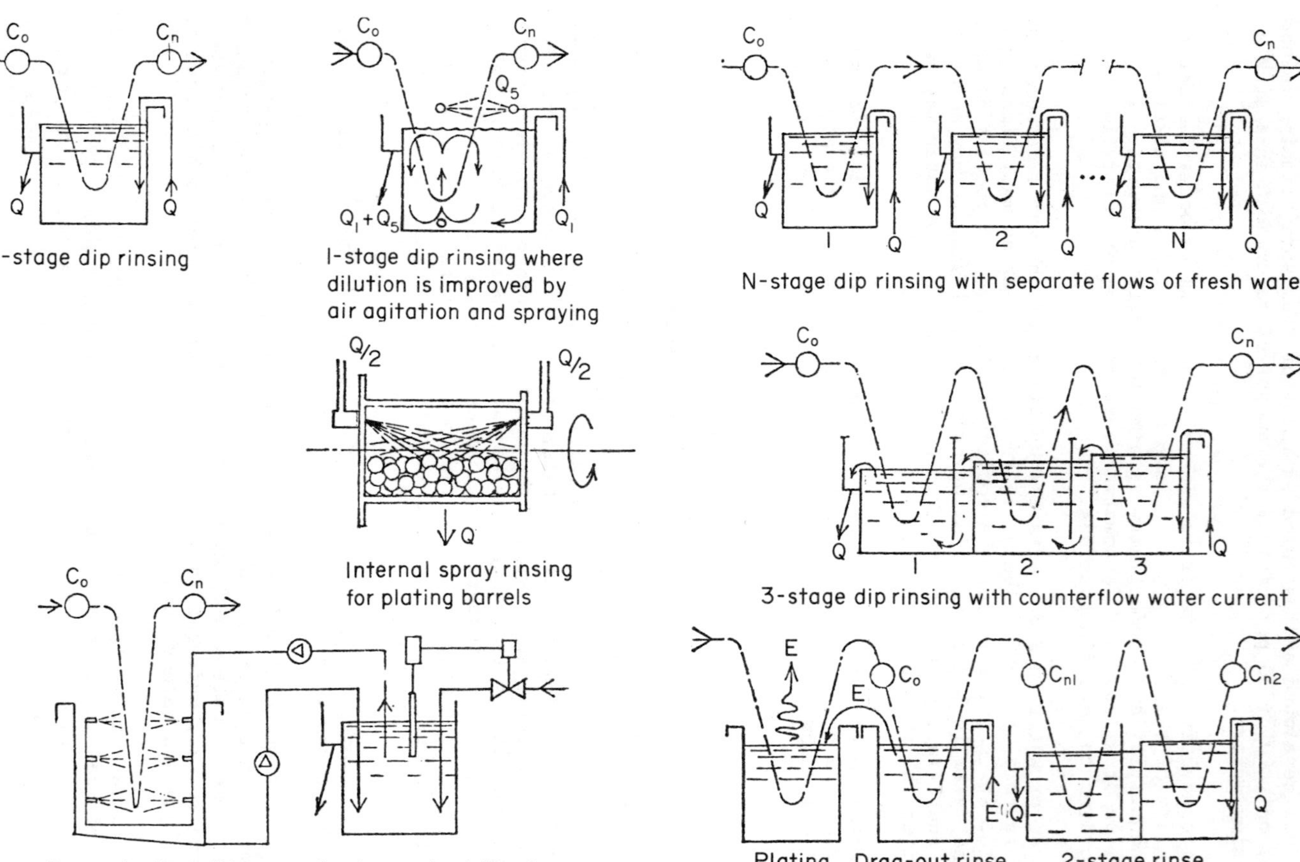

C_o = Concentration of contaminant solution being dragged in.

C_n = Desired maximum concentration after rinsing.
Q = Fresh water flow.
E = Evaporation losses

Fig. 3 Rinsing techniques.

eluate is rendered harmless and non-toxic at the purification plant. The waste is precipitated and dried.

In a galvanic plating plant ion exchangers can be used for the following purposes:

total salt removal from water,

total salt removal from rinsing waste water in a close system,

removal of ions that are harmful to the environment,

recovery of precious metals,

cleaning of process solvents.

In Fig. 4 you can see a plating plant delivered by UPO Ltd. In this plant the nickel- and chromium-contaminated water is purified in an ion exchanger and recycled. Chromium eluate and a percentage of the nickel eluate are rendered electrolytic and returned to the plating bath.

Of the toxic acids used as electrolytes in chromium plating more than 80 per cent may be lost and less than 20 per cent used for the plating itself. By the use of efficient recovery systems these sorry figures can be reversed. Nickel salts are separated from the rinsing water for their intrinsic material value and at the precious-metal bath stage the use of ion exchangers is self-evident.

In the second calculation the use of ion-exchanger equipment in connection with chromium rinsing is examined.

The practical applications of these money-saving systems have been aided by highly advanced automation and also by the UPO organization's policy of developing and delivering the plants as complete units. This has made it possible for UPO to influence all areas of technology in this field and achieve the objects of economy, efficiency and environmental protection.

Calculation 1
$1 = Fmk 3.8

Recovery unit costs for the equipment in Fig. 2.

Consumption figures		
Active carbon	0.001	kg/ per kilo of solvent
Regeneration steam	3.0	kg " " " "
Water	0.005	m^3 " " " "
Solvent	28.0	kg/ per hour
Electricity	0.3	kWh/ per kilo of solvent
Used capacity	0.8	
Efficiency ratio	0.9	
2-shift work, 16 hours per day, 224 days per year		
Energy and raw material costs		
Electricity	0.10	Fmk/kWh
Carbon	10.0	Fmk/ kg
Water	3.0	Fmk/ m^3
Steam	0.035	Fmk/ kg
Solvent	2.30	Fmk/ kg

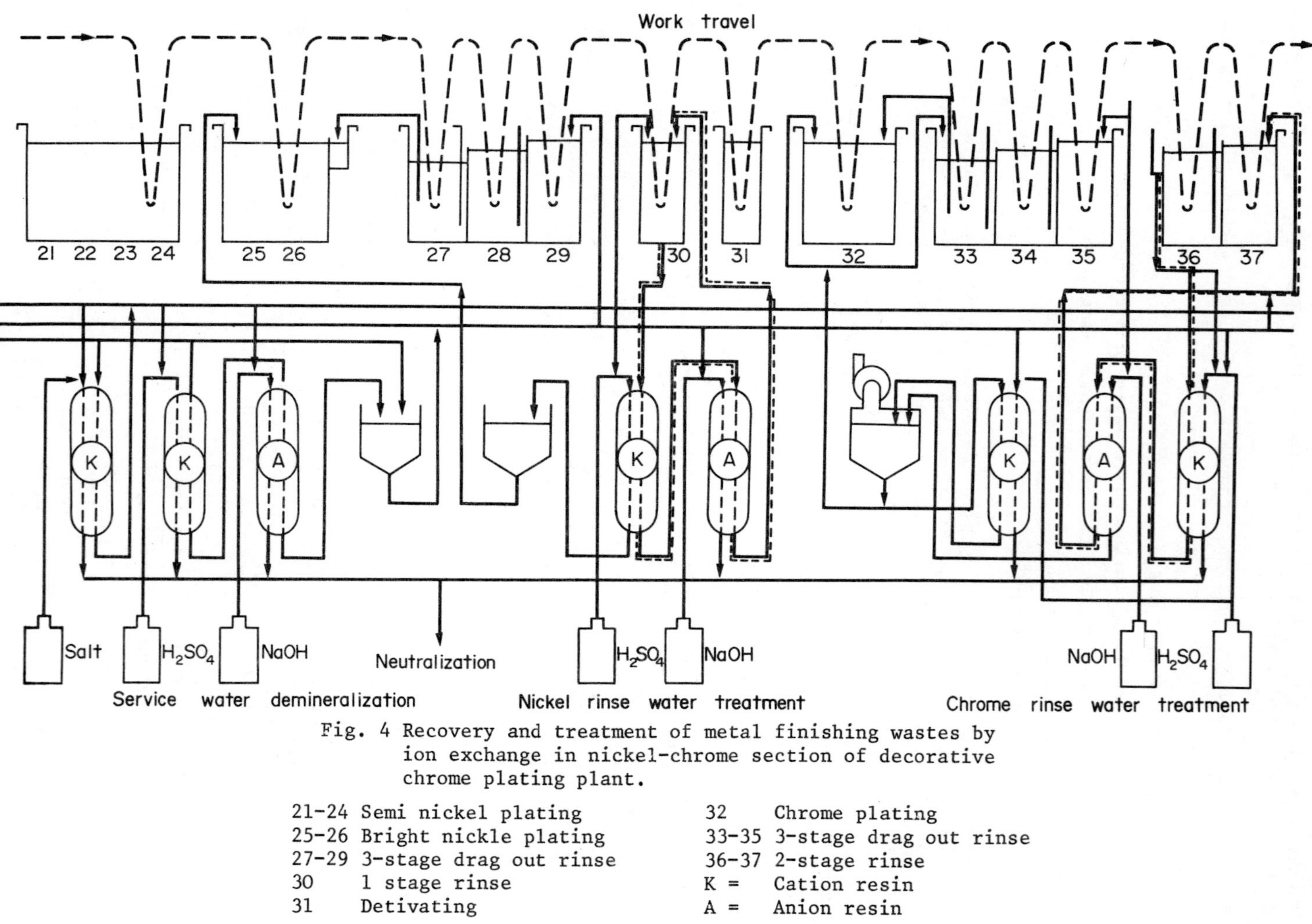

Fig. 4 Recovery and treatment of metal finishing wastes by ion exchange in nickel-chrome section of decorative chrome plating plant.

21-24 Semi nickel plating
25-26 Bright nickle plating
27-29 3-stage drag out rinse
30 1 stage rinse
31 Detivating
32 Chrome plating
33-35 3-stage drag out rinse
36-37 2-stage rinse
K = Cation resin
A = Anion resin

Annual operating costs

Electricity	2880 x 28 x 0.3 x 0.10	=	2420.-
Carbon	2880 x 28 x 0.001 x 10	=	800.-
Water	2880 x 28 x 0.005 x 3	=	1210.-
Steam	2880 x 28 x 3 x 0.035	=	8470.-
Labour	2880 x 28 x 0.3	=	43,200.-
			56,100.-

Investment costs

Regeneration equipment	70,000.-
Water-cooling and circulation equipment	20,000.-
Steam boiler	15,000.-
Installation	20,000.-
Building	15,000.-
	140,000.- (approximately)

Fixed costs

Interest 12%, depreciation 5 years		38,000.- per year

Total costs

56,000 + 38,900	=	95,000.- per year

Annual saving of solvent

0.9 x 0.8 x 225 x 16 x 28 x 2.30	=	167,000.- per year

Calculation 2
\$1 = FmK 3.8

Cost comparison between two methods of handling rinsing water containing chromium

Automatic chromium-plating plant base metal Fe and MS plating capacity	116,000 m^2/per year
Drag out	1 ml/dm^2
Chromium content	350 g CrO_3/l
Yearly operational hours	2500 hours

Alternative A

Single-stage dip rinsing follows chromium plating. Rinsing concentration Cn = 175 ppm. The rinsing water is led to the purification in which the chromic acid is reduced, neutralized and precipitated.

Alternative B

Three-stage drag out rinsing and two-stage counter flow rinsing follows chromium plating. In this method the rinsing water is circulated through the ion exchangers. Both rinsing water and chromic acid are recycled. The purification plant is necessary for the treatment of process liquids and spill water.

Unit Prices	
Electricity	-.10 Fmk/kWh
Water	2.- Fmk/m^3
Hydrochloric acid	-.80 Fmk/kg
Sulphuric acid	-.25 Fmk/kg
Sodium hydroxide	1.50 Fmk/kg
Sodiummetabisulph.	1.40 Fmk/kg
Chromic acid	5.80 Fmk/kg

In alternative A the chromium electrolyte is transferred to the rinsing water in the following quantity:

11,600,000 dm^2/per year x 1 ml/dm^2 = *11,600 litres/year*

of which chromic acid 11,600 litres per year x 350 per litre, = *approximately 4000 kg per year.*

Rinsing water requirement in alternative A is Q = 350 g/1/0.175 g/1 x 11.6 m^3 = *23,000* m^3 *per year.*

In alternative B estimated water consumption totals approximately 1000 m^3 per year.

When the 95 per cent recovery efficiency of the ion exchangers is taken into consideration, chromic acid consumption is according to alternative B: 0.05 x 4000 kg per year = *200 kg per year.*

Operating Costs	*A*	*B*
Raw Material		
Water Chemicals Electrolyte Energy Labour	90,600 Fmk per year	25,800 Fmk per year
Required Investment		
Rinsing bath purification plant, pipe network, etc., installation	65,000 Fmk	

Operating Costs	A	B
Required Investment		
Rinsing baths, ion-exchanger equipment, evaporator, purification plant, pipe network, installation		135,000 Fmk
Fixed Costs		
Interest 12% depreciation 5 years	18,000 Fmk	37,500 Fmk
Total Costs	108,600 Fmk per year	63,600 Fmk per year

Relative chemical consumption	A	B
Chromic acid	100	5
Sulphuric acid	100	20
Sodium hydroxide	100	10
Sodium metabisulphite	100	5

The economy of alternative B is because of the inefficient rinsing techniques in alternative A

In practice, part of the purification-plant investment can be written off against other waste-water treatments.

Experience in Designing a Complex Scheme for Refining and Re-Use of Waste Waters and Creation of a Drainage-free Scheme of Water Supply and Sewerage in an Industrial Enterprise

V.N. Yevstratov and M.I. Kievsky

Ministry of Chemical Industry, Moscow, USSR

1. CONDITIONS OF CREATING A COMPLEX SCHEME

The report deals with information and data on the experience in designing and creating a complex drainage-free scheme of water supply and sewerage in a large chemical plant which is involved in an industrial zone consisting of a town (40,000 citizens), a central heating-and-power plant (CHPP), an industrial base of construction industry, enterprises of food industry and non-ferrous metallurgy, repair bases of different types and other units.

The supply systems of drinking (artesian) and industrial (river) waters as well as sewerage for all units of industrial assembly and town are connected to the chemical plant systems.

This connection is shown in Fig. 1.

There are favourable conditions as to raw materials in the region chosen for the chemical plant construction but the supply of river water is insufficient and conditions are unfavourable for the removal of waste waters into open ponds. The solution of the problem is complicated by the necessity to improve the protection of nature without diminishing economic returns.

2. PREREQUISITES FOR CREATING A COMPLEX SCHEME

In connection with the broad variety of production processes of units in the industrial assembly and chemical plant the kinds of waste waters discharged into sewerage and the nature of pollutants in them are rather different.

The mineral salts, organic and chlorine-containing organic compounds, surfactants, heavy metal salts and petroleum products are the main pollutants in the waste waters. The concentration of these substances varies in the broad range from some milligrammes to dozens of grammes per litre. Pollutants are present in the form of a solution or in an emulsified and suspended state.

Accordingly, it is impossible to discharge waste waters to the ponds, filter fields and sewage farm or to re-use them for industrial water supply without additional refining.

Therefore to solve the problem of possibility of waste water use is to develop the optimal scheme of their refining from undesirable impurities.

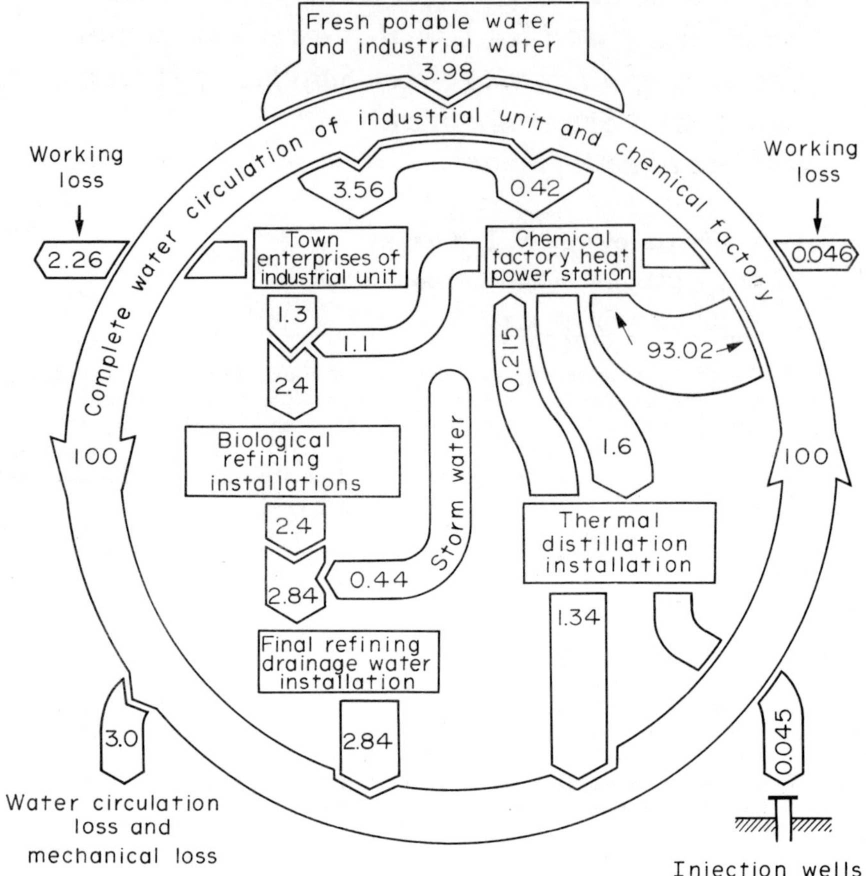

Fig. 1. Balanced scheme of water consumption and use of drainage water in industrial unit and chemical factory.

3. EXPERIENCE IN ORGANISATION OF CREATING A COMPLEX SCHEME

The general designer was in charge of the whole task of creating a drainage-free scheme for water supply and sewerage of the chemical plant and industrial assembly.

The modern achievements in science and technology and prospects in the field of refining and re-use of waste waters were studied and the main directions were outlined for the creation of a drainage-free scheme.

Two main versions were taken as the basis of further work: use of overall waste waters for agricultural sewage farms and re-use of the overall waste waters for industrial water supply at the chemical plant.

Both versions were completely developed. However, the comparison of technical and economic data for both versions of the project showed that the second one was more economically profitable, and the fresh water consumption was less than in the first version.

More than forty highly skilled, specialized Soviet organizations were brought in to carry out research, experimental, investigation, design and other work connected with finding reliable, effective and economically profitable methods of refining and re-use of waste waters of the various enterprises and of the whole chemical plant and industrial assembly.

Coordination, engineering and economic evaluation of all work undertaken were handled by the general designer.

A drainage-free scheme of water supply and sewerage for the chemical plant and industrial assembly was developed, and the construction and step-by-step achievement of the units of the scheme have been accomplished successfully.

The calculations show that the cost of carrying out the investigations will be covered in less than a year.

4. PRINCIPLE OF DEVELOPING A COMPLEX SCHEME

The main principle for developing a complex scheme is the maximum recycling of waste water into the production water cycle without discharge into sewerage.

As it is seen from the "balanced scheme" and Table 1, when the total water cycle of industrial assembly and chemical plant is equal to 100 (in conventional units) the discharge of waste waters into sewerage is 4.44.

The fresh water enters only for domestic and drinking needs (artesian water) and for refilling non-recycling technological and mechanical losses, and water cycle losses (river water) constitute altogether 3.98 units.

The solution of such a problem is achieved by creating local recycling systems that use polluted waters of enterprises with local refining of these waters at certain stages of recycling. (The description of process recycling is not given in the report.)

The industrial waste waters which may not be involved in the local recycling systems of enterprises, depending upon the method of their subsequent refining, belong to the following categories of sewerage:

1. For organically polluted drains.
2. For drains having mineralization up to 3 g/l.
3. For drains having mineralization above 5 g/l.

The domestic drains of the chemical plant, town and of industrial assembly plants enter the sewerage system separately.

There is an all-plant refining system for refining the listed types of waste waters at the chemical plant.

TABLE 1. *Basic data of balanced scheme of water supply and use of waste waters in an industrial assembly and chemical plant (in conventional units)*

Items of water consumption	Before introducing scheme	After introducing scheme
1	2	3
1. Total water cycle of fresh, cooling water and polluted process waters of industrial assembly units, town, CHPP and chemical plant	100.0	100.0
2. Total consumption of fresh process and drinking water	25.14	3.98
including that consumed by chemical plant (without watering cultivations)	22.47	0.42
3. Discharge of domestic, organically polluted and weakly mineralized waste waters for biochemical treatment	19.6	2.4
including that from chemical plant	18.3	1.10
4. Discharge of waste waters into the open ponds (into river)	20.04	—
5. Discharge of strongly mineralized waste waters to desalting plant	—	1.6
6. Supply of mixture of domestic, organically polluted and weakly mineralized waste waters after biochemical treatment and storm waters to additional refining unit	—	1.68
7. Pumping of waste waters that have no perfect refining methods to absorption wells	—	0.045
8. Water use in water cycle, per cent	25.1	95.66
including that at chemical plant, per cent	23.1	97.8

The selection of waste water refining methods for the all-plant system and for local units at the enterprises was determined by the necessity of extracting pollutants from waste waters in such a form that permits the use of extracted sediments and waste materials as an additional output.

Considerable economies are made by the sale of these materials. At the same time, disposal points and slime storage ponds are eliminated, thus reducing the area of land occupied, transport costs, and costs for the neutralization of waste materials.

Overall fresh water consumption at the industrial assembly was reduced by approximately a factor of 7, and wastes discharge for biochemical treatment approximately by a factor of 8 (according to Table 1). Capital investments for the water cycle and for sewerage construction were greatly reduced.

Expenses for construction of the additional systems for refining of waste waters (desalting, additional refining, burial wells) are much less than the reduction of the above mentioned capital investments.

5. ALL-PLANT REFINING WORKS

5.1 *Biochemical Treatment Works*

Waste waters with a mineral content of less than 3 g/l are averaged, neutralized to pH value within the range of 6.5-8.5 and discharged to the buffer pond.

Waste waters containing organic impurities are averaged, neutralized and pumped to biological treatment.

Domestic waste waters undergo mechanical treatment and primary settling then together with waters containing organic impurities pass to biological treatment (air tank-mixers, after-settler) with subsequent chlorination and discharge to the buffer pond.

Storm waters enter the buffer pond without treatment.

After mixing and settling in the buffer pond the water has approximately amount of pollutants (taking into account pollutants which entered with the initial river and artesian water), shown in Table 2.

TABLE 2.

Pollutants and other data	Cl	SO_4	Ca	Mg	Na + K	HCO_3	Cu	Ni	P	Cr	Li Fe B Σ	Dense residue
Amount in mg/l	259	174	177	136	243 266	157	26	3.8	5.6	2.4	0.42	1100

According to a field test the refined water of such composition diluted approximately by a factor of 2 with river water is suitable for irrigating agricultural fields (version 1). However, as was mentioned above, the option of additional refined buffer water and its recycling into the process water cycle was chosen for the actual project.

5.2 *Unit for the Additional Refining of Buffer Water*

The mixture of waste waters from the buffer pond containing pollutants given in Table 2 is subjected to additional refining with the aim of using it to refill the water cycle system of cooling water.

The additional refining is neccesary to extract biochemically non-oxidizing organic substances from waste waters and correction of mineral composition because the accumulation of salts in the system may lead to incrustation and corrosion of the equipment.

The absence of rigidity salts in the additionally refined water permits the elimination of purging of recycling systems and thereby reduction of the drainage of waste water with low mineral content into sewerage, and water consumption for the refilling of systems.

A unit of additional refining provides for the following technological operations: preparation of active microporous anthracite; adsorption of organic compounds on active anthracite; ion-exchange correction of mineral composition; recovery of waste ion exchange resins; preparation of granular nitrogen fertilizers from the recovered solution.

Adsorptive refining is carried out in the fluidized bed of sorbent by a continuous process.

The waste anthracite is fed continuously to the recovery.

Solutions of nitrogen acid and ammonia are used for the recovery of ion-exchange resins. In this case the mixture is formed of nitrates and ammonium salt with concentration of 27-30 per cent.

After salt treatment, granular fertilizers are obtained containing up to 68 per cent of ammonium nitrate.

5.3 Waste Water Desalting Unit

Waste waters with higher mineral content (more than 5 g/l) are subjected to desalting in a multi-chamber evaporating unit after their preliminary physicochemical preparation.

TABLE 3.

Name of pollutant	Na	Cl	Ca	SO_4	Mg	SiO_3	CO_3	HCO_3
Amount mg/l	694.6	1849	1058	1515	61.2	3.85	44.1	380.5

After careful technical and economic comparison of the existing methods of desalting salted waters the evaporation unit was taken for construction at the chemical plant as the most perfected in respect of technology and apparatus.

The condensate from the unit with total salt content of 15 mg/l is used as feeding water at CHPP therefore the first and second stages of chemical treatment are eliminated and the third one is used as a control.

The concentrated brines from the unit are channelled into the preparation of technological raw material and part is buried together with other liquid wastes in absorption wells at a depth of approximately 1.8 km.

6. UTILIZATION OF SEDIMENTS OBTAINED IN THE TREATMENT OF WASTE WATERS AND PRODUCTION WASTES

6.1. The project provides for the preparation of granular organo-mineral fertilizers from the mixture of raw sediments of primary settler and excess active slush by the radiation and heat method.

Investigations are also carried out on separate reprocessing and preparation of fertilizers from raw sediments and of fodder concentrate from excess active slush.

6.2. Granular nitrogen fertilizers are produced from regenerated solutions of the additional refining buffer water unit.

6.3. Commercial sodium sulphate is produced from the mother liquor of evaporating lyes by the counterflow crystallization method.

6.4. Fume gases from limestone calcination ovens are used for the brine carbonation for the purpose of preparing dry ice and liquid carbon dioxide.

6.5. Other gaseous, liquid and solid wastes are also used.

7. RESULTS AND CONCLUSIONS

7.1. The carrying out of the programme of research, investigation and experimental work made it possible to develop a scientifically valid project for the refining and use of waste waters and to construct and introduce a drainage-free scheme of water supply and sewerage for the chemical plant, town and industrial assembly.

7.2. In the Soviet Union, similar investigations and the development of projects have been carried out to create drainage-free schemes for other industrial complexes involving chemical plants and towns.

7.3. Introduction of similar multi-purpose drainage-free schemes are of a great importance in the national economy because they exclude the discharge of waste waters into open ponds, reduces fresh water demands, and greatly improves sanitation and conditions of life.

Moreover, a considerable annual economy is made in terms of lower capital investment requirements for water supply and sewerage units, and of additional product output from the waste materials.

A Review of Non-Waste Technology Problems in Some Major Production Branches

P. Grau

Institute of Chemical Technology, Prague, Czechoslovakia

The key problem of human existence in the future is without much doubt the *generation of energy.*

Energy consumption is one of the most important indicators of technical level attained in various national economics and is also in direct relation to the gross national product. Increase of energy consumption has been of an exponential nature. Electric and thermal energy is mostly generated by burning of *fossil fuels* of all kinds. A tendency exists to use to a greater extent high-quality fuels, such as natural oil or gas. Compared with all kinds of solid fuels, the liquid and gaseous ones are superior in terms of transportation, manipulation, combustion technique and technology including automation, efficiency and pollution.

From the long-term human strategy point of view, the combustion of organic carbon compounds decreases the storage of vital raw materials for chemical and other industries.

Waste products may be classified in three groups - namely, solid and gaseous combustion products and waste heat.

Huge amounts of solid wastes, mostly produced as slurry, are deposited and even in the most successful cases only a very minor part is used as complementary raw material.

Removal of SO_2, SO_3 and NO_x has been solved by a wide spectrum of various processes. Economically acceptable methods of removal of these pollutants from combustion gases will hardly be achieved before the end of the fossil fuel electricity generation era.

Similarly, the excess heat problem has not been economically solved.

Nuclear energy generation will eliminate ash problems and chemical air pollutants. It brings, however, the risk of radiochemical pollution, and increase thermal pollution due to lower thermal efficiency.

New prospects promise high rate reactors, nuclear synthesis reactors and plasma torch reactors.

Hydroenergy (including tidal energy) and *geothermal* energy is of local importance only, as is the case with direct exploitation of *solar* energy.

In connection with the prospective use of hydrogen as an energy carrier, gas engineering will further develop. In this connection also underground coal gasification introduces another example of low-waste production.

Technology of silicates and some other inorganics can effectively use some solid and liquid wastes as raw materials, additives, etc.

This has been demonstrated in the making of bricks, gypsum, cement and glass.

Since many industries produce huge amounts of calcium sulphate, its exploitation, and research and development are of major importance.

The fields of *metallurgy, metal processing of all kinds and corrosion* represent a vast and complex task. Only a few examples may be introduced in this paper.

Mass use of aluminium expands the problems of air pollution caused by open electrolyzers. The harmonization and recovery of chemicals improve the situation but cannot be considered as a final development, which logically tends towards ambient temperature electrolysis.

Metal finishing and galvanic plating bring about another example of how process changes can decrease waste.

There are, however, huge amounts of waste iron hydroxides at various disposal points. Processing of iron and sulphuric acid has been developed. New systems of iron pickling by means of sodium hydride will eliminate the problem in many instances.

Already existing and further exploitation of low-grade metal ores is connected with all kinds of waste production. Hydrometallurgy processing is widely used and in some cases even microbiological processing is under research.

Recovery of metals from scrap, widely used for many years, becomes more complicated. Greater protection of metals against corrosion and modern methods of approach in mechanical engineering, electronic and electrical engineering involve increasingly the use of alloys which are not easy to separate. The collection and classification of *scrap metals* also pose many problems. Due to these facts carbon steel is gradually contaminated by steel alloys and by other metals. Not yet solved is the recovery of metals from many copper alloys.

Production of *plastics* still rapidly increases, and so does the quantity of waste plastic materials. Collection, classification and recovery of this solid waste has been impractical so far. The favoured solution for the near future is to use pyrolysis to produce combustible gas or similar destructive methods. A similar approach can be adopted in disposing of worn tyres.

Pulp production is in general at a low level in non-waste technology development. Indeed, even modern processes based on both sulphite and sulphate technology give a pulp yield of about 50 per cent wood solids, while chemicals recovery reaches about 90 per cent and 45 per cent of the wood solids are burned. A pulp mill attaining these figures is thermally self-supporting. On the contrary, air pollution becomes a serious problem.

Oxygen bleaching decreases wood solids losses by one-quarter. New delignification processes are under research.

It is believed, however, that complementary production, e.g. of feedstuffs, alcohol, magnesium oxide and other materials, can effectively decrease the waste of valuable materials.

The *Food Industry* and its non-waste concepts are closely related to agricultural products. Examples of thin skin potatoes, sunflower seeds, low fat pork, etc., are well known. Difficulties in the treatment, use or disposal of wastes from the food industry are rather rare. Those still existing seem to be solved in the short run. So the main problem remains in the waste of packaging and containers by consumers.

The *chemical industry* produces a variety of wastes, often of a most harmful kind.

Perhaps even more than in other industries the rule of inverse proportion of production capacity to specific waste formation holds true. Naturally, at high capacity even low specific waste formation becomes a real problem.

The methodology of approach towards low-waste production is being developed and includes the following key tasks:

(a) development of effective catalyzers possessing high selectivity;

(b) development of continuous-flow countercurrent apparatus for various processes (extraction, absorption, adsorption, ion change);

(c) development of apparatus for transportation, mixing and separation of dusty, sticky, irregular shaped, non-neutonian and similar "difficult to handle" materials;

(d) exploitation of physico-chemical properties of foams, aerosols and emulsions;

(e) exploitation of reaction of very fine particles;

(f) wide implication of non-classical separation methods such as reverse osmosis, ultrafiltration, dialysis, electrodialysis, etc.

In inorganic chemistry the key high tonnage products are sulphuric acid, superphosphate, phosphoric acid, ammonia, nitric acid, ammonium nitrate, chlorine, sodium hydroxide and soda.

Modern production of sulphuric acid is a good example of non-waste production. Absorption or scrubbing of sulphur dioxide from end gases can decrease its emission down to the order of hundreds of a per cent of produced acid. On the contrary, sulphuric acid production from pyrites is becoming attractive again from the non-waste technology point of view because it can consume up to 30 per cent of waste ferrous sulphate from iron pickling wastes. Burnt pyrites can be successively used in cement production. This offers practically complete use of raw materials in the closed cycle.

Other chemical processes, however, producing sodium sulphate are still typical waste-technology examples, e.g. synthetic fibres, titanium white. Mineral fertilizers can be produced without significant wastes.

Thus silicon fluoride from phosphorus fertilizers can be captured and used in other processes.

Electrothermical production of phosphorus as an intermediate stage in phosphoric acid production offers sludge which can be further transformed into building materials. Similarly, waste gypsum from phosphoric acid production can be used.

Ammonia can be produced without waste; nitric acid is also produced in pressure absorption units practically without significant emissions.

Soda has been produced by the Solvay method, giving a high accompanying production of calcium chloride. Competitive production is based on saturation of sodium hydroxide obtained by the waste diaphragm method with carbon dioxide.

In spite of some doubts regarding the use of asbestos diaphragms, amalgam electrolysis is gradually being limited and decreased. Losses of mercury of up to 100 g Hg/t of the chlorine produced are not acceptable in such a situation that only 80 per cent of the mercury can be removed from the wastes while the remaining 20 per cent, i.e. 20 g Hg/t, of chlorine load the environment.

Hundreds of processes in organic chemistry have been used without paying much attention to the wastes. Many studies are necessary to improve the present situation, especially directed to the discovery of economically feasible reuse or recovery of different waste products. It is clear that only in cases of specialized and medium-tonnage production can waste problems be solved. Low tonnage production, sometimes operated for a short time only, can hardly be without waste.

In organic chemistry, large quantities of inorganic salts are produced. Organic synthesis procedures have been developed in such a fashion that only such reagents are used which form non-toxic and non-objectionable wastes. The following examples can be quoted:

(a) OH: groups are brought into organic molecules principally by oxygenation (synthesis of phenols, naphthols, ethylene-oxides, glycenol, etc.);

(b) similarly oxidation of organic matters classically carried out by various oxidation salts (chromate, permanganate) are replaced by oxygen oxidation or electrochemically;

(c) reduction reagents such as iron chips, zinc, sulphides, sodium, etc., are replaced by catalytic reduction by hydrogen.

CONCLUSIONS

Contemporary science and technology can offer many advances in non-waste technology. Only a very few of them can compete under such circumstances where very low protection of the environment is demanded.

Increased demands for environmental protection together with increased capital and running costs of traditional destructive waste treatment units create more favourable conditions for the implantation of non-waste technology.

Non-waste technology is mostly accompanied by higher energy needs, and a requirement for more qualified personnel and more complicated automation.

Only in very few cases can non-waste technology at present recover really valuable and rare raw materials. In most cases, however, very objectional materials in various wastes are neutralized. Thus, non-waste technology is an excelent tool to protect the environment.

Developing Conservation-Oriented Technology for Industrial Pollution Control

Joseph T. Ling

Vice President, Environmental Engineering and Pollution Control
3M Company 3M Center, St. Paul, Minnesota, U.S.A.

Successful application of a resource conservation-oriented pollution-control technology program throughout a single transnational company has been especially encouraging. It also indicates that on a large scale involving many countries, the rate of industrial conversion to this technology may depend largely on the amount of practical support given by governments.

Strong individual effort has been required in several countries where industrial firms have experimented with conservation-oriented technology, which is a form of non-waste technology. (In this paper, conservation-oriented technology means preventing or reducing pollutants at the source, or utilizing pollutants in a more productive manner.) Even though industrial success in non-waste technology has occurred in limited applications, it has not received much recognition by governments or the general public.

Emphasis has been placed on the traditional removal type of pollution control technology. (This generally means a facility at the end of a production process to remove pollutants before discharge to the environment.) Legislative requirements or the short-term deadlines of recent environmental legislation, particularly in the United States, have forced industry to use removal technology, which is not always the most environmentally efficient method.

The removed residue causes a serious disposal problem and installation, operation and maintenance of removal facilities is a growing expenditure for industry. Various responsible forecasting agencies have indicated that these already substantial costs will rise sharply in the future as environmental regulations become more restrictive around the world. For example, in the USA, the National Research Association estimated this year that the total USA capital investment for pollution control between 1974 and 1983 may reach $263 billion. Thus, the annual cost in 1983 may go as high as $66 billion. This represents over $800 for every US household per year in 1975 dollars. This does not include the impact of new environmental legislation that may be enacted in the next few years. This is a substantial sum even in the USA!

In addition, removal technology consumes large amounts of resources, and the rate of consumption as well as the total cost accelerates exponentially as removal rises to the last few percentage points.

Within industry, the primary objective in management of pollution-control activities is achievement of the highest degree of pollution reduction with the lowest use of human, material and financial resources. Non-waste technology programs appear to be the best means of meeting this objective in many cases.

GOVERNMENTAL POLICIES

However, some governmental policies provide certain tax incentives for installing pollution-removal facilities, but no similar incentive is directed toward preventing or reducing pollution at the source through use of conservation-oriented technology. In many cases, use of conservation-oriented technology also is discouraged by regulations that limit the *concentration* of pollutants emitted, rather than emphasize the *amount* of pollutants discharged to the environment. This is especially true in the recycling of waste streams or use of similar techniques that reduce the total amount of pollutants discharged, even though they increase the concentration of the discharge.

Problems of this nature can be reduced if the technical aspects of pollution control could be recognized by governments as much as the legal and political aspects of environmental legislation.

THE 3M COMPANY PROGRAM

One extensive non-waste technology program recently was implemented by the 3M Company, a large diversified transnational manufacturing company based in the United States. The firm, with nearly 80,000 employees in more than 40 countries, stresses new and improved products. Manufacture of these products often produces pollution-control problems that require special solutions.

Initial results of the 3M program are particularly encouraging because they demonstrate the superiority of this new pollution-control approach over removal technology.

The program was aimed at applying conservation-oriented technology to the company's facilities around the world. It began with the strong support of top management, which was considered essential for successful implementation throughout the firm.

A plan was developed for providing orderly transition from removal technology to the concept and use of conservation-oriented technology. One important aspect was interesting the company's several thousand technical employees in finding ways within their own areas of responsibility to prevent pollution at the source. Appropriate prevention methods include:

1. Product reformulation.
2. Process modification.
3. Equipment redesign.
4. Recovery of waste materials for reuse.

In 9 months the program was introduced in fifteen countries. In the United States, non-waste technology projects eliminated 70,000 tons of air pollutants and more than 500 million gallons of wastewater per year. In addition, the program saved an estimated $10 million in actual or deferred costs associated with pollution control, including energy and raw materials as well as retained product sales.

PROGRAM EXAMPLES

The following examples illustrate the nature of the program, which is called "Pollution Prevention Pays"-or, the 3P program.

"The company developed a new cotton herbicide chemical. The original process emitted a toxic substance and one that caused a strong odor. It also produced 12 pounds of pollutants per pound of product. Using non-waste technology, the laboratory then developed a new process that eliminated the toxic substance and the odor. It also reduced the other pollutants to only 2 pounds of waste per pound of product. In addition, manufacturing costs were significantly reduced.

"Another case involved control and recovery of hydrocarbon solvents, which can contribute to photochemical smog when released into the atmosphere. The firm developed and built a unique inert gas drying process. It features a large oven that operates as a closed system. This prevents hydrocarbon emissions and allows recovery of most of the valuable solvents.

"In a third case, a mercury free catalyst was developed for a resin product to prevent a mercury problem. This made the product more environmentally acceptable and prevented a substantial loss in sales."

These and other successful projects in the firm's non-waste technology program demonstrate what can be done. They have encouraged others within the company to attempt similar achievements. Since this is a continuing program, it is anticipated that eventually most of the organization will adopt this technology.

Rapid progress is not expected, however, because in many cases new manufacturing or recovery methods must be developed, designed and implemented. Some people have blamed technology for causing environmental pollution problems. However, the 3M experience demonstrates an urgent need for the *right* kind of technology-non-waste technology - to solve environmental problems in the most efficient manner.

Since it is impossible to invent new technology to meet arbitrary deadlines, and the investment in new or modified equipment in connection with the non-waste technology program can be costly in many cases, incentives and encouragement from governments could significantly increase the rate of conversion.

This means that both industry and government must work toward greater use of non-waste technology. Industry should be concerned with development and implementation while government should be concerned with developing policies and regulations that promote this concept.

POLLUTANTS ARE MISPLACED RESOURCES

In a sense, many pollutants can be considered misplaced resources. For example, petroleum by-products that once were thrown away are now used to make nylon and other modern clothing materials. These by-products were discarded as waste until someone discovered how to use them for productive purposes. Then they became valuable resources. But it took knowledge (technology) to turn these former pollutants into resources.

This concept can be illustrated by a simple equation:

Pollutants (waste materials) + Knowledge (technology)
= Potential Resources

There are many waste materials that are called pollutants but most of the world's resources are limited and the majority of resources are not renewable. What is needed to complete the equation is knowledge (technology), which is essentially unlimited. Industry is in a good position to provide the leadership needed to increase this technological knowledge.

Our challenge is to develop and apply non-waste technology wherever possible throughout the world so that valuable resources are conserved and the environment is improved for the benefit of future generations. The 3M Company experience indicates that proper application of non-waste technology is possible, productive and profitable.

Both government and industry share a responsibility for creating a much greater awareness of non-waste technology and for putting it to use as the "best available pollution control technology", which is now being required in a growing list of environmental pollution control regulations in many countries.

(The author has indicated a willingness to respond to inquiries about details of the 3M Company pollution control technology program described in this paper.)

The Nordic Organization for Waste Exchange

K.E. Kulander, L-G. Lindfors and E. Lohrdén

Sveriges Industriförbund, Stockholm, Sweden

THE NORDIC ORGANIZATION FOR WASTE EXCHANGE

The project called the "Nordic Organization for Waste Exchange" (NOWE) was initiated by the Federations of Industries in Denmark, Finland, Norway and Sweden, together constituting the Nordic area. It got underway in November 1975 with the Institute for Water and Air Pollution Research (IVL) made responsible for the central secretariat function.

Acting within the Nordic area, NOWE is called upon to furnish information about the supply of and demand for waste or residual products which can be put to productive use or be destroyed on toll at another company. NOWE is also required to furnish advice on waste management and, by actively following up the exchange programme, identify obstacles to using what appears to be recyclable waste. The latter programme can be enlarged to include more penetrating studies where these are deemed urgent.

Under the directives originally laid down for NOWE, the exchange programme is not to concern itself normally with types of waste such as scrap metal, paper and the like, for which an established market exists, nor with wastes of the kind which evidently lack productive value. However, NOWE is to be capable of giving information about the treatment supply for the latter category, a programme that has been enlarged beginning in 1976. The organization consists of national marketing entities, which in Denmark, Finland and Norway are sited on the premises of the Federation of Industries in each of these countries, and in Sweden at IVL. In addition, IVL functions as headquarters for NOWE.

The exchange service is based on a correspondence procedure. Companies wishing to avail themselves of NOWE transmit the appropriate data to their national entity. If desired, they can fill in special forms for this purpose. IVL then compiles the material, which is published in codified form under the auspices of the national marketing entities.

Interest in advertised items is to be notified in writing to the headquarters unit (IVL), which will forward the incoming data to the informant.

A procedure has been devised to assure the original advertiser of reasonable anonymity safeguards. In the absence of special agreement, NOWE will not divulge the identity of the company which stands behind a given exchange notice.

The costs of operating the central marketing entity are underwritten for a 3-year trial period by grants from the Nordic Industry Fund. In addition, it is assumed that the Nordic Federations of Industries will underwrite at least an equal proportion in the form of their own work efforts.

Results of Operations, 1 December, 1973 to 15th December, 1975

During the initial 2-year period 270 items were accepted and advertised, which resulted in 517 notifications of interest. The distribution of different wastes by type is set out in Tables 1-3. As of 1 December, 1975 the result of 147 mediated contacts had been reported. Of these, 27 per cent turned out positively in that wastes were exchanged; 63 per cent turned out negatively. The remaining 10 per cent are identified as not finalized.

Table 1

Statement of Results as at 31 December 1974

Waste type	Ads.	Inquiries	Returned forms	Result			Obstacles in neg. results				
				pos.	neg.	not final	econ.	qual.	int.	legal	misc.
Plastics	37	102	18	4	10	4	1	1	1	0	12
Textiles	30	116	17	6	6	5	1	3	2	0	0
Paper	10	34	5	3	0	2	0	0	0	0	0
Solvents	13	19	8	3	4	1	0	1	0	2	1
Acids	15	1	1	0	1	0	1	0	0	0	0
Inorg. chem.	21	27	8	4	3	1	0	3	0	0	0
Org. chem.	11	7	3	1	0	2	0	0	0	0	0
Slag, sludge	13	26	8	1	4	3	0	2	1	0	1
Miscellaneous	26	30	8	0	5	3	1	1	0	0	3
Total	176	362	76	22	33	21	4	13	4	2	12

Table 2

Statement of Results as at 30 April, 1975

Waste type	Ads.	Inquiries	Returned forms	Result			Obstacles in neg. results				
				pos.	neg.	not final	econ.	qual.	int.	legal	misc.
Plastics	57	142	23	4	15	4	2	3	2	0	8
Textiles	34	124	23	7	11	5	2	5	1	0	3
Paper	13	37	7	4	2	1	1	0	0	0	1
Solvents	16	19	8	3	4	1	0	1	0	2	1
Acids	15	2	2	0	2	0	0	0	0	0	2
Inorg chem.	25	27	12	5	7	0	1	3	0	0	3
Org. chem.	17	9	4	1	2	1	1	0	0	0	1
Slag, sludge	17	31	9	1	7	1	1	2	1	0	3
Miscellaneous	35	34	13	1	10	2	2	2	3	0	3
Total	229	425	101	27	60	14	10	16	7	2	25

Table 3

Statement of Results as at 15 December, 1975

Waste type	Ads.	Inquiries	Returned forms	Result			Obstacles in neg. results				
				pos.	neg.	not final	econ.	qual.	int.	legal	misc.
Plastics	64	158	34	7	24	3	6	6	4	0	8
Textiles	34	139	25	11	13	1	3	3	1	0	6
Paper	14	40	12	5	6	1	2	1	1	0	2
Solvents	18	22	10	4	5	1	0	1	0	2	2
Acids	20	4	3	0	2	1	0	0	0	0	2
Inorg chem.	31	46	18	8	10	0	2	3	0	0	5
Org. chem.	21	15	11	4	6	1	1	1	1	1	3
Slag, sludge	19	42	12	2	8	2	1	2	1	0	4
Miscellaneous	49	51	22	3	17	2	2	3	5	0	7
Total	270	517	147	44	91	12	17	20	13	2	39

It will be seen that one of every four advertised items leads to some form of recycling. This result well squares with the experiences gained by the somewhat older German exchange (VCI-Abfallsbörso).

If the relative interest in each waste type is measured by the number of inquiries received per item and the relative performance by "the number of positive results per item", we get the tabulation shown in Table 4.

Table 4

Waste type	Share (%)	Inquiries/ item	Positive results/ item
1. Plastics	23.7	2.5	0.2
2. Textiles	12.6	4.1	0.4
3. Paper	5.2	2.4	0.4
4. Solvents	6.7	1.2	0.4
5. Acids	7.4	0.2	0
6. Inorganic chem.	11.5	1.5	0.5
7. Organic chemicals	7.8	0.7	0.3
8. Slag, sludge	7.0	2.2	0.2
9. Miscellaneous	18.1	1.0	0.1
	100		

Plastic, textile and paper wastes, i.e. wastes for which a recovery industry is established, account for a substantial proportion of the items advertised even though NOWE tries to avoid items which are deemed to have a well-known market. The study aimed at exploring the feasibility of recovering plastic waste in the Nordic area which NOWE has published (IVL B 209, 1974) has not resulted in any marked decline of the number of plastic items. However, the unalloyed plastic wastes have been increasingly superseded by harder-to-place mixed-waste items. This is also reflected in the result. Apart from a very few exceptions, pure thermoplastic wastes of homogeneous composition are the only plastic type that can be re-used today pending the development of methods for the separation of mixed wastes.

Solvents have been placeable in cases where the amount of pollution has been slight or absent, i.e. when the item can be described as a marketable residual chemical. To qualify for inclusion in this category, however, it will probably be necessary to use less polluted solvents, for instance as washing fluids.

Waste acids have not been placeable, The low interest shown in them has touched off a study which seeks to shed light on the potentials for recycling waste sulfuric acid.

Inorganic and organic chemicals have attracted relatively great interest. These wastes largely consist of marketable residual chemicals. Items that may pose a troublesome "waste type" for a small or medium-sized plant can turn out to command good selling prospects once a waste exchange is operating. Long transport distances do not interpose as deterrent a barrier to this waste type as for others. NOWE has therefore set up a working arrangement with its German counterpart insofar as residual chemicals are also being advertised on the German market. An increasing number of items of the residual chemicals type is also noticeable. Subsumed under the category called *slag and sludge* are metaliferous sludge and airborne particles, coal-dust and the like, some of which have been placeable, notably airborne particles containing zinc. The *"miscellaneous"* category includes waste leather and used packaging material such as cardboard boxes, items that have found interested buyers in most cases.

The reasons stated below have been given to account for a negative result:

1. Economic obstacles: 12%
2. Unacceptable quality: 18%
3. Internally utilized: 8%
4. Legal obstacles: 2%
5. Other reasons: 60%

Included with "other reasons" are competition and cases where the advertiser could not contact an interested party found through NOWE for reasons unknown.

It is remarkable that quality criteria have not posed a bigger obstacle than might be inferred from the foregoing account. That is probably due in large part to the fact that NOWE restricts advertising eligibility to items of relatively high quality, a practice that can be counted on to result in inquiries, as reflected by the fairly great importance attached to prices and the internal utilization of marketable items in a great number of cases.

Legal obstacles, which may chiefly arise in connection with transport across national frontiers, have occurred in only a few cases. For that matter, exchanging wastes across national frontiers has been very seldom considered inasmuch as longer transport distances usually make waste recovery unprofitable.

Central Information System for Problem Wastes

Among the questions taken up in "Disposal of Waste in the Nordic Area", a report presented in July 1974 by a study group under the Nordic Civil Servants Committee for Environmental Protection, is setting up a central information system for problem wastes. Recommendation no. 6 identifies the need for a central unit to gather information about Nordic treatment supply, prices, tariffs and transport regulations.

An agreement has been reached between the aforesaid study group and the Environmental Protection Committee attached to the Nordic Federations of Industries, which called for fitting the information system into NOWE's programme beginning in 1976.

Scope

The assembled information shall encompass:

a continually updated, detailed list of Nordic treatment plants for problem wastes, with specification therein of treatment capacity, intake capacity and prices plus similar data for industry-internal plants with overcapacity;

similar data for certain European plants, especially as regards problem wastes for which Nordic treatment capacity is lacking;

a list of haulage firms;

synoptic information about tariffs and transport regulations.

Although some of the foregoing services have entered into NOWE's functions to date, they have not been explicitly directed towards problem wastes. It has therefore been necessary to collect additional information.

Questionnaire Concerning Recyclable Wastes

In 1975 an information kit was despatched to all companies belonging to the Nordic Federations of Industries, describing in detail the work performed by NOWE. The object was to have this material reach the one or more persons inside the companies which may come to be directly affected by NOWE's services in the course of their employment. Also despatched with the information kit was a questionnaire which asked the companies where appropriate to state the type and annual volume of waste thought capable of having some productive value for another company; the respondents were also asked to state how they took care of the waste as of the time they filled in the questionnaire.

Acceptable returns were received from 164 companies, most of them Swedish.

Table 5 shown below breaks down the number of companies possessing potentially recyclable waste according to type and volume. Treatment methods are set out analogously for each waste type.

Table 5

Number of companies with recyclable waste

	Annual volume (tons)			
	<5	5-50	>50	Σ
Solvent	6	12	5	23
Other chemical	4	6	11	21
Metaliferous waste	3	3	18	24
Plastic	13	13	15	41
Miscellaneous	10	14	31	55
Σ				164

Number of Companies Employing Specified Treatment Method
Solvent Waste

	Annual volume (tons)			
	<5	5-50	>50	Σ
Incineration	2	3	-	5
Deposition	3	1	-	4
Storage		1		1
Recovery	1		4	5
Miscellaneous		7	1	8
Σ				23

Other Chemical

	Annual volume (tons)			
	<5	5-50	>50	Σ
Incineration	2	1	-	3
Deposition	2	-	3	5
Storage	1	1	2	4
Recovery	1	4	2	7
Miscellaneous	1	3	5	9
Σ				28

Metaliferous Waste

	Annual volume (tons)			
	<5	5-50	>50	Σ
Incineration	1	-	-	1
Deposition	-	-	7	7
Storage	2	1	3	6
Recovery	-	2	5	7
Miscellaneous	-	-	7	7
Σ				28

Plastic Waste

	Annual volume (tons)			
	<5	5-50	>50	Σ
Incineration	4	2	3	9
Deposition	3	3	2	8
Storage	2	3	-	5
Recovery	4	9	6	19
Miscellaneous	3	1	1	5
Σ				46

Miscellaneous Waste

	Annual volume (tons)			
	<5	5-50	>50	Σ
Incineration	4	3	4	11
Deposition	1	3	14	18
Storage	4	1	2	7
Recovery	5	1	8	14*
Miscellaneous	1	7	6	14
				64

*Chiefly paper waste.

Programme Considerations and Experiences in Optimizing Industrial Materials Flow and Utilization for a Non-Waste Technology

Jerome F. Collins

Chief, Materials Optimization Branch, Division of Industrial Energy Conservation, United States Energy Research and Development Administration, Washington, D.C.

INTRODUCTION

Industrial experience with non-waste technology can be discussed in two contexts: past and presently evolving.

Past experience has been pragmatic, case-specific, and essentially reflective of internalized cost factors. Managers elected to implement aspects of non-waste technology as they attempted to enhance the profit position of the economic unit under their control.

The present experience is evolving from a reaction to rising materials (and especially energy) prices—again a situation of cost minimization—toward recognition that consideration must be given to planning and acting beyond the immediate interests of the economic unit under the manager's cognizance.

In the United States, one of the ways in which this essentially public benefit consideration can be effected is by means of Federal Government support of technology development of non-waste technology. The Division of Industrial Energy Conservation (INDUS) of the United States Energy Research Development Administration (ERDA) has this as one of its major goals.

This ERDA/INDUS goal is being realized in the form of a two-part program.

A base program structure is being developed by means of materials and process-train systems analyses to understand the energy flows and impacts associated with significant energy intensive commodities and industrial sectors. The objective will be to rationalize - in energy terms—these flows by means of substitute and alternative materials flow and process-train configurations. The second part of the ERDA/INDUS program concerns substantive research and development of more energy efficient unit processes and engineering materials. Both programs will be carried forward to firmly establish the developed technology in the commercial fabric of the United States economy as technically and economically feasible embodiments of non-waste technology. The base program is evolving; the substantive research is further advanced.

EVOLVING BASE PROGRAM

The presently evolving ERDA/INDUS program in non-waste technology is aimed at minimizing and optimizing materials and process-train energy impacts. The basic concept is the reconfiguring of material flows from the generalized resource base or origin to final demand end uses—taking into account the full energy costs of various alternative materials flows and their associated process-trains. The programmatic elements required and obstacles likely to be encountered are discussed in the following sections. A necessary first step is to define for analytic purposes socially necessary standard units of produced goods and services. For example, we may define a standard unit of shelter as X cubic meters of living space comprised of Y_1 tonnes of wood, Y_2 tonnes of brick, cement, clay, etc., Y_3 tonnes of fossil based plastics, textiles, etc.—all with associated life times in tonnes of annualized flows of these materials as "consumed" to provide this standard good. Similarly, standard materials flows for transport, communications, food, public services and so on can be defined.

Following on this is a fundamental consideration of the mix of final demand for goods and services—and the concomitant overall level of materials consumption to be sustained or assumed for analysis. We may choose to accept levels now current for the industrial countries, but two difficulties immediately arise. First, there are considerable differences among these countries, and second, any imputation of even an average of these levels to the developing countries will create an unrealistic global level of demand. Thus, implicit in a supposed value-free technological exercise, will be the necessity to apply essentially subjective inputs.

Explicit recognition is given to energy as the ultimate resource. It is the ultimate and absolute conversion/processing energizer for all commodities. First-law effects concern the inherent energy content of materials and second-law effects concern their utilization in terms of their changing states as they proceed along a flow path from source to end use—and back as a production input in reconstituted or available form.

A related consideration is the problem of differently valued forms of energy, i.e. a joule derived from natural gas is different than a joule from coal. At present, market prices very imperfectly reflect these differential (and implied) values. Thus again, a subjective element is introduced. Former regulatory practices, as expressions of social valuation which were once directed primarily at promoting a benefit on the demand side,are now being redirected toward promoting a benefit on the supply or resource-base side which, however, ultimately redounds to the demand side as the embodiment of a society's values.

Following upon energy is entropy as a basic consideration. Its value is primarily conceptual but with some modification it can be employed as a practical tool of policy and analysis to rationalize materials flows and process-train configurations. The value lies in making explicit that refinements of any and all kinds—in material flows, process-trains, organization, etc.—produce a localized benefit, i.e. an entropy decrease; but at the expense of an entropy increase elsewhere in the larger system. This entropy increase requires a first-law energy input makeup, e.g. fossil fuel, nuclear, or solar energy utilization—the juxtaposition is intended to be provocative. A practical application of entropy concepts will be to define real limits to efficiency and utilization efforts to improve materials, their flows and processing. This requires a modification of basic second-law principles—which are based on reversible, infinitely slow processes—to the finite time processes which obtain in industrial practice.

The sections above lead to consideration of basic data form and their manipulation in analyses. The convenience of standard industrial classifications such as SICs and ISICs must be foregone as units of analysis—except for their advantageous use to obtain and organize economic and energy data as intermediary elements of analysis. These elements are self-limiting intrinsically and act to block reconfiguring among alternative and substitute materials and intermediate goods (and services). Basic information is likely to be obtained or calculated as kWh/tonne. The need to introduce energy-societal valuation as well as the forms of data availability and convenience may indicate the necessity for conversion to kWh per unit-value-added for each flow path.

IMPLEMENTATION

The transfer of technology developed under this program to the commercial world will depend on factors more diffuse and less controllable than those which obtain with conventional equipment or process development. In market economies the perceptions of managers regarding materials flows and process-trains vary with the industry sector. Some industries, in effect, act to maximize their flows of materials and others to minimize them. A convenient differentiating indicator is value-added-per-tonne of product. Its differentiating effects stem from the fact that low VA/tonne industries tend to be in basic commodities and raw materials production while high VA/tonne industries tend to be in goods fabrication. Commodities producers are characterized by high tonnage and energy inputs, e.g. steel production; fabricators are characterized by low tonnage. The application of this indicator can be as a measure of "implementative elasticity", in other words, the likelihood of an industry to respond differentially to materials or energy reconfigurations as a part of some proposed overall or national optimization of materials flows and process-trains. Low VA/tonne industries are likely to be process energy responsive and high VA/tonne industries are likely to be materials flow responsive.

The concept and practice of discounting in evaluating alternatives requires address. In effect the present is weighted to it over the future.

A growing uneasiness is developing with regard to not only the spatial but the temporal negative externalities of present and relatively near-term projects. As more explicit attention is given to future consequences of actions taken now, and as these actions lead to a non-waste technology, the future will tend to be moved up for more favorable consideration in the present.

Perhaps the most potentially socially disruptive consideration stems from the purposeful intent to constrain and possibly to limit man's entrepreneurial nature on a global basis. From time to time and in various local situations it has been necessary to curtail this expansionist tendency of *Homo sapiens*. But there has always been some frontier to be explored—some new opportunity to be developed. It remains to be seen how society ultimately responds to the restraining walls of its own creation.

SUBSTANTIVE RESEARCH

The current substantive research effort of ERDA/INDUS is directed toward the development of new and existing but unproven technologies to improve the efficiency of industrial processes and reduce wastage. Related to this is the development of improved and new engineering materials and the utilization of wastes as fuels or materials inputs.

Effective commercial implementation is to be achieved by the heavy reliance on industry as a research partner in the development effort. Thus, industry's further role is to provide direction regarding economic realities as well as technological feasibility and acceptance of the produced processes or materials.

The ERDA/INDUS program is, in essence, a government-industry partnership with costs generally shared by both on selected projects. The selection of projects for support is based on several factors, including: energy savings potential, cost of the project, technical risk, degree to which the introduction of the new technology can be accelerated, and the assistance to fragmented industries having little or insufficient research and development funds of their own. ERDA/INDUS carefully avoids redundancy with industry efforts. Such programs usually have lower levels of technical risk, high return on investment, and low to moderate cost and will be advanced by the private sector without government support. Six major United States industry categories consume about 70 per cent of the total energy of the US industrial sector. These industries (chemicals, petroleum refining, primary metals, pulp and paper, cement and food processing) are all commodity industries, i.e. they produce materials used in the manufacture of end products. Such industries have several characteristics influencing the introduction of new technologies:

They are production oriented and maximize reliability, capacity and simplicity.

The equipment capital cost is usually extremely high.

They are not R&D oriented and what R&D monies are spent are largely devoted to developing new markets.

Industries in these sectors, in general, are extremely reluctant to introduce new technology—including non-waste technology—unless it has been well proven in a production environment, does not significantly increase capital investment per unit product and indeed has production advantages, and exhibits high internal return on investment. These constraining factors are obviated when such technologies are demanded by statute or administrative requirements to achieve, for example, environmental objectives.

The United States industrial experience with non-waste technology has been relatively limited due to the history of abundant, low-cost energy and materials. The substantive ERDA/INDUS effort in non-waste technology for industry began last year (August 1975) and has been growing constantly since that time. This ERDA/INDUS programme has two basic thrusts: (1) processes with wide industrial application and (2) processes of the energy-intensive industries (of which a few constitute the major items of energy consumption). Key program areas of focus include waste energy and materials recovery and utilization, increased industrial process efficiency, new material development, agriculture and food process efficiency improvement and technology transfer—the function of "selling" the technology, once developed and demonstrated, to the end-user industries.

Some of the key non-waste technology projects that are underway now at ERDA/INDUS are described in the following paragraphs:

Microwave/vacuum grain drying: the use of microwave and vacuum techniques to replace natural gas now used in grain drying

Fuel saving system for paint curing: this system collects the fumes from the paint and incinerates them in zone incinerators on the ovens to provide up to

70 per cent of the heat for curing the paint. This saves energy as well as protecting the environment.

Boiler air/fuel ratio control system: This system utilizes a special flame analyzer and industrial burner feedback controls to optimize the air/fuel ratio on industrial boilers.

Glass conglomerates: a joint program with the US Environmental Protection Agency which conserves energy and cleans effluent gases by preheating the glass bead-feedstock by passing it through the exhaust gases prior to the melting operation.

Waste carbon monoxide utilization: the object of this systems analysis is to identify industrial sources of under-utilized waste CO. A further object is to characterize the CO and its need for clean up as a potential feed stock to replace natural gas in the manufacture of methane and other industrial commodities.

Methane and protein from sulfite pulp mill wastes: the object is to develop a feasible means to desulfurate and breakdown ligno-sulfonates by ozonization. The expected more easily metabolized material will be converted to protein and methane.

Gasoline from pyrolyzed cellulosic wastes: a pyrolysis system is to be optimized for yield and composition suitable for a tandem gasoline synthesis reactor. The synthesis work will concentrate on the selection of catalysts and operating conditions. Significant laboratory quantities of a high octane liquid fuel have already been obtained.

Clean coke process: this project was sponsored by the Office of Coal Research, United States Department of Interior; before this agency became a part of USERDA.

The objective was to utilize high sulphur, high ash, non-metallurgical coal to produce steelmaking coke, chemicals and low-sulphur liquid and gaseous fuels. The process steps undergoing development are: the coal is washed and sized and one stream is carbonized to produce a partially desulferized char. The other stream is oil slurried and hydrogenized to a liquid. The liquid products from both streams are processed further into low sulfur liquid fuels, feedstocks, and recycle oils for further use in the clean-coke process. One of these recycle oils is coked in the carbonizer and blended with another recycle oil and the char. This mix is pelletized and calcined at 1000°C to produce a coke similar to conventional coke.

Key projects planned for initiation in the near future include the following:

High-temperature recuperator: seven development and demonstration projects are planned for recuperators of efficiencies exceeding 70 per cent and applicable to the ambient environments of glass, aluminum remelt, steel reheat, and fume incinerators in the thermal range of 1200°F to 3000°F.

Industrial heat pumps: the development and demonstration of heat pumps which will raise low-grade waste heat streams of 100 - 250°F to useful process levels of 400 - 600°F with a COP of 2.5 to 3.0.

Energy optimized industrial park: the basic concept for this project is the locating of industrial plant in a park complex by order of energy input/output characteristics so that the waste energy is cascaded from one to the other in descending order. The configurations of such parks can and will vary widely depending upon the existing plants, transport facilities, available land, etc., and the designs will also consider potential for using waste materials of some plants as feedstocks for others.

Acetylene production from coal: a 150-kW arc reactor has been developed under a previous federally sponsored program. The reactor successfully converts coal to acetylene as a feedstock replacement for ethylene, now derived from petroleum. The current objective is to develop a scaled-up reactor to provide experience and data on the complete process which will include gas separation and purification.

Conclusion. The rather wide variety of substantive research projects reported above represents a sampling of a greater number now in progress at ERDA/INDUS. Their selection was and continues to be based on planning judgements which reflect the immediate need of the United States industrial sector and the larger economic structure to conserve energy. These projects are largely directed at developing improved unit processes and engineering materials. Some of the ERDA/INDUS projects are to become elements of a materials and process-train optimization effort now evolving. The impact of the fully implemented ERDA/INDUS program will be to move the industrial sector further along the path of a non-waste technology.

No-Waste Salt — No Decontamination: A New Step in the Salt Bath Technology

B. Finnern.

Fa. Degussa, Hanau 1, Federal Republic of Germany

INTRODUCTION

Apart from carburizing, nitriding has become the most frequently applied diffusion method in the heat treatment of iron materials. Now, as before, the treatment in salt baths is of major importance and their use was even extended, because convection results in very uniform-heat transfer conditions and therefore the reaction with the carbon-and nitrogen- supplying substances occurs very rapidly and uniformly. Furthermore, comparing the salt baths with the gas atmosphere, the salt baths dispose of a higher carbon and nitrogen potential as a result of their higher density and therefore their chemical composition and hence their effect changes very slowly only.

Nitriding in cyanide-containing fused salt baths has been known for approximately 40 years and it is applied in more than 1000 plants throughout the world under exactly specified operating conditions, such as the titanium-crucible type with ventilation, i.e. the so-called TENIFER method, for approximately 20 years.

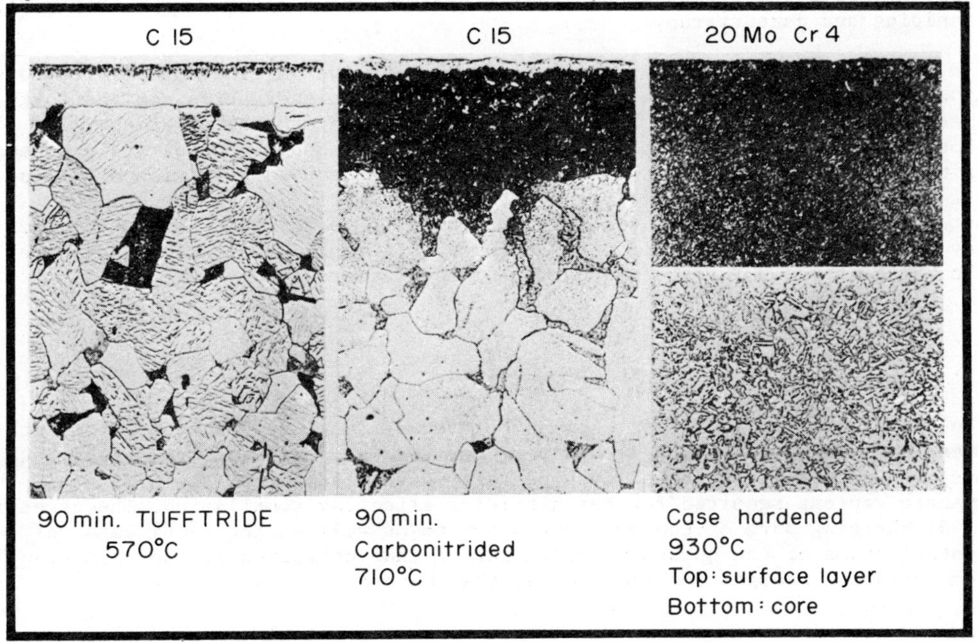

Fig. 1. Marginal zones with different heat-treatment methods.

The advantage of nitriding over carburizing can be explained best by means of Fig 1. When a structural part is treated in a carbon-and nitrogen-supplying atmosphere, it will depend upon the treatment temperature, which one of the two elements will most strongly diffuse into the surface of the structural element. With carburizing, the nitrogen - carbon ratio is 0.1, which is very low, i.e. the nitrogen proportion is very low. With dry-cyaniding under A_1 it amounts to 1:1 and with the TENIFER method it increases to 5:1. Thus, carburizing and dry-cyaniding via A_1 results in a carbon diffusion zone of a concentration and hardness steadily decreasing towards the core. Contrary thereto, dry-cyaniding according to A_1 and the TENIFER method result in two layers of different structures, viz. an external connecting zone (VZ) and a diffusion zone (Diff.-Z.) connected thereto.

The TENIFER method at 580°C results in the generation of sufficiently thick and monophase connecting zones, and as the structure is not transformed, the occurring dimensional changes are lowest with this method. The following properties of the parts are modified by the nitrogen diffusion.

1. Improvement of the antifriction properties combined with a high wear resistance up to operating temperatures of approximately 600°C;
2. Considerable increase of the fatigue strength with respect to bending and torsional stress;
3. Increase of rolling resistance and root resistance with toothed wheels;
4. Increase of resistance against atmospheric, water and seawater corrosion;
5. Avoidance of rubbing corrosion with oscillating stress on positive structural parts.

By virtue of the aforementioned properties, nitriding is much superior to dry-cyaniding and carburizing.

The TENIFER treatment merely implies a change of the marginal properties, and if a higher core strength is necessary because of higher stress of the workpiece, the workpiece would have to be tempered prior to the treatment. It should also be considered that alloy steels are frequently used with casehardening, not because higher core strength might be necessary in view of stress, but because of reduced distortion as the alloy steels may be tempered by hot quenching or oil. When the TENIFER method is used, again unalloyed steels may be used for such workpieces as a result of reduced distortion and hence the price of the workpiece can be influenced favourably.

PHENOMENA IN NITRIDING

The TENIFER method (old type) now as before meets all requirements from technical and commercial aspects. However, more or less daily a part of the bath must be discharged and fresh cyanide salt must be added, so as to maintain the cyanate content required for the nitriding effect by ventilation. The necessity of discharging part of the high-cyanide bath as well as the requirement of decontamination of the quenching media rather than operation of the bath mean some negative charges for the user of this method.

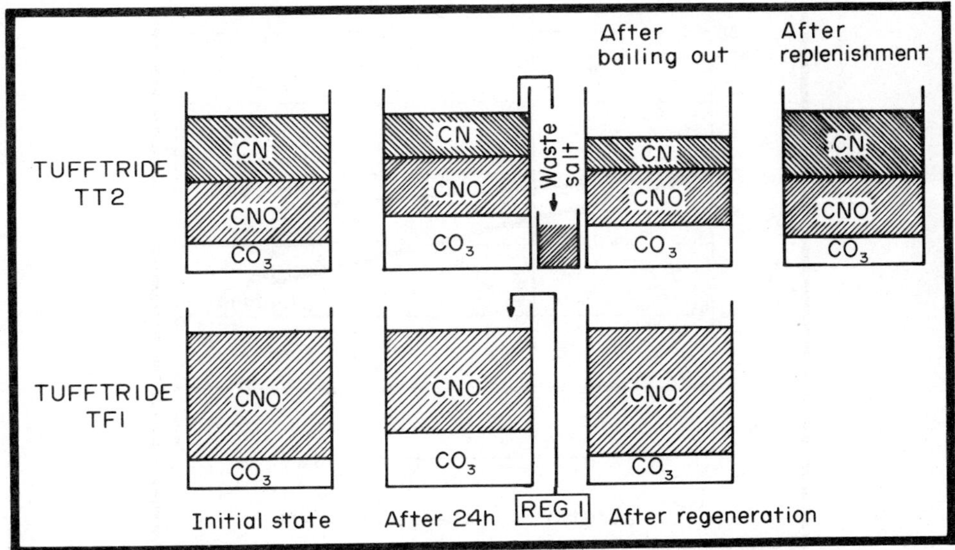

Fig. 2. Completing and regenerating "old" and "new" TENIFER(R) baths.

Therefore, the development of a salt-bath nitriding method not affecting the environment became more and more urgent. The differences between the old and the new methods are schematically shown in Fig 2. The newly developed method implies that cyanate is produced with the carbonate contained in the bath. This product is called Regenerator REG1, which consists of a plastics-like organic compound based on carbon, nitrogen and hydrogen. As a result of addition of this regenerator, the TENIFER bath (new type), which no longer affects the environment does not produce waste salt and the chemicals required for melting, such as TF1 as basic salt and REG1 as a regenerator are of non-cyanide type.

Of course it was required to adapt the composition of the salt bath to this regenerating type of operation. The new TENIFER bath now operates in the sodium - potassium - cyanae - carbonate system of a working range of 33-36 per cent cyanate and a certain adjusted potassium/sodium ratio. During operation of the bath, the cyanate content of the bath decreases slowly as a function of the charge, cf. Fig 3. The carbonate content increases proportionally with the decrease of the cyanate content. Addition of the regenerator results in re-formation of cyanate from the carbonate, which was produced during nitriding. Practice has shown that an REG1 addition of approximately 0.6 - 0.8 per cent of the bath volume per 24 hours is required to maintain a cyanate content of 36 per cent.

Addition of Regenerator REG1 does not result in a change of the bath volume. If the volume is reduced by drag-out, it will be completed by addition of TF1 salt, as can be seen from the schematic Fig. 2.

The operation of the new TENIFER method in favour of the environment does not involve any change of the plants used for the conventional TENIFER method. Treatment now as before is carried out in a massive titanium crucible and ventilation is reduced to half the volume of that of the old TENIFER method.

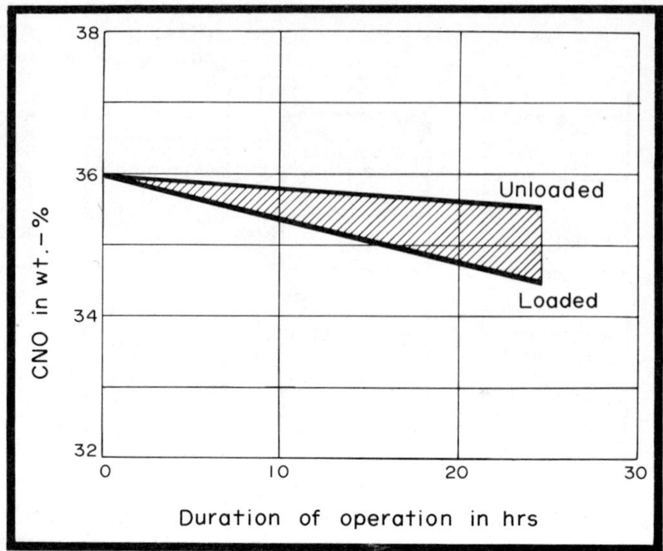

Fig. 3. Influence of operating time on CNO content of the new TENIFER(R) bath.

INFLUENCE OF NITRIDING ON THE WORKPIECE SURFACE

The carbon released by the reaction in the bath is required for the build-up of the connecting zone and stimulates the nitrogen, solubility of which in α-iron is low, to form a connecting zone. This connecting zone, consisting of iron - nitrogen - carbon compound, is produced after short treatment in the surface and consists of the monophase zone ϵ-Fe_xN. Mixed phases of γ-Fe_4N and ϵ-Fe_xN are most frequently generated with treatment temperatures below 550°C.

The build-up of the connecting zone is similar with most steel types and cast iron materials. Therefore, selection of the material can be completely left to the designer's discretion, if only the properties of the connecting zone are important with a particular workpiece. Thus is exemplified by means of the striker forks, Fig 4, from which can be seen that they may be bent from sheet metal or even forged or cast and that all types treated by the TENIFER method have proved to be satisfactory for many years.

As a result of the non-metallic structure of the connecting zone, apart from good sliding and wear properties it does not show the tendency to seize a metallic counter-surface or eventually to gall with the later. This factor is of particular importance for the use of the method with many different types of tools, with which seizing of the cold or warm-flowing material is to be avoided.

The structure of the connecting zone of material Ck15 as well as with steel 34Cr4V and with grey iron GG26 after a 2-hour treatment can be seen from Fig 5. The connecting zone shows porosity in the external marginal zone. With the old TENIFER method such zone is increased by the formation of higher contents of complex iron caused by the high cyanide content of the bath. As a result of the low cyanide content of the new TENIFER bath of approximately 2.5 per cent complex iron will not form in the bath, as can be seen from Fig. 6 as a function of the treatment temperature.

This low cyanide content in the bath results from the reaction between the cyanate and the iron parts to be treated or the sludge in the bath. Practice has now shown that much less porous zones can be generated with said cyanide content of approximately 2.5 per cent than with a lower cyanide content, as can be seen from Fig 7. Furthermore, the structural parts look brighter and cleaner, if the cyanide content is adjusted to the prescribed level by addition of a melting salt during the melting step.

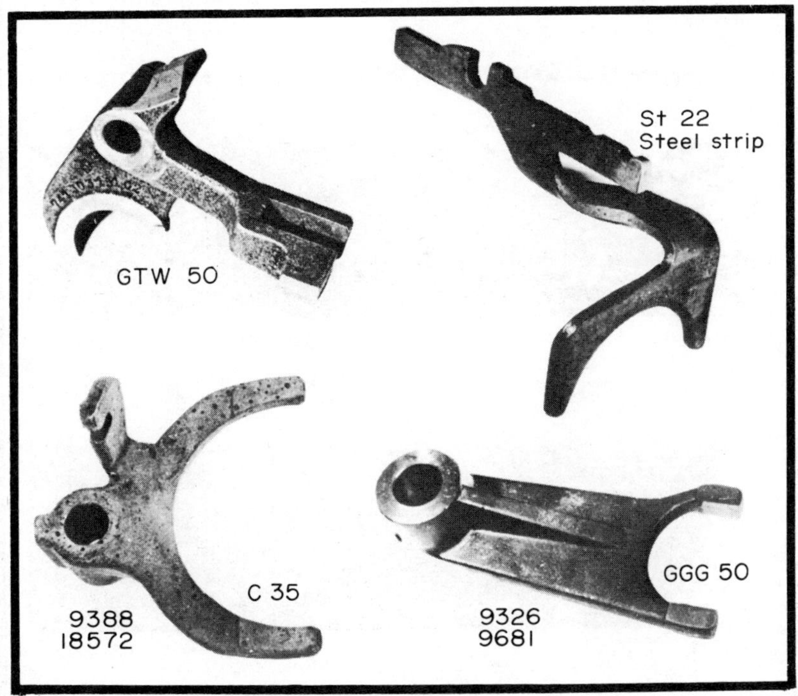

Fig. 4. TENIFER(R)-treated striker forks made of different materials.

The aim of development of the new TENIFER method was to achieve at least the same good properties as those of the old TENIFER method, but the practical application early showed that the nitriding effect of the new bath is stronger. For this reason, many users reduced the treatment time for their structural parts in the TF1 bath by at least one-third, which reduction applies to unalloyed and alloyed steels as well as to grey iron parts.

Practical application of the new TENIFER method in approximately 130 plants for more than 1 year has shown that the salt consumption can be considerably

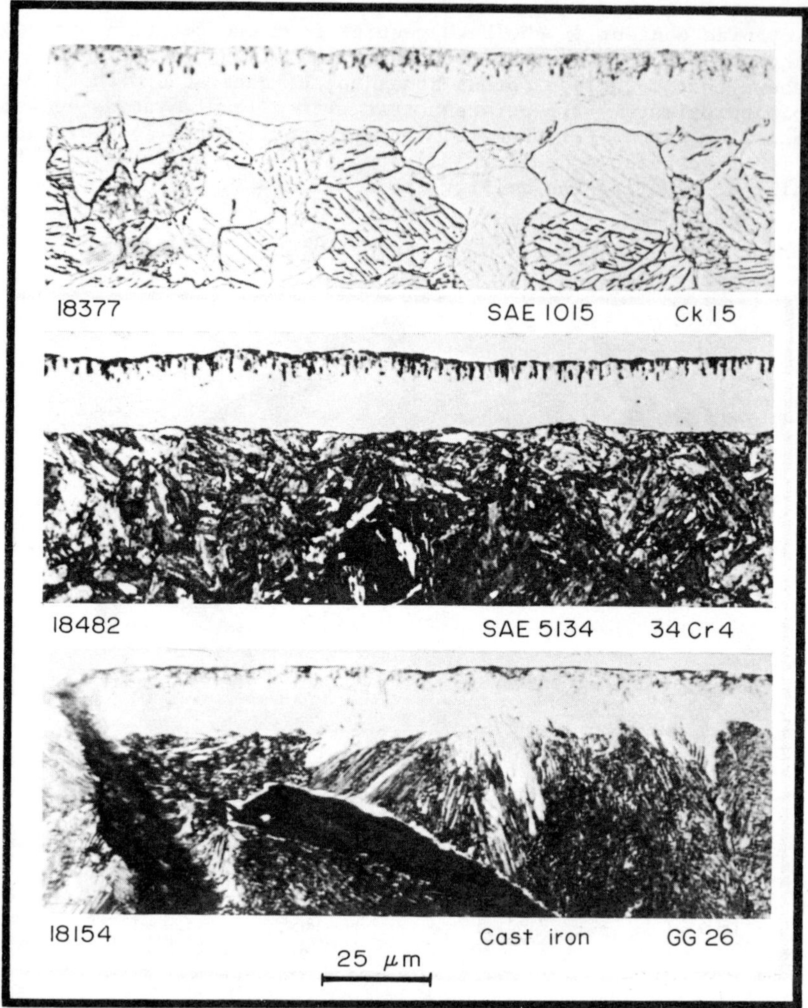

Fig. 5. TENIFER$^{(R)}$-treated materials
Top: Ck15, 90 min, 580°C → salt water 30 min, 300°C tempered.
Middle: 34Cr4, 90 min, 580°C → air.
Bottom: GG25, 120 min, 580°C → air.

reduced, e.g. by avoiding drag-out as a result of the separate addition of Regenerator REG1 and of TF 1 salt for maintaining the bath volume. This is of special importance with the treatment of tools or smooth structural parts and reduces the costs by saving TF1 salt. Apart therefrom, economy of the method is also improved by the fact that the waste-salt removal costs are avoided (bath does not produce waste salt) and that the costs for chemicals required for decontamination of the waste water are considerably reduced.

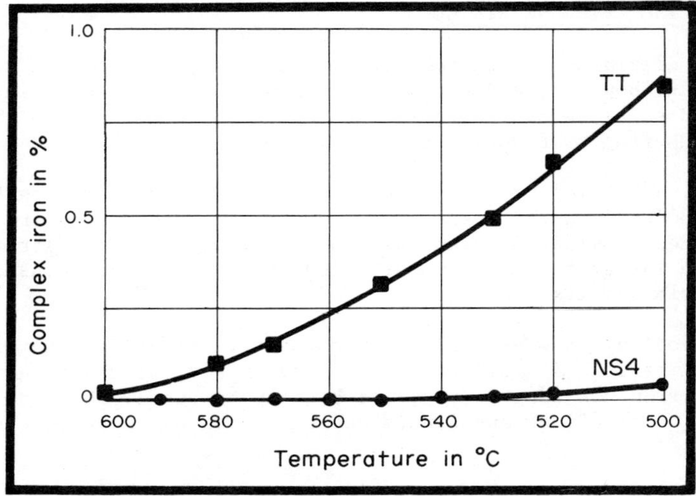

Fig. 6. Influence of temperature on the complex iron content with old and new TENIFER(R) types.

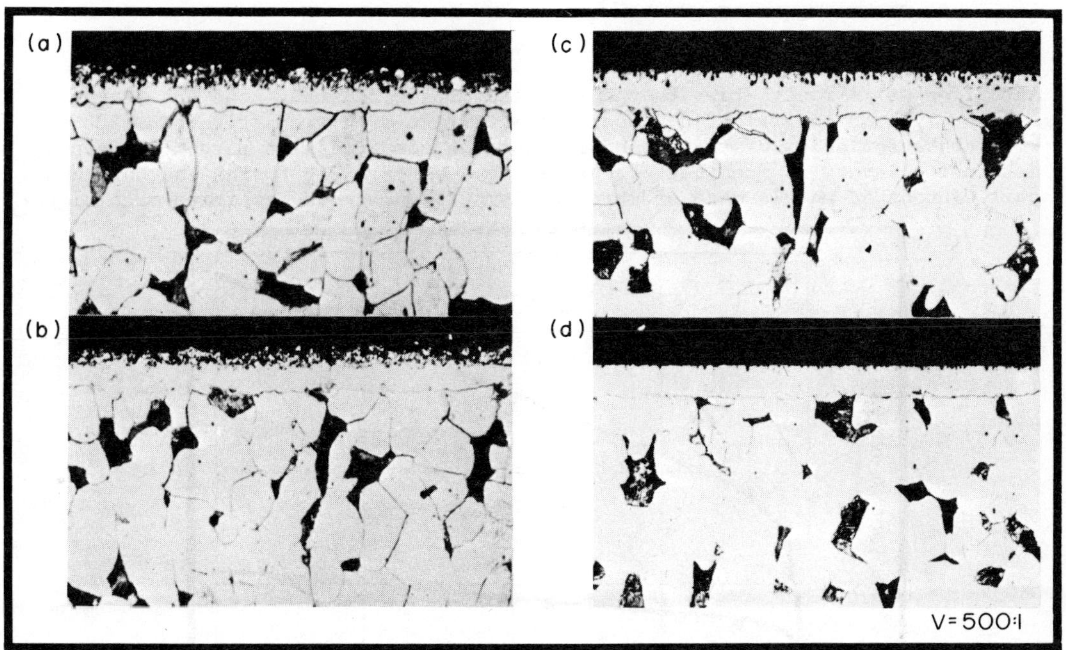

Fig. 7. Connecting zones as a function of the CN content:

(a) 60 min, 580°C; CN = 0.16 per cent; CNO = 32.5 per cent → water.
(b) 60 min, 580°C; CN = 1.12 per cent; CNO = 30.3 per cent → water.
(c) 60 min, 580°C; CN = 2.32 per cent; CNO = 36.7 per cent → water.
(d) 60 min, 580°C; CN = 6.00 per cent; CNO = 35.9 per cent → water.

COOLING AFTER NITRIDING

Cooling after nitriding can be done in the same manner with both the old and the new methods, However, a new cooling bath was developed especially for the TENIFER method so as to destroy the low cyanide contents and the cyanate content of the adhering salt during the cooling period.

This new cooling bath operates on hydroxide basis at an operating temperature of at least 230°C. If parts treated by the TENIFER method only are cooled in this bath, a decontamination plant will not be required. With later rinsing in hot water, the latter becomes strongly alkaline and must be adjusted to the required pH prior to its removal.

This type of cooling, however, also has a positive effect on the distortion of the structural parts and avoids, e.g. cooling in air, oil, nitrogen or vacuum of tools made from hot-forging steel or high-speed steel. The tools cooled in this bath are of excellent appearance and are clean as compared with air or oil cooling.

With cooling in this bath at increased temperature, e.g. 350 - 400°C, high-speed steel tools will obtain a good, brown-black appearance.

PRACTICAL APPLICATIONS

(a) Wear Characteristics

Attention was already drawn to the fact that the connecting zone has extraordinarily good antifriction properties by virtue of its constitution. The wear characteristics of the connecting zone as compared with carbon-containing martensite layers is much better, as can be seen from Fig 8. The casehardened pair of toothed wheels made of material 16MnCr5 and ground exhibits much higher

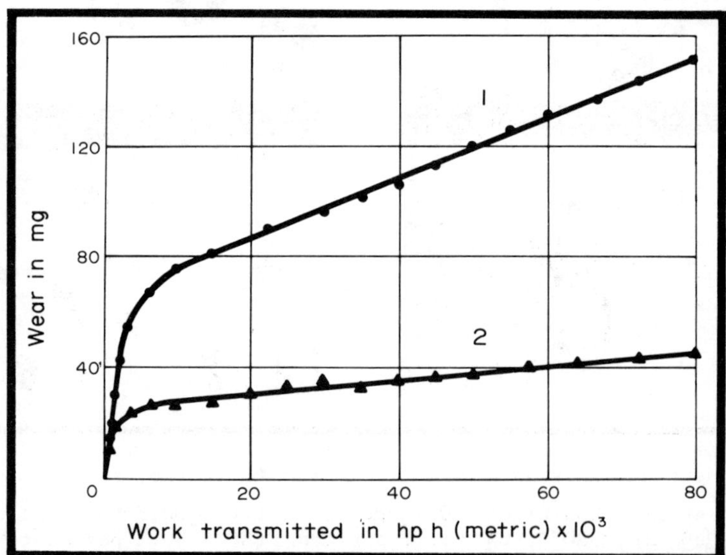

Fig. 8. Toothed wheel wear with casehardened and TENIFER(R)-treated toothed wheels.

wear than the pair of wheels made of material 42CrMo4 tempered and TENIFER-treated for 120 minutes in spite of the higher edge load to which the TENIFER-treated type is subjected.

However, frequently the question arises whether there is still a high wear resistance with two metallic materials with respect to the diffusion zone after the connecting zone has been worn off or whether the structural part will become defective due to wear very soon. The tests, mainly the practical application with crankshafts, cylinder bushes, etc., have shown again and again that the diffusion zone has an extraordinary wear resistance by virtue of the high nitrogen content, which resistance is even comparable with that of carbon-martensite layers.

Comparative wear tests with hard-chrome deposits have shown that the connecting zone of a structural part is superior to hard-chrome deposits. This applies particularly to sliding surfaces for shaft packings, wherein the packings cut the hard-chrome deposit very soon, whereas the connecting zone is hardly damaged after the same period of use.

As can be seen from Fig 9, the TENIFER treatment proved to be very satisfactory even with measuring tools as compared with previous hard-chromium plating or application of hard metals, and thus the manufacturing costs of said tools were reduced considerably.

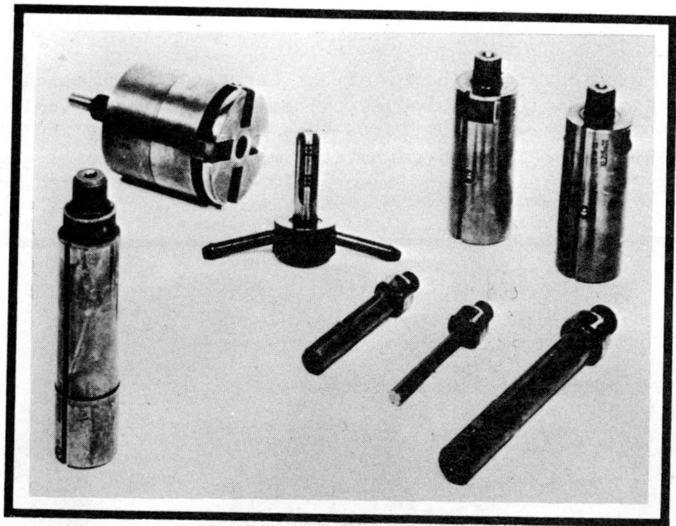

Fig. 9. Pneumatic test mandrels, TENIFER$^{(R)}$-treated.

To exemplify the wear resistance, TENIFER treatment of actuator sleeves is also referred to. Such sleeves for high-speed diesel engines with dry or wet sleeves or even for low-speed stationary and ship's diesel engines are treated by this method to a considerable extent. This TENIFER treatment is carried out to attain a good internal wear resistance with respect to the piston rings and thus to achieve lives of approximately 500,000 kilometres. The nitrogen diffusion avoids burning of the piston rings with first use of the engines and finally seizing of these rings. New actuator sleeves externally prevent cavitation occurring with various engine types by virtue of the connecting zone.

Manufacturer	Motor	Material	Preparatory treatment	TUFFTRIDE® time	Tappet / Follower arm
Porsche	P	Hard cast iron	-	3 hrs	Steel tappet case-hardened
Audi NSU Auto-Union	P	1045	Induction hardened	2 hrs	Hard cast-iron tappet
Chrysler France	P	Alloyed cast iron	Induction hardened	3 hrs	Steel follower arm TUFFTRIDE® treated
Steinmetz/ Opel	P	Alloyed cast iron	Induction hardened	3 hrs	Hard cast-iron tappet
Hanomag	D	1045	Hardened + tempered	2 hrs	Hard cast-iron tappet
KHD	D	1045	Induction hardened	3 hrs	Hard cast-iron tappet
MTU	D	8620	Carburised + hardened	2 hrs	Hard cast-iron tappet
Saviem	D	4140	Hardened + tempered	2 hrs	Hard cast-iron tappet
Humber	D	Hard cast iron	-	3 hrs	Hard cast-iron tappet
Ford Europe	D	Alloyed cast iron	-	3 hrs	Hard cast-iron tappet

P = petrol engine D = Diesel engine

Fig. 10. Material and treatment time of TENIFER(R)-treated camshafts.

It is interesting to note that combinations between the classical heat treatments such as casehardening and induction-hardening and the TENIFER method are used more and more. Fig. 10 illustrates this development by the example of camshafts of very different Otto carburetor engines and diesel engines. Camshafts made of steel or cast iron inductively hardened or casehardened are

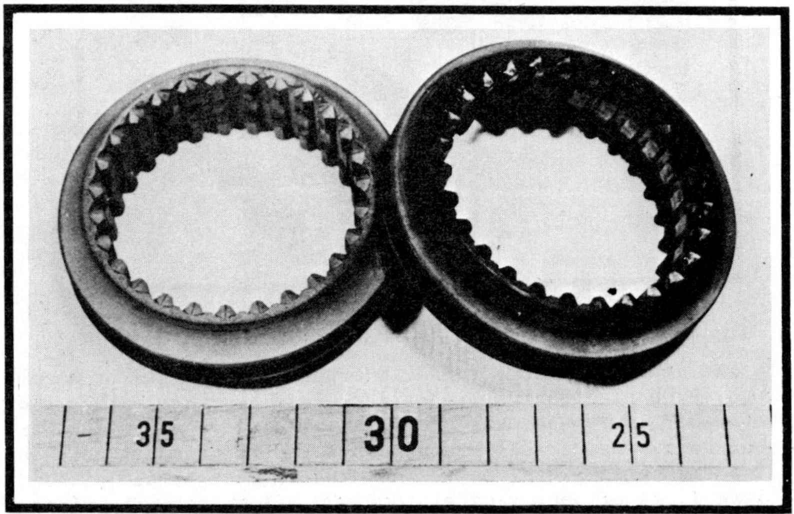

Fig. 11. TENIFER-treated clutch sleeves made of steel Ck45N, 90 min, 580°C → water and control lugs inductively hardened.

started after this heat treatment at approximately 580°C and are finally TENIFER-treated for 2 hours after grinding. This example shows very clearly that wear does not always and only depend upon absolute hardness and upon the depth of the hardening zone but that good sliding properties of the surfaces, i.e. hard cast-iron plungers on the cam surface in this case, are most important.

Casehardened and TENIFER-treated parts, as investigated by Niemann and Rettig, also exhibit the highest limit-seizing load capacity as compared with tempered or casehardened parts.

In this connection, it is to be noted that an increase of the surface roughness as a function of the initial state is attained by the TENIFER treatment. Therefore post-lapping of the nitrided surface may become necessary. This will be necessary particularly in case it is combined with a much softer countermaterial such as bronze, aluminium, white metal or plastics. This lapping procedure is very short and is realized by lapping linen of 360 grain size, but it can often be done by brushing with rotating steel brushes.

When carbon-containing steels are used, induction hardening can also be realized after the TENIFER treatment. When hardening at medium frequency, the connecting zone will be destroyed in most cases, while it can be conserved when hardening at high frequency, because the latter means rapid heating.

The example of synchronous rings for gears, Fig. 11 shows that such a heat treatment combination results in a more economic production of the respective parts as compared with case-hardening, in which the parts have to be hardened on a mandrel because of the higher dimensional changes. Simultaneous change from the previous alloyed steel to material Ck45 resulted in a further economic advantage.

Even pivot pins made of material 42CrMo4 and tempered to approximately 85 kg_f/mm^2 and TENIFER-treated for 2 hours with subsequent induction hardening lose their previous seizing tendency as a result of the nitrogen martensite with surface hardnesses of HRc = >63 and HV1 = >900 kg_f/mm^2.

(b) Corrosion Resistance and Avoidance of Frictional Oxidation

As a result of the generation of the connecting zone in the surface of the workpiece, the corrosion resistance of cast-iron materials and steels is increased considerably. The connecting zone should be sufficiently thick and as homogeneous as possible. For example, test probes on the sea bed in the area of the mouth of the Weser river and of the island of Helgoland, which were placed 2 years ago, proved to be highly satisfactory. This protection against corrosion relates to atmospheric, water and seawater corrosion. With different corrosive media only tests can show whether sufficient protection is provided by the connecting zone. This applies particularly to the so-called corrosion-resistant steels, the anticorrosive properties of which are reduced due to the connection between nitrogen and chromium, which results in chromium nitrides in the marginal zones. The extremely bad wear resistance of these steels, however, is considerably improved by the TENIFER treatment. As to the newly developed cooling bath, comparative corrosion tests have shown that workpieces cooled in this bath have better anticorrosive properties than parts cooled in water or other media after the TENIFER treatment, which can be seen from Fig 12.

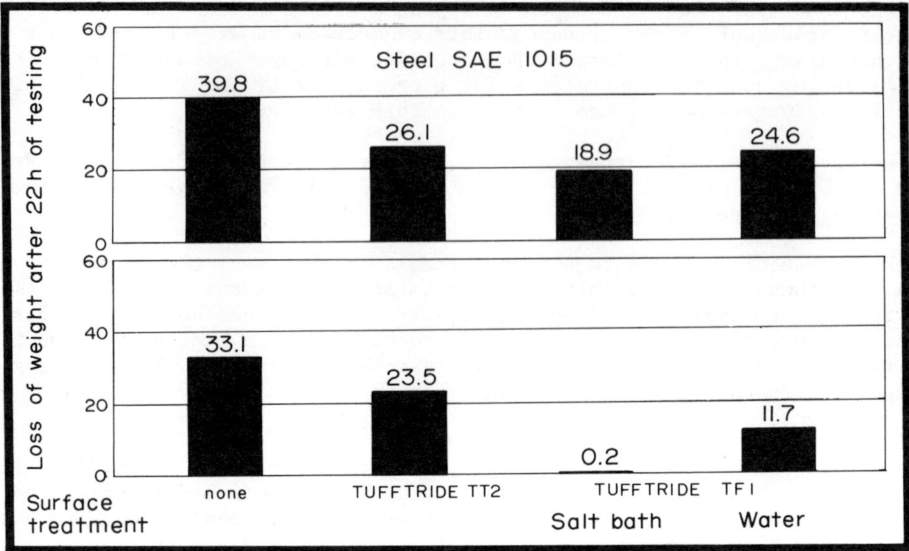

Fig. 12. Seawater test on TENIFER(R)-treated steels.

Frictional oxidation occurs between two positive surfaces exerting a relative motion under pressure. A typical example is the directing levers with cone-type connection, which are shown in Fig 13. This frictional oxidation is avoided by formation of the connecting zone.

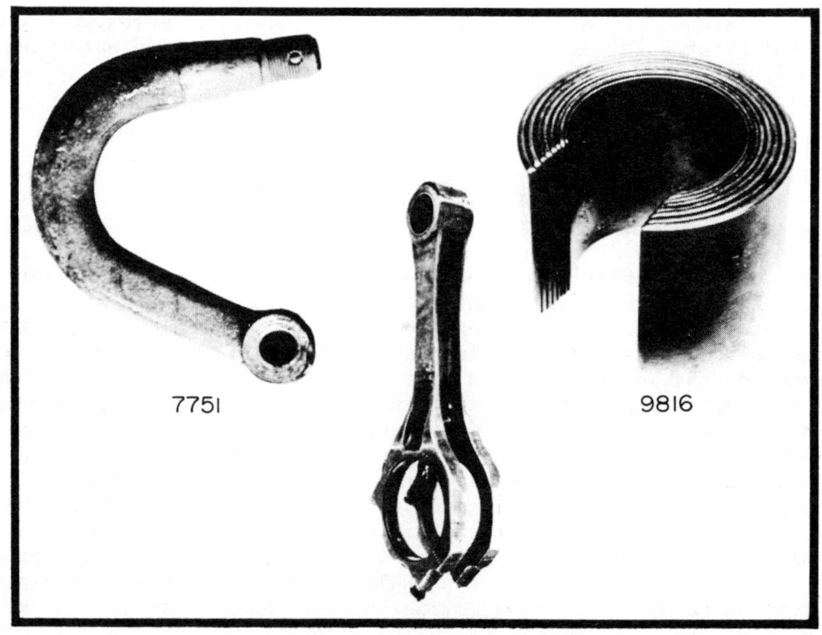

Fig. 13. TENIFER(R)-treated directing levers.

(c) Treatment of Toothed Wheels

As already shown, the wear resistance of the tooth flank is considerably increased by the TENIFER treatment of toothed wheels. Furthermore, the nitrogen diffusion results in such an increase of the root strength that it is higher than that of casehardened toothed wheels. In practice, however, these values cannot be realized because TENIFER-treated toothed wheels will fail prior to attaining such root strength as a result of the formation of pittings on the tooth flank, i.e. the criterium of TENIFER-treated toothed wheels is the rolling strength, tooth fractures are not to be reckoned with.

Fig 14 shows the rolling strength K_D for toothed wheels made of material C45N and 42CrMo4, which are tempered, as a function of the treatment time. The values attained with both materials practically are the same, which applies to the old as well as the new TENIFER method, as latest tests have shown. As these test results were attained with precision master wheels and manufacturing and assembly faults, etc., are not considered, the attained values can be realized in practice only for K_D values of approximately 2.5 - 3.0 kg_f/mm^2 and nitriding cannot be applied for toothed wheels subjected to highest stress, e.g. in vehicle construction. However, in the fields of machine, crane and vehicle construction there are sufficient applications for the TENIFER method with respect to toothed wheels, as shown in Fig 15 illustrating internally toothed wheels. With such internally toothed wheels the two connected wheels may generally be TENIFER-treated too, but the driving pinion must be casehardened because of the high-rolling strength required.

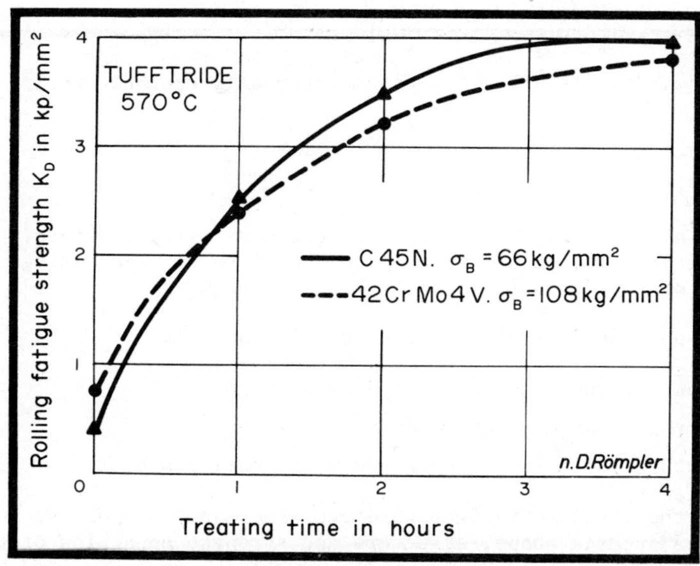

Fig. 14. Influence of TENIFER(R) treatment on rolling strength of the tooth flank.

(d) Increase of Fatigue Strength

The fatigue strength is increased by the nitrogen content of the diffusion zone. Hardness and diffusion depth of this layer are substantially influenced

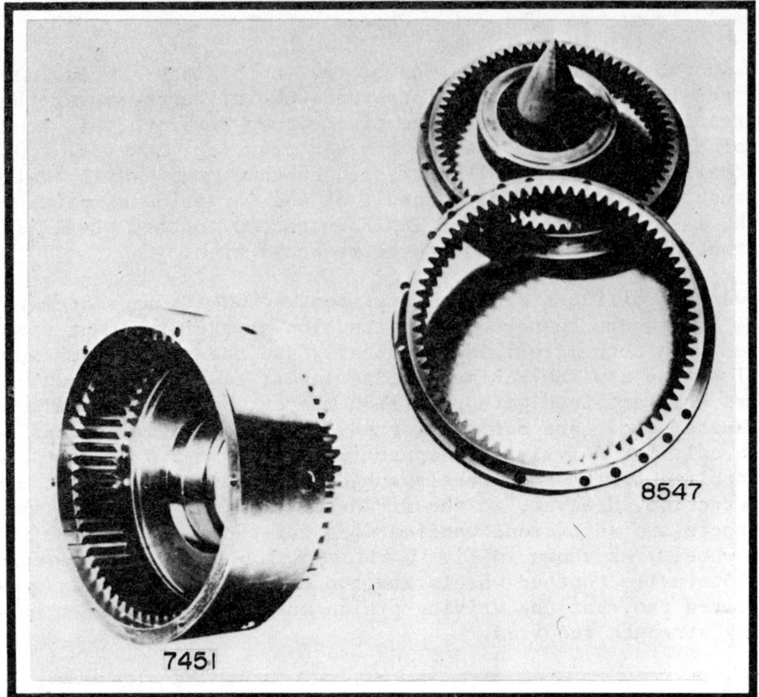

Fig. 15. TENIFER(R)-treated internally toothed wheels.

by the alloy elements in the steel, i.e. the more alloy elements, particularly chromium, are present, the lower will be the nitrogen penetrating depth, while the hardness of the diffusion zone will be the higher.

For this reason, the fatigue strength is influenced by

(a) the depth of the zone influenced by nitrogen,

(b) the nitrogen content in the diffusion zone and

(c) the state of solution of nitrogen with unalloyed steels.

Furthermore, the absolute fatigue strength also depends upon the known influencing factors such as shape, structure and strength condition of the workpiece involved.

The fatigue strength of notched samples made of material C15 with a shape factor of $\alpha_K = 2$ after treatment in the old and new TENIFER baths is shown in Fig. 16. It can be seen that the samples treated in the old and the new TENIFER baths at 570°C and for 2 hours practically have the same fatigue strength. On the other hand, the increase of the treatment temperature to 580°C results in an increase of the fatigue strength.

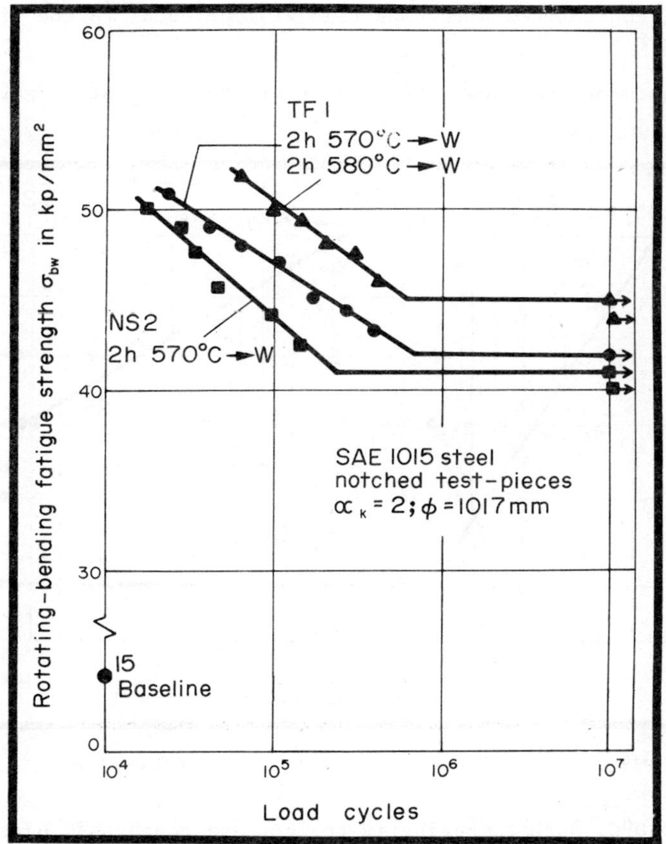

Fig. 16. Revolving fatigue strength of notched samples made of C15; α_K = 2, diam. = 10/7 mm.

Tests with crankshafts have shown that shafts made of material Ck45 showed higher values when they were treated in the new bath rather than in the conventional bath for 90 minutes. With tests with niobium-alloyed C45 in the new cooling bath at 230°C, the fatigue strength of crankshafts was in the same range as that of series shafts, as can be seen from Fig. 17. However, it is essential that the concentricity error was within the prescribed limit below 0.1 mm after cooling in the new cooling bath and the shafts had not to be straightened.

TREATMENT OF TOOLS

(a) High-Speed Steel Tools

One of the earliest applications of salt-bath nitriding is the treatment of metal-cutting tools made of high-speed steel. To avoid embrittlement of the cutting edges because of the high content of nitride-forming alloy elements, only short nitriding times are applied. They depend upon the edge shape and generally range between 1 and 10 minutes at 570 or 580°C. The tool is to be preheated well so that the nitriding time will be equal to the time of immersion in the bath.

After this treatment such tools frequently are tempered at temperatures bet-

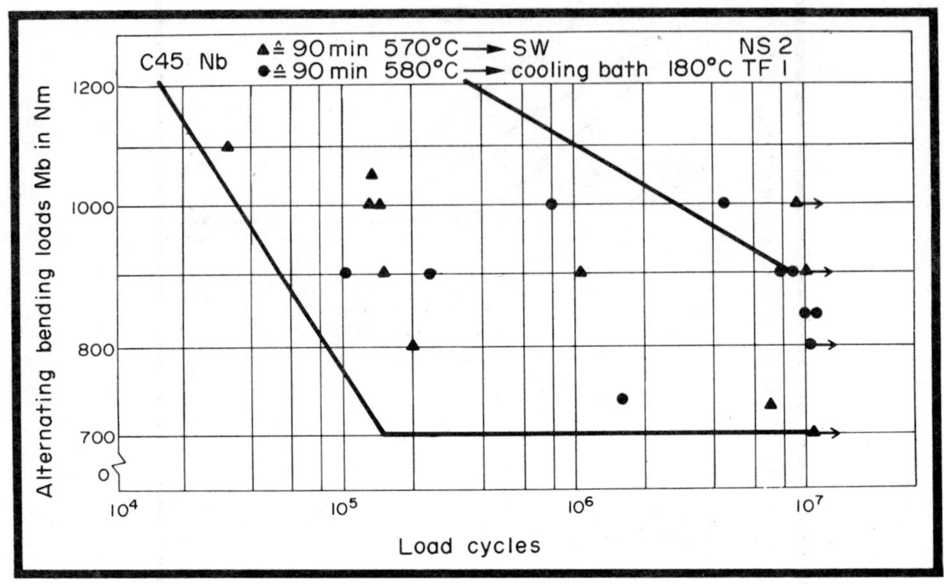

Fig. 17. Fatigue strength of BMW crankshafts.

ween 350 and 540°C. Such tempering is to increase toughness, particularly of the cutting edges. Tempering is done in both AS140 and annealing salt baths or in vapour-type tempering furnaces. Fig. 18 shows a series of cutting tools after TENIFER treatment. To obtain an advantageous appearance of such parts, the tools may also be cooled in the new cooling bath at temperatures between 350° and 400°C and they will be of brown to black appearance according to the material quality and surface finish.

The nitriding effect on such tools can be best proved by testing the surface hardness with HV1. Fig. 19 shows the surface hardness of high-speed steel DMo5 after a treatment in the old TENIFER bath and in the new TF1 bath. It came out that Vickers surface hardnesses increased by approximately 100 to 150 are obtained by treatment in the new bath so that the treatment time for such tools can also be decreased for this reason.

(b) Treatment of Jet Molds for Thermoplastic Materials

These tools still are often casehardened, but distortions due to casehardening often are excessive with large and complicated molds. Therefore, jet molds made of alloyed steel (20MnCr5) or of pretempered alloyed heat-treatable steels (34Cr4, 42CrMo4) are TENIFER-treated with good success. With high requirements as to stability of shape low-tension annealing at approximately 620°C prior to the last metal-cutting treatment may become necessary. Such tools will be TENIFER-treated for 90 to 120 minutes after preheating to 350°C.

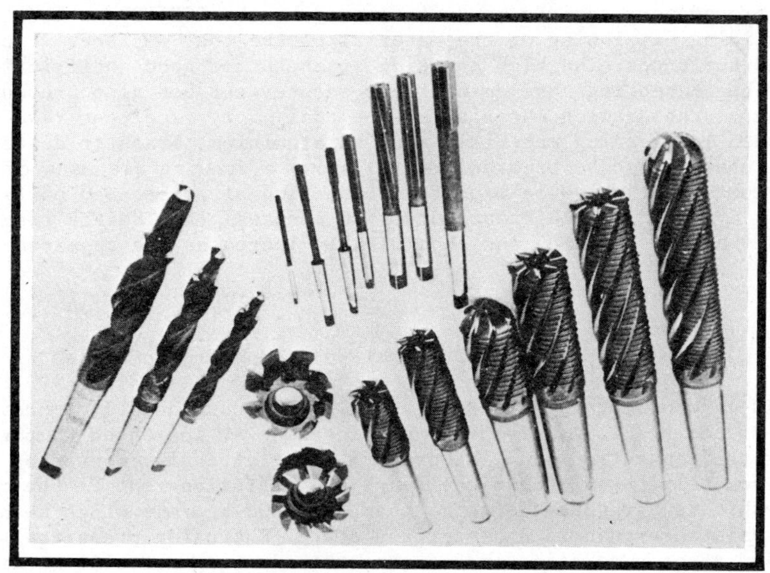

Fig. 18. TENIFER-treated metal-cutting tools made of high-speed steel.

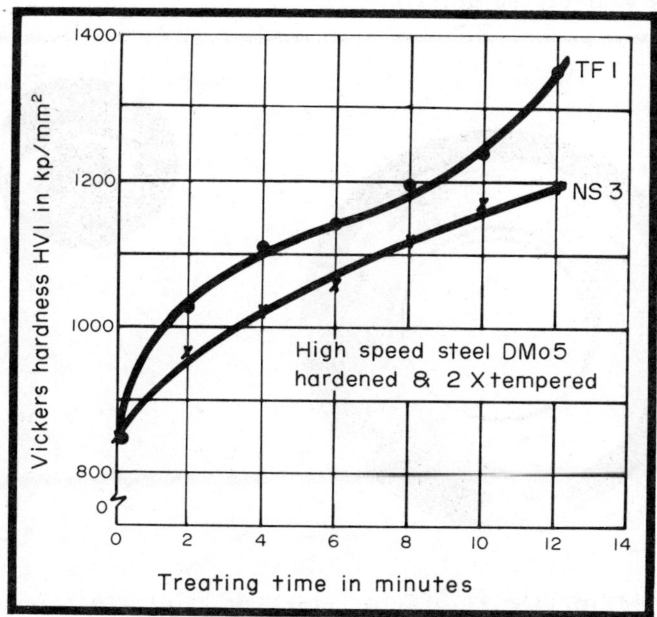

Fig. 19. Influence of old and new TENIFER(R) types on surface hardness of DMo5 as a function of the treatment time.

(c) Cold Working Tools

These tools for non-cutting transformation, such as drawing, bending and rolling, also proved very satisfactory if treated by the TENIFER method. The work involves sliding or flowing of the material on the tool surface, which results in wear and build-ups. The high abrasion strength and good antifriction properties of the connecting zone do not only reduce wear but also prevent adhesion and welding of the drawing material on the tool. Very excellent results are obtained when lubricating materials such as aluminium, brass or austenitic plates or tubes are to be treated. Fig 20 shows a drawing die made of 12 per cent chromium steel, which is used for production of cup-shaped parts from thin sheets after a TENIFER treatment of 90 minutes. The TENIFER treatment both reduces the welding tendency and considerably increases the appearance of the drawn cups.

(d) Die-cast Metal, Extrusion Press and Forging Tools

These tools made of hot-forging steel will have a considerably improved life by virtue of the stability of the connecting zone at increased temperatures. Even after the connecting zone is worn off after extensive operation due to thermal decomposition or mechanical wear, the diffusion zone of the materials involved exhibits hardnesses of more than HV = 900 kg_f/mm^2 which mean much more abrasion resistance than in non-nitrided state. Extrusion press tools will be disassembled after a certain time of use, the adhering material remainders will be removed and the tools will be machined and nitrided again. Press discs could be used again even after ten post-treatment operations. The increase in volume due to nitrogen adsorption often is advantageous with later nitriding, as it counteracts the wear of the profile.

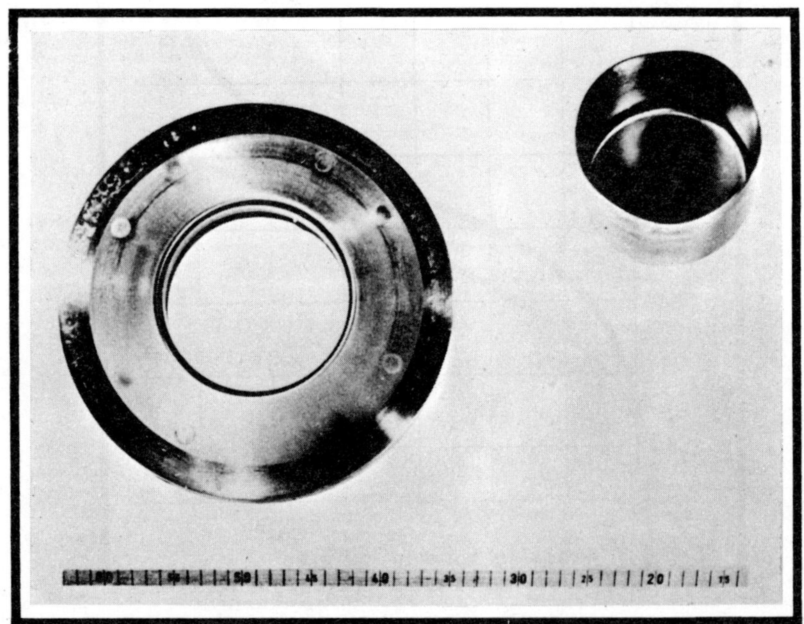

Fig. 20. TENIFER$^{(R)}$-treated drawing ring made of 12 per cent chromium steel (hardened).

Fig. 21. TENIFER(R)-treated aluminium extrusion press tools made of material 2343/2344.

Fig 21 shows TENIFER-treated extrusion press tools for the processing of aluminium.

The treatment in the new TENIFER bath excellent from environmental aspects generally takes 2 hours for such tools and in this case, too, cooling in the new cooling bath at increased temperature has shown particularly advantageous.

SUMMARY

The TENIFER method is characterized by many types of application and, considering the possibility of conversion to the new TENIFER method, which does not affect the environment and which is combined with the possibility of cooling in a bath destroying minor remainders of cyanide and cyanate, many advantages of economic nature are attained so that the conditions for extended application of the nitriding technology according to the TENIFER method are given.

The Design of Non-Waste Technologies Taking the Example of a Lignite Transformation Complex in the German Democratic Republic

W. Kluge

Institute of Energetics, Leipzig, German Democratic Republic

The raw lignite from the opencasts is the most important primary energy in the GDR. From this raw lignite, upgraded energy carriers (electric energy, coke, gas) and raw materials (tars, phenols, sulphur, etc). are produced in different transformation steps. The territorial concentration of these transformation steps in complexes of factory plants is particularly economical. The largest complex processes 100 000 t raw lignite daily (see Fig. 1.).

In fulfilment of our law for the protection of the environment and of our principle of seeking the most economic use of natural resources, e.g. water and energy, the workers of this complex have developed a nearly closed circulation system for treating the liquid and solid waste products. This system is discussed below as an example of the design of non-waste technologies (see Fig.2). In the GDR the available water quantity is low in relation to the population density and requires the multi-step utilization of water in the large industrial plants and complexes.

The waste water with the highest load is produced in the gas works in the form of gas liquor (1). This liquor is freed from its dusty and tarry ballast in the settling pond (A1) and is purified together with the process waste waters from the coking plant (2) extractively in the Phenolsolvan plant (ph) and biologically in an activated sludge plant (Bio). About 60 per cent of the treated water (3a) is used in the wet-type dust collection of the briquetting plant and is there loaded with coal dust (4). In the settling pond (A2) the water is clarified mechanically (5) and rinses the ash from the power station (i) and the gas works (g) into residual pits of the opencasts (6). There the ash settles. The waste water (7), which is practically free of harmful matter, is treated with the ferruginous water from the drainage of the opencast in a neutralization plant (N) and then discharged as pure water (9) into a river. On the lower reaches of this river an internationally known nature reserve, the "Spreewald", and the capital of the GDR are situated. In spite of the complex and further large thermal power stations in the upper reaches of the river every sports fisherman may see for himself the biological life in the river at these locations. Now more than 20 per cent—all biologically treated waste waters (3b) at latest till 1980—are sprayed on to agriculturally used areas after dilution with 5 parts of cooling water. The spraying is done both on agricultural areas to meet the nitrogen demand and on areas which were formerly devastated by open-cast mining and which shall be recovered for further utilization in agriculture or forestry by the legally and practically well-organized process of recultivation. An essential aspect of multi-step water utilization within the complex is the fact that no continuous enrichment with harmful matter occurs. The waste waters from the gas works and the coking plant, highly loaded with harmful matter, are purified in several stages and the water leaving the complex is practically free from any harmful matter.

Decisive in non-waste technologies is the conversion of the mostly solid or liquid substances removed from the water into utilizable or at least environmentally harmless products. The most important consequential measures are described below.

In the settling pond A1 two products are formed: a coarse-grain product (a) which is sold as a fuel and a fine-grain thixotrope product (b) which is processed together with briquette breeze (c) in the mixer (M) into a high-quality fuel and burnt in the power station. In the settling pond A2 the coal dust is separated with the help of a ferric hydroxide sludge as coagulant and is used also as a fuel (f) after filtration.

The relief gas (I) produced by the gas purification in the Rectisolan plant is purified in a Sulvosolvan and Claus plant (5c) and thereby elementary sulphur is recovered. The purified gas (II) is used as fuel for the coking plant and the power station.

The xylitol content and the Fe-concentrate are separated from the ash of the power station and are sold. A land cultivation effect is gained by the filling of the residual pits by means of ash rinsing. In an increasing degree the filter ash (i) is used to still greater effect as a raw material (k) for the recultivation (lime content 20-30 per cent) and as a binder for the stabilization of overburden dumps and slopes and for road construction.

Finally it should be pointed out that conformity with the requirements of environmental protection and the economy of the complex have been attained. The multiple utilization of the water not only results in pure waste water, but also reduces the cost of fresh-water use considerably. The energetic utilization of the waste products results in an annual saving of 900 000 t raw lignite, equivalent to a 2.7 per cent increase in the energy transformation efficiency of the complex.

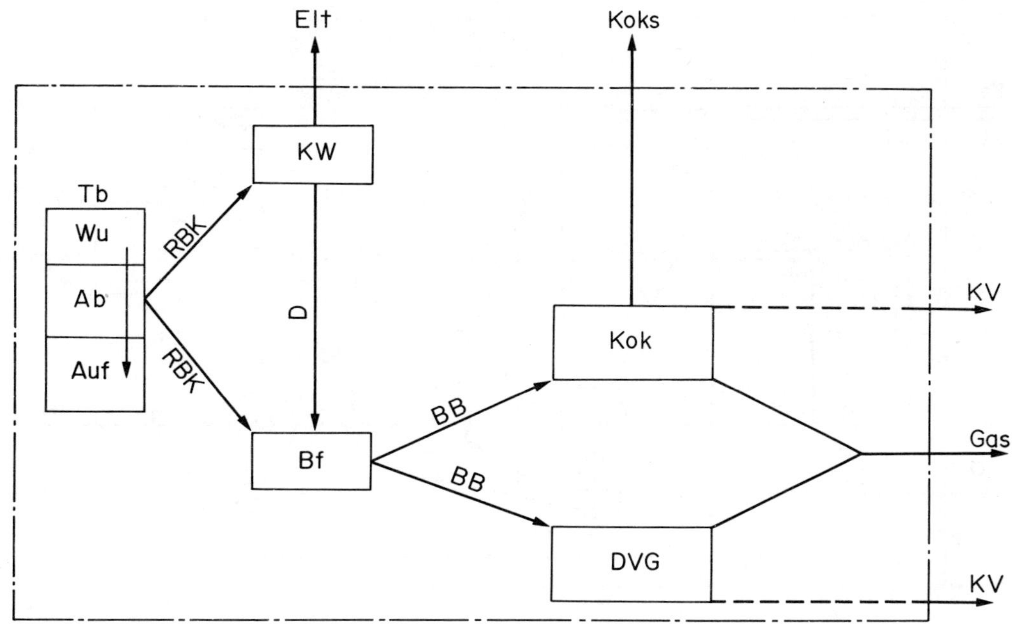

Fig. 1. Main production line of the complex. Tb, opencast; WU, area of recultivation; Ab, area of mining; Auf, area of development; KW power station; Bf, briquetting factory; Kok, coking plant; DVG, high-pressure gasification plant (gas works); RBK, raw lignite; BB, lignite briquette; D, steam; Elt, electric energy; Koks, lignite high-temperature coke; Gas, town gas; KV, coal transformation products (phenols, tars, sulphur).

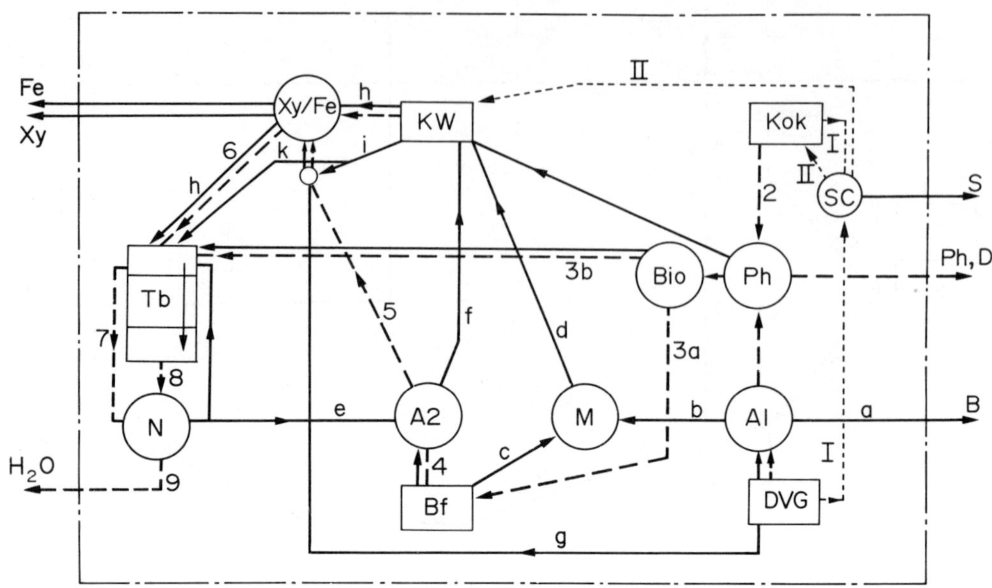

Fig. 2. Circulation scheme of the main solid and liquid waste products. Continuous lines, solid waste products; slotted lines, liquid waste products; dotted lines, gaseous waste products; Bf, KW, Kok, Tb, DVG (See Fig. 1); S, sulphur; Ph, phenols; D, fertilizer (NH_3-water); B, fuel; H_2O, purified waste water; Fe, magnetite concentrate; Xy, xylitol; A1, A2, settling ponds; Ph, phenol; Bio, biological waste-water purification; SC, Sulvosolvan and Claus plant; M, mixer; Xy/Fe, xylitol and magnetite removal; N, neutralization plant; 1, gas liquor; 2, process waters from the coking plant; 3a, 3b, biologically purified water; 4, water with coal dust; 5, flushing water for power-station ash removal; 6, water with ash; 7, clarification water; 8, water from the drainage of the opencast; 9, purified waste water; a, course-grained fuel; b, fine-grained fuel; c, briquette breeze; d, mixed coal product; e, iron oxide sludge; f, filter cake (fuel); g, gas works ash; h, wet ash; i, electric filter ash; k, ash for ash utilization; I, relief gas; II, fuel gas.

The Iron and Steel Industry: Pollution Control and Recycling

Y. Hellot

Mineralogical Department, Ministry of Industry and Research, Paris, France

INTRODUCTION

The iron and steel industry have always been classed among the industrial sectors which are the biggest consumers of water, one of the branches of industry most prone to the emission of various pollutants into the atmosphere, and finally, as a major source of solid wastes.

With respect to water, total requirements for the French iron and steel industry at the present time come to about 5 billion cubic meters per year which, for an annual production of 26 million tons of steel, represents a requirement in the order of 200 cubic meters per ton of steel. In actual practice, due to multiple internal recyclings in an average amount of about 70 per cent total water use only amounts to 1.5 billion cubic meters per year, This water is not consumed. The better part (97 - 98 per cent) is discharged into the natural environment together with its added wastes (matter in suspension, matter in solution, or toxic substances).

In the area of air pollution, the iron and steel industry is particularly identified with the emission of particles, which are more or less fine, of iron oxide dusts in the case of sintering plants and especially so in the steel mills. Coking plants are a major source of highly varied emissions (particles of coke and carbon, ammonia, phenols, sulfur compounds and the like) and all of the reheating furnaces for pig iron or steel burning fuel are sources of emissions of sulfur compounds.

As regards the production of solid wastes, the iron and steel industry is without doubt the branch of industry which is the major offender. To produce a ton of steel in an integrated plant, no doubt twice as much solid waste is produced, especially in the blast furnaces (slags) and to a lesser extent in the hot and cold rolling mills.

Efforts on the part of the steel industry to reduce pollution, and especially water pollution, did not begin yesterday. Certain remarkable achievements were made possible due to the presence in this industry of some extremely favorable circumstances. In the area of water, for example, water quality for processing is such that it can in most cases be easily reached by subjecting effluents to some basic technical procedures. This allows the recycling of effluents benefiting from the advantages in such cases : better use of raw materials, savings in water and a reduction in the charge in the drain.

Results obtained by such means are impressive since the difference between a factory operating on an open system and a plant which has extended to its maximum these principles, and is thus working on a closed system, varies between

200 m^3 of water per ton of steel for the one and 3 m^3 for the other.

In the same manner, in the field of air pollution, a steel mill using oxygen only emits 120 mg/Nm^3 of dust if effective redusting is installed, whereas in mills which could be qualified as "old", such as with Thomas converters, dust emissions are possible at a level of 150 g/Nm^3.

Examples such as these of specialized technology proper to a process, or of recycling, are taken from among recent developments in the iron and steel industry in France : in sintering plants, in blast furnaces, in coke plants and in steel mills.

In conclusion, we will be reviewing the status of the iron and steel industry as regards results in the area of air and water pollution.

COKING PLANTS: WATER POLLUTION

In coking plants, water serves three main uses:

- the cooling of ovens, gas and ancillary facilities;
- the extinction of the coke on drawing the charge;
- the treatment of gas and the separation of tars, and "ammoniacal water" is obtained.

Cooling water entails large volumes, but since the water is not polluted, it can be recycled.

The extinction of coke is in general carried out wet, through spraying of about 1 m^3 per ton of coke. Of this 0.5 m^3 is evaporated and the remaining 0.5 m^3 contain suspended matter, phenols and ammonium salts. After simple decantation, this water is recycled with the addition of an amount to replace the evaporated water. Ammoniacal water is still used in some cases to supply this additional water, but this approach should be prohibited because it leads to a conversion of water pollution into air pollution. In an integrated plant, it is possible to make advantageous use of a system run-off which, for example, is polluted with suspended matter, for the purpose of additional water.

"Ammoniacal water" comes from the treatment of gas and water made up of carbon. In addition to the matter in suspension, it also contains organic matter (BOD_5 and COD from 5 to 20 g/l), ammonia (5 to 10 g/l), phenols (2 to 4 g/l) sulfides, cyanides, sulfocyanides, tars and oils. The high pollution and toxicity of this water is to be emphasized. Different treatment techniques (biological treatment) which have been perfected remove the better part of the organic matter and phenols, but none or very little of the ammonium compounds (free or bonded salts). After treatment, this water is then discharged in the natural environment. This should no longer be tolerated. Studies performed in France have developed a specialized process proper to these methods which enables the recycling of the better part of this water.

The process is as follows:

- decantation and separation of the tars;

displacement of ammonium chloride with sodium;

a pre-distillation of alkalinized water thus enabling steam to draw off the free ammonia or ammonia freed with CO^2, HCN and H^2S, and to recover at the bottom of the column the phenols and other organic matter;

the water so "distilled" is then concentrated in multiple stages and a very slightly laden distillate is recovered. This distillate then passes over active carbon which gives good-quality water which can be recycled in the factory. The carbon is not regenerated;

ammonia can as appropriate be injected in the sulfate plant;

ammonia sulfate is a by-product of the gas in the coking plant;

or it can be injected into the Phosam plant if there is one;

the concentrated, phenol-containing residue can be treated by liquid - liquid extraction, such as, for example, by a solvent like benzol, which is a by-product of the coking plant. The solvent is regenerated in an evaporator.

The investment cost for a unit called upon to treat 500 m^3 of ammoniacal water per day (a coking plant producing 2000 t/day or 700,000 tons of coke per year) is 8 million francs (as of 1975). Compared to the total investment for a coking plant of that capacity—600 million francs—this represents therefore 1 per cent. As to operating costs, they would be 11 francs/m^3 of treated water, or 2 to 3 francs per ton of coke, not including any possible profitable use of the ammonia and the phenols.

This technology therefore makes it possible to discharge into the environment only water which is very slightly polluted and in reduced quantities.

COKING PLANTS: AIR POLLUTION

The production of coke is a source of pollution to the atmosphere throughout all of the preliminary steps and the coking of the carbon: preparation, charging, coking, drawing the charge, extinction, screening and handling.

The quantity of dust emitted by a coking plant is approximately equal to 0.9 kg/ton of coke produced, plus handling which in itself represents 1.5 kg.

Significant progress has been made in the extinction of coke (it should be noted in this respect that dry extinction making use of an inert gas is the most satisfactory approach because it practically eliminates any emission of dust or waste water) where an 80 per cent reduction in emitted dusts is achieved by positioning at the top of the towers louvers with slats inclined at 45°.

As regards filling the ovens, palliatives do exist but they represent a very costly investment and are expensive to operate.

For drawing the charge of coke, an interesting approach is taking shape at the present time. It consists of building a room against the battery of ovens in which the drawing takes place and in which the emitted fumes are accumulated under the roof, from where they are drawn off with a rapid flow rate together with a final dedusting.

SINTERING PLANTS: WATER POLLUTION

Recently constructed plants do not cause any direct pollution of water either because the dedusting systems operate entirely dry, or because the waste-water system for wet dedusting is closed (decantation, flocculation, and eventual cooling).

AIR POLLUTION

The preparation of ore and coke, the production of a briquette, and its cooling can produce 5 - 40 kg of relatively large grain dust per ton of briquette.

However, all of these plants are equipped with installations for dust control using dedusters whose average yield can be estimated at 95 per cent. For example, in a modern plant with an area of 400 m^2, with 2 million Nm^3/hour of output to be treated, the main emissions are reduced from 5 g/Nm^3 to 150 mg/Nm^3, for an efficiency of 97 per cent. Due to recent techniques, a sintering plant overall pollutes the air very little with particles.

Nevertheless, the release of sulfur dioxide (coming from the sulfur in the coke) is another source of pollution. At the present, more emphasis is given to a dispersion of the smoke by injecting gas at high speed in very high stacks than to removing the sulfur from the smoke. Still, investigations into the removal of the sulfur are in progress and pilot plants are operating on one-fifth of the output to be treated.

Blast Furnaces

Formerly, blast furnaces were considered, and justly so, as major polluters due to gas (very noxious gas, because it contains 25 per cent CO) which it released together with dust, and this up to 40 kg/m^3.

This gas was very soon collected as a source of energy and it was completely used. To make this possible, the dust had to be removed, from 6-18 kg/Nm^3 to 10 mg/Nm^3. This was carried out with a multi-stage dedusting system, the last of which was wet (dust-pots, multicyclones, venturi washers). This gas wash water was very heavily laden with suspended matter and slightly charged with cyanides, fluorides and traces of heavy metals.

Simple techniques using decantation, flocculation, with cooling, recycle this water without any problem, nevertheless, draining off 1 - 5 per cent is necessary in order to reduce the concentration of salts dissolved in the water (hardness, chlorides).

Sludge coming from decantation is recovered, but unfortunately the content in zinc is in some cases too high to allow recycling in the plant. In Japan, treatment in rotary furnaces at a high temperature gives pre-reduced pellets which can be used in blast furnaces or in steel making.

Water from the cooling systems is recycled without any apparent difficulty.

When a ton of pig iron is produced, some 700 to 800 kg of slag on the average are produced. Slag brings together such residual elements as gangue, fluxes and fuel ash (CaO, MgO, SiO^2, Al^2 and O^3). Slag coming from a blast furnace is cooled by two methods.

These slag cooling methods are as follows:

- by granulation with water, which is a sudden cooling on contact with water. The product has a vitreous structure. It is used as a hydraulic binder (cement or road mixtures);
- by slow cooling of the liquid slag which gives a crystallized slag which can be used as rail ballast, artificial gravel for roads and aggregate for concrete.

Profitable uses for this by-product have developed a great deal over the past years and it can be estimated that some 80 per cent of the produced tonnage is put to profitable use. The remaining 10 per cent is still dumped on the slag heaps.

When the factory carries out the granulation of the slag, this takes place in basins with a filtering bottom. This water circuit is closed on the condition that it is strictly managed and that there is supervision to avoid any overflow of floating particles.

The blast furnace, therefore, is one area of the factory which is the source of very little pollution.

Nevertheless, it still remains to deal with cleaning out the blast furnace cast-houses where dust is still emitted in abundance. Several installations (in Japan for example) are in operation which provide satisfaction, but power requirements are high which implies high operating costs.

Steelmaking

Pig iron is an alloy of iron and oxidizable non-metallic elements (silicium, carbon and phosphorus mainly). It is converted into steel by the oxidations of these metalloids through various methods:

- conversion mills using the Bessemer and Thomas process (old converters) and the oxygen or air enriched in oxygen blast process (LD, OLP, BOP processes);
- open hearth mills (Martin furnace and electric furnaces).

In every case, gas from refining consists of carbon monoxide and carbon dioxide. They contain in addition fine particles of iron and iron oxides which are called "red smoke" due to their color. Depending upon the process employed, there are great variations in the amounts of CO and CO^2 and particles (50 - 150 g/Nm^3). Thomas converters produce a great deal smaller quantity of gas and particles than, for example, oxygen furnaces in which secondary air is drawn in at the converter nozzle in order to ensure total combustion of gas, which increases greatly the volume of gas to be treated.

Dedusting of the mills is carried out in all of the new plants. By contrast, this is not at all the case in the old plants (Thomas and Martin furnaces, for example, which is speeding up their disappearance). This is carried out in some cases (electric furnaces for example).

Collecting and dedusting gas can be considered to be well developed in the case of the oxygen blast furnace. Depending upon the case, the gas is collected (collection without combustion) or they are burned with partial or total recovery of energy in the form of steam.

An example of a recent installation is described below. This facility was designed and built in keeping with several objectives:

to reduce to a minimum the volume of gas to be collected;

to recover the gas containing a major proportion of the carbon monoxide so that its high heating power can be used in the steel furnaces;

to dedust the gas to the maximum;

to recycle all (right up to the drain) of the water in the gas wash water system.

A description of the installation is as follows:

collection of the gas in a skirt cooled by water circulation;

cooling of the gas in a hood designed as a water-jacket, then in the stack;

two-stage wet dedusting: first in a saturator with water spray, where gas fumes are supersaturated with moisture, then in a purifier with six successive stages of multiple venturis with little loss of pressure;

transfer through a constant-speed blower and discharge into the atmosphere by way of a torch stack.

This installation was installed in a steel mill with two 250 ton converters, with a maximum flow of oxygen at the nozzle of 1 000 Nm^3/min (blast time varies from 18 to 20 minutes), in order to treat a maximum output of gas of 192,000 Nm^3/hour at a temperature of 1500°C in the hood and 850°C in the processing tower, which enables a production per charge of 6.5 tons of low-pressure steam (hood) and 8.5 tons of high-pressure steam (in the processing tower).

This dedusting system results in a concentration of less than 120 mg/Nm^3/per charge on the average (the manufacturer can guarantee 70 mg/Nm^3).

As could be seen, the dedusting is carried out wet (which is obligatory for reasons of safety as regards explosion, with the exception of electric furnaces) The volume of water employed varies between 5 and 20 m^3, but a lot of power is entailed, which implies high costs.

The volume of water employed per ton of steel varies between 5 and 30 m^3. This water is heavily laden with suspended matter (2 - 10 g/l, or 15 - 40 kg/ton of steel), very fine particles of iron, iron oxide, calcium carbonate, and oxides of manganese, magnesium and aluminium.

The wash-water system includes the following:

a pre-decanter for the sludge water, removing the largest particles, combined with a classification;

an air cooler;

a decanter-flocculator (circular boom decanter with automatic extraction of sludge);

storage tanks for water purified before recycling;

treatment of sludge containing 65 per cent iron, successively by filtration over a filter under vacuum with a rotary drum, and by drying in a tunnel kiln.

This results in water containing less than 50 mg/l of matter in suspension, which is recycled with a drain, and in pellets of iron which are recycled in the plant.

The establishment of these procedures markedly limits water and air pollution. There remains, nevertheless, a secondary source of dust emissions. These are to be found in the pouring from the ladles, charging the crucibles, which are carried out in the cast house. The collection and treatment of these emissions is carried out in Japan, and is under study in France. It is certain that the volumes of air to be treated will be considerable, as will the expenditures for power.

We might also cite the example of the LWS process, which is a blast process which blows from the bottom air enriched in oxygen into the Thomas converters. This process for the conversion of pig iron into steel is carried out by separate and simultaneous blowing from bottom to the top through the base of the converter, in double tuyères, and under high pressure (8 - 12 bars) of an oxidizing gas (pure oxygen); and of a cooling liquid protecting the pipes (mazout for example). The OBM (or Q-BOP) is very similar. These two processes greatly reduce (from half to one-third) the amount of red smoke as compared to converters with vertical injection. Emitted gas remains *very* abundant and the collection-dedusting system described above, without gas combustion, can now be carried out. It is enough to note that French production in 1976 using this process will be approximately equal to 4.5 million tons of steel (about 15 per cent).

In the case of the open-hearth mills (Martin and electric process) it is certain that dedusting of emissions from the Martin furnaces is anticipated in only a few cases since the process is going out of style.

By contrast, dedusting in electric furnaces (arc furnace) is carried out in all of the new furnaces and this will be progressively the case for all of the furnaces already in operation. Two types of smoke collection are to be distinguished:

collection in the furnace through a "4th hole" installed in the dome (about 1 000 $Nm^3/h/t$);

collection under the roof as concerns in particular fumes spread about (about 10 000 $Nm^3/h/t$).

Dedusting at the 4th hole is performed by a venturi washer. This results in the following:

dust contents of 10 - 15 g/Nm^3;

discharge into the air of 100 mg/Nm^3;

flow density of 400 - 600 Nm^3/t.

specific power consumption of 8 - 12 kWh/t.

Dedusting at the 4th hole + hood is carried out with a dry filter. This results in the following:

dust contents of 1 - 2 g/Nm^3;

discharge into the air of 20 mg/Nm^3;

flow density of 3000 - 5000 Nm^3/t;

specific power consumption of 8 - 13 kWh/t.

Dedusting costs can be estimated at 15 per cent of the total steel furnace investment (including water treatment). Dust can still be recovered with or without pelletizing and reintegrated in the production process.

It is to be noted that the composition of these dusts varies greatly, but median values could be cited based on a furnace producing special-purpose steels:

TiO^2	0.1 per cent	Mn	4 per cent
MgO	3 per cent	Cu	0.6 per cent
Pd	2 per cent	Zn	12 per cent
F	very low	Fe	28 per cent

Rolling Mill Furnaces

Steel cast in an ingot mold, in order to be cooled as little as possible, is reheated in Pits oven, then hot rolled to become billets or slabs; it is once again cooled and reheated in ovens to be rerolled, thus taking its final shape for the market: various sections, bars, beams, sheet, rails, wire.

These successive reheatings take place in ovens heated with gas from the blast furnaces, coking plants, natural gas or mazout. These fuels generate little dust, but they do cause pollution due to the sulfur dioxide (SO^2). This is eliminated with the smoke through the stacks which enables a satisfactory dispersion.

Hot Rolling

The steel ingot, after reheating, undergoes an initial rolling to spread it out, then with or without reheating, a second rolling is performed in a roll train which transforms it into various flats or sections (final products).

At the end of each reheating, an iron oxide skin (calamine) forms on the surface of the metal, which must be removed. Water in hot rolling has three uses:

cooling the trains, ovens and ancillary facilities (motors);

removal of the calamine from the rolled product;

cooling of the laminated product and the housing, which is performed with the removal of the calamine.

This system becomes loaded up with calamine (iron oxides), oils and grease. It contains up to 25 - 30 kg of suspended matter per ton of steel, within 10 - 20 m^3.

This water can be recycled, and this is the case more and more often. For this purpose, a treatment system is used as follows:

scale pits or gyrocyclone;

decantation-flocculation in a circular boom decanter with the addition of multiple electrolytes and the trapping of oils by scraping with the scraper boom;

filtration on rapid sand filters (the level of matter in suspension is thus reduced to less than 50 mg/l and hydrocarbons to less than 5 mg/l);

cooling with an air cooler;

recovery tank and recycling to the rollers;

finally, decantation or cycloning (water cyclons) of the filter wash water.

All of the sludge and iron oxides which are recovered are removed and returned to the production process (sintering or steelmaking).

Considering the great diversity of rolling trains for hot rolling and treatment techniques, it is difficult to give any figures on the investment cost. It is still possible to estimate that the average investment for a plant producing 1 million tons of steel per year is in the order of 10 million francs (as of 1976).

Cold Rolling

Cold rolling is only a very minor source of air pollution. By contrast, it is a major source of water pollution due to matter in suspension, and especially due to soluble oils.

Depending on the rate of reduction carried out on the rolled product, use is made of either soluble oils (at 1.5 - 4 per cent oil) or a mixture of vegetable or animal oils in suspension in water.

These oil systems are closed, but a drain is necessary (1 m^3 per ton of steel approximately) containing 0.2 kg of suspended matter; 0.6 - 0.2 kg of BOD_5; 0.2 - 0.6 kg of COD; and 1 kg of oil. The separation ("breaking") of the water oil emulsion requires a physico-chemical process, then decantation and flotation. The oils are burned, but the waste water is discharged. As regards the oil suspensions after pH adjustment, they can be treated directly by flotation or electro-flotation.

Pickling

Pickling of steel is carried out in the iron and steel industry either with sulfuric acid, hydrochloric acid or nitrofluoric acid (stainless steel). There are two kinds of waste: a continuous discharge of rinsing water and a discontinuous discharge from the used baths of acid containing 30 - 40 g/l of acid and 100 - 130 g/l of iron.

An example of hydrochloric acid pickling merits being mentioned because, as soon as the installation is sufficiently large, the bath can be regenerated in a grid reactor so as to recycle the HCl and produce iron oxide in grains or powders which can then be marketed. Rinse water can be used in part in this installation and any residual pollution which is produced can be held to a strict minimum.

CONCLUSIONS

The examples given of processes which make it possible to reduce pollution produced by the iron and steel industry show that considerable improvement has been achieved in France. Nevertheless, such progress is necessary and it must be pursued even further in such areas as steel furnaces (secondary dust emissions), coking plants (air pollution), cold rolling mills (waste water) and in the entire plant as regards the discharge of wastes.

It is difficult to make a distinction between the control of air and water pollution, especially as regards investments. Nevertheless, for a new factory, a two-thirds to one-third allocation between the protection of air and the protection of water could be advanced. In the case of an integrated plant, the total percentage of the investment devoted to pollution control could be evaluated at 5 - 10 per cent, with significant variations from plant to plant (25 per cent for a steel-making plant, but 2 - 3 per cent for rolling mills).

The Outlook for Progress and Technological Methods in a Paper Industry Confronted with Environmental Problems

P. Monzie

Centre Technique du Papier, Grenoble, France

INTRODUCTION

The paper industry, which discharges large amounts of organic and inorganic matter into the environment, has the reputation of a polluter.

Statistics available for the years before this industry installed its first water-purification plants attributed 20 per cent of France's industrial pollution to paper and pulp manufacturers.

It should be emphasized that the two branches do not share equal responsibility for this pollution; in fact, pulp manufacturers are estimated to produce 90 per cent and paper mills only 10 per cent of the total figure. This report will therefore deal mainly with pollution problems related to pulpmaking.

It is noteworthy that a profound change in the orientation of pollution abatement studies took place during the years 1965-1970.

The original idea was to treat effluents and various wastes discharged from the mills, but application of these techniques on an industrial scale revealed the extent of the problem and the threat posed by such solutions to the long-term profitability of manufacturing processes. Pulpmakers gradually became aware of the need to find a more rational, more economical solution based on a change in the manufacturing principle itself. Subsequent studies were therefore oriented towards developing new technologies and better, cleaner manufacturing processes. This report is largely devoted to defining these new trends and determining the progress made by research and development efforts in this area.

Because of the diversity and complexity of interrelated pollution problems, and for the sake of clarity, it is essential to give a brief description of the most sophisticated paper production line available and to point out the origins of various wastes generated by the process.

1. WHERE POLLUTION ORIGINATES

Figure 1 is a diagram showing a combined pulp and paper production line in its most highly developed form. Types and origin of wastes are indicated by a series of arrows connected to blocks at the bottom of the diagram.

The various stations are as follows:

1.1. *Woodyard - Barking*

Wood is delivered to the mill in the form of logs. The percentage of unbarked wood receipts, whether roundwood, sawmill waste (edgings) or smallwood, is constantly increasing. Residue collected when woodyards are cleaned, which is done periodically, makes up 0.5 - 2 per cent of the wood used in pulp production. In France, where annual consumption totals about 107 million steres, such residue represents a loss of 25,000 to 100,000 tons of raw materials. Only a portion of this residue (about 30 per cent) can be fed into mill boilers to generate heat.

Barking. The roundwood is barked, then chipped, screened and stored. These processes produce significant quantities of solid wastes (bark or sawdust) totalling about 12 per cent of the processed hard- or softwood.

Losses therefore total 600,000 tons/year for an annual consumption figure of 5 million tons of wood.

To these solid wastes should be added the pollution load contained in water used in wet barking processes. Some experts estimate that suspended solid content of this water makes up 25 per cent of total suspended matter for the entire mill.

1.2. *Pulping*

Wood may be pulped by mechanical or chemical means or by a combination of the two (semichemical and chemi-groundwood pulps).

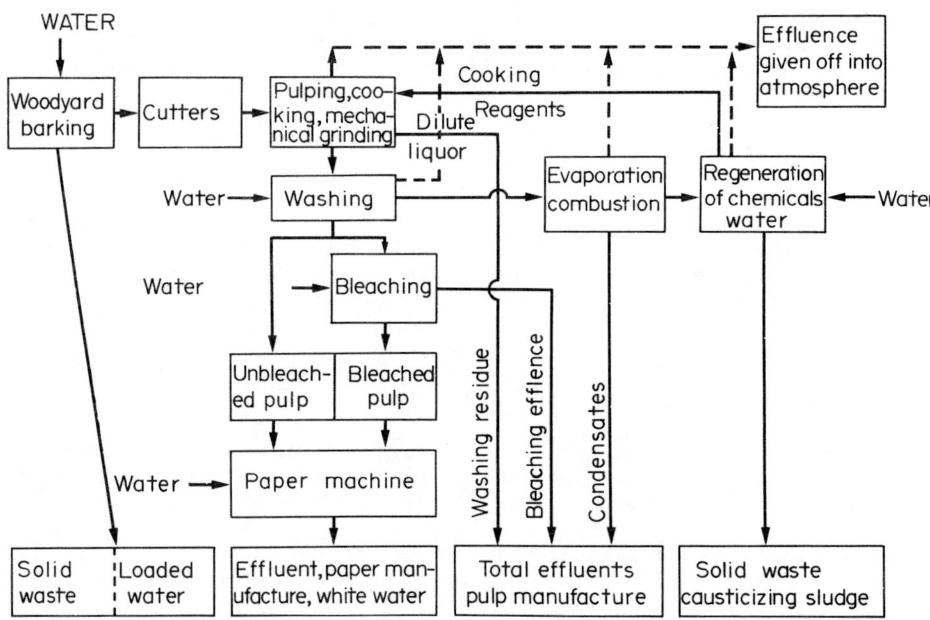

Fig. 1.

Table 1 shows estimated pollutant discharge in terms of suspended matter, oxidizable matter and equitox for various types of processes: bleached or unbleached pulps, paper and cardboard.

1.2.1. *Mechanical Pulp*

Pollutants originate in the woodyard and grinding mill. They are made up of suspended fibers and fines as well as materials in solution or in colloidal suspension.

TABLE 1

GLOBAL POLLUTION

	Total suspended matter g/kg of product	Oxidizable matter* g/kg of product	Inhibitors equitox†
Semichemical and strawpulp			
without liquor recovery	50	290	(1)
with liquor recovery	40	90	(2)
Sulfite pulp			
without liquor recovery	60	450	(1)
with liquor recovery	50	250	1.60
Unbleached kraft pulp			
without liquor recovery	20	240	(1)
with liquor recovery	10	40	0.21
Bleached kraft pulp			
without liquor recovery	50	290	(1)
with liquor recovery	40	90	0.35
Kraft paper and board	10	10	(2)
other types of paper and board	30	10	(2)
manufacture of paper and board from mechanical pulp. including pulpmaking	30	10	(2)

$$\text{*Oxidizable matter} = \frac{2BOD_5 + 1COD}{3}$$

† Equitox level in 1 m^3 of effluent equals 100/C150, where C150 is the concentration, expressed in percentage of volume, at which 50 per cent of the daphnids used in the tests are paralyzed after 24 hours.

(1) Measured, not global pollution.
(2) Non-toxic.

1.2.2. *Semi-chemical Pulp*

In addition to woodyard effluents, which create problems for all types of pulps, semi-chemical pulpmaking is beleaguered with the problem of cooking liquors which are difficult to regenerate.

There are three possible cases:

(a) Large mills (producing more than 400 tons/day) using sodium sulfite which burns their cooking liquors and regenerates the cooking agent.

Residual pollution in this process is caused by:

evaporation condensates,

water from the washing stage.

There is no plant of this type operating in France.

(b) Plants using ammonium sulfite and which destroy their cooking liquors by combustion without recovering chemicals. Sulfur is given off as SO_2 in flue gases and ammonia is destroyed by combustion.

Pollution in such mills comes from condensates and water from the washing plant.

One such unit is currently operating in France.

(c) Units producing less than 100 t/day and using sodium sulfite or ammonium sulfite, for which neither recovery nor elimination of chemicals would be economical but which are now forced to contemplate this solution or to install expensive water-purification equipment.

1.2.3. *Chemical Pulp*

1.2.3.1. *Sulphate pulp*. Except for a few low-output units producing pulp for special papers, these mills all have capacities exceeding 300 t/day and are all equipped with combustion-regeneration plants.

Pollution is caused by three types of effluents:

"washing residues", i.e. a very dilute black liquor (concentration of less than 10g/l);

evaporation condensates;

bleaching effluents.

The last of these poses the most critical pollution problem: because of the fact that they are very dilute and contain chlorides, the dark colored bleaching effluents, whose BOD and COD are very high, cannot be cycled back into the recovery system. They are therefore discharged into rivers, and it is estimated that they are responsible for between 50 and 70 per cent of the pollution produced by kraft mills.

Kraft mills are also confronted by a significant air-pollution problem due to the formation of volatile, bad-smelling sulfur compounds of the mercaptan type also found in evaporation condensates.

1.2.3.2. *Sulfite pulp*. The following cases are possible:

Mills using calcium bisulfite, whose cooking liquors can be eliminated by combustion but not regenerated. In most cases, these effluents, along with water from the washing plant, are purely and simply discharged into rivers, and that is one reason why this process is tending to disappear. In older mills, effluents from the cooking plant may represent up to 95 per cent of total pollution.

Mills using soluble bases: ammonium, sodium, magnesium, which burn their liquors and recover cooking agents (in the case of sodium and magnesium).

In addition to pollution caused by water from the washing plant and by bleaching effluents, evaporation condensates from sulfite cooking liquors are worse pollutants than those from the kraft process. The sulfite process frees large quantities of volatile organic substances: acetic acid, aldehydes (furfural), methanol, and this results in condensates with high BOD and COD.

1.3. *Estimated Cost of Anti-pollution Treatment*

At the present time, anti-pollution regulations are aimed mainly at chemical pulp mills.

Bleached kraft pulp effluents. Kraft pulp accounts for more than 85 per cent of chemical pulp production. Normal production figures for kraft pulp, whose output is constantly increasing at the expense of sulfite pulps, reach 30 million tons, of which:

6 million are produced in Western Europe, and

4.5 million in Scandinavia.

Estimated average pollution figures for this industry are as follows:

BOD : 50 kg/t of pulp,

COD : 200 kg/t of pulp,

Color : 200 kg/t of pulp,

SS : 30 kg/t of pulp.

Recent studies show that in countries such as France and the United States, where strict regulations are enforced, cost of pollution-abatement measures total 40 francs/ton of paperboard, most of the burden of which is borne by pulp manufacturers.

According to an OECD report drafted in 1973, the future of the pulp industry would be placed in jeopardy if pollution-abatement costs exceeded 4 per cent of the cost of pulp, i.e. 80 francs/t on the basis of present costs.

An evaluation carried out in the United States for the application of the Federal Water Pollution Control Act Amendments of 1972 showed that:

in 1977, implementation of the most efficient pollution-abatement techniques will cost bleached kraft pulp mills from 40 to 60 francs/ton;

to arrive at zero pollutant discharge in 1985, a bleached kraft pulp mill will need to spend more than 200 francs/ton of pulp.

These statistics bring to evidence the fact that pollution-abatement costs may place a heavy burden on the paper industry and justify current efforts toward finding and developing more economical solutions.

2. PROBLEMS INVOLVED IN THE UTILIZATION OF SLASH AND OTHER FOREST RESOURCES

In view of the high cost of raw materials, and particularly those of the wood used in papermaking, the pulp industry must increasingly orient itself toward the utilization of various forest residues and other raw materials of vegetable origin.

Studies have therefore begun to develop means of using:

sawdust and various waste chips, residues from wood-conversion industries;

the whole tree;

thinnings, etc.

Solutions must first be found to the problems of high bark content and heterogeneity which set these materials apart from roundwood chips. The above examples may be quoted to emphasize the fact that interindustrial cooperation can offer favorable solutions to problems of waste disposal. More should be done in this direction, in particular as regards finding new outlets for certain industrial waste products (mud, bark).

2.1. The Whole Tree—Smallwood

An intensive effort is being made to develop uses for these cheap raw materials as a supplement to conventional resources. Wood chipped on the felling site is mixed with other furnishes in the production of bleached semichemical and kraft pulps.

Research and tests are being conducted on methods designed to cut down bark content, either by submitting chips to a preliminary mechanical treatment or by improving cooking and bleaching techniques, or by using new processes, all with a view to increasing the amount of waste materials used in pulp furnishes. The development of oxygen processes to replace more pollution-prone kraft pulping techniques may be one solution to the problem, since oxygen has a strong effect on bark.

2.2. *Sawdust*

Use of sawdust has been developed in particular in the United States and Canada, where 5000 tons of kraft pulp are produced daily from sawdust.

In France, development of this raw material has come up against problems of uniformity and fineness of sawdust particles. It is recommended that sawdust be used in combination with chips, and special (rotary or screw) digesters should be used to facilitate circulation of this raw material. Furthermore, pulps obtained from these materials are generally of mediocre quality, partly because their fibers are too finely separated and greatly damaged. Current studies aim at using sawdust in the production of chemi-groundwood pulps which serve as furnishes for some unbleached papers and cardboards. It should be recalled that sawdust output for France totals more than 500,000 tons per year, which would go a long way toward meeting needs.

2.3. *Annuals*

Resources available in an agricultural country such as France are considerable. Of a total 30-35 million tons of straw (wheat, barley, oats, etc.) produced each year, an estimated 10 to 12 million tons cannot find buyers. If basic storage and price stability problems could be solved at a national level or by means of local agreements between producers and industry the market for straw in the paper industry would undoubtedly make new gains.

With regard to the technological problems which have discouraged pulp manufacturers from making use of this raw material, we have every reason to believe that they will be solved in the near future thanks to research programs aimed at promoting innovative technologies and processes involving both lower costs and less pollution.

3. WASTE TREATMENT—OUTLOOK FOR DEVELOPING NEW LOW-POLLUTION PROCESSES

Pollution abatement problems should be attacked from two directions:

- waste should be treated and pollution diminished by optimizing operation of conventional installations;
- new processes and techniques should be developed to cut down volume and level of pollution.

Only by combining the advances made in these two areas will it be possible to decrease, not to mention eliminate, all forms of waste products (solid, liquid, gas) without jeopardizing the profitability of these installations.

3.1. *Waste Treatment*

3.1.1. *Woodyard and Barking*

Woodyard residues and bark are likely to increase, since consumption of all wood grades and of residues from wood conversion and logging operations is expected to pick up.

Bark. The solution rests in perfecting dry barking techniques to furnish more combustible bark and cut down the pollutant discharge from wet barking units.

Attempts have been made to use bark in the following ways:

- as fuel, in the form of "bricks";
- as a component of building boards;
- as compost in combination with urban waste;
- for by-products (polyphenols or resins);
- as an adsorbent (in powdered form);
- as activated charcoal; a 150 t/d unit of this type is operating in Japan;
- to make coverings for protection against cold, drought, weeds.

3.1.2. *White water*

Waste water from the paper machine contains suspended solids comprising fibers or fiber fragments and various agents used to manufacture paper (pastes, resins, fillers of organic and mineral origin).

Efforts are being made to cut down the volume of water to be treated by intensively recycling effluents.

Unfortunately, only products such as building panels and some cardboards lend themselves to closed-circuit recycling. In the other product categories there are problems both at the operating and at the quality level.

Reusable fibers must be separated from other, finer materials which cannot be cycled back directly into the production circuits, and this requires special, repetitive cleaning phases.

The development of new—specifically dry—sheet formation techniques, use of new furnishes combining cellulose and synthetic fibers, finally, the improvement of filler retention on the wire cloth. These are the fields of research which may lead to a reduction in waste water from the paper machine.

3.2. *New Pulpmaking Processes and Techniques*

3.2.1. *For Chemical Pulps*

We are again dealing with the kraft process which is tending to replace other chemical processes.

As mentioned previously, most of the effluents originate in the bleaching plant. These effluents are too dilute and have too high a chloride content to be cycled back and eliminated in the liquor combustion-recovery plant.

For this reason, research has followed two main currents:

- reduction of effluent volume;
- reduction of chloride content or total elimination of chloride to be replaced by new, non-chlorine reagents according to new bleaching sequences to be developed and adapted to each case (integrated or non-integrated mill, semibleached pulps).

3.2.1.1. *Cutting down effluent volume*

(a) Chlorine gas bleaching. Laboratory studies have shown that in recycling effluents from various bleaching stages, it is possible, for a CEDED type sequence to cut down the amount of water from 30 m^3/t to 6 m^3/t of pulp, without any effect on the quality of the treatment.

Such recycling is only partially satisfactory, since it is accompanied by an overconsumption of reagents representing about 4 kg of soda and 25 kg of ClO_2 per ton of processed pulp. This example confirms the efficiency of high density chlorine gas treatment and the advantages inherent in making more widespread use of this method, particularly with relation to research on new sequences.

(b) Displacement bleaching. Of very different design, this new technique, developed by Kamyr, uses the diffuser principle. Pulp flows at a consistency of 10 per cent from the bottom to the top of a cylindrical bleaching tower into which the bleaching agents are fed successively and without mixing. A 120 t/d pilot plant of this type is operating in Finland. Displacement bleaching lasts 1½ hours instead of 4 hours and cuts down water consumption from 30 to 18 m^3/ton for an EDED sequence. Power consumption is also cut in half (65 kWh/t of pulp).

3.2.1.2. *Changes in existing processes*

(a) Oxygen bleaching. Fig. 2. shows where oxygen treatment fits into the bleaching line and illustrates the advantages of such a treatment. Conventional bleaching processes call for chlorination and extraction treatment of unbleached pulp. The resulting effluents contain about 80 per cent of pollutant discharge which is directly proportional to the lignin content of the unbleached pulp. Because of their high chloride content, these effluents cannot be recycled.

"Oxygen bleaching" involves a special pressurized lignin-removal treatment between the unbleached washing stage and the conventional bleaching stage: lignin is partially dissolved and the liquors, containing 50 to 70 per cent of the pollutant discharge from the bleaching stage, are recycled, thus reducing final pollution discharge by as much.

The possibility of eliminating the chlorination/extraction stage after a period of transition has been contemplated, and this would further cut pollution. This process is currently being used on an industrial scale by a French manufacturer of bleached hardwood kraft pulp.

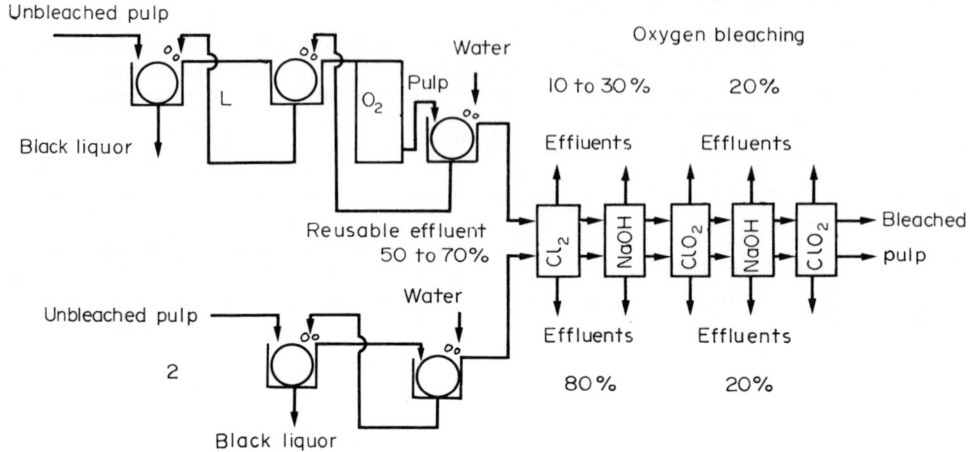

Fig. 2

(b) Soda-oxygen cook. A new, low-pollution process is presently being developed to replace the kraft process. The advantage of the soda-oxygen cook is that it eliminates ill-smelling sulfur compounds and the drawbacks of sulfide (corrosion).

An American mill will be implementing the two-stage soda-oxygen process in the near future.

Studies are being carried out in France in particular with a view to adapting this process to local tree species (both hardwoods and softwoods) and to developing a simpler, one-stage pressurized process involving less capital outlay.

(c) New bleaching sequences. APS sequences. A number of modified sequences based on the conventional CEDED schedule have been laboratory tested under the name APS (anti-pollution sequences). We will not go into a description of the four alternatives suggested to date. We will simply repeat that the principle consists of replacing all or part of the chlorine by ClO_2 in the first stage, which would mean:

a reduction of effluent coloration and BOD;

avoiding the drawbacks of high temperatures (loss of pulp quality);

a reduction in chloride content.

Furthermore,

acid and alcaline effluents are collected and treated separately;

alcaline effluents, whose volume is decreased, are cycled back into combustion-regeneration circuits;

acid effluents are resin-treated and regenerating liquors are treated with the recycled alcaline effluents. Acid effluents are rejected after treatment to eliminate chlorides and other impurities.

Rapson process. The various alternatives of this process, which was developed by Professor Rapson of the University of Toronto, call for recycling all effluents to the regeneration plant. The recommended bleaching schedule uses mainly ClO_2 and counter-current recycling in order to cut down on the quantities to be treated. The chlorides used to regenerate ClO_2 are successively eliminated than recovered by means of various steps as listed briefly below:

evaporation of white liquor. The concentrated NaOH + Na_2S solution precipitates NaCl crystals;

crystallization of green liquor;

elimination of NaCl-enriched dust from the electrostatic precipitator;

Clean sequences. The study of new sequences in which no chlorine compound is used represent a big step toward very low-pollution mills with almost fully closed-circuit systems.

This research relates to the use of reagents such as O_2, O_3, H_2O, etc., and to the development of the appropriate sequences where possible making use of gas treatment whose advantages were described elsewhere.

(d) Other processes. The most rational and the most complete solution to all forms of pollution, whether attributable to BOD, COD, coloration or any other criterion likely to come into play in the future, is to close the circuits entirely.

However, it should be noted that if we are to arrive at closed-circuit systems, it will be necessary to develop additional treatments to purify the circuits of micro-pollutants (in particular trace elements) whose build-up might hinder the successful operation of the installation. Among these specific treatments whose applications are either already being studied or in the process of being devloped for effluents from the extraction stage, the following two are worthy of mention:

Ultrafiltration treatment. The advantages of ultrafiltration treatment for pulp mill effluents have been recognized in the course of the past few years.

However, the membranes used in the process have to date evidenced a tendency to react with alcaline effluents.

Rhone Poulenc's new, more resistant membrane, whose pH is high, has made it possible to effectively eliminate colored organic substances from alcaline extraction effluents, which are the main color culprits in the bleaching process.

Tests carried out by the Centre Technique both in the laboratory and in mills have shown that 85 per cent of all color and 70 per cent of COD may be eliminated at 10 per cent concentration.

By using an appropriate washing sequence, it was possible to an average flow of 3.5 m^3/day per m^2 of membrane.

A techno-economic study carried out on this process in a mill producing 500 t/d of kraft pulp showed that the ultrafiltration treatment competes well with other existing processes.

Synthetic resin treatment. It is also possible to eliminate color by means of synthetic ion or adsorbent resins.

The ionic resin process was perfected by a Swedish company and is currently being implemented in a 300 t/d kraft pulp mill. Two other installations were built in Japan for mills producing 500 t/d. Colored substances remaining on the resin are then washed out with an alcaline solution originating in the mill circuits and after evaporation are burned in the recovery boiler.

The adsorbent resin process is currently still at the pilot study stage. It is slightly less effective in decoloring and reducing COD than its ionic resin counterpart. Furthermore, while the basic process is the same—fixing of the color in an acid medium, then elution in an alcaline medium—because of their adsorbent effect, these resins have no affinity for inorganic chlorides.

3.2.2. Semichemical and Chemigroundwood Pulps

Residual liquors from these processes contain very few organic and inorganic substances and generally have little recovery value. The major problem is therefore to find the threshold of profitability for treatment plants.

For conventional concentration - combustion - regeneration techniques the profitability threshold is generally thought to correspond to a 350 t/d output.

For a large number of mills whose production falls below this level, the only solution to pollution problems lies in developing new, less cumbersome combustion units than conventional ovens and boilers.

It would also be advisable to develop more flexible techniques, better adapted to treating liquors from processes currently being perfected.

3.2.2.1. Fluidized bed combustion/reduction. This technique would seem to meet the above criteria, as shown by the studies carried out in France by the Centre Technique du Papier in collaboration first with CERCHAR then with Société HEURTEY within the framework of a study program partially financed at the start by the French Ministry of the Environment.

A colloqium held recently in Stockholm confirmed the need to develop new technologies, not only to meet the needs of average-sized mills, but also to replace boilers which will be subjected in future to stringent safety standards and therefore difficult to maintain in operation.

In addition, the outlook for development of new production units in underdeveloped countries (bagasse and straw pulps) will create potential markets for this process.

It is noteworthy that this process entails eliminating dilute liquors without additional need for fuel oil, which in the context of the energy crisis could be advantageous for small units.

3.2.2.2. *Other pulping processes.* Research aimed at using reagents which would not require regeneration and at cutting down investment burdens on average-sized installations is continuing. Such research would increase the profitability of such mills.

The following are good examples:

carbonate cooks with or without oxygen, or particular interest in processing annuals;

magnefite cooks which simplify regeneration in advance;

more recent studies on carbonate-sulfide cooks.

CONCLUSION

This brief look at waste-treatment problems and possible economically viable solutions to such problems reveals the importance of past and future research efforts.

It is expected that by 1985 the pulp mill will be operating on a series of entirely new principles:

Firstly, at the supply stage, increasing use will be made of forest rejects and agricultural residues. The pulp and paper industry will thus be able to meet ever-increasing wood requirements without overconsuming forest resources and spoiling park areas, both of which are vital to the quality of life and a healthy ecology.

From a technical standpoint, new processes and technologies will change the face of future pulp mills. These units will be more compact, and the disappearance of air pollution will mean better environmental conditions in the immediate vicinity of the mill.

Large storage areas and waste treatment plants will have diminished in size if they do not disappear entirely.

In order to attain goals in this area, it is necessary:

to develop close cooperation on an international scale;

to maintain close contact between national government - local industry and research agencies.

Only by combining all these efforts can the environment be protected in everyone's interest and the paper industry develop means of making better use of a constantly renewed source of raw materials.

Would it not be advisable to also look beyond this goal to the potential offered by chemical conversion of cellulose into a high polymer of recognized quality and to lignin which is still not being used to its best advantage, and finally to by-products of the process: sugars, acids, alcohols, etc.?

Perhaps the time has come to prepare for the advent of these new processing industries which offer an unending source of carbon.

Non-Waste Production of Bleached Kraft Pulp

W. Howard Rapson* and Douglas W. Reeve**

**University of Toronto, Toronto, Canada*
***ERCO Envirotech Ltd., Toronto, Canada*

Global production of bleached kraft pulp, a high brightness, high strength cellulosic pulp with excellent papermaking properties, is approximately 60 million tons per year. With each ton produced, 100 to 200 cubic meters of contaminated effluent is normally discharged, bearing a substantial load of suspended solids, and dissolved solids which consume oxygen, are highly coloured and are highly toxic to aquatic and marine biota.

In the processes for conversions of logs to bleached kraft pulp, four major effluents are normally formed. The logs are first debarked, commonly with large volumes of high-pressure water, producing an effluent high in suspended solids. The wood is chipped and then cooked at high temperature and pressure with regenerated sodium hydroxide and sodium sulfide. This process decomposes and dissolves the interfiber binding and when the pressure is rapidly released, free fibers are formed. The spent cooking liquor is partially removed from the brown pulp by countercurrent washing. Typically 92-98 per cent of the spent cooking liquor is recovered, concentrated by evaporation and burned. However, some liquor is lost from the washing process giving rise to up to 30 per cent of the biological oxygen demand (BOD) discharge from the mill. Further, on evaporation of the spent liquor, volatile components are stripped out of the liquor and become concentrated in the condensed vapours or condensate. The methanol and malodorous sulfur compounds in the condensate contribute up to 30 per cent of the BOD discharge of the mill.

The fourth major effluent is produced on bleaching the highly coloured brown pulp. In the multistage bleaching process the pulp is treated alternately with chlorine and/or chlorine dioxide and with sodium hydroxide. Washing is employed after each treatment. The effluent from the bleach plant has the largest volume and contains up to 50 per cent of the BOD and most of the colour discharged from the mill.

Thus, bleached kraft pulp mills typically have four major water-polluting effluents: from the wood room; from spent cooking liquor left in the partially washed unbleached pulp; the spent cooking liquor evaporator condensates; and the bleach plant effluent. In addition, leakage, spills and wash-ups make a significant contribution. Such water pollution can only be eliminated either by a series of increasingly costly effluent treatments, or by recycling all polluted water within the mill.

Primary effluent treatment settles out a large part of the suspended solids.

---

*University of Toronto, Toronto, Canada.

†ERCO Envirotech Ltd., Toronto, Canada.

Secondary treatment biologically oxidizes part of the dissolved and suspended solids, but still leaves 10-20 per cent of the BOD in the final effluent. Furthermore, this effluent usually still contains all the colour, a large part of the toxicity, and often unacceptably high suspended solids. It also contributes nutrients, nitrogen and phosphorus, which must be added to the secondary treatment to allow the micro-organisms to grow. Disposal of the sludge from these processes by burial or incineration is difficult and costly. In addition, in those instances where colour and/or toxicity effluent standards are imposed, which will be more prevalent in the future, the cost of effluent treatment is greatly increased.

To eliminate the need for partially effective external treatment the following steps can be taken. Debarking is accomplished without water or with recycled water from which the suspended solids are removed and burned. The unbleached pulp-washing system must be highly efficient, operating on only recycled water and discharging no effluent. All contaminated condensates must be steam stripped and recycled. Finally, bleach plant effluent must be recovered.

The possibility of eliminating water pollution by bleached kraft pulp mills by recovery of bleach plant effluent, the closed-cycle concept, was conceived in 1965, and the first paper suggesting specific steps for accomplishing this goal was presented in 1967. The bleach plant effluent and evaporator condensates are used for pulp washing and in preparation of the cooking liquor so that no fresh water is added in pulping or chemical recovery. The spent bleaching chemicals and the organic matter dissolved during bleaching thus get into the pulping chemical recovery cycle. All the organic matter is burned and the spent bleaching chemical, sodium chloride, leaves the furnace with the pulping chemical smelt. The inorganic smelt is dissolved and then causticized to form "white" liquor for re-use in pulping.

There are several consequences of this procedure. All process water used in the bleach plant eventually ends up in the chemical recovery system. All chlorine in the bleaching chemicals ends up as sodium chloride in the chemical-recovery system, and must be removed at the rate at which it is introduced. Since there is no purge for salt in the normal chemical-recovery system, it was necessary to develop a process for salt recovery and control.

The greatest problem was recovery of sodium chloride, of high quality, in sufficient amounts and at reasonable cost while controlling the recirculating burden of sodium chloride to tolerable levels. Through 4 years of intensive research at the Department of Chemical Engineering of the University of Toronto, a process was devised. Further research, to refine the process, was supported for an additional 2 years by the Committee for Pollution Abatement Research (CPAR) of Environment Canada. Part of this program was a brief large-scale trial of the process, using equipment available in an operating kraft-mill. In 1972 ERCO Envirotech Ltd. was formed to complete development of the salt-control process. Two years were required for pilot plant and laboratory studies and to develop the remaining process and equipment design technology required.

Thus, a new chemical process has emerged. By this process, sodium chloride is removed from kraft pulp mills, a key to allowing recovery of bleach plant effluent. In the process, regenerated cooking liquor, white liquor, is concentrated in multiple-effect evaporators crystallizing sodium carbonate and sodium sulfate, unregenerated pulping chemicals, which are recycled to the system. The partially concentrated white liquor is further concentrated and cooled to crystallize sodium chloride. The concentrated white liquor, stripped of sodium chloride, contains all the active chemicals required for pulping and so, after dilution, is applied to fresh wood chips. The sodium chloride crystals are leached and washed countercurrently to remove impurities. The sodium chloride may then be used to regenerate bleaching chemicals: sodium hydroxide, chlorine and chlorine dioxide.

Considerable development has also been required in bleaching technology, particularly with respect to decreasing the effluent volume. The individual components of this technology have been tested by years of commercial practice. However, in order to design and operate a bleach plant at 20 cubic meters, total "effluent" per ton of pulp, the maximum which can be recovered, an entirely new design philosophy is required. ERCO Envirotech Ltd. has developed the required bleach plant design which minimizes steam and fresh water consumption and maintains pulp quality and efficient utilization of bleaching chemicals.

In a parallel development program, J.A. Histed of CIP Research Ltd., Hawkesbury, Ontario, under the sponsorship of the CPAR program of Environment Canada, has confirmed the feasibility of such bleach plant water re-use by extensive laboratory experiments.

Substitution of 70 per cent of the available chlorine by equivalent chlorine dioxide in the first stage of bleaching is used in the closed-cycle mill and has the following advantages compared with all chlorine. The pulp is stronger, cleaner and is more stable towards yellowing with age. The yield of pulp is increased by about 1 per cent because the chlorine dioxide does not degrade and dissolve the carbohydrates as much as chlorine does. High temperature (60-70°C) can be accepted in the chlorination stage. The amount of sodium hydroxide required for bleaching is decreased to half of that required with all chlorine and it follows that the amount of sodium chloride produced per ton of pulp is decreased by 50 per cent, substantially decreasing the amount of salt that must be removed by the salt-recovery process.

When 70 per cent of the chlorine in chlorination is replaced by chlorine dioxide, a large, effluent-free chlorine dioxide plant is required. In fact when 70-80 per cent chlorine dioxide is used in the first stage, all the oxidizing chemical required in the bleach plant can be supplied by modern chlorine dioxide plants as developed by ERCO Industries Ltd., of Toronto, Ontario. Water required to absorb chlorine dioxide and chlorine is minimized by absorbing both in the same solution.

In a mill in which water and all chemicals are recycled, if nothing escapes, the sodium chloride removed must balance stoichiometrically with the sodium hydroxide, chlorine and chlorine dioxide used in the process. If it does not, the chemical inventory will increase or decrease. It also follows that the amount of sodium chloride removed is exactly that required to produce the bleaching chemicals. The sodium chloride may be electrolyzed to produce sodium chlorate and also sodium hydroxide and chlorine. The sodium hydroxide would be used directly in the bleach plant. The chlorine would be burned with hydrogen and resulting hydrochloric acid combined with the sodium chlorate to make chlorine dioxide. Theoretically, in an entirely closed mill the chemical inventory should be constant, only electricity being consumed. In practice, there will be some sodium, potassium, chloride and sulphur introduced with the wood and some loss of solid particles to the atmosphere, although this must be minimized.

The closed-cycle mill will require none of the traditional kraft makeup chemical, Na_2SO_4. With spills, screen room effluent and bleach plant effluent eliminated and losses to the atmosphere minimized, hardly any pulping chemical makeup will be required. Sodium makeup can be provided by purchased caustic. Sulphur makeup, in many mills, will be provided by purchased sulphuric acid used for tall oil acidulation. Thus the need for handling another chemical, Na_2SO_4, is eliminated.

The bleach plant uses only pulp machine white water for washing and a minimum of fresh water enters the bleach plant through chemicals or other sources. Instead of the typical fresh water requirement of 100 m^3/ton pulp, the bleach plant requirement is virtually zero. Similarly, in the closed-cycle mill no fresh water is added in brown stock screening or washing which typically eliminates 40 m^3 fresh water makeup per ton of pulp. Thus treated process water requirements are greatly decreased. It should be noted that 50 m^3/ton pulp additional cooling water is required, for the salt-recovery process although in most cases this water would require no treatment.

In the closed-cycle mill an increased fraction of the organics dissolved on pulping and all the organics dissolved in bleaching are recovered and burned. This will increase steam production by at least 0.3 tons per ton pulp. In addition, countercurrent washing and other features in the bleach plant design can result in a saving of up to 2.0 tons steam per ton pulp. Against this saving of 2.3 is a cost of 1.1 tons steam for evaporation of white liquor.

The capital cost of building a new bleached kraft mill, completely eliminating water pollution, is only about 5 per cent more than building the same mill with no effluent treatment. This must be compared with 10 per cent or so additional capital usually required to provide primary and secondary treatment, sludge disposal and colour removal.

The operating cost (including the cost of capital) of a closed-cycle mill, taking into account the increased yield of pulp on wood, the energy saving, fibre saving and the lower water cost, would likely be less than that of present mills without effluent treatment. The operating cost of a new mill with primary treatment, secondary treatment, sludge removal and colour removal is about 5 per cent higher than that of a mill without effluent treatment. Yet all the colour, much of the toxicity and about 10 per cent of the oxygen-consuming matter still remain in the effluent after treatment. To decrease the pollutant load further by external treatment would increase both capital and operating cost exponentially. Therefore, it is very much cheaper to eliminate water pollution by internal recycle of process water.

With the components of closed-cycle mill technology demonstrated by commercial practise, in some aspects, thoroughly examined in pilot plant studies, in other aspects; full scale, commercial demonstration remains. It is close at hand. The world's first installation of a bleached kraft pulp mill, designed to recover all bleach plant effluent, is, at the time of writing, under construction at Thunder Bay, Ontario, Canada, and is expected to startup in 1976. The Great Lakes Paper Co. Ltd., in August 1975, announced it had chosen to implement the ERCO Envirotech Ltd., closed-cycle concept at a cost of 8 million dollars. This eliminates the need for installation of effluent-treatment facilities for the 250,000 ton per year market pulp mill. All process streams containing suspended solids, BOD, colour and toxicity will be re-used within the mill and only cooling water, free of contaminants, will be discharged into the nearby river.

The design of the mill is the result of the close collaboration of Great Lakes Paper, the general consultants E & B Cowan Ltd., and ERCO Envirotech Ltd., with the Federal and Ontario Ministries of the Environment playing an important part in the overall environmental protection system.

Another important factor in the realization of the non-waste bleached kraft pulp mill is a $1,158,000 cost-sharing contract with Environment Canada. Under the terms of the Development and Demonstration of Pollution Abatement Technology program, the contract requires Great Lakes to document and evaluate the new process and insures that the technology will be freely available to other interested Canadian companies.

This development qualifies as "non-waste" technology in a number of ways. Firstly it evades investment in pollution abatement add-on facilities that do only part of the job. It also evades the wasteful use of receiving water as a disposal sink for waste materials. Mainly, of course, it reduces chemical usage which saves both mineral and energy resources.

BIBLIOGRAPHY

1. W.H. Rapson, The feasibility of recovery of bleach plant effluents to eliminate water pollution by kraft pulp mills. Pulp and Paper Mag. Can. 68, No. 12, T635-T640 (1967).

2. D.W. Reeve and W.H. Rapson, The recovery of sodium chloride from bleached kraft pulp mills. Pulp and Paper Mag. Can. 71, No. 13, T274-T280 (3 July 1970).

3. D.W. Reeve and W.H. Rapson, Kraft bleach plant effluent elimination. Part II. Equilibrium data for sodium chloride recovery by white liquor evaporation. Pulp and Paper Mag. Can. 74, No. 1, T19-T27 (1973).

4. D.W. Reeve, J. Lukes, K.A. French and W.H. Rapson, The effluent-free bleached kraft pulp mill. Part IV. The salt recovery process. Pulp Paper Mag. Can. 75, No. 8, T293-296 (1974).

5. D.W. Reeve, Ph.D Thesis, Sodium chloride recovery from kraft pulp mills. University of Toronto, 1971.

6. D.W. Reeve, G. Rowlandson and W.H. Rapson, The effluent-free bleached kraft pulp mill. Part VIII. Bleach plant renovation and design, presented at the International Pulp Bleaching Conference, Chicago, Ill., 2-6 May 1976.

7. W.H. Rapson, C.B. Anderson and D.W. Reeve, The effluent-free bleached kraft pulp mill. Part VI. Substantial substitution of chlorine dioxide for chlorine in the first stage of bleaching, paper presented at the Alkaline Pulping Conference, Williamsburg, Va., 16 Oct. 1975.

8. W.H. Rapson and D.W. Reeve, Can. Patent 915,361, 28 Nov. 1972.

9. W.H. Rapson and D.W. Reeve, Can. Patent 928,007, 12 June 1973; US Pat. 3,740,308, 17 June 1973.

10. W.H. Rapson and D.W. Reeve, Can. Patent 923,256, 27 Mar. 1973.

11. W.H. Rapson and D.W. Reeve, Can Pat. 915,362, 28 Nov. 1972; U.S. Pat. 3,740,307, 19 June 1973.

12. D.W. Reeve, US Patent 3,950,217, 13 April 1976.

13. D.W. Reeve, The effluent-free bleached kraft pulp mill. Part VII. Sodium chloride in alkaline pulping and chemical recovery, presented at the Technical Section, Canadian Pulp and Paper Association, Montreal, Canada, 27-30 Jan. 1976.

Réduction de la Charge de Pollution de l'Eau Provenant d'une Usine de Pâte au Sulfate Blanchie

P. Lieben

Direction de l'Environnement, OCDE, Paris, France

L'OCDE a publié en 1972 un rapport sur "l'application des techniques avancées à la lutte contre la pollution dans l'industrie des pâtes et papiers".* Ce rapport décrivait un certain nombre de nouvelles techniques internes visant à réduire les déchets de fabrication déversés avec les effluents de l'industrie papetière, et responsables de la pollution de l'eau. On tentait également, dans ce rapport, de comparer les résultats et le coût de ces nouvelles techniques, combinées de différentes manières, avec le traitement externe traditionnel des eaux usées.

Bien que cette étude ait souffert de la généralisation des estimations et de la disparité des données de base, elle montrait cependant que nombre des mesures internes considérées étaient capables, dans certaines limites, de réduire les déchets de fabrication et la charge polluante à un coût inférieur à celui des mesures externes, pour une même diminution de DBO_5.

Cependant, pour parvenir à une très forte réduction de la pollution, il semblait nécessaire de combiner les mesures internes avec le traitement externe.

Depuis la publication du rapport ci-dessus, les techniques se sont développées de façon continue, et tous les coûts ont augmenté, mais pas nécessairement dans les mêmes proportions.

Le présent rapport examine la même usine que le rapport précédent (500 tonnes par jour de pâte au sulfate blanchie), et les mêmes types de mesures, internes et externes. (Le procédé Rapson-Reeve ne sera pas étudié ici; il fera l'objet d'une étude séparée.) La technologie adoptée est celle qui prévaut aujourd'hui, et les couts (investissements et exploitation) sont basés sur les prix pratiqués actuellement en Suède.

Lorsqu'on examine les résultats qui ressortent de cette étude il convient de garder à l'esprit que les coûts peuvent varier d'un pays à l'autre. Il faut également souligner que de nombreux facteurs locaux dont il n'est pas possible de tenir compte dans ce type d'étude générale (espace disponible, problèmes de paysage, disposition de l'usine, etc.) peuvent jouer un rôle décisif dans le choix des techniques à utiliser.

---

*Ce rapport est disponible sur demande au Secrétariat de l'OCDE, Direction de l'Environnement, 2, rue André Pascal, 75775 Paris Cedex 16. Des copies seront également mises à la disposition des participants au Séminaire sur les principes et la création de techniques et de systèmes de production sans déchets.

Il existe un certain nombre de techniques internes (Tableau 5) capables de réduire la charge polluante provenant de différentes unités de production, pour une dépense inférieure à celle des traitements externes. Ces techniques permettent de réduire d'environ 50% la charge polluante dans le type d'usine considéré. Il semble cependant que, au stade actuel de développement de la technologie, une combinaison de mesures internes et de traitement externe des effluents soit nécessaire pour atteindre le niveau de réduction de la pollution fixé par la législation dans de nombreux pays. La manière dont cette combinaison est réalisée dépend de circonstances particulières et de la structure des prix localement.

Déscription de l'usine hypothétique

Cette usine hypothétique produit 500 tonnes de pâte au sulfate blanchie par jour, à partir de résineux. Elle utilise l'équipement conventionnel normalement employé dans les usines construites en 1960-1965. La situation d'ensemble de l'effluent est:

Débit d'effluent: 240 m^3/tonne de pâte.
Matières en suspension: 33 kg/tonne de pâte.
DBO_5: 45 kg/tonne de pâte.

On admet que le rejet maximum autorisé pour l'usine équivaudrait à:

Matières en suspension: 5 kg/tonne de pâte.*
DBO_5: 5 kg/tonne de pâte.

Ecorçage [1.1]. L'écorçage par voie humide est employé, et l'effluent est tamisé. Les rejets de tamisage sont déshydratés et pressés. L'écorce (environ 120 tonnes en sec par jour) est brûlée dans un incinérateur séparé. L'effluent combiné des tamis des filtres et des presses est dirigé vers un bassin de sédimentation en terre, petit et peu efficace, puis est rejeté. La quantité d'effluents déversés qui est de 10 m^3 par tonne de pâte, contient 14 kg de matières en suspension et 3 kg de DBO_5 par tonne de pâte.

Cuisson et lavage [1.2]. La cuisson se fait dans des lessiveurs discontinus. La vapeur produite est condensée dans un condenseur. On admet que la quantité de condensats qui se forment est de l'ordre de 1 m^3/tonne de pâte. La thérébentine est séparée des condensats de dégazage dans un décanteur. La pâte écrue est lavée dans des filtres à contre-courant. Le taux de récupération de la matière sèche est estimé à 97%. (Beaucoup d'usines existantes ont des taux plus bas en raison d'un lavage insuffisant.)

Epuration et raffinage [1.3]. Le système d'épuration fonctionne sans recyclage et la quantité d'effluents provenant de cette source est de l'ordre de 100 m^3/tonne de pâte.

Blanchiment [1.4]. La séquence du blanchiment est CEDED et la blancheur de 90 G.E. Les effluents des phases D et E_2 sont utilisés partiellement dans les phases C et E_1 et les effluents de ces stades partent à l'égoût. La

---

*Il est reconnu que les boues biologiques ne permettent pas d'atteindre ce niveau de matières en suspension; aussi ce chiffre se rapporte aux matières en suspension fibreuses (boues primaires).

quantité d'effluents acides est de 60 m^3/tonne de pâte, et la quantité d'effluents alcalins est de 40 m^3/tonne de pâte.

Evaporation [1.5]. L'évaporation s'effectue dans un évaporateur à 5 effets. La quantité de condensats est d'environ 6 m^3/tonne de pâte, dont environ 4 m^3/proviennent des effets d'évaporation et 1,5 m^3/du condenseur de surface, après le dernier effet. L'eau d'étanchéité et les condensats résiduaires provenant de la pompe à vide, en même temps que les fuites, représentent environ 1 m^3/tonne de pâte.

Caustification [1.6]. Les effluents de la caustification (y compris le four à chaux) représentent 5 m^3/tonne de pâte et peuvent contenir des traces de boue de chaux.

Chaudière de récupération [1.7]. Les divers effluents de four sont de l'ordre de 13 m^3/tonne de pâte et peuvent contenir des matières en suspension et des matières dissoutes.

TABLEAU 1

Effluents d'une usine de pâte au sulfate blanchie de 500 t/jour

	Effluents		Matières en suspension			DBO_5		
	m^3/jour	m^3/tonne	mg/l	kg/jour	kg/tonne	mg/l	kg/jour	kg/tonne
1. Ecorçage	5000	10	1400	7000	14	250	1500	3
2. Cuisson et lavage	500	1	-	-	-	4000	2000	4
3. Epuration et raffinage	50 000	100	140	7000	14	130	4500	9
4. Blanchiment acide	30 000	60	40	1200	2,4	170	5000	10
alcalin	20 000	40	40	800	1,6	125	2500	5
5. Mise en pâte	2000	4	250	500	1,0	-	-	-
6. Evaporation	3500	7	10	35	0,1	2000	7000	14
7. Caustification	2500	5	20	50	0,1	-	-	-
8. Chaudière de récupération	7500	15	10	75	0,1	-	-	-
Total	120 000	240	138	16 650	33,3	188	22 500	45

La pollution peut être réduite soit par des mesures externes seulement, c'est-à-dire le traitement des effluents, soit en appliquant diverses mesures internes à l'usine, destinées, à éliminer au moins une partie des rejets de la production, complétées autant que nécessaire par le traitement externe des effluents. Les valeurs économiques présentées indiquent seulement l'ordre de grandeur des investissements et des coûts d'exploitation qu'entraîne l'application de ces mesures. Les prix sont donnés en Couronnes Suédoises; ils correspondent aux premieres mois de 1976.

EMPLOI DE MESURES EXTERNES

Lorsque l'on se borne à employer des mesures de traitement externe des effluents, l'objectif à atteindre (5 kg/tonne) suppose tant un traitement primaire qu'un traitement secondaire. Les effluents de l'usine doivent être divisés en deux catégories. Les rejets 1-6 (Tableau 1) sont envoyés au traitement. Les rejets 7-8 peuvent être déversés directement (en admettant que l'effluent N° 7 ne contienne pas davantage que 20 mg/l de matières en suspension). La méthode de traitement choisie pour cette étude est le système par boues activées.

Clarificateur primaire [2.1.1]

Les divers effluents seront mélangés dans un réservoir, neutralisés et pompés vers le clarificateur primaire. Celui-ci est du type cylindrique à alimentation centrale (Ø 62 m, vitesse ascensionnelle 1,5 m/h) et le résultat du traitement est le suivant (Tableau 2).

TABLEAU 2

	Volume	matières en suspension		DBO_5	
	m^3/jour	mg/l	Tonnes/jour	mg/l	Tonnes/jour
Débit d'entrée	111 000	150	16,5	205	22,5
Débit de sortie	110 000	40	4,5	185	20,2
Boues	600	20 000	12,0	3800	2,3

Traitement secondaire [2.1.2]

En vue d'arriver au taux désiré de traitement - 5 kg/tonne soit 2500 kg/jour - les quantités suivantes doivent être enlevées au cours du traitement secondaire:

DBO_5: 17,7 tonnes/jour (87,5%).
Matières en suspension: 2,0 tonnes/jour.

La concentration permise est de 22,5 mg/l

Le dispositif de traitement secondaire consiste en:
- addition d'éléments nutritifs et bac de mélange,
- bassins d'aération: 15 bassins (1000 m^3 chacun),
- clarificateurs secondaires: 6 décanteurs à alimentation centrale (Ø 34 m, vitesse ascensionnelle 1,1 m/h),
- installation de traitement des boues (y compris boue primaire): 3 épaississeurs de boue et centrifugeuses, equipement de transport de la boue.

Aucune précipitation chimique pour l'enlèvement du phosphore n'est prévue.

Resultat final obtenu [2.1.3]

Le résultat final obtenu est présenté dans le Tableau 3.

TABLEAU 3

	Volume	Matières en suspension			DBO$_5$		
	m^3/jour	mg/1	tonnes/ jour	kg/ tonne	mg/1	tonnes/ jour	kg/ tonne
Débit d'entrée	111 000	150	16,5	33	205	22,5	45
Débit de sortie	111 000	15	1,6	3	20	2,2	4,5

La production de boue est d'environ 165 tonnes/jour. Les matières sèches représentent 27 tonnes/jour et la concentration après centrifugation est de l'ordre de 16,5%.

La boue est dirigée vers une incinération séparée où elle est mélangée à l'écorce et brûlée.

Coûts du traitement externe [2.1.4]

Investissements. Les investissements pour une installation de traitement des effluents, telle qu'elle a été décrite ci-dessus, sont à peu près les suivants (les coûts varient naturellement selon les pays):

Traitement primaire:	*'000 C.S.*
- bâtiments	2000
- machines installées	3375
- tuyauteries, instruments, etc.	1575
Total	6950
Traitement secondaire:	*'000 C.S.*
- bâtiments	6580
- machines installées	13 180
- tuyauteries, instruments, augmentation de la capacité d'incinération d'écorce, etc.	3490
Total	23 250
Préparation du site	1500
Total des investissements	31 700
Charges de capital par an (9%, sur 10 ans)	4920

Coûts d'exploitation. Les coûts annuels d'exploitation sont approximativement les suivants (ils varient selon les pays):

Traitement primaire:	*'000 C.S./an*
C.S. 0,009/m^3 d'effluent (350 x 110 000 x 0,009)	346
Traitement secondaire:	
- C.S. 400/tonne de DBO_5 appliqueé/jour (350 x 20,2 x 400)	3828
- manutention des boues: C.S. 45/tonne de boue enlevée (350 x 6,1 x 45)	96
- frais directs de main-d'oeuvre*	640
- entretien et réparations: 4% des investissements	1268
Total du traitement secondaire	4832
Ensemble des coûts d'exploitation	5178
Coûtes totaux	
Total par an ('000 C.S.)	10 098
Total/tonne de pâte (C.S.)	57,70

EMPLOI DE MESURES INTERNES

Si l'on examine les possibilités de réduction de pollution au moyen de

*Frais directs de main-d'oeuvre: C.S. 80 000/homme an.

mesures internes, il faut se rappeler que quelques-unes des techniques avancées ne sont pas encore largement utilisées. On admet cependant que toutes ces techniques ont une chance raisonnable d'être appliquées commercialement dans un avenir proche.

Ecorcage [2.2.1]

L'effluent provenant de l'écorçage pourrait être éliminé en passant à l'écorçage à sec. On admet que la quantité de bois (pin) utilisée est: 2900 m^3, comptés en plein, avec écorce, par jour (500 tonnes/jour x 5,1 m^3, compté en plein par tonne x le facteur d'écorce 1,13). Le bois arrive en longueurs de 4 m. L'écorçage est effectué dans un cylindre sec, Ø 3,8 m et 1 = 54 m.

Estimation des coûts

Investissements:	*'000 C.S.*
- bâtiments et fondations	3000
- équipement installé	12 000
Total	15 000
Coût en capital par an (9% sur 10 ans) coût d'exploitation:	2325
Coûts d'exploitation	*'000 C.S./an*
- coût supplémentaire de la main d'oeuvre, par comparaison avec l'écorçage par voie humide	néant
- entretien	néant
Economie due au rendement accru*	760
Coûts totaux:	
Total des coûts additionnels par an ('000 C.S.)	1565
Total des coûts additionnels par tonne de pâte (C.S.)	8,95

Cuisson et Lavage [2.2.2]

Les condensats seront pompés vers une colonne de stripping. Dans cette colonne les composés du soufre et le méthanol qui se trouvent dans les condensats peuvent être enlevés et envoyés au four à chaux pout y être brûlés. La plus grande partie de la DBO provenant des condensats peut ainsi être éliminée, et les condensats peuvent être utilisés, par exemple, au lavage.

---

*Rendement accru: Prix du bois à l'usine = C.S. 150 par m^3 plein avec écorce.
Rendement accru = 0,5%.

La même colonne de "stripping" sera aussi utilisée pour les condensats d'évaporation.

Les pertes survenant au cours du lavage peuvent être réduites en ajoutant un filtre de lavage de plus. Le taux de récupération atteindra ainsi 98,5% et les pertes du lavage seront de l'ordre de 6 kg de DBO_5 ou 10 kg de Na_2SO_4. La dilution augmentera également.

*Estimation des coûts**	*'000 C.S.*
Investissements:	
- cout d'investissement pour un filtre de lavage nouveau d'une capacité de 500 tonnes/jour, installé, compte tenu des coûts de construction des bâtiments correspondants	5500
Coûts de capital par an (9%, 10 ans)	850
Coût d'exploitation:	*'000 C.S./an*
Coûts d'exploitation additionnels	
- énergie et entretien	761
Economies	
- récupération plus importante de produits chimiques (13 kg Na_2SO_4 par tonne de pâte, à C.S. 0,30/kg)	682
- énergie	385
Total des économies	1067
Economies nettes de coûts d'exploitation	306
Coûts totaux:	
coût additionnel par an ('000 C.S.)	544
coût additionnel par tonne de pâte (C.S.)	3,10

Epuration [2.2.3]

L'usine dont il s'agit utilise des quantités importantes d'eau de dilution (100 m^3/tonne) pour l'épuration.

En adoptant un système normal *en circuit fermé*, la quantité d'eau peut être ramenée à environ 50 m^3/tonne ce qui, en conséquence, réduira le volume total des matières en suspension dans l'effluent d'épuration de 50%. L'emploi d'un système en circuit fermé n'affecte pas la quantité de DBO rejetée, qui restera la même que celle qui provient du lavage de la pâte, soit 6 kg de DBO_5 par tonne.

Dans ce cas, outre la fermeture du circuit d'eau, la méthode modifiée

---

*Les coûts du "stripping" sont compris dans les chiffres figurant sous la rubrique "évaporation" (voir ci-après).

d'*épuration à chaud* pourrait être utilisée, du fait que l'on envisage une augmentation de la capacité de lavage. La consommation d'eau fraîche pour l'épuration peut être réduite à environ 8 m^3/tonne de pâte. L'effluent d'épuration est utilisé au lavage après chauffage dans un réchauffeur.

De la sorte, l'effluent d'épuration peut être éliminé, mais la charge polluante provenant du lavage va avec la pâte au blanchiment, ce qui accroît le coût du blanchiment.

Estimation des coûts

Investissements	*'000 C.S.*
- système en circuit fermé	5500
- système d'epuration à chaud	-
Total	5500
Coûts de capital par an (9%, 10 ans)	850
Coûts d'exploitation	*'000 C.S./an*
- coût accru des produits chimiques dans le blanchiment	100
Coûts totaux	
total des coûts additionnels par an ('000 C.S.)	950
total des coûts additionnels par tonne de pâte (C.S.)	5,40

Blanchiment [2.2.4]

On étudiera sous cette rubrique deux étapes:
- réduction de la consommation d'eau, et
- réduction de la DBO provenant du blanchiment.

La réduction de la DBO pourrait être obtenue par un échange d'ions, par blanchiment à l'oxygène, ou par lavage à contre-courant.

Réduction de la consommation d'eau [2.2.4.A]. L'installation de blanchiment consomme quelque 100 m^3 d'eau/tonne; 60 m^3/tonne sont rejetés après la phase de chloration et 40 m^3/tonne proviennent de la phase alcaline. On admet que l'eau sera recyclée comme suit:

(i) Une moitié environ, soit 20 m^3/tonne, des effluents de chloration sera utilisée au lieu d'eau fraîche pour la dilution au sommet de la tour de chloration. Cinq m^3/tonne de cet effluent seront utilisés pour diluer la pâte dans le cuvier de pâte écrue.

(ii) En utilisant les effluents des dernières phases dans les rampes de lavage, l'économie d'eau fraîche utilisée dans les filtres de blanchiment se montera à 25 m^3/tonne de pâte.

A la suite des modifications ci-dessus, la quantité totale d'effluents décroîtra de 50 m^3/tonne et ces effluents seront les suivants:

30 m^3/tonne d'effluents acides,
20 m^3/tonne d'effluents alcalins.

Il n'y aura pas de changement important dans l'ensemble des eaux usées. On admet que la DBO est également divisée entre les deux effluents.

Estimations des coûts

Investissements	*'000 C.S.*
- système de recyclage	450
Coût en capital par an (9%, 10 ans)	70
Coûts d'exploitation	*C.S./tonne de pâte*
- coût supplémentaire de produits chimiques	10
- réduction des dépenses de vapeur, et récupération de fibres	11
- économies de coûts d'exploitation	1*
Coûts totaux	
Total des coûts additionnels par tonne de pâte	*néant*

Réduction de DBO par échange d'ions [2.2.4.B.1]. L'effluent de la phase d'extraction alcaline peut être dirigé vers un échangeur d'ions. L'éluat contient seulement de petites quantités de chlorures et peut être évaporé et brûlé. Compte tenu de l'expérience acquise dans des opérations à échelle industrielle, la charge en COD peut être réduite de 60%, la charge en DBO de 20 à 40%, et la couleur (lignine) de 90%.

Des améliorations récentes de ce système indiquent que les effluents de la phase C peuvent également être traités de la même manière, moyennant quelques adaptations, et conduire à une réduction supplémentaire de pollution. Il est cependant difficile de prévoir exactement la réduction de DBO qui en résulterait.

Après avoir mis en oeuvre les mesures décrites ci-dessus pour réduire les quantités d'eau employées à l'usine de blanchiment, les hypothèses suivantes concernant les effluents de blanchiment semblent raisonnables (Tableau 4).

TABLEAU 4

	Eaux usées (m^3/tonne)	Matières en suspension (kg/tonne)	DBO_5 (kg/tonne)
Effluents acides	30	2	8
Effluents alcalins	20	1	7
Total	50	3	15

*On n'a pas tenu compte de l'économie réalisée par la réduction de la quantité d'eau industrielle à traiter.

Après traitement des effluents alcalins par échange d'ions, le total des effluents du blanchiment sera:

- débit: 50 m^3 tonne
- matières en suspension: 2 kg/tonne
- DBO_5: 10,5 kg/tonne

Estimation de coût

Investissements ('000 C.S.)	24.000
Coûts en capital (9%, 10 ans) (C.S./t. de pâte)	21
Coûts d'exploitation par tonne de pâte (C.S.)	10
Coût total par tonne de pâte (C.S.)	31

Réduction de la DBO par le blanchiment à l'oxygène [2.2.4.B.2]. On admet qu'une phase O_2 sera ajoutée avant la série C - E - D - E - D de blanchiment. On estime en outre que les effluents provenant des différentes phases seront:

	DBO_5 (kg/tonne)
- pressage de la pâte au stade O_2	3
- stade O_2, sortie du réacteur	8
- phase C	2,5
- phase E	3,5
Total	17

En admettant un rendement de 80% au lavage après la phase O_2, les rejets provenant de cette installation de blanchiment seraient:

	DBO_5 (kg/tonne)
- stade O_2	2,5
- phase C	2,5
- phase E	3,5
Total	8,5

La quantité totale d'eau usée du blanchiment est estimée à 50 m^3/tonne.

Estimation des coûts

Investissements	*'000 C.S.*
- stade O_2 et lavage final	45 000
- changements nécessaires dans l'usine	8000
Total	53 000
Coûts de capital par an (9%, 10 ans)	8268

Coûts d'exploitation	*C.S./t. de pâte*
- produits chimiques, électricité, vapeur et entretien	35

- économies dans la consommation d'autres produits chimiques (Cl_2, NaOH et ClO_2)	50
Total des gains d'exploitation	15
Coûts totaux	
Total des coûts additionnels par tonne de pâte (C.S.)	32,00

Evaporation [2.2.5]

Les condensats provenant de l'évaporation peuvent être divisés comme suit:

- le condensat du premier effet est utilisé pour l'alimentation en eau de la chaudière;
- les condensats du second et troisième effet et du condenseur primaire (2,5 m^3/tonne) sont relativement propres et peuvent être rejetés comme tels;
- les condensats du quatrième et cinquième effet (2 m^3/tonne) sont utilisés pour la caustification;
- les condensats du condenseur de surface sont épurés dans une colonne avec les condensats de soufflage; les gaz de colonne de stripping sont brûlés dans le four à chaux; les condensats épurés (2,5 m^3/tonne) sont mis à l'égout.

La liqueur blanche faible provenant de la caustification sera utilisée comme eau servant à assurer l'étanchéité des pompes à vide, et est recyclée à la caustification.

En améliorant le rendement du lavage (voir "cuisson et lavage" ci-dessus) on a également augmenté la dilution. Ceci postule une capacité d'évaporation additionnelle. On suppose qu'un effet de plus sera ajouté, disposé en parallèle. Le condensat en surplus (0,5 m^3/tonne) est envoyé au collecteur. La quantité d'effluents résultant de l'évaporation sera ensuite réduite à environ 5 m^3/tonne, dont 3 m^3/tonne peuvent être rejetés sans aucun traitement. Les 2,5 m^3/tonne restants contiennent environ 4 kg de DBO_5/tonne de pâte (condensats allant au stripping).

Estimations des coûts

Investissements	*'000 C.S.*
- un nouvel effet installé	1000
- colonne de stripping, échangeurs de chaleur, tuyauteries	2000
Total	3000
Coûts en capital par an (9%, 10 ans)	465
Coûts d'exploitation	*'000 C.S./t. de pâte*
- évaporation	660
- stripping	2100
Total	2760

Coûts totaux

Coûts additionnels totaux par an ('000 C.S.)	3225
Coûts additionnels totaux par tonne de pâte (C.S.)	18,40

TENTATIVE D'EVALUATION ECONOMIQUE

Le Tableau 5 et les Figs. 1-3 constituent un résumé concis des mesures décrites plus haut, destinées à diminuer la charge polluante des eaux usées.

Même si les données économiques fournies ne sont que des estimations très imparfaites, qui peuvent varier d'un pays à l'autre, il est possible de les utiliser pour se faire une idée du coût des diverses mesures utilisables pour atteindre la limite des 5 kg de DBO_5 par tonne dans le cas de cette usine hypothétique de 500 tonnes par jour.

En étudiant les estimations indiquées ci-après, on doit se rappeler qu'elles se rapportent à la situation d'une usine existante qui est déjà équipée de manière classique. Dans le cas d'une nouvelle usine à construire, on pourrait réaliser des économies importantes en substituant à la place des équipements traditionnels quelques-unes des nouvelles techniques examinées dans le présent rapport, voire de nouveaux procédés de fabrication.

Emploi exclusif de mesures externes

Sans aucune mesure interne à l'usine, et en effectuant un traitement primaire et secondaire des effluents, le coût total de réduction de pollution dans les effluents à 5 kg de DBO_5/tonne de pâte serait de:
C.S. 57,70 par tonne de pâte.

Emploi de mesures internes a l'usine

Cas N° 1. En utilisant l'écorçage à sec (2.2.1), en améliorant la capacité de lavage (2.2.2), en réduisant la consommation d'eau pour l'épuration (2.2.3) et le blanchiment (2.2.4.A), et en traitant les condensats (2.2.5), la quantité d'effluents peut être ramenée à 100 m^3/tonne et cet effluent contiendra environ *9 kg de matières en suspension et 25 kg de DBO_5 par tonne*.

Les investissements, pour ces mesures internes à l'usine, seront d'environ C.S. 24 millions et le coût total environ C.S. 30,35 par tonne de pâte.

Les investissements en ce qui concerne l'installation de traitement externe des eaux usées, nécessaire pour arriver au chiffre de 5 kg de DBO_5/tonne,

TABLEAU 5

Résumé des mesures examinées, Résultats estimés et coûts

Mesure	Réduction de DBO_5 kg/t. pâte	Réduction de DBO_5 tonnes/jour	Investissement millions C.S.	Coût par tonne de pâte C.S.	Coût par tonne de DBO_5 enlevé C.S.
1. Traitement externe boues activées	40	20	31,7	57,70	1440†
As	7	3,5	7,8	14,70	2100‡
2. Mesures internes					
2.1. Ecorçage à sec	3	1,5	15,0	8,95	3000
2.2. Lavage amélioré	3	1,5	5,5	3,10	1000
2.3. Epuration à chaud	6	3	5,5	5,40	900
2.4.A. Réduction d'eau au blanchiment	-	-	0,5	-	-
2.4.B.1. Echange d'ions	4,5*	2,25	24,0	31,00	6900
2.4.B.2. Blanchiment à l'oxygène	6,5*	3,25	53,0	32,00	4900
2.5. Traitement des condensats	14	7	3,0	18,40	1300

*Amélioration importante en ce qui concerne la couleur, également.

†Traitement externe seul.

‡Traitement externe combiné avec mesures internes (cas n° 2 dans "tentative d'évaluation économique").

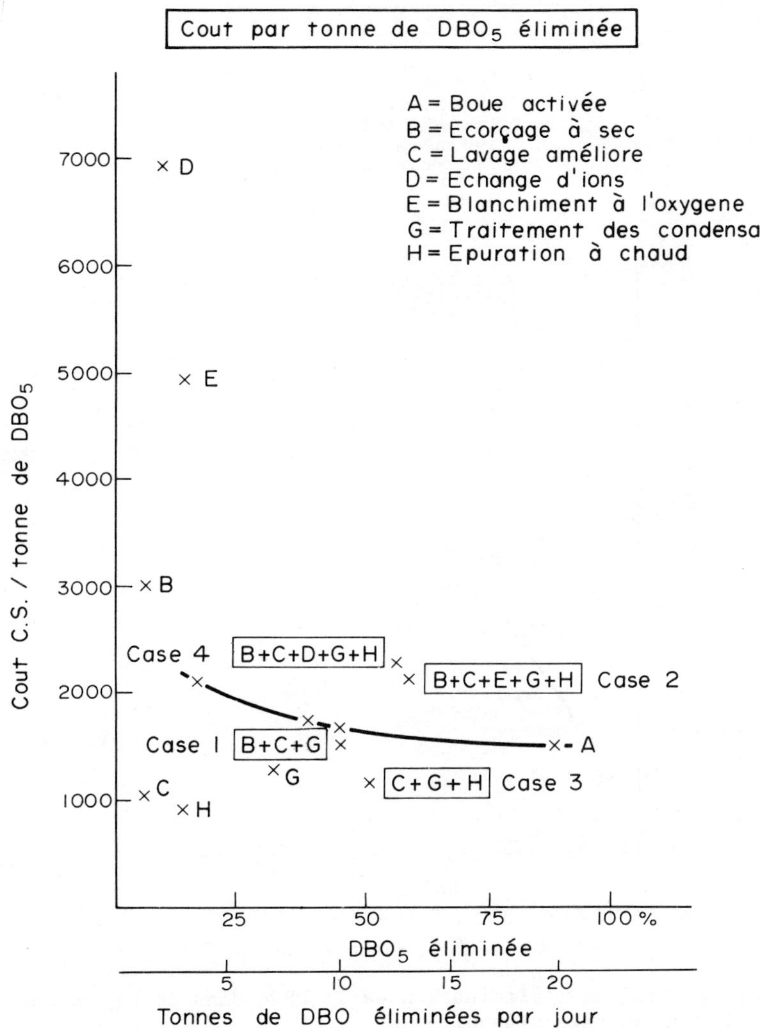

Fig. 1. Cout de l'elimination de la DBO_5 dans le cas d'une usine de pate au sulfate blanchi.

sont de l'ordre de:
C.S. 19,7 millions.

Les coûts d'exploitation seront de:
C.S. 2 800 000 par an.

Le coût total du traitement externe est ainsi de:
C.S. 5 876 000 par an, ou
C.S. 33,60 par tonne de pâte.

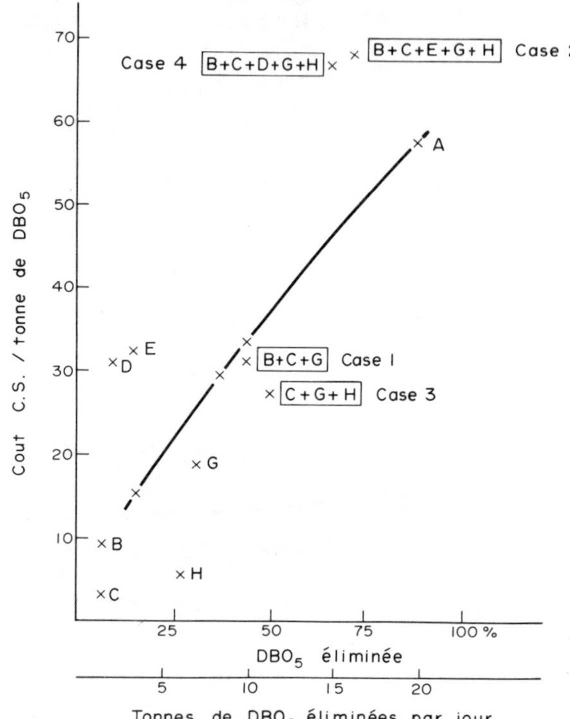

Fig. 2. Coût de l'élimination de la DBO_5 dans le cas d'une usine de pâte au sulfate blanchi.

Le coût total, pour atteindre l'objectif de 5 kg de DBO5/tonne en appliquant les mesures ci-dessus est ainsi d'environ:

C.S. 64,00 par tonne de pâte.

Il n'y aura pas d'amélioration importante en ce qui concerne la couleur de l'effluent.

Cas N° 2. En utilisant les techniques susmentionnées, en combinaison avec un système d'épuration fermé (2.2.3) et le blanchiment à l'oxygène (2.2.4.B.2), la quantité d'effluent peut être ramenée à environ 50 m^3/tonne et la *DBO_5 à environ 12 kg par tonne de pâte.* Les investissements prévus pour ces mesures sont d'environ C.S. 82,5 millions, et le coût total par tonne de pâte de C.S. 67.85.

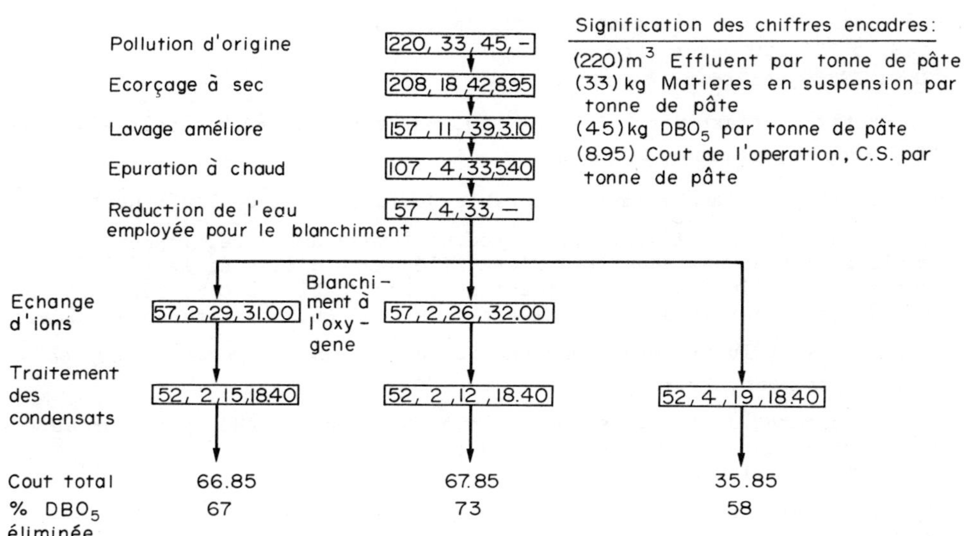

Fig. 3. Presentation schematique des mesures descrites.

Pour enlever de l'effluent de l'usine 7 autres kg de DBO_5 par tonne ou 3,5 tonnes de DBO_5 par jour, il faut compter un investissement d'environ:

C.S. 7,75 millions

pour les installations de traitement externe. Les coûts d'exploitation seraient de l'ordre de:

C.S. 1 355 000 par an

et le coût total du traitement externe de l'effluent sera donc de l'ordre de:

C.S. 2 564 000 par an, ou

C.S. 14,70 par tonne de pâte.

En conséquence, le coût total de cette variante sera environ:

C.S. 82,50 par tonne de pâte.

La couleur de l'effluent de l'usine sera fortement réduite.

Cas N° 3. Si l'on continue à employer l'écorçage par voie humide, mais en améliorant le lavage (2.2.2), en fermant le système d'épuration (2.2.3), en employant moins d'eau pour le blanchiment (2.2.4) et en traitant les condensats (2.2.5), les investissements nécessaires sont d'environ C.S. 14,5 millions et le coût total s'établirait à C.S. 26,90/tonne.

En ce cas l'effluent contiendrait *22 kg de DBO_5 par tonne,* soit *11 tonnes DBO_5/jour.*

Les investissements nécessaires pour enlever 17 kg de DBO_5 par tonne, ou 8,5 tonnes de DBO_5 par jour, au moyen d'un traitement externe, seraient approximativement de:

C.S. 17,8 millions.

Les coûts d'exploitation sont estimés à:
C.S. 2 440 000 par an.

Le coût total de ce traitement externe serait donc d'environ:
C.S. 5 220 000 par an, ou
C.S. 29,80 par tonne de pâte.

Le coûts total pour arriver à 5 kg de DBO_5 par tonne de pâte serait ainsi de:
C.S. 56,70 par tonne de pâte.

Il n'y aurait pas d'amélioration sensible de la couleur de l'effluent.

Cas No 4. En utilisant l'écorcage à sec, un lavage amélioré. l'épuration à chaud, la réduction des quantités d'eau, l'échange d'ions et le traitement des condensats, l'investissement total peut être estimé à C.S. 53,5 millions et le coût total serait approximativement de C.S. 66,85 par tonne de pâte.

Dans ce cas, l'effluent aurait une charge en *DBO_5 de 15 kg par tonne, soit 7,5 tonnes par jour.*

Les investissement nécessaires pour enlever 5 tonnes de DBO_5 par jour au moyen de traitements externes seraient de:
C.S. 12,7 millions

Les frais d'exploitation seraient d'environ:
C.S. 1 900 000 par an.

Le coût total du traitement externe est donc d'environ:
C.S. 3 880 000 par an, ou
C.S. 22,20 par tonne de pâte.

Le coût total pour arriver à 5 kg de DBO_5 par tonne de pâte serait donc à peu près:
C.S. 89,000 par tonne de pâte.

La couleur de l'effluent de l'usine *sera fortement réduite.*

QUELQUES TECHNIQUES NOUVELLES

L'étude de cas qui vient d'être présentée est un exemple de certaines techniques, et leur coût d'application, capables de réduire les déchets provenant de l'industrie papetière.

Il convient cependant de garder à l'esprit que, si ces techniques ont pour effet de réduire les quantités de matières en suspension et de substances dissoutes dans l'effluent de l'usine, beaucoup d'entre elles nécessitent une utilisation accrue d'autres ressources, notamment les ressources énergétiques.

Lorsque l'on examine les technologies sans déchets dans l'industrie papetière il convient de noter que certains développements importants intervenus récemment ont également pour effet de réduire les quantités de déchets, soit (a) en trouvant des utilisations aux déchets, soit (b) en augmentant le rendement en matière de produits finis.

(a) Les procedés industriels nécessitent de l'énergie, très souvent sous forme de températures élevées. Une grande partie de cette énergie est dissipée avec les eaux de refroidissement, les réfrigérants et les fumées; elle se transforme en chaleur perdue. Les déchets d'un procédé peuvent cependant être valorisés ailleurs; c'est le cas également des rejets thermiques.

Une étude réalisée récemment en Suède a montré que, en utilisant des techniques éprouvées, une grande quantité de chaleur perdue pourrait être récupérée; ceci s'applique notamment à l'industrie des pâtes et papiers.

Cette étude a montré que l'industrie papetière suédoise dispose de 24 unités de production (sur un total de 100) qui pourraient fournir de la chaleur à un système voisin de chauffage urbain, dans des conditions intéressantes. Les investissements nécessaires pour cette récupération de chaleur sont estimés à 69 millions de Couronnes suédoises (soit US $16 millions), résultant en une économie annuelle de 70 000 m^3 de mazout. Au prix de 400 Couronnes suédoises par m^3 de mazout, et en utilisant dans les calculs les coûts normaux d'exploitation, l'amortissement de ces investissements serait réalisé sur trois ans environ.

(b) Tout récemment encore le papier-journal standard avait un poids de 52 g/m^2, et était constitué de 20% de pâte au bisulfite non blanchie et de 80% de pâte mécanique. Dans ces conditions, 1 kg de bois (poids sec) produisait 16,7 m^2 de papier-journal.

Le poids a maintenant été abaissé à 48, 8 g/m^2, et on a montré que ce papier-journal pouvait être fabriqué en utilisant uniquement de la pâte thermomécanique (au moins pour ce qui concerne les bois résineux scandinaves). Un kilo de bois (poids sec) peut donc aujourd'hui produire 19,7 m^2 de papier-journal.

En d'autres termes, avec les techniques anciennes un journal de 64 pages de dimensions standards (A_2) nécessitait 0,479 kilos de bois; aujourd'hui, le même journal ne consomme que 0,406 kilos de bois. Dans le cas d'un journal tirant à 100 000 exemplaires, cette économie atteint 2555 tonnes de bois (poids sec) ou 6500 m^3 (volume plein, sans écorce) par an pour 350 numéros.

Ces données seraient évidemment modifiées si on considérait également la récupération des vieux papiers.

Displacement Bleaching

Johan Gullichsen

Arhippainen, Gullichsen & Co., Engineers and Consultants, Helsinki, Finland

INTRODUCTION

About 100 million tons of chemical wood pulp is produced in the world today. The quantity is forecasted to increase to 160 million tons before 1985.

More than 70 million tons is produced by the kraft process, of which an increasing portion is bleached.

About 48-52 per cent of the wood is dissolved in the kraft cook. Ninety-nine per cent of the dissolved wood substance and the chemicals used can be recovered and burned after evaporation. The recovered energy covers practically all the energy required by the manufacturing process.

The unbleached product cannot as such be used for bright paper qualities, but has to be bleached further.

Since roughly 5-10 per cent of the unbleached product is dissolved during bleaching it is understandable that bleaching causes a considerable stream pollution if the effluents containing the dissolved organic substance plus the chemicals used are sewered without treatment.

According to present pollution regulations, bleach plant effluents cannot be sewered without treatment. The treatment is both complicated and expensive due to the unfavourable conditions of high quantities and low concentrations of dissolved compounds.

THE CONVENTIONAL TECHNIQUE

Figure 1 shows a flowsheet of the present conventional technique developed in the 1930s.

The pulp is treated in five to six consecutive stages with oxidant and caustic soda. Chlorine, hypochlorite and chlorine dioxide are the most common oxidants used. The pulp is washed with water on a drum filter after each of the treatments and pumped after chemicals addition to the following stage. The operational principle of a drum filter includes an internal liquid circulation which amounts to roughly 100 times the pulp throughput. When this and the pumping energy needed to transport the pulp from stage to stage is considered, the fairly high electric energy requirements in conventional bleaching are understandable.

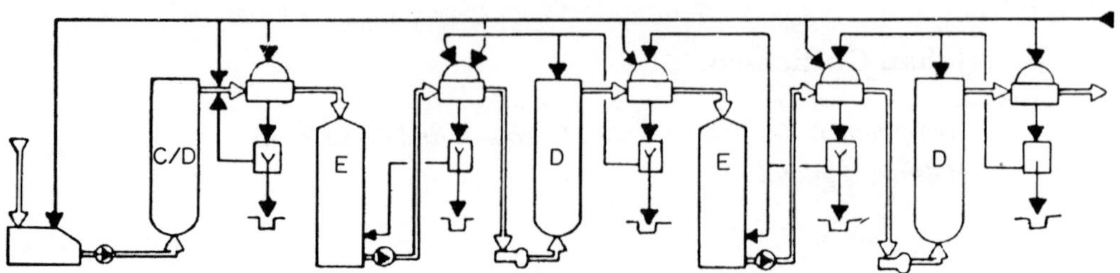

Conventional bleaching, sequence CEDED

Fig. 1.

Because of the long retention times needed (1-4 hours) for each of the treatments and the current consistency levels (10-15 per cent), the reaction towers required are huge.

The corrosiveness of the chlorine chemicals used demands that the equipment is built out of expensive acid-proof material as special-quality stainless steel, hastelloy C or titanium.

Both the operating and capital costs are thus high in bleaching.

The price difference between bleached and unbleached pulp does not generally cover the excess cost of bleaching and bleaching has in fact for the last 10 years been justified with marketing reasons.

The present stream pollution regulations calls for a considerable reduction, preferably complete elimination of bleach plant effluents. This is basically possible to achieve by closing the liquid circuits and through application of the counter-current washing principle. The problem is, however, that corrosive compounds as acid chlorides will accumulate in the circuits to a level higher in order of magnitude than presently, which means that the capital expenditures of bleaching will increase considerably since more expensive corrosion resistant materials has to be used.

THE NEW TECHNOLOGY

The aforementioned factors plus the increasing costs of energy calls for a new technique which features reduced use of materials of construction, reduced space requirements as well as reduced effluent volumes and energy demand.

A positive development took place already in the early 1960s, when Kamyr Ab of Sweden developed the continuous diffuser.

The continuous diffuser is a displacement washing device which can be placed in the bleaching tower and operates without the dilution circuit typical for the drum washer. Since the dilution circuit and the transportation of pulp to the washer was eliminated a substantial saving in pumping energy was achieved. The basic difference in water use between a drum washer and diffuser system is demonstrated in Fig. 2.

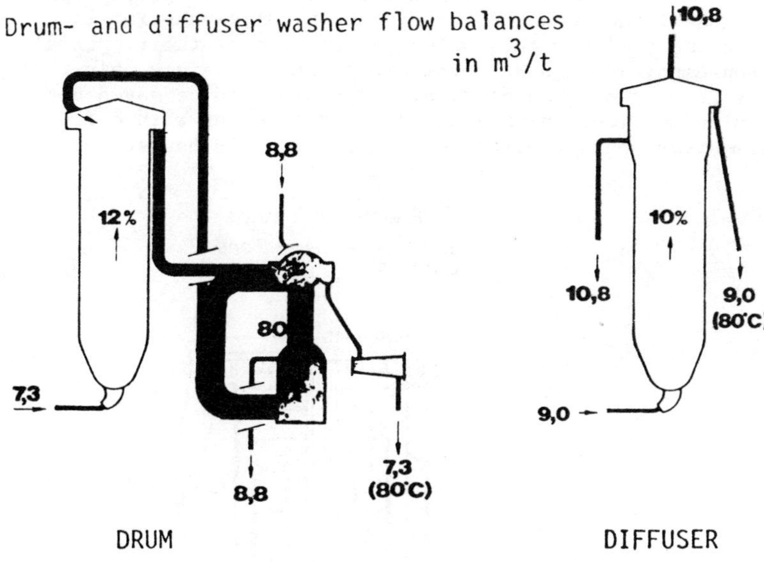

Fig. 2.

The basic process with huge retention towers remained, however, the same and the savings in space and materials use were minor.

DISPLACEMENT BLEACHING

Professor Howard Rapson of the University of Toronto, Canada, made an interesting discovery in 1965. He observed that if bleaching agents are displaced through a pulp mat, instead of mixed into the pulp, very rapid bleaching can be performed.

Considerable research to evaluate the benefits of the method was started in a number of laboratories throughout the world.

It was concluded that the high bleaching rates were achieved through continuous displacement of the rate controlling reaction product diffusion layer formed around the fibre and that maximum active chemicals concentration was maintained at the fibre surface throughout the whole reaction period.

The method was called dynamic bleaching. Since no equipment to meet the requirements of the new method was available at the time the new process remained a theoretical curiosity.

In the year 1970 the idea to combine the continuous diffuser with the principles of dynamic bleaching was born. Since it was proven that a bleaching stage could be performed in 5-10 minutes it was felt that several displacement stages of the diffuser type could be built in the same tower and that the whole multistage process could be carried out in a tower smaller in size than any one of the conventional bleaching stages. If the liquid circuits could be completely closed, the demand for active chemical exhaustion in a stage could be eliminated and the chemical concentrations set at their optimal levels. The power consumption would be considerably reduced since pulp transportation from tower to tower would be eliminated. Heat would be saved both through the closed circuits applied, which means less heat losses with effluents and because intermediary contact with the atmosphere would be reduced.

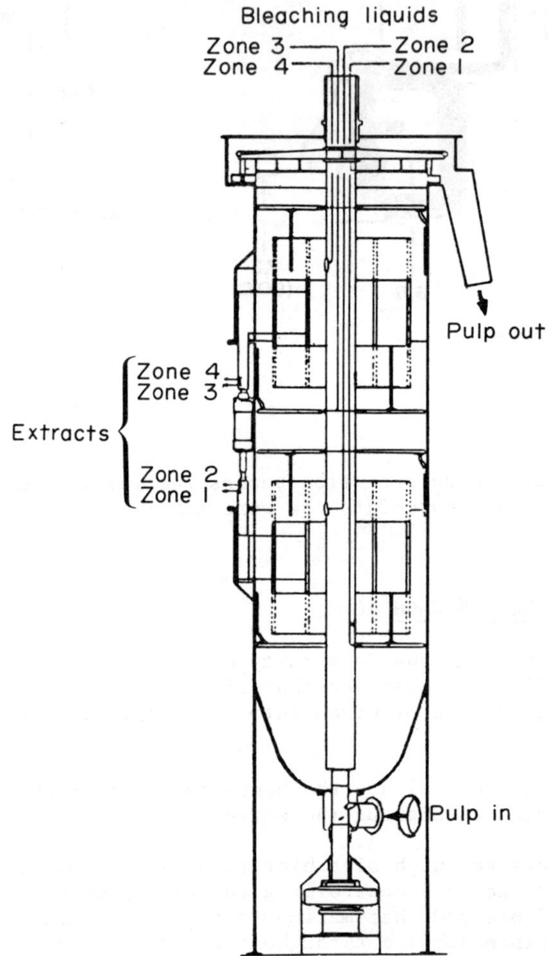

Fig. 3.

Preliminary tests results generated in 1970 with a small pilot plant installation were so convincing that Makyr Ab decided to build a four-stage pilot plant with a capacity of 120 tons per day.

Ths Scandinavian research and development funds SITRA in Finland, STU in Sweden and Utvecklingsfondet in Norway agreed to support the project mutually with partial financing.

The laboratory work was commenced in 1971 and the pilot plant started up in early 1972. The pilot plant was built as a by-pass production line to Kymmene Ab´s birch kraft bleach plant at Kuusanniemi Kraft Pulp Mill in Kuusankoski, Finland.

The assembly drawing in Fig. 3 shows the diffuser arrangement in the four-stage pilot plant.

The process flowsheets of Fig. 4 demonstrate bleaching sequences tested on chlorinated and semibleached birch kraft pulp.

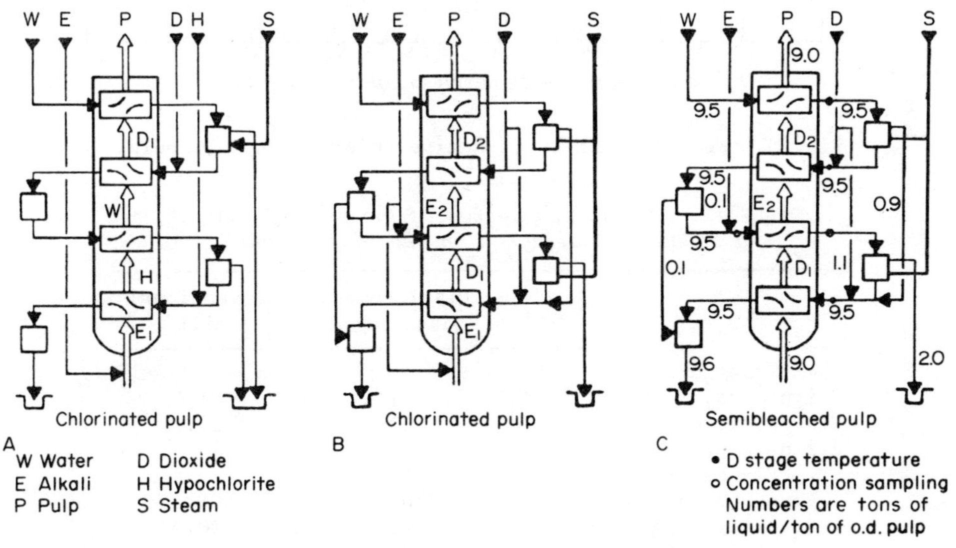

Fig. 4.

Figure 5 summarizes the main testing results in terms of pulp brightness versus net chemical consumption in comparison to what the mill is achieving with their conventional process. Corresponding strength property data are presented in Table 1.

The results show that the displacement bleaching process gives a product with properties that compare favourably with conventionally bleached pulp.

A comparison based on the results, process simulation and design as shown in Table 2 demonstrates the benefits of displacement bleaching.

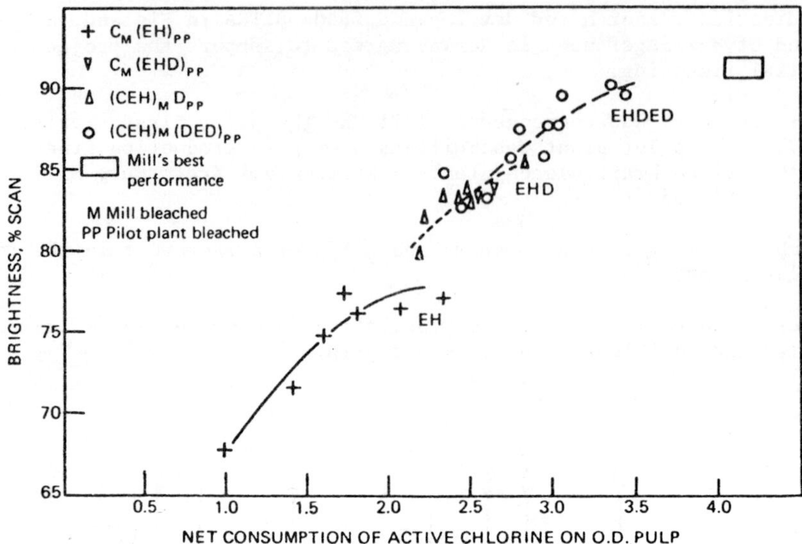

Fig. 5. Net consumption of active chlorine on O.D. pulp.

TABLE 1 *Typical properties of birch kraft pulps displacement bleached in the pilot plant**

	Displacement pilot plant	Conventional mill
Semibleached		
Brightness, % SCAN	71.0	71.0
Viscosity, cp	22.5	22.0
Strength properties PFI-beater		
Revs. to 35° SR	3100	3300
Breaking length, km	10.4	10.0
Burst factor	79	80
Tear factor	76	69
Bulk, cm^3/g	1.18	1.15
Fully bleached		
Brightness, % SCAN	89.4	91.5
Viscosity, cp	26	22
Strength properties PFI-beater		
Revs. to 35° SR	3400	3700
Breaking length, km	10.2	9.4
Burst factor	75	75
Tear factor	82	76
Bulk, cm^3/g	1.20	1.20

*Brightness was measured with the Elrepho instrument. Viscosity was measured with standard method SCAN C 15 and converted to Tappi T 230, 0.5 per cent CED viscosity.

TABLE 2

	Bleaching method		
	Displacement	Conventional	
		1960s	1970s
Relative space requirement	0.25	1	1
Water consumption, m^3/t	15	100	40
Heat consumption, Mcal/t	300	900	720
Electric energy demand, kWh/t	65	150	150

This development was achieved by the end of 1973 and Kamyr started to look for customers willing to buy full-scale prototypes. Three installations have been sold so far:

Eastex Inc., Evadale, Texas, USA,
Weyerhauser Company, Plymouth, North Carolina, USA,
Kymin Osakeyhtiö, Kuusankoski, Finland.

The sequencies applied, the capacities and start-up dates are shown in Table 3.

TABLE 3

Customer	Sequence*	Capacity, t/d	Start-up year
Eastex	(C)EDED	500	1975
Weyerhauser	(C)EDED	500	1976
Kymin	(CEH)DED	170	1977

*The stages in brackets are conventional.

The first installation did start up in November 1975. The experience so far is very positive and the expectations have been met.

The flowsheet in Fig. 6 representing the Eastex installation shows how a displacement bleach plant can be combined with a conventional chlorination stage.

SPIN-OFF DEVELOPMENT

The pilot-plant activities have been continued until late 1975. The work has been concentrated on evaluation of a number of spin-off ideas. Such are:

1. Multistage counter-current washing in one tower.

2. Pulp sulphonation in conjunction with brownstock washing.
3. Capacity escallation by using a twin diffuser as a single stage.
4. High-density chlorination.

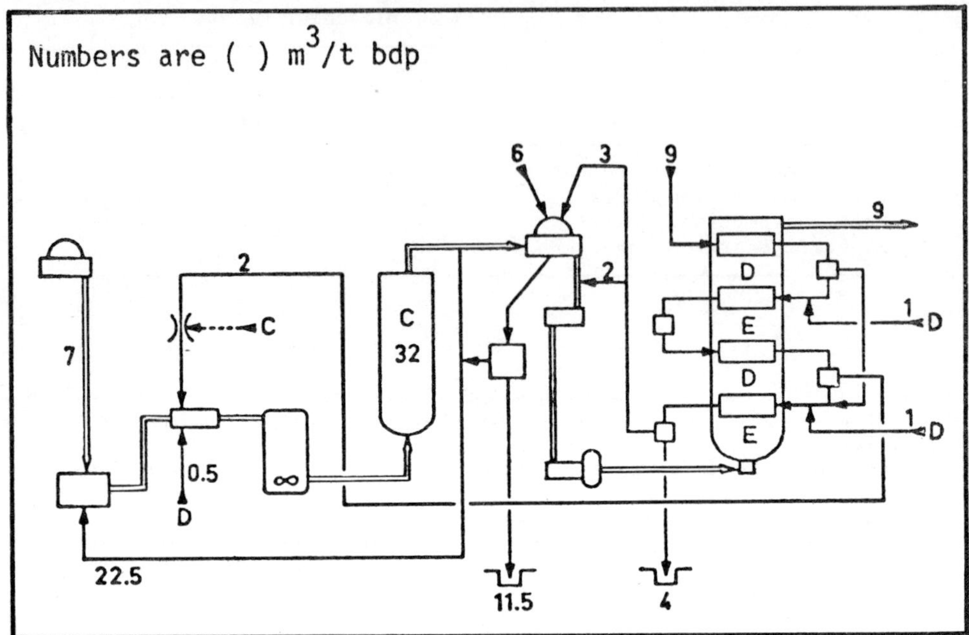

Fig. 6.

Out of these 1, 3 and 4 have already reached commercial realization. The Eastex bleach plant is preceded by a multistage brownstock washing tower. Kymin Osakeyhtiö has converted the four-stage pilot-plant to a two-stage brownstock washing unit and the first commercial high-density chlorination installation will start up in 1977.

The most interesting spin-off development is definitely the high-density chlorination since it is a logical complement to displacement bleaching. Figure 6 shows how much equipment is needed to carry out a conventional chlorination. Conventional chlorination is traditionally carried out at a low pulp consistency (3-4 per cent) while all other stages including displacement bleaching operate at 10 per cent or higher consistencies.

A pulp suspension of 10 per cent consistency is not a stable visco-elastic body. The suspension is fluidized when it is exposed to a power forceful enough to break the fibre network. The process is reversible which means that the stable network will re-form when at rest.

Gas can be dispersed into such a fluidized suspension and the portion non-soluble in the liquid phase is trapped within the fibre network as small bubbles and dissolved into the liquid at a rate corresponding to the consumption in the liquidphase.

The higher consistency makes temperature control economically feasible.

A set of two specially designed mixers which utilize the fluidization-dispersion principle has been tested at the pilot-plant.

The results not only demonstrate the benefits of the principle but also proved that the chlorination equipment becomes much simpler and that good process control is easier to achieve with this principle than in conventional processes. Figure 7 shows a flowsheet of a future displacement bleach plant including high-density chlorination. The flow balance shows that further reductions in effluent volumes can be achieved; from 15 m^3/ton of pulp as in Fig. 6 to about 10 m^3/ton of pulp in Fig. 7. This quantity is small enough to meet, as far as bleaching is concerned, the requirements of the "effluent-free pulp mill" recently discussed in the literature.

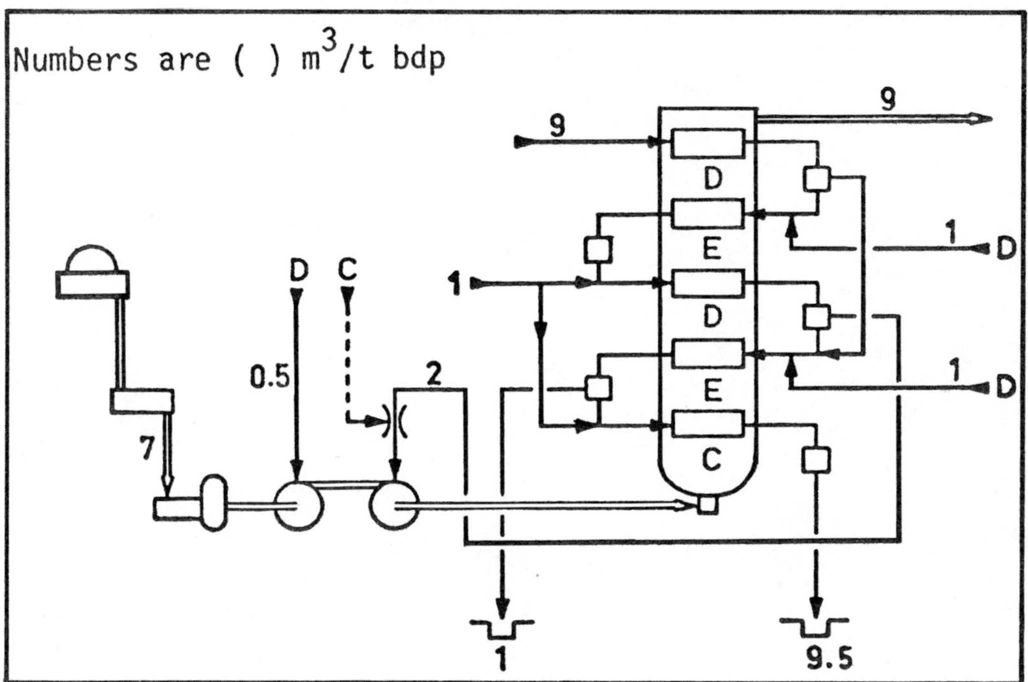

Fig. 7.

LITERATURE CITED

1. W.H. Rapson and A.B. Andersson, *Tappi*, 49 (8): 329 (1966).
2. International Paper Company, personal communication (1971).
3. I. Palenius and T. Laxén. The Finnish Pulp and Paper Research Institute, personal communication (1971-1973).
4. J. Gullichsen, *Tappi*, 56 (11):78 (1973).
5. J. Gullichsen, *Tappi*, 57 (10):111 (1974).
6. W.H. Rapson, *Pulp Paper Can.* 68, No. 12, T635 (Dec. 1967).
7. W.H. Rapson and D.W. Reeve, *Tappi*, 56, No. 9, 112-115 (Sept. 1973).

Biological Method for Purifying Kraft Pulp Mill Condensates

Ilpo Vettenranta

Enso-Gutzeit Osakeyhtiö, Paper Division, Ihatra, Finland

The main characteristics of kraft pulp effluent are suspended solids, biological oxygen demand (BOD) and the toxic effects and bad odours caused by organic sulphur compounds.

Suspended solids are reduced by sedimentation in mechanical raked clarifiers. BOD-load is reduced by biological methods, i.e. activated sludge, trickling filter, or aerated lagoons.

Steam stripping is the conventional method for purifying the poisonous malodorous gases and condensates. However, this method has the drawback of high operation costs. For example, an average kraft mill producing 100,000 tons of pulp yearly would have to spend 1 million Finnmarks (Fmk) for the steam energy required to purify its condensates by the steam stripping method.

The greatest part of the bad odours and toxic sulphur compounds of a kraft pulp mill effluent is concentrated in the condensates from the evaporating plant and the digester room. When they find their way into the water courses, these condensates have an adverse effect on all aquatic life. However, the most noticeable effects appear in the fish which are higher up in the food chain. In addition, the condensates give the water a noticeably bad odour.

The bad odours of the gases and condensates are caused chiefly by hydrogen sulphide and the organic sulphur compounds methylmercaptan, dimethylsulphide, and dimethyldisulphide.

DEVELOPMENT OF ENSO-BIOX

In 1970, when Enso-Gutzeit's Research Centre began to study the possibilities of treating condensates biologically, Dr. Onni Koistinen, the inventor of the Enso-Biox method, had the following views:

The malodorous sulphur compounds are the same compounds which exist naturally in all decaying processes.

Nature has its own biological methods for eliminating odours.

Centralizing, intensifying and controlling these natural biological processes should be the goal of the research and development work.

The laboratory units

For the first trials in 1970, a biological filter of 25 litres capacity was built. The filtering medium for this filter was some softwood bark which had been pretreated with the nutrients necessary for biological activity. The condensate to be treated was allowed to trickle down through this medium while air was blown up through the medium countercurrent to the liquid flow.

When planning the first trials, Dr. Koistinen had a theory that the bark material was able to bind the nutrients so that they would not be washed out of the filter during the filtration. The results of the trials appeared to support the theory. There was no need to add nutrients continuously into the incoming effluent. Optimum conditions were found for a strain of bacteria which not only tolerates but requires toxic sulphur compounds for its growth. On 7 December 1970 we applied for a patent for the process. Results of laboratory unit 1970 are shown in Table 1.

TABLE 1

Enso-biox purifying effect (patent application 1970)

	Feeding (mg/l)	Passed through (mg/l)	Purifying (%)
H_2S	7.7	0	100
CH_3SH	20.8	0.5	97.6
$(CH_3)_2S$	28.0	0.2	99.3
$(CH_3)_2S_2$	37.6	0	100
BHT_5	1334	148	89

The research group

A work group led by Dr. Onni Koistinen, a biochemist, began to study more about the biological filter and how it might be used in practice to purify condensates. During the course of the study, several persons made research for the group at different times. Contributions towards the research have been made by a microbiologist, by a student preparing his thesis and by a biochemist. Enso-Gutzeit's Research Centre has provided modern equipment and back-up support for the research work.

The pilot plant studies

The laboratory size unit was used to determine the loading conditions required to give optimum cleaning results. Together with the laboratory scale studies some pilot plant studies were made at our Tainionkoski mill.

A rather large number of interdependent parameters were considered for the pilot plant studies. However, because the range of variation for each parameter which could be investigated was large, it was realized that a full investigation could not be completed in a very short time. In fact, a study of all the possible parameters would take a year or more to complete. Pilot-plant studies were made also at our Uimaharju mills. The results of the studies are shown in Tables 2 and 3.

TABLE 2

Enso-biox purifying effect. Pilot-plant equipment, Tainionkoski

	Feeding ($g\ S/m^3$)	Purifying (%)
H_2S	444	100
CH_3SH	791	100
$(CH_3)_2S$	7.3	97
$(CH_3)_2S_2$	31.3	66
Smelling sulphur compounds total	1274.0	99
MeOH		31
Acetone		53

TABLE 3

Enso-biox purifying effect. Pilot-plant equipment, Uimaharju

	Feeding ($g\ S/m^3$)	Purifying (%)
H_2S	116.7 - 208.0	95 - 100
CH_3SH	86.9 - 292.0	90 - 98
$(CH_3)_2S$	2.5 - 8.2	36 - 88
$(CH_3)_2S_2$	47.2 - 126.4	27 - 91
Smelling sulphur compounds total	1250 - 1710	88 - 98
MeOH		15 - 50
Acetone		42 - 91

During the research work, it was found that the biochemistry of organic sulphur compounds was nearly an undeveloped area of science. However, some graduate-level work on the biofilter has been done in the Department of General Microbiology at the University of Helsinki. During their investigations, researchers at the university isolated some of the micro-organisms which effectively oxidize reduced sulphur compounds to harmless forms. The researchers determined that these micro-organisms were members of a special group of sulphur bacteria which can be found living in soil, ditches, mud and turf. After isolating and identifying the bacteria, the researchers studied and determined the optimum conditions for the use of the bacteria with bark as a filter medium in a biofilter.

A decision to build a mill-size unit at Uimaharju

In order to study methods for constructing the biofilter, it was decided to build a trial mill size unit at Uimaharju. Uimaharju was chosen as the site for the first mill size unit for two reasons. The first reason was a matter of scale. Because the Uimaharju kraft pulp mill was a new mill using the continuous digesting method, the amount of condensates to be handled was reasonable. The second reason concerns the plans made by the National Board of Waters for purifying Uimaharju mill's effluents. The Board requested that facilities for purifying Uimaharju mill's condensates be installed before the summer of 1976.

The planning and construction work for the Uimaharju biofilter started in autumn 1974 and was finished in autumn 1975. The Uimaharju filter consists of two different units, a liquid filter and a gas filter. Both units use a filtering medium of softwood bark specially treated with a nutrient mixture which is required to support the continuous biological process. The liquid unit at Uimaharju has been divided into two parallel filters each of which is 14 metres in diameter and 200 m^3 in volume. After being purified in the liquid units, the condensates are clean and safe enough to be released into the water system. Because the air which is blown through the liquid units becomes contaminated in the process, it must be purified in the gas filter unit before being released into the atmosphere. This gas filter is similar in construction to the liquid filter but is a little larger with a diameter of 20 metres. The results from the operation of this mill size unit have agreed very well with our expectations (Table 4).

TABLE 4

Enso-biox purifying effect. Mill plant equipment, Uimaharju

	Feeding ($g\ S/m^3$)	Purifying (%)
H_2S	160.6	100
CH_3SH	338.6	100
$(CH_3)_2S$	9.8	77
$(CH_3)_2S_2$	66.6	41
Smelling sulphur compounds total	575.6	93
MeOH	2670	60
Acetone		67

One of the incentives for the development of the biofilter technology was the fact that the earlier known steam stripping methods required much energy for purifying the condensates. Because the results depend upon the conditions at each mill, a direct comparison between the biofilter and the steam stripping methods cannot be made in general. However, in the case of the Uimaharju mill study, it has been estimated that the costs of constructing a steam stripping unit would have been 1.5 to 2 times the cost of building the biofilter unit. Furthermore, it would have cost 10 times more to operate a steam stripping unit than it costs to operate the biofilter unit.

We intend to continue to do development work with the mill scale biofilter at Uimaharju in order to broaden the application for the filter.

Packaging Alternatives for Wine

W.P. Fornerod

Institute TNO for Packaging Research, Delft, Holland

INTRODUCTION

The mobility of the average Dutchman has very much increased in the past decade. Specifically during the holiday periods, enormous streams of traveling Dutchmen can be seen in Holland's neighbouring countries. This mobility has had a very strong impact on the behaviour and the habits of the people. One consequence is that wine has become very popular. Estimated wine consumption per year now is 1.4×10^8 litres, meaning a *per capita* consumption per year of more than 10 litres. Of course this is not yet as high as, for instance, in France (some 70 litres) and not as high as the milk consumption in Holland (some 50 litres), but nevertheless a notable quantity.

WINE PACKAGING SITUATION IN HOLLAND

The majority of the wine is imported in barrels and bottles. With very few exceptions the wine is packed in non-returnable glass bottles. This situation is quite different from the situation, for instance, in France, where at least 20 - 30 per cent of the wine is bottled in returnables. The glass bottle consumption for wine packaging amounts in Holland to some 70,000 tons.

Several alternatives, diminishing the total tonnage, have been proposed. One very attractive alternative is a type of Blok-pack package (a paper-aluminium laminate pack). The packaging material is a paper-aluminium laminate, coated with polyethylene.

To be able to compare the non-returnable bottle with this non-returnable laminate pack, it is necessary to compare the energy/resource consumption and the environmental impacts.

RESOURCE CONSUMPTION

A typical one litre glass bottle consists of 575 g of glass. The aluminium cap weighs 1.908 g. Inside the cap is a corklayer with a polyethylene paper lamination: 0.41 g cork, 0.027 g polyethylene film and 0.058 g paper as lamination. Furthermore, a paper label is applied, weighing 1.5170 g.

The laminate-pack, consisting of a paper/aluminium/polyethylene laminate, weighs 30.6 g. The paper component (inclusive of printing ink) weighs 21.916 g. The aluminium foil weighs 2.895 g. (For this particular wine use, the aluminium foil weighs twice as much as the aluminium foil used for this type of pack for milk.)

The two-sided polyethylene coating weighs in all 5.789 g.

Table 1 gives the expected resource consumption of each of the package types, if all the consumed wine in the Netherlands were packed in either the glass bottle or the laminate-pack.

TABLE 1

Comparative resource consumption assuming the Dutch wine consumption of 1.4 x 10^8 *litres were entirely packed in non-returnable glass or completely in the laminate-pack*

	Non-returnable glass bottle (tons)	Laminate-pack (tons)
Paper	220.5	3068.24
Glass	80,500	-
PE	3.78	810.46
Cork	57.4	-
Aluminium	267.12	405.3
Total	81,048	4284

The total weight of the potential refuse quantity diminishes from approximately 81,000 tons to 4,000 tons a year, if all the wine is packed in the paper-aluminium laminate packs instead of in non-returnable glass.

The energy consumption of the two types of packages as between glass bottles and laminate-packs is as seen in Table 2.

TABLE 2

Energy consumption on the same assumption as in Table 1. Unit of energy is t.o.e. (ton oil equivalent)*

	Non-returnable glass bottle	Laminate-pack
Paper	264.6	3681.9
Glass	33,005	-
PE	7.6	1629.1
Cork	-	-
Aluminium	2003.4	3039.8
Distribution†	156.8	22.4
Total t.o.e.	35,437.4	8373.1

*Unit of energy equal to 1 ton of a standard crude oil (1 t.o.e. equals 10 Gcal).

†Energy needed to transport the weight of the packages of the packed wine to the retail shop. This figure is dependent on the distribution patterns and systems in the country.

It is clear that glass requires more than four times as much energy as the laminate-pack. It is, of course, possible to express the energy consumption in any energy unit. We choose tons oil equivalent, because it visualizes the tons of oil involved.

We can imagine that one would like to compare the energy consumption with the national energy consumption, which is expressed in the energy unit kilowatt hour (kWh). We therefore repeat Table 2 in Table 3 with energy units 1000 x kWh.

TABLE 3

Energy consumption as given in Table 2.
Unit of energy MWh (1 MWh = 1000 kWh)

	Non-returnable glass bottle	Laminate-pack
Paper	3.08	42.81
Glass	383.78	-
PE	0.088	18.94
Cork	-	-
Aluminium	23.30	35.35
Distribution	1.82	0.26
Total	412	97.4

WHAT CAN BE DONE WITH THESE DATA?

To decide what is better from the viewpoint of environmental protection, we refer in the first instance to Table 1.

The total household refuse is of the order of 3×10^6 tons a year in Holland.* That means in Holland a saving of some 2.6 per cent (in weight) can be reached. As the disposal of refuse costs Dfl. 250,000,000 (US $90,000,000)* a year, the community can save Dfl. 6,400,000 (US $2,300,000) per annum, if ordinary wine is packed in a laminate-pack type of package.

Table 3 reveals that Holland can save some 315,000 kWh of energy consumption per annum, if it changes to a laminate-pack type of package for wine. As the total per annum consumption is 7×10^5 MWh, this means a saving of 0.5 per cent.

And lastly, the pollution of water and air by the production of the packaging material. Going back to Table 1 one can obtain general comparative figures by setting the pollution of the production of 80,000 tons of glass against the pollution of the production of 3000 tons of paper plus 800 tons of polyethylene.

*Figures SVA, Amersfoort.

Unfortunately, the emission data up till now are not sufficiently consistent to produce information without omissions.

Generally it can be stated that glass production and polyethylene production are rather "clean". The production of paper gives rise to some problems at the pulp-making stage. This pollution is a water pollution problem which can be rather easily (but expensively) removed at the fabrication plants.

Other factors to be considered before deciding to change the package

The consumer/housewife nowadays is motivated to help in preserving the environment. Therefore it is highly likely that due to this attitude, many used bottles will be disposed of separately to facilitate recycling. This may save resources and diminish the amount of glass refuse. Recycling of the paper-aluminium laminate, on the other hand, is still not possible, so that all the used packs would be part of the household refuse.

Another way to diminish waste is the change to returnable bottles. In this respect, wine has the advantage, that its used bottles are much easier to clean, for instance than used milk packages or used paint packages. It must, however, be emphasized that such a change is not merely a technical problem. The distribution of wine in Holland is completely adjusted to non-returnables. Hence, wine in returnables will involve a totally structural change of the distribution pattern and that much will be required of the housewife in effort and energy.

Another important aspect is of a marketing nature. Before deciding on a non-glass package for wine, it is necessary to know whether it will be accepted. In Holland the consumer prefers even ordinary wine in glass bottles. Pinard-wine is sold in glass bottles and in laminate-packs in self-service shops. The consumer appeared to prefer the glass-packed wine, even though this is slightly higher priced than the same wine in the laminate pack.

Moreover, it must be taken into account that not all wine varieties permit non-glass packs. It is difficult to visualize a wine cellar with highly-prized varieties in laminate packs.

CONCLUSION

As this case study is a comparison between two types of packaging it does not suggest that one of the two packages is *the* environmental packaging solution for wine, because other alternatives (returnable bottles, non-returnable PVC., etc.) must as well be studied before a final decision can be taken. Although the environmental impact of packaging appeared to be a very minor part of the total environmental impact caused by society, it does *not* mean that efforts to decrease the impact are not useful.

If one considers the impacts of separate activities of society, it will appear that all these separate activities have minor impacts. If, however, an effort is made to lessen the impact of each activity, the total saving is considerable. Hence, it is advisable not to postpone measures to lessen such impacts, if the reason is *only* that the particular activity (makes a very small contribution to the overall impact.

ANNEX

For the calculation, the following assumptions are made:

1. *Conversion table for different units of energy*
   1 t.o.e. (ton oil equivalent) = 10 Gcal = (10^7 kcal) = 11,76 kWh.

These conversions are only valid in comparing energy *contents*. In *generating* electricity, much more than one ton of oil is necessary to produce 11.76 kWh because of the losses in energy in the production.

2. Energy consumption of packaging material, including resource extraction, production, transportation of the raw materials.

1 ton PE	= 2.01 t.o.e.	= 23.4 x 10^3 kWh
1 ton aluminium	= 7.5 "	= 87.2 "
1 ton glass bottles	= 0.41 "	= 4.77 "
1 ton tinplate	= 1.2 "	= 13.95 "
1 ton paper	= 1.2 "	= 13.95 "

3. 1000 litres gas/diesel oil ≃ 0.8 t.o.e. ≃ 9.4 x 10^3 kWh.

The Recovery of Glass in Switzerland

Yves Maystre

Director of the Institute of Environmental Sciences of the Federal Polytechnic, Lausanne, Switzerland

From the point of view of the protection of the environment against pollution and degradation, a system of production without waste holds no advantage if it automatically entails a system of consumption with a high level of waste.

The glass industry (bottles, flasks, etc.) is given here as an example to illustrate this point. Figure 1 represents a glass-factory in this system.

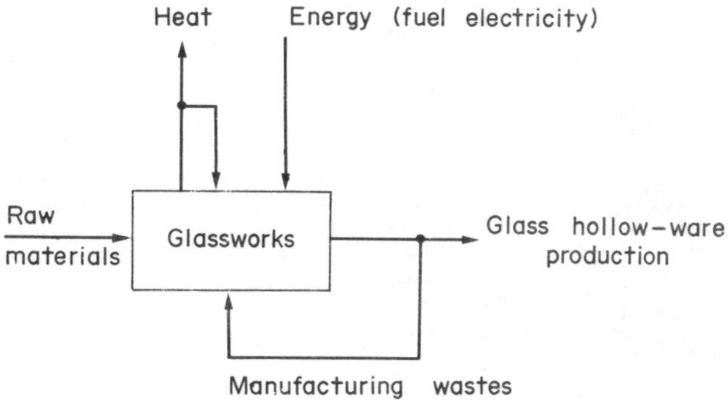

Fig. 1

In order to optimize this so-called "system of production without waste", it is necessary to obtain the highest possible rate of recirculation from the wastes into production and in energy.

The recirculation of the wastes caused during production sets neither a technical nor an economic problem: the remelting of glass in the glass-factory's furnace is easily executed and the energy consumed in this process is only a fraction of that consumed in melting the equivalent quantity of raw materials.

The profitability of the glass-works is the objective of the owners, managers of this industry: however, this objective holds no interest from the point of view to the protection of the environment, as the production of "non-returnable" glass entails a waste of resources and an ever-increasing source of pollution. (In the case of glass, the squandering of resources outweighs the rate of pollution, but the contrary is true for a large number of artificial products.)

In considering the system represented in Fig. 2, one can see that it includes production and consumption and that, in consequence, the objective would be here to minimize the "net production" of the system, instead of to maximize the production as is the case for the glass-factory.

In minimizing the net production, one equally minimizes the consumption of resources, so as to find the most interesting technology from the point of view of the environment. In Fig. 2 the cycles of recirculation of Fig. 1 are included in the box "glass-factory".

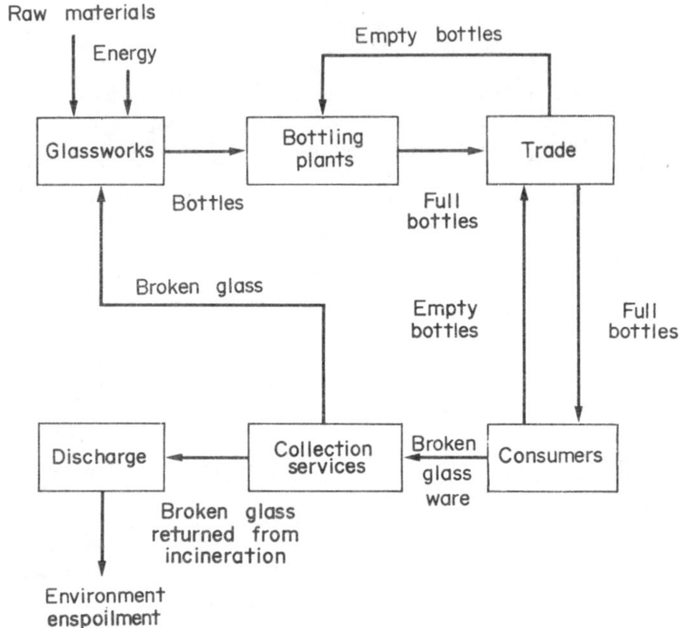

Fig. 2

There are two possible cycles for recirculation:

1. The recirculation of cullet: this permits the glass-factory to retain its own volume of production.

2. The recirculation of the bottles. This requires the creation of a new industrial branch - the bottle cleaners: it will reduce the production output of the glass-factory, it requires the transport of empty bottles by the retailers, the consumers returning their empty bottles to the retailers when buying full bottles.

Figure 3 presents the quantity of glass bottles which would have been found in Switzerland in 1974 in the two extreme considered cases and disregarding imports and exports. Allowing 8 uses a year, on an average for returnable bottles, the production capacity of the glass-factory would be fifteen times less in the case of recirculated bottles than that of the recirculation of cullet. The reality lies between these two extremes, as illustrated in Fig. 4. The figures for production, elimination and recirculation of glass in

Switzerland for 1974 are presented in this figure.

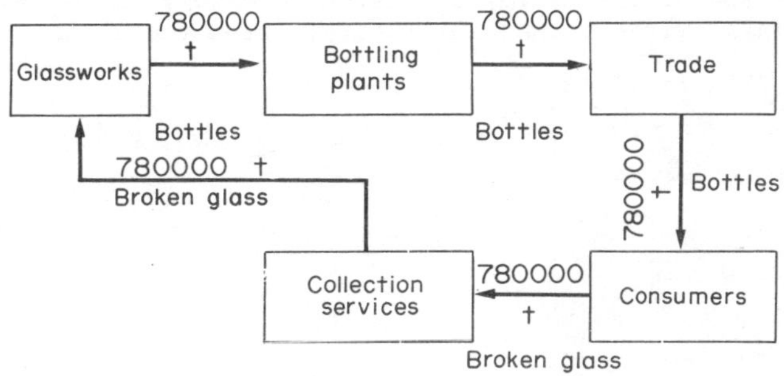

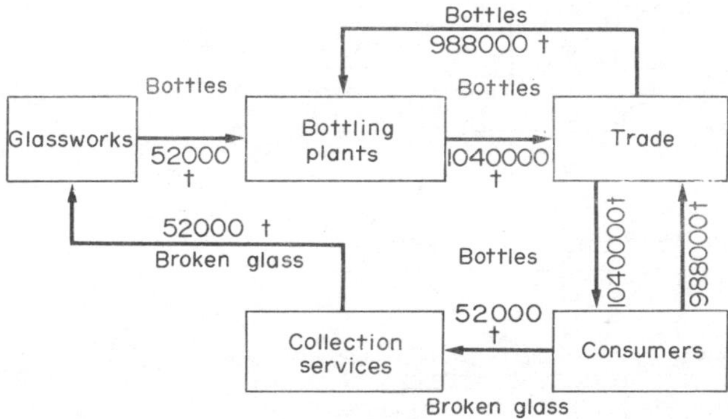

Fig. 3.

Eighty per cent of the glass produced is eliminated with the household refuse. The recycling flow between bottlers and the glass-factories represents 16 per cent of the production, while the recovery through municipal collection represents 4.5 per cent of the production. Given the unsteady state of the system at the considered time, certain figures cannot be given with more precision than $\pm$ 10 per cent.

The optimization of the "production and consumption" system for glass can be based on different criteria, which in turn can lead to very different results. From the environmental point of view the main parameter to consider is energy, an expensive resource and the generator of thermical pollution and of SO_2 and NO_x. In comparison, neither the raw materials of glass manufacturing nor the wastes from glass are parameters of importance.

Among all the operations of transport, stocking, manufacture and treatment, the fusion of the glass is by far the most important consumer of energy.

Figures 5 and 6 give the theoretical basis for the calculation of energy requirements. In Fig. 5 the results obtained for the energy of fusion are compared for the case of green-glass, with the results obtained from an empirical formula established by a French author, Mr. Levêque. It can be noted that the compatibility is good between this empirical formula and our thermodynamic calculation of the energy of fusion necessary in the manufacture of 1 ton of green-glass, without the heat recovery from the gases.

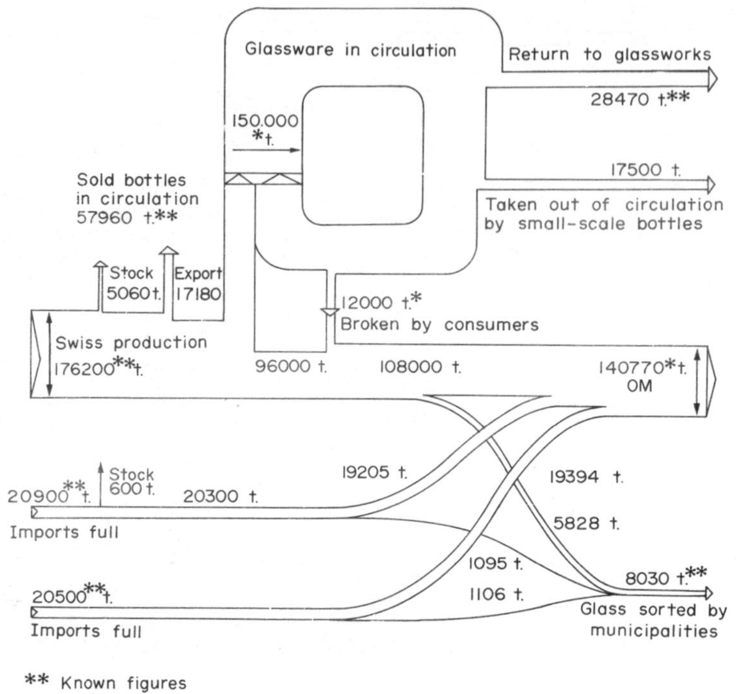

Fig. 4.

Figure 6 presents the results of the calculation for the theoretical energy of fusion necessary in the fusion of three common qualities of glass made in Switzerland from the most recent thermodynamic data. This graph indicates the quantity of energy needed for the fusion of 1 ton of glass for different percentage of cullet introduced as raw material in the furnace, with heat recovery from the gases.

In reality, the efficiency of a glass-factory furnace plays a major part in the actual calculation of energy. This efficiency has varied considerably in the last few years.

Figure 7 shows the evolution of the energy necessary for manufacturing 1 ton of glass in relation to the efficiency of fusion furnaces, as shown by different authors. The heterogeneity of the figures concerning the recycling of glass, which appears in the literature, is due, in a large part, to the differences in furnace efficiency adopted by the various authors. These results are presented simultaneously with the probable evolution of the efficiency up to 1980. By then, the efficiency will probably exceed 40 per cent. The furnace mentioned in our study is the one of St-Prex which already has an interesting efficiency.

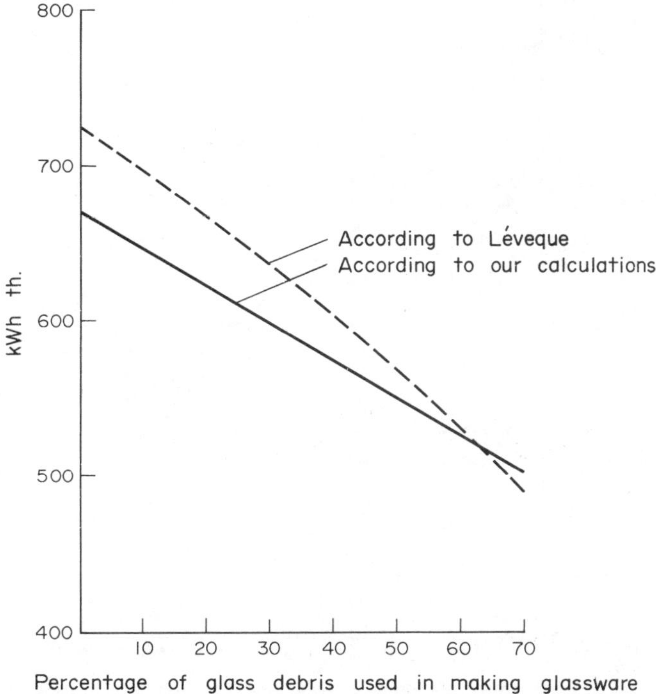

Fig. 5.

Figure 8 illustrates the influence of the furnace efficiency and of the percentage of cullet on the energy consumption. In 1980 the efficiency will exceed 40 per cent, which, compared to a furnace in 1970, represents an economy of energy equal to 33 per cent. This economy almost corresponds to that obtained by introducing 80 per cent of cullet into a furnace built in 1970.

In order to take into account other processes, we have calculated the consumption of energy per ton of the given material. Figure 9 sums up the different basic facts necessary to establish the overall balance of energy of the recovering of glass, from the extraction and transport of raw materials to the elimination of the wastes. One can see that the most important item is the fusion: this explains the interest of the glass-manufacturing industry for furnaces with a high efficiency.

Having assembled the necessary facts, it is possible to calculate the consumption of energy involved in the manufacture and consumption of the glass. The calculations have been made for a capacity of 100 l of liquid. After inquiry, we allowed an average weight of the non-returnable glass equal 600 g per liter and an average weight of 800 g per liter for the returnable bottles (which must be more resistant) with 20 recirculations per bottle.

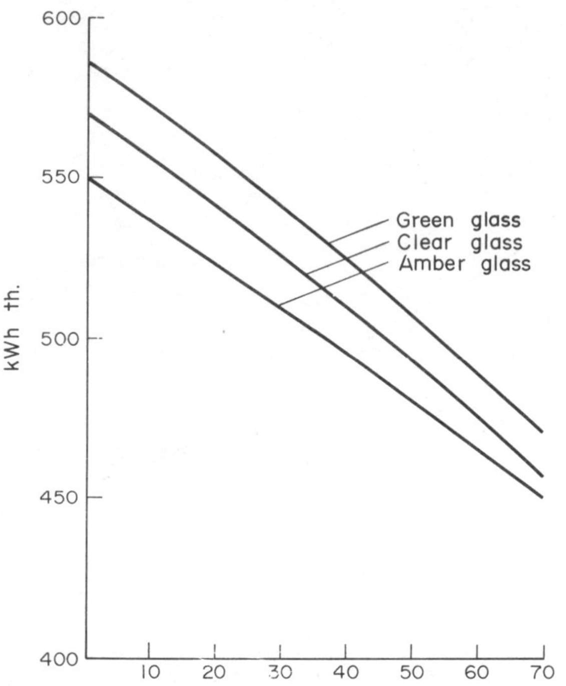

Fig. 6.

Figures 10 and 11 show the corresponding costs of energy. The overall balance of energy for manufacturing the quantity of non-returnable glass corresponding to a capacity of 1000 l is shown in Fig. 10. The electric and thermal kWh are added separately and the conversion made on the basis of:

1 el. kWh = 3 th. kWh (efficiency of nuclear power plants) or

1 el. kWh = 1445 th. kWh (overall efficiency of the production of electricity in Switzerland in 1974).

From this one can ascertain that the economy of energy is obtained above all with the fusion. A fair economy of energy is also obtained with the manufacture of sodium carbonate, while all the other processes together represent less than 20 per cent of the total consumption of energy. An economy of energy of 50 per cent is possible when using 100 per cent cullet as raw material.

In Fig. 11 the balance of energy is drawn up for returnable bottles allowing 20 recirculations per bottle.

If one is to compare Figs. 10 and 11 correctly, one must allow for the same capacity on the market at a given time and consider a period of time equal to the average life of the bottles in recirculation.

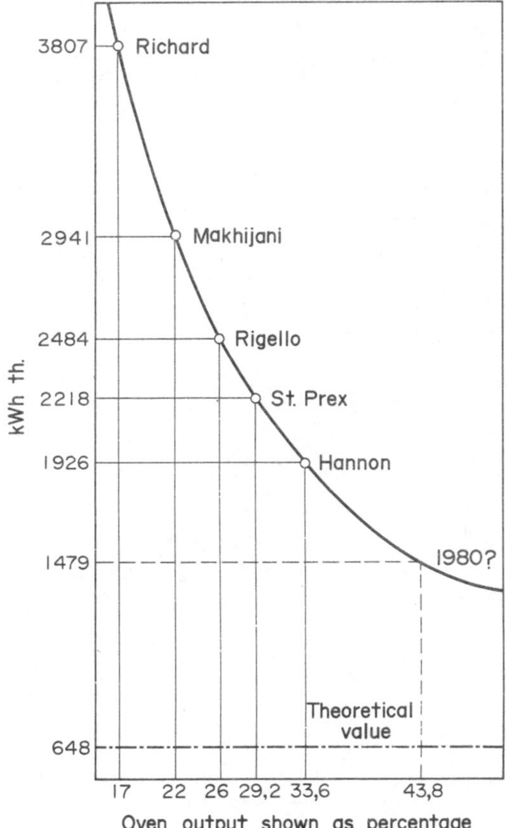

Fig. 7.

Taking 1000 l as the capacity permanently on the market: in the case of the non-returnable glass, it would be necessary to manufacture 20 times 1000 l for the considered period.

In the case of the returnable glass, it would be necessary to manufacture a capacity of 1000 l - but clean and distribute 20 times these 1000 l - i.e. 20,000 l.

Figure 10 being for 1000 l, Fig. 11 must therefore be calculated for one-twentieth of 1000 l for the manufacture and 1000 l for the cleaning and distribution.

As one can see, from the only point of view of energy (considered as the most representative parameter of the protection of the environment in the case of glass) the recirculation of bottles (returnable bottles) is by far more economical than non-returnable glass, as the ratio of energy consumption =

2259.27 : 368.62, i.e. approximately 6. Even in raising to 100 per cent the rate of recycling the cullet from non-returnable bottles or broken returnable bottles, the ratio of energy consumption is 4.

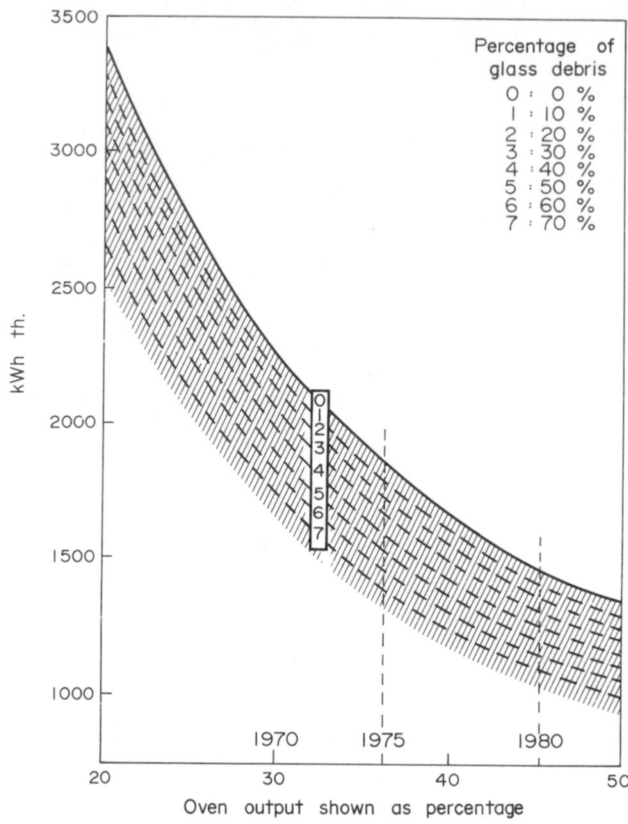

Fig. 8.

Now, one can refer to Fig. 3, the ratio of energy consumption for the two systems is - according to the preceding calculations - of 1157.58 : 295.21, i.e. approximately 4.

The graph of Fig. 12 indicates the distribution of the energy consumption according to the partners in the process. One can see that the glass industry represents 90 - 95 per cent of the total consumption: it is, therefore, the principal beneficiary and the principal interested party in the recycling of cullet and who should bear all the supplementary expenses resulting from the separate collection of cullet. The graph of Fig. 12 illustrates the balance of energy for the case of non-returnable glass. The operations are grouped in four main stages. For three of these four stages, the energy consumption drops as the percentage of cullet increases. For the transport of raw materials, the energy consumption increases with the amount of cullet. This is due to the fact that the collected cullet recovered, has then to be also transported by lorry to the glass-factory, inducing an heavy consumption of energy.

Raw materials	kWh el./to	kWh th./to
Manufacture of soda Na_2CO_3		2907
Extraction, crushing of lime	38	53,5
Production processes	**kWh el./to**	**kWh el./to**
Glass melting (0% of glass debris)		2234
Bottle making	265,7	
Municipal collection of glass debris		50
Transport of bottles		83,3
Collection of household waste		50
Waste destruction by burning	55	
Distribution of returnable bottles (Kwh/1000 l)		118
Washing of returnable bottles (")		100
Transport	**kWh el./to km**	**kWh th./to km**
Railway transport	0,119	
Pipeline transport		0,073
Waterborne transport		0,130
Truck transport outside built-up areas		1,02
Truck transport in built-up areas		1,256

Fig. 9.

The model shown here has its limits of interpretation and does not allow for a categorical interpretation such as "all non-returnable glass must be forbidden" or "heavy taxes should be added on non-returnable glass". In fact, the calculations carried out here are relative to average values and not to the marginal values in different local situations. It is evident that the energy consumption for each stage studied would be radically modified if the size of flow varied as much as shown in the diagram of Fig. 3.

Accordingly, the above calculated values can be considered as marginal values, as long as one remains around the average quantities. It is possible, however, to perceive the general tendency: the recirculation of bottles reduces the quantity of energy consumed per unit and the recycling of cullet reduces it also, but to a lesser extent. The graph of Fig. 13 has been drawn from the four extreme values taken from Figs. 10 and 11. It offers a first approach to the examination of possible alternatives. For example, increasing the amount of cullet from 10 to 60 per cent, while keeping a rate of 30 per cent for the recirculation of returnable bottles, would produce the same economy of energy (1600 - 1100 = 500 th. kWh per 1000 l of content in circulation) as

Activite	Taux de recyclage (kWh)							
	0%		25%		50%		100%	
	Electr.	Autres	Electr.	Autres	Electr.	Autres	Electr.	Autres
1. Manufacture of soda Na_2CO_3 (Solvay)		389,30		291,98		194,65		-
2. Extraction, crushing of limestone ..... lime	3,04	4,28	2,28	3,21	1,52	2,14	-	-
3a. Transport of sand Nemours-Vallorbe	23,59		19,17		11,79			
3b. Transport of sand Vallorbe-St-Prex	2,56		2,08		1,28			
4. Transport of lime La Sarraz-St-Prex		1,31		0,98		0,65		
5a. Transport of Dolomie Gorlago-Simplon	0,67		0,50		0,34			
5b. Transport of Dolomie Simplon-St-Prex	0,61		0,46		0,31			
6. Transport of soda Zurzach-St-Prex	4,46		3,35		2,23			
7a. Transport of phonolite Düsseldorf-Bâle	2,18		1,64		1,09			
7b. Transport of phonolite Bâle-St.Prex	0,68		0,51		0,34			
8a. Transport of fuel Libya-Genoa (Ship)		22,62		20,42		18,02		12,44
8b. Transport of fuel Genoa-Swiss frontier (pipeline)		3,57		3,23		2,84		1,96
8c. Transport of fuel Swiss frontier-Collombey (pipeline)		0,40		0,36		0,32		0,22
8d. Transport of fuel Collombey-St-Prex (CFF)	0,71		0,64		0,57		0,39	

No.	Item								
9.	Collection of glass debris from the municipalities				4,5		12,0		27,0
10.	Transport of glass debris			0,36	13,09	0,96	34,91	2,17	78,55
11.	Glass melting		1340,41		1210,36		1067,83		737,23
12.	Bottle making	159,42		159,42		159,42		159,42	
13.	Transport of bottles		49,98		49,98		49,98		49,98
14.	Collection of glass in the municipal rubbish		30,0		25,50		18,0		3,0
15.	Burning of municipal rubbish	33,0		28,05		19,80		3,3	
16.	Energy wastes in the slags		83,72		71,16		50,23		8,37
	Total	230,92	1925,59	218,46	1694,77	199,65	1451,57	165,28	918,75
	Total (1 kWh el.=3 kWh th.)	2618,35		2350,15		2050,52		1414,59	
	Total (as per production outline in Switzerland)	2259,27		2010,44		1740,06		1157,58	

Fig. 10. Waste glass - energy consumption linked to the production of a quantity of green glass corresponding to 1000 l of the content.

	Activité	Taux de recyclage (kWh)							
		0%		25%		50%		100%	
		Electr.	Autres	Electr.	Autres	Electr.	Autres	Electr.	Autres
1.	Manufacture of soda $Na_2\ CO_3$ (Solvay)		25,95		19,47		12,98		
2.	Extraction, crushing of limestone .... lime	0,20	0,29	0,15	0,21	0,10	0,14		
3a.	Transport of sand Nemours-Vallorbe	1,57		1,28		0,79			
3b.	Transport of sand Vallorbe-St-Prex	0,17		0,14		0,09			
4.	Transport of lime La Sarraz-St-Prex		0,09		0,07		0,04		
5a.	Transport of Dolomie Gorlago-Simplon	0,04		0,03		0,02			
5b.	Transport of Dolomie Simplon-St-Prex	0,04		0,03		0,02			
6.	Transport of soda Zurzach-St-Prex	0,30		0,22		0,15			
7a.	Transport of phonolite Düsseldorf-Bâle	0,15		0,11		0,07			
7b.	Transport of phonolite Bâle-St-Prex	0,05		0,03		0,02			
8a.	Transport of fuel Libya-Genoa (Ship)		1,51		1,36		1,20		0,83
8b.	Transport of fuel Genoa-Swiss frontier (pipeline)		0,24		0,22		0,19		0,13
8c.	Transport of fuel Swiss frontier-Collombey (pipeline)		0,03		0,02		0,02		0,01
8d.	Transport of fuel Collombey-St-Prex (by railway)	0,05		0,04		0,04		0,03	

9. Collection of glass debris from the municipalities				0,07		0,57		1,57
10. Transport of glass debris			0,03	1,15	0,07	2,60	0,15	5,51
11. Glass melting		89,36		80,69		71,19		49,15
12. Bottle making	10,62		10,62		10,62		10,62	
13. Transport of bottles		3,33		3,33		3,33		3,33
14. Distribution of bottles		118,0		118,0		118,0		118,0
15. Washing		100,0		100,0		100,0		100,0
16. Collection of glass in the municipal rubbish		2,0		1,7		1,2		0,2
17. Burning of municipal rubbish	2,2		1,87		1,32		0,22	
18. Energy wastes in the slags		5,58		4,74		3,35		0,56
Total	15,39	346,38	14,55	331,03	13,31	314,81	11,02	279,29
Total (1 kWh el. =3 kWh th.)	392,55		374,68		354,74		312,35	
Total (as per Swiss production outline)	368,62		352,05		334,04		295,21	

Fig. 11. Glassware in circulation - energy consumption linked to the production of a certain amount of glassware corresponding to 1000 l of content (20 circulations; 800 g/litre).

a rate of 22 per cent in the recycling together with a rate of recirculation of returnable bottles of 50 per cent.

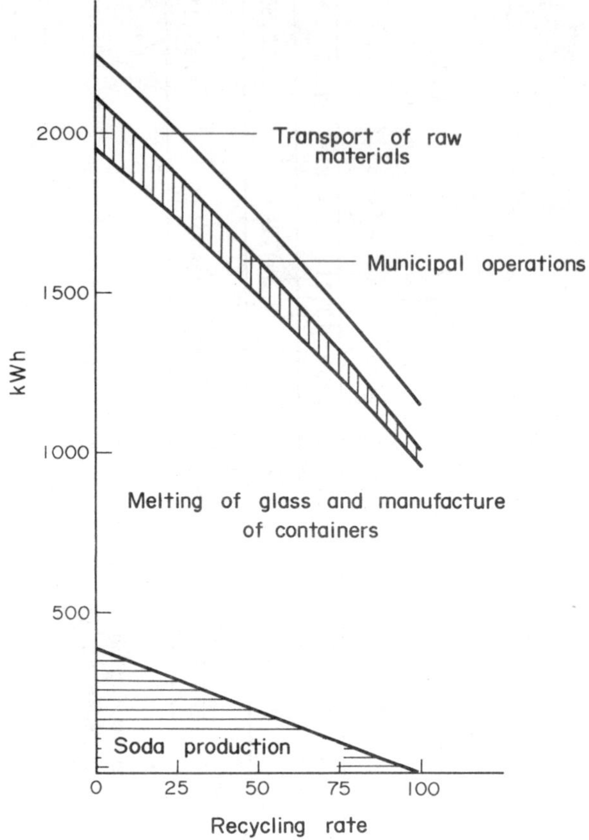

Fig. 12.

In an analysis of the situation, further considerations must be taken into account:

1. The capacity of production of the existing glass-factories. (If they are operating at a fraction of their capacity, the marginal consumption of energy will be very different from the calculated averages.)

2. The possibility of setting up systems of recirculation of cullet and of returnable bottles.

3. The possible effects of the competition with other types of containers.

4. The problems of imports and exports.

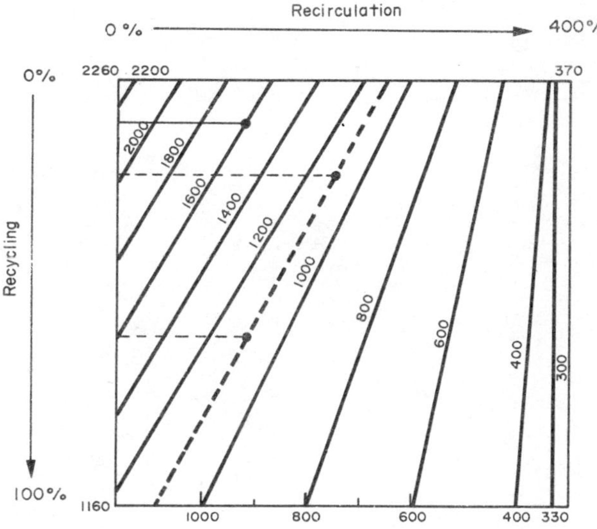

Fig. 13.

In conclusion, the point reached by the present study permits one assumption to be made, without specifying to what extent it will be true: the recycling of cullet and the recirculation of bottles must both be encouraged, as they represent a decrease of wastes and of the consumption of natural resources for the system of production and consumption of glass.

The Status of Non-Waste Technology in the U.S. Steel Industry

Arthur H. Purcell

TIP, Inc., Washington, D.C., USA

INTRODUCTION

For every twenty units of energy consumption in the United States, more than one unit goes for production of iron and steel. Stated more dramatically, iron- and steelmaking uses an amount of energy greater than one-sixth of US energy imports. This tremendous resource intensiveness has prompted much concern over how, through innovative waste reduction technologies, the country's iron and steel industry can reduce its dependence on our limited energy resources and still produce its needed flow of materials and products.

Steel* presents both difficult problems and great promise for non-waste technology (NWT) efforts, in material as well as energy areas. In all major phases of steel and steel-product manufacture, use, and discard, NWT applications are possible:

- Basic steel-producing processes
- Material design
- Product design and use
- Product disposal and reuse

BASIC STEEL-PRODUCING PROCESSES

The current use of the term "energy conservation" has obscured to a certain extent the problem facing the steel industry: energy is not readily available in its desired forms. Some fuels are relatively plentiful while others are scarce. The challenge before the steel industry is then two-fold; total energy conservation and conservation of scarce fuels.[1]

The major innovations of the past few years that have had significant impact on energy and material intensiveness in steel production have been electric furnace steelmaking and continuous, or strand, casting of steel shapes. More recently, direct reduction has been introduced. While estimates vary in regard to energy intensiveness of direct reduction, it is a process which can be linked to electric furnace and continuous-casting operations to provide what probably represents the best working system of waste minimization in basic steelmaking processes. Electric furnace steelmaking can cut steel furnace energy requirements by well over 50 per cent; for example, in the United States an average of 32.9 billion joules is required to produce a ton (metric) of raw steel from the basic oxygen furnace or open-hearth furnace, while electric furnace steelmaking, which can utilize large scrap charges, requires only

*For simplification, we shall use "steel" to refer to iron and steel.

about 14.5 billion joules per ton. Direct reduction is a process by which pelletized iron oxide charged into a reducing gas produced by steam re-forming natural gas is converted directly into metal. Fuel economy close to that of electric furnace steelmaking has been reported for direct reduction.[2] Direct reduction is amenable to small increments of capacity and, on a per net ton of product basis, requires lower capital equipment costs than conventional systems.[3] Limitations on natural gas supplies have, however, minimized feasibility of widespread adoption of direct reduction systems.

The yield from liquid steel to cast shape using continuous casting is about 10 per cent greater than that with conventional ingot casting; in addition, significant energy is saved in continuous-casting operations since casting steps can be eliminated and shapes poured directly.

Recovery of waste heat in steelmaking represents a virtually untapped source of non-waste technology. It has been estimated that as much as 88 per cent of the heat energy required to produce finished steel is wasted. Practical recovery rates are on the order of 35 - 60 per cent, depending on the waste heat source.[4] Elliot[2] has reported:

> "It would appear that the heat available for recovery might be between 0.3 and 1.5 million Btu (.335 billion joules and 1.7 billion joules/ton) of low-temperature heat per ton of product in ore agglomeration processes, possibly 2 million Btu of low and intermediate temperature heat per ton of coke in coking (2.24 billion joules per ton), as high as 3 million Btu of low temperature heat per ton of pig iron (3.3 billion joules per ton), between 1 and 3 million Btu of low-temperature heat per ton (1.1 and 3.3 billion joules per ton) of steel in the steel-making process, and up to 1 million Btu of low-temperature heat per ton of steel in casting operations. . . . With a capture efficiency of 50 per cent, approximately 10 per cent of the total energy consumed currently in producing finished steel could be recovered."

An attractive possibility in basic steelmaking energy recovery is utilization of blast-furnace gas for the production of methanol, a clean fuel which could be utilized directly in the steelmaking process. The chemical reaction involved is a simple one:

$$2CO(g) + 3H_2(g) = 2CH_2OH(\text{liquid})$$

The process can be carried out with various catalysts at temperatures and pressures in the vicinity of 250°C and 100 atmospheres.

The use of nuclear reactors, particularly the high-temperature gas reactor, to produce energy in steelmaking that would otherwise have to come from conventional supplies has received much attention. Many view nuclear steelmaking as the best way to enhance direct reduction applicability, since heat from the reactor can be utilized to re-form the natural gas used in direct reduction, as well as generate electricity for electric furnaces. Important technological, economic, social and environmental and safety problems stand in the way, however, of adoption of nuclear steelmaking in the near future.

Application of computer technology to steelmaking holds important potential for process energy savings in steelmaking. Because a variety of fuels are used in steelmaking, there often is significant waste in determining the proper mix. Battelle Research[5] and a number of steel companies[6] have used computer techniques to develop systems for optimizing fuel mix. Computer simulations have been developed which can delineate where changes in operating practice, in fuels utilized within given facilities, and changes in product scheduling will result in a better utilization of internal fuels while minimizing the usage and cost of external fuels. These simulations can be used to develop rules by which fuels are to be alloted for usage in various process stages.

Material Design

The two major materials design areas in steel where NWT principles best apply are in the corrosion and strength areas. While estimates vary considerably, it is clear that corrosion of steel accounts for significant material lost annually; some projections indicate that between 20 and 40 per cent of the steel produced annually goes toward replacing corroded steel. A study done for the Ministry of Technology a few years ago indicated that the economic implications of corrosion accounted for about 3½ per cent of the British gross national product. This same report claimed that corrosion losses could be cut by over 20 per cent through proper application of existing corrosion-inhibiting design technologies. These technologies include: coatings, substitution, cathodic protection, and passivation, as well as thermal treatment to reduce or eliminate stress risers in steel forms, which are areas particularly susceptible to corrosion. The technology of coatings is one which will probably increase in importance; for example, if coated steel can be made as resistant to corrosion as stainless steel, the chronic chromium shortage problem can be alleviated, and economic and energy savings realized.

One very promising future direction in corrosion research is that of ion implantation. In this process, ions of widely ranging elements such as argon, chromium, platinum and even ytterbium can be implanted in precisely controlled fashion on the surface of steel and other metals to effect formation of stable, probetive coatings. The details of the surface chemistry are poorly understood in ion implantation; the process apparently can alter electrical properties at the surface, as well as promote non-crystallinity of surface film formation. The US Bureau of Mines at its College Park laboratories is developing methods of implanting chromium ions on plain steel to produce a stainless material. Because ion implantation uses exceedingly small amounts of material and is controllable to precise tolerances, it presents a bright future direction in corrosion control.

New energy technologies may present a potential hazard in efforts to reduce material waste through corrosion. Magnetohydrodynamics, coal gasification, nuclear fusion, and geothermal all involve environmental conditions which lead to extreme material corrosion. Should any of these energy options be adopted on a large scale, significant loss of steel and other materials through corrosion will take place unless corrosion-inhibiting technologies of a sophisticated nature are developed.

Important questions of tradeoffs between energy and material savings will also complicate the corrosion research picture; generally, building corrosion resistance into materials design means increasing energy intensiveness. Hence, the question of short-term energy losses versus long-term materials gains is an important one in non-waste technology applied to the area of corrosion.

The stronger a steel is, the less material that has to be used for a given application. Hence, programs aimed at improving strength properties of steel has been conducted for several years. A great number of high-strength steels have been developed for applications on all major steel product levels. An important new direction in ultra-high-strength steel development has been microalloy steels, which many consider the wave of the future in structural steels. Utilizing very small amounts of alloying elements which form second-phase compounds, microalloy steels have exhibited yield strengths more than twice that of medium carbon steel. 0.05 Niobium steel, for example, in which niobium carbonitrides are finely dispersed, has demonstrated yield strength on the order of 551.6 meganewtons per square metre.[7] As with corrosion-resistant steel, high-strength steel usually has an energy cost which must be weighed against material benefits.

PRODUCT DESIGN AND USE

NWT design of products made of steel (and other materials) has four major components:

*Minimizing materials and energy requirements of fabrication
*Minimizing materials and energy requirements of product use
*Maximizing durability and repairability of product
*Maximizing reuse or recyclability of discarded product

In the United States automobile manufacture is a highly significant example of how technological efforts to reduce waste in a steel product industry can lead to reduction in resource use. Due mainly to the continuing pressures of the "energy crisis" and soaring fuel costs, the auto industry has begun to produce and sell on a large scale automobiles which weigh considerably less, and use proportionately less steel, than the conventional full-size "family car" that was a trademark of this country before the 1973 OPEC embargo. For example, one major manufacturer, Chevrolet, has introduced the "Chevette" which weighs 45 per cent less than its standard, full-size "Impala" model. This lighter weight allows economies of fuel needed to operate the car often approaching twice that of conventional vehicles.

The American auto industry, the single largest user of steel in the United States, has learned that, if given the choice, the public is interested in buying products which represent minimization of waste. This is a market situation that was totally unanticipated just a few years ago.

In the United States the manufacture of throwaway beverage containers accounts for more than 1 per cent, or over seven quintillion joules, of annual national energy expenditures; a significant fraction of this is for steel containers. The materials and energy waste inherent in the throwaway system has been of great concern to the public. Most estimates show that energy intensiveness could be reduced by about 50 per cent if we replace the throwaway system with an all-returnable system. State governments are under increasing pressure to enact laws to ban throwaway containers as resource and environmental conservation measures. By the time this paper is presented, in fact, the people of four states (Michigan, Massachusetts, Maine and Colorado) which entail more than 8 per cent of the population of the country, will have decided through the election process whether their governments should enact such bans. Both industry and labor are very concerned by this situation; they feel that laws forbidding throwaway containers will lead to severe industry economic dislocations and employment losses. One can company official has recently stated that "tens of thousands" of jobs could be lost in the United States if the country bans throwaway containers.[8]

At least one industry-- American Can Company --has embarked on a program of non-waste technology aimed at drastically reducing the energy and material demands of steel containers. According to a recent American Can report,[9] the company is committed to producing a two-piece throwaway steel can by 1980 with energy intensiveness only about 2 per cent greater than that of a glass bottle which is reused ten times (2627.2 joules per 0.348 liter container vs. 2574.4 joules per 0.348 liter container). While not giving details of its plan, American Can states that this reduction in energy will be accomplished through four steps:

Source reduction: reducing material usage by 15-20 per cent.

Resource recovery: a level of 50 per cent recycle through municipal waste management systems is assumed to have been attained.

New process technology in container manufacture: the use of water base and/or ultraviolet cured coatings to minimize energy consumption.

Reduction of paperboard use for container packaging.

The company has cited technological non-waste precedents in its work, including a reduction of steel and tin used in its standard 303x406 vegetable can of 25 per cent between 1952 and 1974. It is interesting to note that by slightly necking the top of its standard 0.348 liter beverage can, American Can claims savings of 12 per cent on lid material.

The area of designing durability and repairability into steel and other products has been poorly treated in the United States. In general, technological trends have run counter to this concept. The above example of the beverage container industry points to a philosophy of making non-durable goods less resource intensive instead of making more durable goods; consumers have generally accepted this philosophy. American automobiles, once easy to repair, are now difficult and costly to repair. The combination of technical complexities of devices and high labor costs has generally prompted manufacture of appliances and other goods with little regard to repairability.

The attitude of American consumers is changing on the durability and repairability question; this change can be fostered through increased NWT efforts. One promising method involves life-cycle costing procedures for consumer products. In this system, long-range energy, resource and environmental implications are factored into the price of goods, thus allowing consumers to judge the true costs of products.

Through proper design applying NWT principles, the possibility of re*using* and re*cycling* discarded steel products can be greatly enhanced. Direct reuse of steel discards for purposes other than those of original design has generally been difficult; discarded steel products generally are too corroded or worn to be amenable to secondary use. This contrasts with products of other materials. Discarded rubber tires, for example, have a variety of direct reuse applications in both modified and unmodified form; these include highway crash barriers, boat bumpers and asphalting mix constituent. Glass jars can be reused many times as storage receptacles after their original purpose is fulfilled. Similarly, many plastic containers are reusable. Steel product designs generally make use of them in discarded form difficult. Oil drums are one of the few widespread steel products which find secondary uses, generally as trash receptacles. With imaginative efforts, however, it should be possible to make reuse of steel products more widespread. Hence, this NWT area of steel is wide open for development.

While re*use* of steel products has proved difficult, re*cycling* of them is relatively easy, provided they are designed with recycling in mind. The success of recycling of discarded steel products is directly dependent on whether non-ferrous materials can be easily separated from them. Steel beverage cans, for instance, have not been readily amenable to recycling because of their lead/tin solder content and, in many instances, aluminum tops.[10] The impurities, in small concentrations, can cause severe metallurgical problems. In the steel furnace, the lead can cause reactions leading to refractory damage. Tin can cause "hot shortness" and lead to poor-quality steel. Aluminum oxidizes and, in continuous-casting operations, the oxide can collect and clog molten metal flow. Presently, design plans geared toward replacing solder with polymeric cement and eliminating bi-metal cans are going into effect by the major can makers. In most areas of steel product design, however, recyclability is inadequately, or not at all, considered. Design for recyclability is thus another major area in which NWT principles need to be vigorously applied.

As outlined, reuse and recycling represent important areas of NWT potential for steel. Estimates of how much obsolete scrap is currently recycled on an annual basis in the United States vary considerably, with about 25 per cent representing the lowest probable fraction. The steel industry has always used a certain amount of scrap in its open hearth and basic oxygen furnaces; the advent of the electric furnace, which can produce less energy intensive steel, has greatly increased industry capacity to recycle, since it can take scrap charges of at least 70 per cent. Research geared toward efficient recovery of steel from junk automobiles, a major source of scrap steel, has increased. Several institutions, including the US Bureau of Mines and Prohler Industries, report development of low-polluting methods of removing (through thermal means) the considerable organic fraction from auto wastes and thereby producing scrap steel readily amenable to recycling.

Waste steel represents the heaviest non-organic fraction of the massive municipal wastes this country generates. Because they can be presorted or separated magnetically from combined wastes, considerable attention has been directed toward recovery of them. Most recovery systems are still in experimental or demonstration stages. Significant controversy exists at present over the merits of high technology separation vis-à-vis low technology separation; the latter entails preseparation through manual or other low-energy means, while the former entails large-scale, often rather energy-intensive bulk separation systems.

The important role that recycling can play in reduction of energy, materials and environmental impacts has been clearly documented. A recent report by Purcell and Smith[11] has reviewed a number of major studies assessing the quantitative impacts of recycling and confirmed that steel recycling, particularly when the electric furnace method is used, entails reduction in energy, environmental and material impacts ranging from modest to large, in virtually every category measured. For example, steel recycling in the electric furnace can reduce particulate air pollutants by nearly an order of magnitude, water effluent discharges by a factor of 4, solid wastes by a factor of nearly 50, and energy consumption by more than twice.

CONCLUDING REMARKS

Non-waste technology in the production of steel shapes and products has both important problems and significant prospects. Challenges in areas ranging from waste furnace heat recovery to development of corrosion-resistant materials to imaginative design for reuse to implementation of high-efficiency recycling systems are evident.

It is clear that energy, environmental and material benefits are possible in direct proportion to the degree to which NWT principles can be applied. One vital component in the success of efforts to reduce waste in our steel and other industries will be assistance from the consumer sector in accepting and promoting NWT principles. In that regard, it will be instructive to analyze the results of referenda in four states of the United States on 2 November 1976; voters will have the opportunity to accept or reject the non-waste concept of banning by law one-way throwaway beverage containers. In making their decisions, they will have to weigh vital energy, environmental and economic factors associated with the measures. The results of the voting may well indicate the future direction that non-waste technology will take in this country in the coming years.

REFERENCES

1. G.K. Sigworth, Fuel and energy conservation in the steel industry. Proceedings, Annual Meeting, AIME, New York, 16-20 Feb. 1975.
2. J.F. Elliot, Energy for steel production - prospects and options. Proceedings, Annual Meeting, AIME, New York, 16-20 Feb. 1975.
3. T.F. Barnhart, Energy requirements for various steelmaking processes. Proceedings, Annual Meeting, AIME, New York, 16-20 Feb. 1975.
4. American Iron and Steel Institute, "Energy Conservation in the Steel Industry," Report on 84th General Meeting, New York, Technical Session, 26 May 1976.
5. Battelle Memorial Institute, *Energy Use by the Steel Industry of North America*, 1971.
6. W.R. Gray, J.D. Fekete and M.I. Tarkoff, A steel plant energy model. *Iron and Steel Engineer*, 51, No. 11, p. 54.
7. M.E. Fine, Northwestern University, private communication.
8. TIP, Inc., *Citizens and Waste*, Report on the 1975-76 Workshop Series, June 1976.
9. *Resource Recovery and Source Reduction*, Fourth Report to Congress, US Environmental Protection Agency, Office of Solid Waste Management Programs, 1976.
10. A.H. Purcell and F.L. Smith, "Energy and Environmental Implications of Materials Alternatives: An Assessment of Quantitative Understanding", Resource Recovery and Conservation (to be published).

The Status of Non-Waste Technology in the US Packaging Industry

W. David Conn

Assistant Professor of Environmental Planning, University of California, Los Angeles, USA

INTRODUCTION

The packaging sector has received a great deal of attention as a possible target for the introduction of non-waste technology due to:

its rapid growth rate (historically);

its high consumption of certain materials (see Table 1);

its high consumption of energy (the production of raw materials for packaging alone consumed some 5 per cent of total energy used by US industry in 1971);[1]

its major contribution to the solid waste stream (see the latest figures in Tables 2 and 3).

Efforts to introduce non-waste technology can be divided into two categories, namely "waste reduction" and "resource recovery". Time and resources have not permitted a complete survey of the status of all activities falling into these two categories; instead, this paper will provide some examples of significant trends and developments in the US packaging industry.

TABLE 1

Packaging material consumption in relation to total material consumption, 1971

Material	Packaging (10^3 tons)	Total packaging and non-packaging (10^3 tons)	Packaging percentage
Paper	27,700	58,652	47.2
Glass	11,100	14,900	74.2
Steel	7255	87,038	8.3
Aluminum	757	5074	14.1
Plastic	2900	10,000	29.0

Sources: American Paper Institute, *The Statistics of Paper*, Washington, 1972; American Iron and Steel Institute, *Shipments of Steel Products*, Washington, D.C., 1972; Arthur D. Little, *Incentives for Plastic Recycling and Reuse*, Boston, 1972; US Department of

Interior, *Minerals Yearbook*, Aluminum Chapter Reprint, 1971, Washington, D.C., 1973; The Aluminum Association, *Aluminum Statistical Review, 1971*, New York, 1972.

TABLE 2

Post-consumer net solid waste disposed of, by product category, 1971-1974. (As generated wet weight, in millions of tons)

Products	1971	1972	1973	1974
Non-food product waste:				
Newspapers, books, magazines	10.3	10.9	11.3	11.5
Containers and packaging	41.7	45.2	46.9	45.7
Major household appliances	2.1	2.1	2.1	2.2
Furniture and furnishings	3.2	3.3	3.4	3.3
Clothing and footwear	1.2	1.2	1.3	1.3
Other products	18.4	19.5	20.5	20.4
Total non-food product waste*	76.9	82.1	85.4	84.5
Food waste	22.0	22.2	22.4	22.6
Yard and misc. organics	25.9	26.3	26.9	26.9
Total*	124.8	130.6	134.8	134.5

Source: Office of Solid Waste Management Programs, Resource Recovery Division, and Franklin Associates, Inc. Data revised 25 June 1976.

* Totals may not agree with details due to rounding.

WASTE REDUCTION IN THE PACKAGING SECTOR

Waste reduction (formerly known as source reduction) is defined by the US Environmental Protection Agency (EPA) as the "prevention of waste at its source, either by redesigning products or by otherwise changing societal patterns of production and waste generation".[2] Two approaches to waste reduction are particularly applicable in the packaging sector. These are: (i) the use of less materials and/or energy per unit of product; and (ii) the re-use of packaging.

(i) *The use of less materials and/or energy per unit of product*

Competitive market forces, accentuated by the recent sharp increases in the prices of many raw materials, have caused many companies to seek new ways of reducing resource use. For example, the standard half-pint paperboard milk container (in which approximately 11 per cent by volume of milk was sold in the US in 1973) has been redesigned; the new container with a smaller base has been estimated to use 31 per cent less paper and 16 per cent less low-density polyethylene than the old container.[3] The introduction of the new container in several parts of the country is leading to savings in paper and polyethylene (and the environmental impacts associated with their production), to a reduction in the post-consumer waste stream, and to dollar savings for the purchasers and fillers of the cartons (which may be passed on to consumers). The paperboard milk container (and other milk containers made of glass or rigid plastic) are facing competition from the flexible polyethylene pouch which is already used to package more than 60 per cent of fluid milk in Canada. An earlier attempt (some 2-3 years ago) to market pouched milk in the US was largely unsuccessful, but some dairies are now supplying it to schools, where the resulting substantial reduction in solid waste generation is seen as an important advantage.[4] A resource and energy profile analysis of several different milk-container systems is currently being conducted for the EPA.

Intense competition between the producers of steel, aluminum, glass and plastic containers for food, beverages and other products is causing a trend toward increasingly lightweight cans and bottles. A major development has been the shift from three-piece to two-piece steel cans. The new cans have no side or bottom seams and may have thinner walls than the traditional three-piece cans, thus significant materials savings are possible, as well as energy savings in making and transporting the lighter containers. Steel savings of as much as 20-30 per cent have been claimed in trade journals, although unsupported by published evidence. In practice, the difference in the weight of steel actually incorporated into the cans may be much smaller, although additional savings result from the fact that more prompt (and readily recyclable) scrap is created in two-piece can-making.

There are potentially both advantages and disadvantages for companies introducing two-piece steel cans. The advantages include:

- savings in the costs of steel and tin-plate;
- easier quality control and reduced possibilities of leakage;
- elimination of the fear of lead contamination from some side seams;
- savings in labor costs due to a more highly mechanized process;
- savings in transportation costs due to lighter containers.

On the other hand, the disadvantages include:

- high capital cost of two-piece can-making line compared with three-piece line;
- reduced flexibility in changing can size and/or label (since the new lines are best suited to high-speed production of a limited range of containers);
- a possible problem of over-capacity resulting from the too-rapid introduction of more productive two-piece lines (each replaces at least two three-piece lines, and a glut of used can-making equipment has been reported);[5]
- lower suitability of two-piece can for food products as these require a stronger container than carbonated beverages (internal pressure of beverage adds strength to the can and permits a thinner wall).[5]

In practice, the two-piece can is making significant inroads into the beer and soft drink container markets, but is progressing more slowly in the market for food containers (where it is generally being introduced not to replace existing lines but rather to equip new installations).[5]

In response to the developments occurring in the steel can industry, attempts are being made to save materials and energy in the manufacture of aluminum

TABLE 3

Post-consumer residential and commercial solid waste generated and amounts recycled, by detailed product category, 1974.
(As generated wet weight, in thousands of tons)

Product category	Gross discards	Material recycled		Net waste disposed of		
		Quantity	Per cent	Quantity	% of total waste	% of non-food product waste
Durable goods:	14,610	280	2	14,330	11	17
Major appliances	2250	100	4	2150	2	3
Furniture, furnishings	3300	0	0	3300	2	4
Rubber tires	1760	180	10	1580	1	2
Miscellaneous durables	7300	0	0	7300	5	8
Non-durable goods, exc. food:	27,685	3275	12	24,410	18	29
Newspapers	10,170	2145	21	8025	6	10
Books, magazines	3770	300	8	3470	3	4
Office paper	6375	830	13	5545	4	7
Tissue paper, incl. towels	2200	0	0	2200	2	3
Paper plates, cups	570	0	0	570	-	1
Other non-packaging paper	1240	0	0	1240	1	1
Clothing, footwear	1300	0	0	1300	1	2
Other misc. non-durables	1970	0	0	1970	1	2

Containers and packaging:	51,290	5540	11	45,750	34	
Glass containers:	12,070	275	2	11,795	9	14
Beer, soft drink	5990					
Wine, liquor	1705					
Food and other	4375					
Steel cans:	5920	60	1	5860	4	7
Beer, soft drink	1565					
Food	3470					
Other non-food	885					
Aluminum:	905	50	6	855	1	1
Beer, soft drink	475	50	11	425	-	1
Other cans	65	0	0	65	-	-
Aluminum foil	365	0	0	365	-	-
Paper, paperboard:	27,380	5155	19	22,225	17	26
Corrugated	14,720	3710	25	11,010	8	13
Other paperboard	6360	790	12	5570	4	7
Paper packaging	6300	655	10	5645	4	7
Plastics:	2990	0	0	2990	2	4
Plastic containers	480	0	0	480	-	1
Other plastic packaging	2510	0	0	2510	2	3
Wood packaging	1850	0	0	1850	1	2
Other misc. packaging	175	0	0	175	-	-
Total non-food product waste	93,585	9095	10	84,490	63	100
Add: Food waste	22,600	0	0	22,600	17	27
Yard waste	25,500	0	0	25,500	19	30
Misc. inorganic wastes	1900	0	0	1900	1	2
Total	143,585	9095	6	134,490	100	159

Source: Office of Solid Waste Management Programs, Resource Recovery Division, and Franklin Associates, Inc. Data revised 25 June 1976.

cans (all of which are two-piece). For example, a new 12-oz aluminum beer can has recently been announced which weighs only 27 lb per 1000 cans compared with 31 lb for its immediate predecessor.[5]

A trend toward lighter containers is also evident in the glass container industry. One development is a new process (introduced by Kerr) which permits manufacturers to tightly control the distribution of glass in new containers, thereby enabling their strength to be enhanced in critical places while the use of material is minimized.[6] Another development is the use of plastic coatings on glass containers, permitting their strength to be maintained with less glass; for example, a coated quart bottle weighs about 15 oz compared with 19 oz for an uncoated non-refillable bottle of the same capacity. The use of coatings on glass containers also permits a reduction in the use of secondary packaging, giving further savings in materials and energy. Almost 90 per cent of the larger (48 oz and 64 oz) soft drink bottles are now coated.[6]

Another method of reducing waste in the packaging sector is by shifting from smaller to larger packages. For example, according to information supplied to the EPA, the 7-oz returnable glass bottle requires about twice as much glass per ounce of soft drink as the 32 oz size.[7] There currently seems to be some trend away from smaller bottle sizes (e.g. 12 oz and 16 oz) towards larger sizes (e.g. 48 oz and 64 oz) for soft drinks, although at the same time it must be reported that the small 7-oz container for beer is increasing in popularity.[6]

(ii) *The re-use of packaging*

The issue of refillable versus non-refillable beverage containers has captured much attention in recent years. It is not within the scope of this paper to review the arguments for and against legislation to encourage the use of refillables (many of which arguments are based on little more than guesses of what the likely impacts of legislation might be); rather, the purpose is to report on present trends and indicate how these are affecting industry.

To date, legislation requiring mandatory deposits on beverage containers has been enacted in two states (Oregon and Vermont) as well as a few localities; in addition, some private organizations (such as Cornell University and the Yosemite Park and Curry Company) are employing deposit systems on a trial basis.

In Oregon,* according to the latest available information, [8,9]

- beverage container litter has decreased by an estimated 66 per cent;
- non-refillable containers have declined to 12 per cent of the soft drink market and 6 per cent of the beer market;
- sales of beer and soft drinks have neither declined below the level of the

---

*The Oregon legislation, in effect since 1 October 1972, requires a minimum 2-cent refund to purchasers on return of "certified" containers of beer, malt beverages, and carbonated soft drinks, and a 5-cent refund on the return of all other containers for those beverages. Certified containers are defined as those used by, and accepted for reuse by, more than one manufacturer. In addition, the law outlaws the sale of flip-top or pull-tab beverage containers.

year prior to enactment nor have increased as rapidly as in previous years;
the price of beer and soft drinks to the consumer has been lowered *on the average* (beverages in refillables being less expensive than those in non-refillables);
refillable soft drink containers are being returned at a rate between 93-96 per cent and refillable beer containers at a rate of about 85 per cent;
the law has decreased container manufacturing and canning industry profits substantially due to the transition from non-refillable to refillable containers;
significant job losses have occurred in the container manufacturing and canning industries;
significant numbers of jobs have been created in the brewing, soft drink, and retail sectors of the economy;
the number of non-local beers sold in the state has dropped from 29 to 9 since passage of the law.

Complete data are not yet available for Vermont,* but the following trends have been reported:[10]

beverage container highway litter decreased by about 67 per cent between 1973 and 1974;
beer sales dropped by about 10 per cent in the first year, but then began to climb again;
the prices of beer and soft drinks have risen, although the increases can be at least partially explained by increases in the prices of materials and labor;
there has been a considerable shift toward the use of refillables for soft drinks, and some shift for beer;
return rates of the order of 87-92 per cent are being achieved;
increases in employment have been reported by several soft drinks distributors and beer wholesalers;
no significant sales or employment impacts have been experienced by container manufacturers (whose sales volume in Vermont is relatively small).

With the passage of mandatory deposit legislation thus far restricted to a few states and localities, the impacts on industry nationwide have been relatively small. Non-refillable containers previously destined for places where laws are now in force have been diverted to other, unrestricted markets. Even the effects noted above for Oregon and Vermont cannot necessarily be attributed solely to the passage of mandatory deposit laws; for example, a recent study was unable to find evidence that implementation of the Vermont law led to the following year's drop in beer sales. On the contrary, the study found that a drop in tourism (due to the gasoline shortage and an abnormally bad skiing season) as well as some overstocking by distributors prior to the law's enactment provide more plausible explanations for the observed statistics.[11] Nevertheless, the industries potentially affected by mandatory deposit legislation have mounted vigorous campaigns to have such legislation defeated whenever introduced.

---

*The Vermont legislation, in effect since 1 September 1973, requires a deposit and refund of a minimum of 5-cents on the purchase and return of all containers for beer, malt beverages, and carbonated soft drinks. The law also requires that a handling charge equivalent to 20 per cent of the deposit be paid by the distributor to the retailer.

Some shift toward the refilling of beer containers has been reported in a region where it is not required by law. Five north-west breweries who have begun their own buy-back recycling and re-use programs have apparently achieved an average return rate exceeding 40 per cent for their previously "non-returnable" bottles, and the programs have been found to more than pay for themselves. The saving in bottles has led to a corresponding reduction in the number purchased from the suppliers, although as brewery sales overall are increasing, the demand for bottles is steadily increasing also.[12]

Even in the absence of mandatory deposit legislation, some distributors offer soft drinks in refillable bottles. It is difficult to obtain firm data on return rates, but there are several reports of figures exceeding 96 per cent.[13,14] With rates as high as these, a refillable system not only leads to significant savings in materials and energy, and to a decreased negative impact on the environment, but it can also be very profitable financially for the distributor.

Although most attention has been focused on the re-use of beverage containers, other packages and packaging accessories can be, and are being, increasingly re-used. These include crates and pallets made of wood or plastic, as well as corrugated/cardboard cartons.

RESOURCE RECOVERY IN PACKAGING SECTOR

There are three approaches to resource recovery affecting the packaging sector, namely: (i) source separation; (ii) mechanical separation; and (iii) energy recovery.

(i) *Source separation*

Source separation is the setting aside of recyclable waste materials at their point of generation by the generator. Separation is followed by the transporting of these materials to a secondary materials dealer or directly to a manufacturer. The packaging items most commonly recovered in this manner are corrugated paper and metal cans, although small quantities of other items such as glass cullet are also recovered.

Used corrugated containers are discarded primarily from commercial and industrial sources. Most of the country's major supermarket chains now separate corrugated, as do many auto assembly plants and other commercial/industrial establishments. However, the price paid for the reclaimed material has varied considerably: in 1973 it rose to $60/ton at the mill in most parts of the country, but in the following year it dropped to less than half this level.[2]

The aluminum industry has mounted a major effort in the past few years to encourage the return of aluminum cans to recycling collection centres, where a bounty is paid on the weight of aluminum brought in. Reynolds Aluminum, which launched its recycling program in 1967, has now established Reynolds Aluminum Recycling Co. as a separate "profit center", and is increasing the number of recycling centers to its ultimate goal of 85 centers and 150 mobile units. According to the company, the quantity of recycled scrap received now equals between one-third and one-half the aluminum used in cans that it manufactures each year. Each pound of scrap received is said to save 4 lb of bauxite and 95 per cent of the energy required to process the virgin ore.[15,16] Overall in the industry, the Aluminum Association claims the quantity of metal reclaimed annually has risen from 26,500 tons in 1972 to 87,000 tons in 1975 (although these figures probably include aluminum from other products such as foils, cooking trays, etc.).

According to the Glass Container Manufacturers Institute, source separation of glass cullet by the public has increased significantly over the past few years, from 188,000 tons annually in 1971 to nearly 313,000 tons in 1975. However, the quantity involved is a small proportion (perhaps 2-3 per cent) of the total amount of glass discarded by consumers.

(ii) *Mechanical separation of materials*

Although there has been a major thrust in recent years to develop mechanical systems for separating the components of mixed post-consumer wastes, so far there are few plants operating continuously, and this method is not yet a major factor in recovery (although it is expected to be in the future). Even ferrous separation (which is most readily accomplished by magnetic means) is not widespread; in 1975 EPA reported having knowledge of only twenty-five cities recovering ferrous from mixed waste, although at least eighteen additional facilities were then planned.[2,17] In 1973 less than 2 per cent of the steel cans discarded were recycled.[2]

Currently, the copper precipitation industry consumes more than half of the scrap steel cans recovered from waste; however, consumption in the de-tinning industry (which not only upgrades the scrap to a high grade of steel, but also recovers a valuable resource - tin) is increasing. Efforts are being made to improve processes technologically to increase scrap marketability (e.g. by removing contaminents such as tin, aluminum and lead).

(iii) *Energy recovery*

Various technologies for recovering energy from the solid waste stream are currently in operation or under development, notably incineration with heat recovery, pyrolysis and combustion of a shredded, pelletized or wet-pulped fuel separated from the refuse. The paper and plastics used in packaging are significant contributors to the heat content of the waste stream; for example, plastics typically have calorific values in the region of 15,000-20,000 Btu per pound compared with an average value of 4500 Btu per pound for mixed municipal solid waste. Another method of energy recovery that is currently being developed is methane recovery from sanitary landfills. Paper undergoes decomposition and contributes to gas formation in landfills, but plastics do not. So far, energy recovered from solid waste has not been significant in meeting national energy needs, but its significance is expected to grow in the future.

WASTE REDUCTION, RESOURCE RECOVERY AND THE US PACKAGING INDUSTRY

Reviewing current trends and developments, it appears that waste reduction and resource recovery are not high priority activities in the packaging industry at the present time, except where companies can use them to gain a market advantage (or escape from a market disadvantage). Attempts are being made within the industry to reduce resource inputs per unit of output since this reduces costs and improves a company's competitive position, but efforts by others (e.g. government agencies) to encourage re-use (and thereby cut down on the need for new packaging) are on the whole being vigorously opposed. On the other hand, resource recovery is being actively promoted, both because it is clearly compatible with an increase (rather than decrease) in the amount of packaging produced, and also because it is in itself a growth industry.

REFERENCES

1. Gordian Associates. Energy consumption for six basic materials industries. US Environmental Protection Agency Contract No. 68-01-1105, 1973. (Unpublished data.)

2. US Environmental Protection Agency, Office of Solid Waste Management Programs, *Resource Recovery and Waste Reduction; Third Report to Congress*. Environmental Protection Publication SW-161. Washington, US Government Printing Office, 1975. 96 pp.

3. International Paper Company. (Unpublished data.)

4. Whatever became of milk-in-a-pouch in the States? *Modern Packaging*, 48 (10): 6-8, Oct. 1975.

5. A face-off in metals. *Modern Packaging*, 49 (6): 20-23, June 1976.

6. Personal communication. P. Corcorn, R. Powell and G. Teiteldaum, Glass Packaging Institute, to W.D. Conn, University of California.

7. Personal communication. Glass Container Manufacturers Institute and Brockway Glass Company to E. Claussen, US Environmental Protection Agency, Office of Solid Waste Management Programs.

8. Applied Decision Systems, Inc. *Study of the Effectiveness and Impact of the Oregon Minimum Deposit Law*. Final report presented to Oregon Legislative Fiscal Officer and Department of Transportation, Oregon Division of Highways, Oct. 1974.

9. Personal communications from various industry sources to J.H. Skinner and N. Humber, US Environmental Protection Agency, Office of Solid Waste Management Programs.

10. M. Loube, *Beverage containers:* the Vermont experience. Environmental Protection Publication SW-139. Washington, Environmental Protection Agency, 1975. 16 pp.

11. C. Stern, *et al.* *Impacts of Beverage Containers Legislation on Connecticut and a Review of the Experience in Oregon, Vermont, and Washington State*. University of Connecticut, Department of Agricultural Economics, Mar. 1975. 181 pp.

12. M. Westerman, Olympia sets recycling pace for Northwest breweries. *Beverage World*, May 1976.

13. Personal communication. G. Norton, Royal Crown/Dr. Pepper, to W.D. Conn, University of California.

14. Personal communication. P. Chokola, Chokola Beverage Company, to W.D. Conn, University of California.

15. Recycling nets more than goodwill. *Modern Packaging*, 48 (11): 8, Nov. 1975.

16. Personal communication. D. Gautschy, Reynolds Aluminum Company, to W.D. Conn, University of California.

17. Resource Technology Corporation, *Solid Waste Processing Facilities*. Technical Report 103701, Rev. A. New York, American Iron and Steel Institute. May 1974.

Non-Waste Technology: The Case of Tyres in the United States

Haynes C. Goddard

Solid and Hazardous Waste Research Division, Environmental Research Center, US Environmental Protection Agency, Cincinnati, Ohio, USA

I. INTRODUCTION

What is the tire problem? Essentially, it is a large supply of extremely durable and non-degrading material that cannot be landfilled well without expensive and energy-demanding shredding first. As traditional methods of reclaiming and re-using discarded tires are suffering declines, the problem is becoming more severe. The policy problem derivative from this situation then is to find solutions to the scrap tire surfeit.

The concept of non-waste technology as defined by the organizers of this seminar indicates a concern for the rational, or synonymously, the efficient use of resources. This discussion paper seeks to apply these concepts to the scrap tire problem in the US, focused on tires as solid wastes, and not on the problems of air and water pollution occurring in tire fabrication.

There exist two aspects of the concept of efficient resource use that can aid us in understanding the sources of the scrap tire problem, and in the pursuit of rational policy responses to it. These aspects are termed "engineering efficiency" and "economic efficiency".

Engineering efficiency is the familiar concept of employing resources in order to obtain maximum input for a given output, or conversely, minimum input for a given output. Adoption of this concept as a principle leads to the search for the least-cost combination of the factors of production for any level of output.

Economic efficiency is much less widely understood as a concept that is importantly related to the pursuit of rational resource use. It refers to those decision rules that affect the organization and distribution of a society's output, and to the use and eventual discard of materials. Specifically, it refers to those decision rules that lead societies to choose optimal levels of output, and to choose optimal characteristics of that output. In practical terms, optimality takes the form of maximum net benefits of resource use accruing to individuals and to society, where net benefits are defined as total benefits minus total costs of that resource use.

The balance of this paper is devoted to an explanation of how these concepts are of use to us in understanding and correcting the tire problem in market economies, such as that in the US, and to a brief review of past, present and expected future research on the problem that has been conducted in the US.

II. NON-WASTE TECHNOLOGY FOR TIRES AND EFFICIENT RESOURCE USE

Engineering efficiency

Waste of material can and does occur at each stage of the materials cycle - from extraction to fabrication to distribution to use to discard. Since materials are scarce, there is an incentive to minimize this materials loss, subject to the minimum constraint that increased resource recovery or prevention of loss does not lead to increased costs.

In market economies characterized by the profit incentive, the stimulus to prevent materials losses and to recover resources in the manufacturing process is typically strong enough to stimulate firm and plant managers to indulge in a continual search for new technologies and organizational forms that will reduce costs and/or raise labor productivity. This is the concept of engineering efficiency and in the US the tire industry constantly searches for cost-reducing technologies. While this search applies to all inputs, not just materials, these cost-cutting technologies do have substantial impacts on reducing materials waste. As an example, the widespread adoption of computerized tire construction equipment has made it possible to control tire specifications more closely with the result that tires are less likely to be built with too much material, that is, unnecessarily exceed the specifications, and similarly, that they are less likely to be built with too little material, leading to rejection or premature wear and/or failure which is also wasteful.

The point of this discussion is, then, that if governments seek to maintain appropriate incentives for managers, they can be expected to behave in a manner fully consistent with the concept of efficient resource use, or non-waste technology. This writer concludes, that in the US at least, more attention must be devoted to understanding how the concept of economic efficiency can be utilized to improve the social use of materials in tires. Problems of materials waste in tire manufacture are less important at this time.

Economic efficiency

This concept leads us to focus on the question of whether or not the economy is led by current markets to produce and use the quantities and qualities of tires that are consistent with what we understand to be economic efficiency. In other words, are the quantities and the distribution of tire durabilities produced those which lead to maximum net benefits for this use of these materials, principally petroleum and natural gas? We can begin to see a negative answer to this question by briefly mentioning the nature of the problem.

In the US, discarded tires have become a disposal problem for essentially four reasons:

1. There has been a growth in automobile ownership which has led to an increased supply of used tires.

2. The real prices of petroleum and natural gas have declined secularly (until 1974), which has made new replacement tires more attractive relative to retreaded tires. This has had the effect of increasing the supply of used tires, and simultaneously decreasing the demand for them.

3. There persists an aversion to retreaded tires that began during World War II, when quality was low, which lessens the demand for retreading.

4. Recent changes in tire design (radials) have lessened the demand for reclaimed rubber for use in the construction.

An obvious initial interpretation of this situation is that the market mechanism assigns inappropriate prices to those resources which must be devoted to the management of waste tires. Thus, there seems to be an inappropriate structure of incentives facing all those economic units which make decisions concerning tire design, fabrication, use, re-use and discard. This interpretation is reinforced when one examines the relative states of technology in new tire construction vs. retreading. The fact that tire-retreading technology is relatively primitive is reflective of a possible market failure, if we recognize that the very direction and pace of technology development is greatly influenced by the structure of markets and the prices these markets assign to resources.

Indeed, there is a market failure -- that is, no prices are assigned to tire management at a critical junction in the tire life cycle. Specifically, there is levied no residuals charge on the waste tire generator. Thus the usual necessary condition for the maximization of net benefits of resource use is violated, for tire users will tend to purchase tire characteristics (durability, retreadability, etc.) for which the marginal net benefit to them is higher than it is to society. Stated more familiarly, the social marginal opportunity cost exceeds the private marginal opportunity cost. This is not to suggest that the appropriate policy is in fact a residuals charge per tire, as this would lead to surreptitious and indiscriminate dumping and littering. The general idea is that because the cost of disposal is not made explicit to users and fabricators of tires, there is a tendency to fabricate lower durability tires than is technically feasible, which leads to less retreading (low-durability tires are unsuitable for retreading), resulting in a greater production of new tires than would otherwise occur. This is not to say that private markets as they currently function have not responded to a demand for increased tire durability. Early tires lasted barely 9000 km, and now exceed 90,000 km. Rather, the point is that the lack of a residuals charge has biased the distribution of tire durability (as measured by production and sales) toward lower average durability as opposed to higher. This then increases the supply of unretreadable tires, and also lessens the demand for retreaded tires.

Finding an effective way to make the costs of waste tire management explicit to waste tire generators with appropriate allowances for those who buy retreaded tires will tend to have the following waste-reducing effects.

1. It will tend to alter the structure of demand in favor of greater tire durability and away from unretreadable tire-casing construction.

2. It will tend to increase the demand for retreaded tires.

3. It will tend to reduce tire discard, and will result in less new tire production for the replacement market.

4. As new tire production for the replacement market declines, it will tend to squeeze tire manufacturer's profits and motivate them to find process waste-reducing technologies, that is, it will stimulate

searches for increased engineering efficiency.

The exact nature of such an incentive mechanism has yet to be explored, but it would appear that a deposit system with appropriate adjustments for tire condition at the time of return would be appropriate.

If such a mechanism can be developed and made to work acceptably, it may do much to contribute to the goals expressed in the concept of non-waste technology. The mechanism is appropriately labelled as one which seeks to encourage economic efficiency by administratively generating a set of prices to function where the market does not or does so only weakly and/or unsatisfactorily. If these prices are properly set, they would tend to improve resource allocation in the tire industry as a result of equating the full range of marginal benefits and costs, a usual necessary condition for net benefit maximization.

In sum, placing scarcity prices on resources where none existed before will tend to alter production processes to produce less waste, will tend to increase average tire durability, and will tend to reduce the overall flow of tires to the environment. What it will not do is stimulate the recovery of the tire materials in one form or another for use in other activities (although higher energy prices will tend to do this). In the US, virtually all research activity sponsored by government has been on the latter activity, and little attention has been paid to the kinds of economic efficiency considerations which would correct tire market failure.

III. RECOVERY OF TIRE MATERIALS

An aspect of none-waste technology for tires that has not yet been discussed is tire recovery or recycling of tires in forms other than retreading. There are a number of candidate technologies for recycling. These include rubberized asphalt, incineration for energy recovery, carbon black recovery, pyrolysis for recovery of gases and oils, the manufacture of products from reclaimed rubber, such as rubber mats, and artificial reef and breaker construction.

Research on these alternatives relates to improvement in both engineering and economic efficiency. Some research has been devoted to developing and/or improving new or current technologies, and some with demonstrating the net benefits of technologies in order to create viable markets for recycled tires.

As mentioned earlier, the traditional markets for scrap tires have been in a state of decline for a number of years. Sales of reclaimed rubber have declined as tire manufacturers substitute natural rubber or SBR (styrene-butadiene rubber) for the sidewall compounds to meet the more stringent flexing requirements of radial tires, and as plastic has been substituted for rubber in a number of products, such as mats.

Undoubtedly, the secular decline in the real price of petroleum has been an important factor in the reduced demand for scrap tires. Although we do not have direct measures of this impact, the recent increase in the sale of retreaded tires is a response to higher petroleum costs, and we can expect that higher petroleum prices will further this effect, as well as make other uses of scrap tires more attractive.

Whether tires are retreaded or not, they must eventually become scrap. The problem then is to find that use (or simple disposal) which represents the least cost of management (if costs exceed the prices of recovered materials) or the greatest net benefit (if benefits exceed costs).

A recent and relatively short analysis of the relative benefits of scrap tire re-use alternatives was conducted by the author for the purpose of providing some guidance on what direction continued research in this area should take. The results of this analysis are summarized in Table 1.

TABLE 1

Summary of Relative net benefits of scrap-tire use alternatives (per tire, circa *1970-74)*

Alternative use	Net benefits ($)	Rank
Retreading	14-20	1 or 2
Road-building uses		
Seal costs	9.30 - 48.60	1 or 2
Joint and crack filler	4.35	3
Asphaltic concrete	N.A.	
Strain-relieving interlayer	N.A.	
Emulsions	N.A.	
Hydrogeneration	0.14 (?)	4 or 5
Energy recovery	0.14	4 or 5
Carbon black feedstock	-.62 to -.02	6
Destructive distillation	slight profit	?
Artificial reefs	N.A. (costs from 0.35 to 4.14)	?

Source: see reference.

As a consequence of this analysis, given that there is a fair amount of industry research being conducted on retreading, it was decided to pursue an investigation of the technical and economic performance of rubberized asphalt, particularly seal costs. Experimental sections of this material are being laid around the country this year and next, and their performance will be monitored until 1980.

In addition, the US EPA is sponsoring a more complete analysis of the benefits and costs of alternative tire designs and uses, where the net benefits of longer-lived tires (durability) are explicitly treated. A preliminary analysis showed the net benefits of increased durability to be sensitive to the discount rate, although changes in some incorrect parameter values used in the earlier work may change this.

Virtually no work has been performed on developing and analyzing the implications of alternative policy instruments for improving the economic efficiency of tire markets. Mechanisms such as tire deposits with appropriate credits need to be researched, so that the net benefits of such less capital intensive techniques of waste-tire management can be compared with the more capital intensive ones that are frequently proposed, such as energy recovery, carbon-black recovery, etc.

It is the area of incentives and economic efficiency that tends to be the most neglected in environmental research, and it is this area that is likely to have some of the highest payoffs in non-waste technology.

REFERENCE

1. Haynes C. Goddard, An economic evaluation of technical systems for scrap tire recycling, U.S. Environmental Protection Agency (EPA-600/5-75-019), Cincinnati, Ohio, 1975.

Two Examples of Low Emission Technologies in the Pulp and Paper Industry

E. Jochem

Fraunhoffer-Gesellschaft, Karlsruhe, Federal Republic of Germany

This report initially describes the importance of production, structure and emissions of the pulp and paper industry in the Federal Republic of Germany. Subsequently, low-emission manufacturing methods, one for each industry, which are just now being developed in the Federal Republic and which are promoted by the Federal Government, will be explained.

1. PRODUCTION AND STRUCTURE

The production figures of the pulp and paper industry of the Federal Republic of Germany developed very differently in the past 25 years. While the paper industry could increase its production output by more than 400 per cent, i.e. to approximately 6.4 million tons a year in the years from 1950 to 1973, the production of the pulp industry only increased by approximately 50 per cent, i.e. to 720,000 tons a year in the same period (cf. Table 1).

Compared with world production, the pulp production of the Federal Republic of Germany is relatively unimportant. It consists of approximately 75 per cent pulp and approximately 25 per cent artificial-fibre cellulose. Thus, a great proportion of the paper industry's demand for pulp must be imported. The imported proportion of the raw material used for the paper produced in the Federal Republic of Germany amounts to approximately 35 per cent.

TABLE 1

General survey (1973)

Product	Production	Proport. of world production (%)	Import balance (million t)	Sales (million DM)	Factories
Cellulose	0.72	0.8	1.9	500	14
Paper, cardboard	6.36	4.0	2.0	5560	280

Source: Refs. 1 and 2.

The waste-paper proportion of the complete cellulose material used in paper production amounts to 48 per cent; this proportion, which even exceeds the rate of Japan, which disposes of little raw materials, presumably will not exceed 50 per cent provided that the needs and markets do not change considerably.

The cellulose generated, in fourteen factories is produced exclusively according to the bisulphite technology, since sulphate-type cellulose factories could not be installed in the densely populated territories of the Federal Republic of Germany because of the bad smell of their exhaust gases. While three of the fourteen cellulose factories producing approximately 10,000 tons a year and therefore are to be considered as relatively small production units, a medium production capacity of 75,000 tons a year is calculated for the remaining factories.[3]

The output of approximately 280 factories of the paper industry, which comprises a few big companies and a large number of medium-sized firms, varies between 1000 and 350,000 tons a year. This structure results from the history of the industry and in this respect is similar to the French, Italian and English paper industries.

2. EMISSIONS OF THE PULP INDUSTRY

In an international comparison, the pulp industry of the Federal Republic of Germany, which requires an average of 220 m^3 waste water per ton of pulp, has a low specific waste-water rate. These waste-water quantities occur in four sections of the production: in the pulping shop, in the bleachery, the by-product generation and the pulp dewatering. Pulping and bleaching result in 80 per cent of the waste water and the highest proportion of emissions.

The CSB emissions, which consist of dissolved organic wood substances, viz. 10 per cent thereof, have their origin in the vapour condensates of the evaporating plants, by which the sulphite liquors from the *pulping shop* are evaporated (cf. Fig. 1). With the present coverage degree (related to CSB) of the sulphite liquors from the pulping shop, which amounts to approximately 70 per cent, the remaining liquor causes an average emission of 340 kg CSB per ton of pulp; this value corresponds to approximately 60 per cent of the current CSB emissions of the pulp industry.[3]

The waste waters from the *bleachery* contain approximately 150 kg CSB per ton of pulp, which means a contribution of approximately 20 per cent to the CSB emission.[3] The most important load on the waters from the bleachery, however, does not reside in the CSB emissions but in the relatively toxic chlorine compounds, which currently amount to an average of 22 kg per ton of bleached pulp and to approximately 13,800 tons per annum.[4]

Finally approximately 14,000 tons of solid materials are emitted by the West German pulp factories each year in the category of treatment of sewage effluents.

To be able to draw a comparison between the indicated harmful materials, water standards are applied for determining their importance. To this end, a relative water load R_w is defined.[4]

$$R_w = \frac{M_i}{V \cdot WSD_i} \qquad \text{(equation 1)}$$

wherein M_i are the annual emission quantities of the emission component i,

V is the estimated medium river volume, which receives the emissions in the Federal Republic (V = 1.35 billion m^3)[4], and

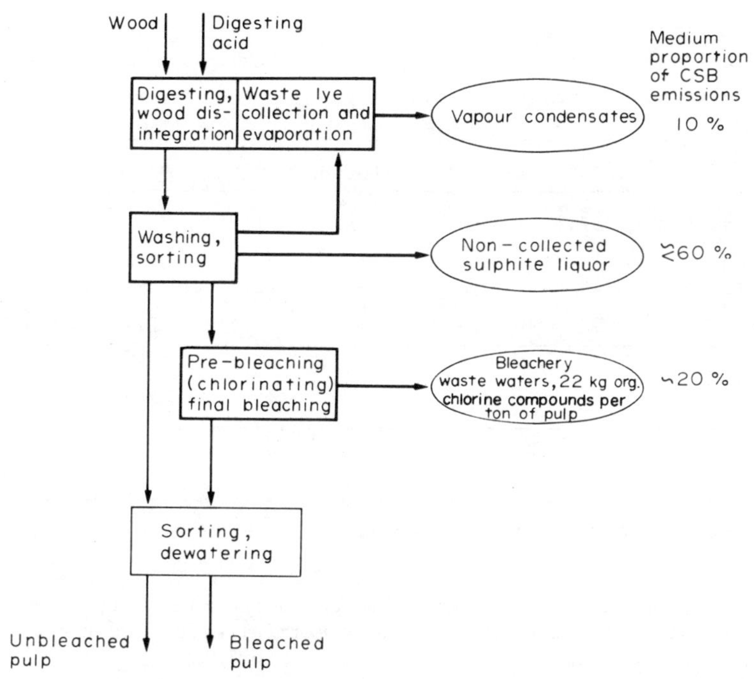

Fig. 1. Crucial points of emission in pulp production. 2 - Wood. 3 - Digesting acid. 4 - Digesting, wood disintegration. 5 - Waste lye collection and evaporation. 6 - Vapour condensates. 7 - Medium proportion of CSB emissions. 8 - Washing, sorting. 9 - Non-collected sulphite liquor. 10 - Pre-bleaching (chlorinating), final bleaching. 11 - Bleachery waste waters, 22 kg org. chlorine compounds per ton of pulp. 12 - Sorting, dewatering. 13 - Unbleached pulp. 14 - Bleached pulp.

WSD_i is the water standard of the emission component i, in this case:

- CSB: 15 mg per l[6]
- org. chlorine compounds: 0.1 mg per l[5]
- chloride: 250 mg per l[5]
- removable materials: 0.1 ml per l[6]

Thus the relative water load R_w indicates how many times per year the river volume would have to be replaced in order to secure water-quality standards against emissions. This determination makes it possible to combine different emission components and to estimate the importance of the separate crucial components of emission with respect to their influence on the environment or the urgency of new low-emission technologies. Assuming a replacement of the river volume of 80 to 100 times per year,[22] it occurs that solely the emissions of the pulp industry constitute an impact on river water quality close to the tolerable limits (cf. Table 1).

TABLE 1
Relative load on waters from the pulp industry (1973)

Section of emission	Emission quantities 1000 t p.a.			R_w			ΣR_w
	CSB	org. Cl	chloride	CSB	org. C*	chlr.	
Vapour condensates	37.5	-	-	1.85	-		1.85
Non-collected sulphite liquor	245	-	-	12.1	-		12.1
Bleachery	93	13.8	35	4.6	102	0.01	106.61
Total of solid materials		14.0			4.7		4.7
Total of pulp industry							~128

*Assumed conversion factor: 0.022 kg/l.

As can be seen from Table 1, for the time being the bleachery rather than the non-collected sulphite liquor has the most harmful effect on the waters, with the organic chlorine compounds alone making approximately 75 per cent of the total load from pulp production.

The importance of organic chlorine compounds from the pulp bleachery can be demonstrated by means of a cost estimate for the procuring of water from the bank filtrate of the lower Rhine; the costs of drinking-water preparation would be reduced from approximately DM 0.7 to DM 0.6 per m^3, if the organic chlorine compounds from the pulp factories connected to the Rhine (including the French ones) were avoided and hence the active-carbon adsorption of drinking water preparation were considerably reduced.[7]

3. BLEACHING OF THE SULPHITE PULPS WITH HYDROGEN PEROXIDE

Since more than 80 per cent of the pulp production in the Federal Republic is bleached and in future surface waters will have to be increasingly applied as drinking-water sources, the development of a low-emission bleaching method is of very high importance, if the other possibility of applying an (expensive) post-treatment method for bleachery waste waters by adsorption or similar processes is disregarded.

Therefore it is the main aim of a low-emission bleaching method to substitute chlorine as a bleaching agent in pre-bleaching as far as possible so as to avoid the emission of the relatively toxic organic chlorine compounds.

While hydrogen peroxide has already been applied with the lignine-conserving bleach of groundwood pulp, the production of half stuffs and pulps of medium brightness as well as in the last sequences of a multi-stage bleach, hydrogen was not yet applied in the multi-stage bleaching of sulphite pulps and more particularly in the first stage thereof until 1974. Since the use of peroxide appeared successful in the first bleaching stage as a result of laboratory tests,[8] tests on a large technical scale were started in 1974 in a pulp factory of the Federal Republic of Germany and these tests were financially supported by the Federal Ministry for Research and Technology.

4. RESULTS ALREADY ATTAINED BY THE FACTORY TESTS

The industrial tests were carried out in existing plants (200 tato); the reaction parameters of the peroxide stage (P_E) could always be adapted to the degree of decomposition in the digesting shop. The operating temperatures of the P_E stage ranged between 80° and 85°C without resulting in any difficulties and recognizable bleaching agent losses.

With leaf wood sulphite pulps (e.g. beechwood - rayon - pulp), which was digested more softly, the use of chlorine could be dispensed with completely. It was possible to bleach to brightness degrees of more than 90 per cent MgO by means of the following sequences:[1]

P_E - H - D - H,

P_E - H - D - or

P_E - D - H.

With harder pine wood sulphite pulps (e.g. spruce-paper pulp) chlorination could not be dispensed with completely; only 1 - 2 per cent of chlorine was applied rather than the usual 5 - 7 per cent. The paper pulp qualities obtained by the sequences

(C) - P_E - H - D - H,

(C) - P_E - H - D,

(C) - P_E - D - H,

(C) - P_E - H

did not suffer any stability loss as compared with the qualities obtained with types conventionally bleached with the two stages C-E.

Furthermore, the tests have shown that approximately 80 - 90 per cent of the CSB emissions from bleacheries occur in the waste waters of the pre-bleaching stage (P_E or (C)-P_E). Since the bleachery waste waters of the peroxide bleach do not contain any or only few chloride ions, they can be evaporated without problems either for recycling or for combustion without any difficulties. Such a preparation or collection of the bleachery effluents of the C-E stage was not possible because of the high chloride-ion concentration (corrosion).

Apart from the considerable reduction of the CSB emissions by 80 - 90 per cent and the chlorine compounds by 90 - 100 per cent, a reduction of the specific water consumption by 60 - 80 per cent, i.e. to 30 - 40 m^3 per ton of pulp, could be reckoned with for bleaching because of the modified water circuit.

As a result of first cost estimates, the bleaching costs of the new process are approximately DM 20,00 to DM 30.00 per ton higher (without recovery of the alkali components). Considering that at least DM 50 per ton had to be paid for treating the waste water from a conventional bleach by physical and chemical methods (with a much worse degree of elimination), the peroxide bleach appears to be a low-emission and cost-saving method with good prospects for the future, particularly if the lower fresh-water consumption and the waste-water tax to be paid with the beginning of 1981 are considered.

The development activities with regard to the peroxide bleach will be terminated in mid-1977.

Other activities focus the development of a chlorine-free bleach for harder sulphite cellulose, e.g. bleaching sequences with peroxide and chlorine dioxide,[8,10] with peroxide and peracetic acid[8] or with hydrogen (under pressure) and peroxide.[11-14]

5. EMISSION OF THE PAPER INDUSTRY

Similar to the pulp production, the specific waste water rate of the paper industry of the Federal Republic of Germany, which amounts to approximately 50 m^3 per ton of paper, is on a relatively low level if compared internationally. As a matter of fact, the values are very different for the separate factories and types of products, the individual values range between 820 m^3 per ton (special papers) and no waste water (papers from waste paper).[3]

The charges on waste waters from paper production, on the one hand, are a function of the raw materials applied (groundwood pulp, cellulose, waste paper, rags) and their preparation and, on the other hand, are due to materials which are added to the wet section of the paper machine (pigments, fillers, resins, starch, aluminium sulphate, refining chemicals).

Amounting to less than 12 kg CSB per ton, the specific charge of dissolved organic materials with paper production waste waters is much lower than with cellulose production. The content of removable materials amounts to 425 litres per ton on an average. These emission data already consider the current standard of waste-water treatment in the paper industry of the Federal Republic of Germany, for which a mechanical and chemical cleaning of approximately 50 per cent of the production and an additional biological clarification of approximately 20 per cent is indicated.[3]

The efficiencies of the mechanical or mechanical and biological clarification technologies as well as of the mechanical and biological treatment in Table 2 show that, it is true, the removable materials were sufficiently eliminated by a 100 per cent mechanical and biological treatment, which, however, does not apply to the CSB emissions with an average degree of elimination of approximately 55 - 70 per cent (according to the type of production).

TABLE 2

Table 2. *Degrees of elimination and application of final waste-water treatment* (1973/74)

Final waste-water treatment	Degree of elimination						Appln. in F.R. of Germany
	Removable materials			CSB content			
	Medium	Min.	Max.	Medium	Min.	Max.	
Mech. or mech./chem.	0.97	0.94	0.99	0.26	0.06	0.46	0.50
Mech./biological	0.998	0.99	0.000	0.55	0.50	0.72	0.17

Sources: Refs. 3, 4 and 15.

The annual emission quantities of the paper industry of the Federal Republic of Germany, which are calculated from the data of Table 2, likewise were expressed as relative water charge in line with equation (1). Since even with an assured 100 per cent mechanical and biological post-treatment of the factory waste waters nearly 30,000 tons of CSB would be emitted, in the last years the question was discussed whether the water circulation should be closed for paper qualities other than simple packing papers, too.

TABLE 3

Relative load on water by the paper industry (1973)

Emission field	Emission quantities Removable solid mat.	1000 t p.a. CSB	R_W Rem. sol. mat.	CSB	R_W
16 factories for simple packing papers	0	0	0	0	0
Paper industry total	59.3	75.0	20	3.7	23.7

Paper production with closed water circulation and without any emission of harmful materials (cf. Table 3) is a very recent development and today it is substantially limited to waste-paper processing factories.[16-19]

6. COMPLETELY CLOSED CIRCULATION IN PAPER PRODUCTION

With closed circulation, any waste water which arises is used anew without or with corresponding clarification, i.e. only the water lost by evaporation and the losses due to the moisture of discharged solid residues and the final product have to be replaced by fresh water.

Figure 2 shows a simplified block diagram of an example of closed circuit. In this case primary and secondary circuits are operated without stores or material and liquid separation, i.e. the dilution from the paper machine and the material disintegration is carried out by filter water. In the tertiary circuit, the waste water is clarified by different methods (e.g. mechanical and chemical clarification, in the circumstances rubble filters, adsorption and ion exchange) to such an extent that the water from the clear-water reservoir (Res. 2) can be dispended to the places which would be conventionally fed with fresh water.

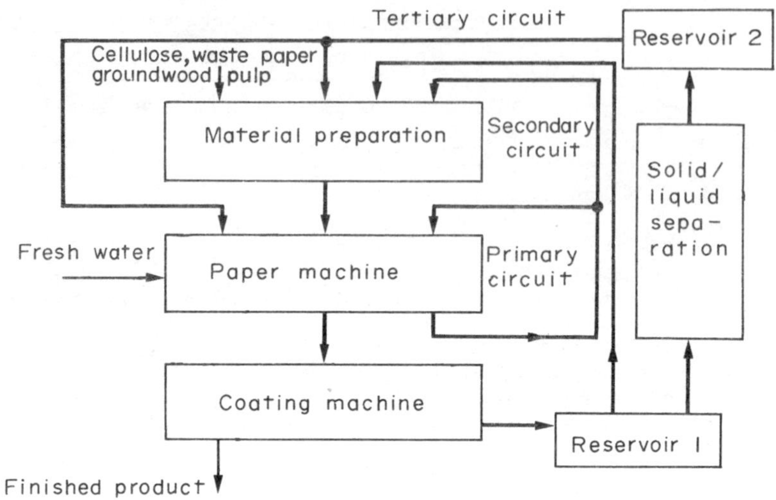

Fig. 2. Block diagram of a closed water circuit in paper production. 2 - Tertiary circuit. 3 - Cellulose, waste paper, groundwood pulp. 4 - Material preparation. 5 - Secondary circuit. 6 - Reservoir 2. 7 - Solid/liquid separation. 8 - Fresh water. 9 - Paper machine. 10 - Primary circuit. 11 - Coating machine. 12 - Finished product. 13 - Reservoir 1.

The advantages of closed water circuits with paper production are as follows: very low specific water consumption, increased material yield, no emission as well as only unimportant amounts of sewage sludges, since the thick materials can be recycled to the material preparation.

As the problems of closed circuits with manufacture of high-quality papers still are considerable (e.g. salt deposition, sludge formation, smell development, residual colours, corrosion), it was decided in 1975 to start a long-term research programme of 10 years, which is to systematically investigate the problems in laboratories and on minor industrial scale and which is to elaborate solutions. The Federal Government supports these activities, which are primarily carried out at the "Wasser- und Abwasserforschungsstelle des Verbandes Deutscher Papierfabriken" with the "Institut für Papierfabrikation der Technischen Hochschule Darmstadt", so far over a period of 4 years with the sum of DM 1.7 millions.

7. SOME RESULTS OF THE RESEARCH PROGRAMME

Table 4 shows some properties of circulated water in a closed circuit.

TABLE 4

Content	Range	Unit
Chlorides	200- 3200	mg Cl^- per litre
Sulphates	350- 3200	mg SO_4^- per litre
BSB_5	2000- 8000	mg per litre
CSB	4000-12000	mg per litre
H_2S	e.g. 225	mg per litre

The observed high concentrations of cations and anions, which are introduced into the system through semi-materials, auxiliary materials and fresh water, cause increased corrosion, which should be restricted by the selection of corresponding materials, if possible.

First laboratory tests with pinewood sulphite cellulose or waste paper show that no notable decrease of the tearing length of the laboratory sheet could be observed in spite of strong *salt concentration*. However, a distinct stability reduction occurs after artificial ageing (in a drying cabinet at an air temperature of 150°C for 16 hours) with high ion concentrations. Similar observations were made with respect to the degree of remission (degree of whiteness) of the sulphite pulp. With waste paper, the pressure required for corrugating and upsetting was not influenced by high ion concentrations.

Furthermore, enormous high BSB_5 and CSB values (*concentration increase*) were observed with completely closed water circuits. Laboratory tests have shown that the concentrations of the two sum parameters for organic substances approximate a balance value, i.e. the finished product receives a higher CSB load than with the open system.

Comparative investigations with respect to *micro-organisms* in water circuits of some waste-paper processing factories have shown only an unimportant higher germ charge in the closed circuit as compared with the open system. Only the anaerobic total number of colonies was higher by one order.

The results of the laboratory tests were widely confirmed by the factory results of a West German tissue factory, which has a nearly closed circuit with a water consumption of approximately 1 m^3 per ton. To avoid full exchange of the circuit water in the event of colour changes between white and coloured qualities, the circuit water was subjected to a bleach with chlorine-containing bleaching agents, if necessary, which agents widely disinfected the water as a by-effect and thus rotting and glut formation were avoided.[21]

According to the test and investigation results already attained, the possibility emerges that 20 per cent of the paper production can be done without discharge of waste water, while a further 70 per cent might considerably reduce the water circuits so that a total reduction of the water consumption by 60 - 70 per cent is feasible.

The emissions, particularly of organic contents, will be reduced to a similar extent (50 - 70 per cent) without consideration of the increased application of the clarification of the residual waste water.[4]

To what degree the waste water quantity and emissions can be further decreased under the condition of uniform product quality and economically possible costs on a long-term basis, can only be estimated in the course of the oncoming years on the basis of the research results and the practical experience to be expected.

The investment and operating expenses of the circuit technology are estimated to amount to 30 - 100 per cent of the investment and operating costs of the open systems with downstream residual waste water clarification. Increasing water prices in future and the introduction of the waste-water tax in 1981 will increase the pressure for efforts to introduce circuit technology.

REFERENCES

1. Stat. Bundesamt, *Statistisches Jahrbuch*, Wiesbaden, 1975.

2. Verband Deutscher Papierfabriken, "Papier 1973 - Ein Leistungsbericht", Bonn, 1974.

3. G. Rinke, L. Göttsching et al., Gutachten über einzel- und volkswirtschaftliche Auswirkungen des geplanten Abwasserabgabengesetzes auf die Papier- und Zellstoffindustrie. *Studie i.A. des Bundesministers des Innern*, Darmstadt, 1975.

4. E. Jochem and J. Wiesner, Beurteilung von F&E-Projekten im Bereich Umweltfreundliche Technik. *Studie i.A. des Bundesministeriums für Forschung und Technologie*, Karlsruhe/Frankfurt, 1974.

5. Internationale Arbeitsgemeinschaft der Wasserwerke im Rheineinzugsgebiet, IAWR-Grenzwerte. *Gas-Wasser-Abwasser*, 53 (6), Sonderdruck, 1973.

6. Bundesrat, Gesetz über Abgaben für das Einleiten von Abwasser in Gewässer, Drucksache 373/76, Bonn, 1976.

7. J. Reichert and M. Fischer, Systemanalytische Arbeiten auf dem Gebiet der Wasserreinhaltung. *Studie i.A. des Bundesministeriums für Forschung und Technologie*, Karlsruhe, 1976.

8. Patentschriften der DEGUSSA, DBP 2219504 und DBP 2219505 sowie Offenlegungsschrift 23 27 900.

9. H. Krüger and W. Traser, Möglichkeiten des Ersatzes von Chlor durch Wasserstoffperoxid beim Bleichen von Sulfitzellstoff. Vortrag vor der Drei-Länder-Kommission des Umweltausschusses der Zellstoff- und Papierindustrie BRD, Österreich, Schweiz am 1.9.1975 in Linz.

10. N. Liebergott, *Pulp and Paper Mag.Can.* 72, 109-117 (1971).

11. H. Makkonen and M. Ranua, Sauerstoffbleiche der Sulfitzellstoffe. *Das Papier*, 29 (10A), V25 - V32 (1975).
    H. Makkonen, M. Pitkänen and I. Laxen, Vortrag zu Int. Pulp Bleaching Conference, 3-7 June, Vancouver, B.C., Canada (1973).

12. Christensen, P.K., *Norsk Skogindustrie*, 28 (2), 39 - 41 (1974).

13. Kymmene AB, Kuusankoski, Deutsche Auslegungsschrift P 2040 763.0-45.

14. Ch.E. Farley and M. Gruyson, American Cyanamid Corp., US Pat. 3 719 552.

15. H.-L. Dalpke and L. Göttsching, Über den spezifischen Abwasseranfall und die spezifische Schumtzfracht von Papierfabriken. *Wochenblatt für Papierfabrikation*, 102 (19), 721 - 730 (1974).

16. W. Brecht and H.-L. Dalpke, Geschlossene Wasserkreisläufe in einer altpapierverarbeitenden Papierfabrik. *Wochenblatt für Papierfabrikation*, 100 (16), 579 - 585 (1972).

17. W. Brecht, H.L. Dalpke and F. Börner, Untersuchung der geschlossenen Wasserkreislaüfe in einer weiteren altpapierverarbeitenden Papierfabrik. *Wochenblatt für Papierfabrikation*, 101 (4), 1 - 7 (1973).

18. H. and O. Widmer, Geschlossener Wasserkreislauf in einer Papier- und Kartonfabrik. *Wochenblatt für Fabrikation*, 100 (23/24), 930 - 932 (1972).

19. W. Brecht, H.L. Dalpke and F. Börner. Geschlossene Wasserkreisläufe in weiteren altpapierverarbeitenden Papierfabriken. *Wochenblatt für Papierfabrikation*, 102 (7), 223 - 234 (1974).

20. H.L. Dalpke, Besondere Schwierigkeiten bei der vollständigen Kreislaufschliessung. *Das Papier*, 29:(6) 236 - 240 (1975).

21. L. Göttsching and H.L. Dalpke, Chancen und Risiken der Wasserkreislaufschliessung in Papierfabriken. Vortrag (17.6.1976) Baden-Baden (wird im Oktober 1976 im "Vortragsheft" von *Das Papier* veröffentlich).

22. R. Keller, Wasserbilanz der Bundesrepublik Deutschland. *Umschau*, 73(3), 73-78 (1971).

Treatment and Preparation of Dusts and Sludges in the Steel Industry

M. Haucke and W. Theobald

Verein Deutscher Eisenhüttenleute, Düsseldorf, Federal Republic of Germany

The production of iron and steel involves generation of more or less exclusively anorganic residues, which primarily consist of non-usable residual slags, used refractory linings, rubbish, scale, dusts and sludges (Fig. 1). The total quantity of waste materials from crude iron, crude-steel and rolled-steel production can be assumed to amount to approximately 29 million tons for a crude-steel production of 50 million tons.

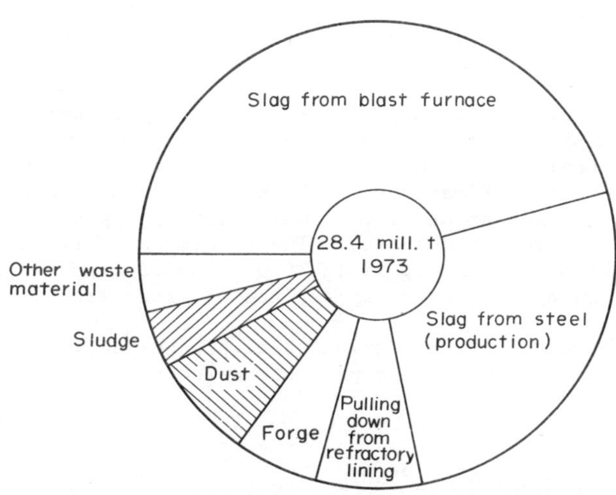

Fig. 1. Waste materials from steel production in the Federal Republic of Germany, 1973.

The big waste quantities, insufficient dump surfaces, consideration of the requirements of landscape design, the physical and chemical properties of the waste, which are favourable for further use of the above materials, and the recognizable market capacity led to the establishment by the steel industry of the technical and organizational conditions necessary to produce and sell by-products developed from residues in recent years. Most important is the use of blast-furnace slag, of which 10.73 million tons were produced in 1975 and which was combined with slag taken from dumps as the basic material for construction materials, street construction materials and fertilizers. The smelting plant sand produced by water granulation of liquid blast-furnace slag amounted to 3.3 million tons and thus constituted a considerable proportion of utilized blast-furnace slag, which was primarily used for the production of cement. Minor quantities are applied in street construction. Apart from this utilization exemplified by blast-furnace slag and from distribution of slags without economic use for secondary construction purposes, the in-plant utilization of waste materials results in a reduced waste quantity to be deposited in dumps. For example, 2.4 million tons of low-phosphorus steel-plant slag were used as base carrier in crude-iron production in 1973.

An investigation carried out in 1973/74 first showed a sure quantity figure for the production and destination of the waste materials in the steel industry:

Waste materials: 573 kg/ton of crude steel, 100 per cent.
Sale and distribution: 320 kg/ton of crude steel, 55.9 per cent.
Own utilization: 148 kg/ton of crude steel, 25.9 per cent.
Deposit: 53 kg/ton of crude steel, 9.3 per cent.

Thus, the total quantity of waste materials was reduced from 573 to 53 kg per ton of crude steel by virtue of sale, distribution and own utilization so that only approximately 10 per cent of the waste materials from the production in blast-furnace, steel-factory and rolling mill plants had to be deposited. The remainder of 9.1 per cent not included in the list is composed of distributions to waste-disposal companies, temporarily stored blast-furnace and steel-mill slags and waste materials removed by other steps such as combustion.

The materials most difficult to handle from aspects of waste management and waste removal undoubtedly are the dusts and sludges, which are exclusively generated by the operation of plants for keeping air and water clean, with the exception of unimportant remainders. The proportion of dusts and sludges is shown in the third quadrant of Fig. 1, which covers approximately 11 per cent of the total waste material quantity. The deposited dusts and sludges ascertained amounted to approximately 13 kg TS per ton of crude steel and are contained in the shown deposit of 53 kg per ton of crude steel.

The dusts and sludges occurring during steel production will now be described in a sequence corresponding to the production sequence.

BLAST-FURNACE COKING PLANTS

The companies of the steel industry established outside the water-management districts of the state of North-Rhine Westphalia started to use biochemical methods for cleaning the waste waters from coke production at the beginning of the sixties. The first activation plant of this type was built with Metallhuttenwerke Lübeck GmbH, which developed the "Lübecker Becken" (Luebeck basin), which later became known for the treatment of municipal sewage water. The next step was the use of sewage plants, in which the activation basin is combined with the final clarifying basin to form a circular structural body. Another clarification plant of a coking plant applied surface ventilators, use of which has the advantage of a reduced current consumption and reduced investments, but also the disadvantage of higher space requirements and higher cooling in winter as compared with the aforedescribed type of construction (Fig. 2).[1,2] Nowadays, the sewage corresponding to approximately 30 per cent of the production of blast-furnace coking plants is treated by biochemical methods.

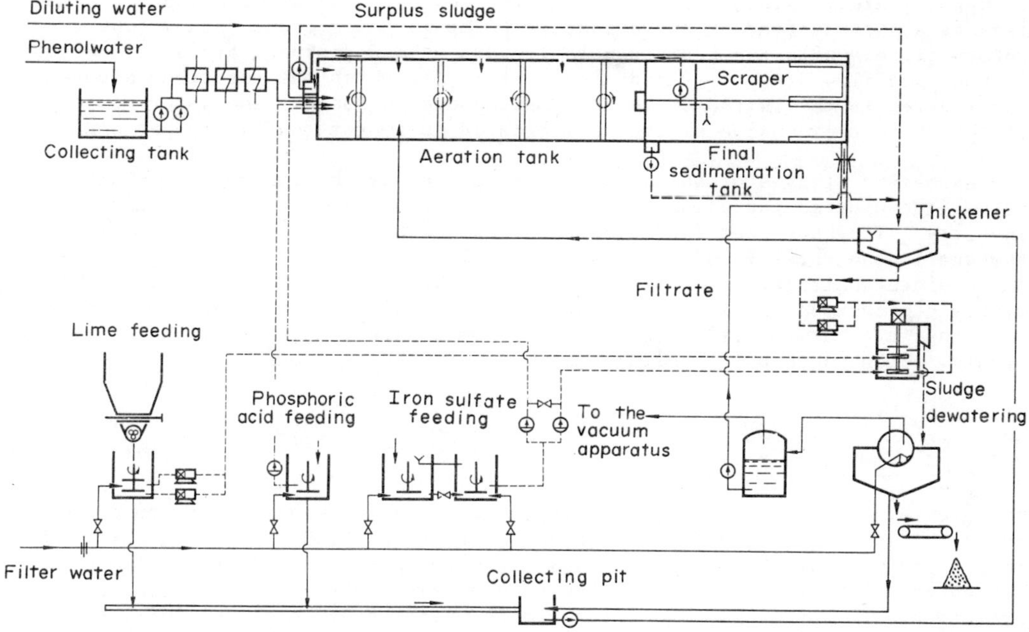

Fig. 2. Biochemical plant for the treatment of coking plant sewage.

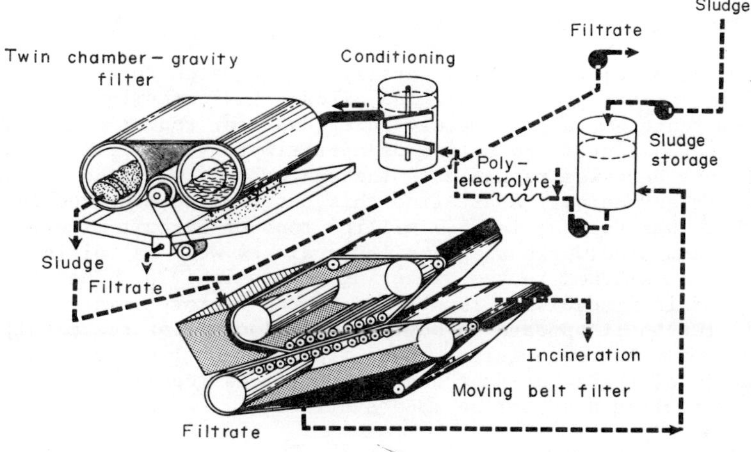

Fig. 3. Drainage of excessive sludge from coking-plant sewage treatment.

The highly loaded one-stage activation plants for clarification of coking plant sewage produce approximately 0.15 - 0.45 kg TS sludge per kg of phenol decomposition according to the particular operating conditions so that the excessive sludge currently produced in activation plants amounts presumably to approximately 25,000 m^3 per annum. Thickeners equipped with rabble mechanisms of a surface feed of approximately 20 kg TS per m^2d and vacuum-type rotary filters with the addition of lime are used for sludge drainage. Figure 3 shows the flow diagram of a drainage plant, by which the described sludges are treated in the United States of America and in Japan. The water content of the drained excessive sludge is indicated to amount to 80 - 85 per cent.

The excessive sludges of the sewage-treatment plants of coking plants are primarily burnt as the sludges from the activation plants of the chemical industry. They have not installed their own combustion furnaces because of the small quantities involved; the excess sludges for the time being are fed on to sinter belts or are applied with the coking coal. The solutions applied widely depend upon the specific factory; investigations in other countries too show that a treatment of excess sludges from sewage treatment plants of coking plants, which meets all requirements, has not yet been found.

SINTERING PLANTS

The growth of crude-iron production and the increased requirements to be met by big blast furnaces, which now produce more than 10,000 t/d, have resulted in a rapid worldwide increase of capacities for producing lumps of fine ores, concentrates, mill scale and flue dust in recent years. When building agglomerating plants, primarily sintering plants were chosen in the vicinity of ironworks, while pelletizers were preferred in the vicinity of ore mines or in the import harbours. Induced-draught sintering plants were exclusively built in the Federal Republic of Germany. The dusts of such plants are grouped according to their origin, viz.

dusts of mechanical treatment and of transport of the material and of the sinter and

dusts in the exhaust gas of the sinter belt.

The total amount of sinter dusts generally is estimated to amount to 1 - 2.5 per cent of the sinter production. The already indicated investigation showed a medium occurrence of approximately 14 kg dust per ton of sinter and 550,000 tons of dust for a sinter production of approximately 40 million tons, and 99 per cent of this dust was returned to the sinter belt with the returned materials after screening. Apart from this dust, which occurred in the same plant and which was used again, 5.7 million tons of circuit materials such as blast-furnace and steelworks slag, rolled scale as well as rolling mill dusts and sludges were returned to the sintering plants in 1973. Therefore, the sintering plant is of decisive importance in the factory's material circuit, and for this reason it appears to be the central plant of the metallurgical waste management. Raw materials such as iron ore and lime are replaced by the utilization of these residues and dump surfaces required if such utilization were not applied need not be made available.

Recycling of waste materials in the sintering process faces requirements of keeping the air clean.[3] The electrical gas cleaning of sintering plants lately proved to be that portion of the plant, which imposes limits to the recycling aimed at by waste management or which at least strongly affects recycling.[4] The surface glow fires, which repeatedly occurred in dry electric filters of the exhaust gas cleaning of sintering plants and which resulted in distortions of the deposit plates and destruction of the internal equipment of the filters, are generally explained by the use of oil and grease-containing rolled scale and the corresponding deposit of hydrocarbons and their cracking products on the filter discs. For this reason, it was proposed some years ago to equip sintering plants with cyclones and downstream washer (Fig. 4). The mechanical preliminary cleaning additionally would facilitate the circuit of the separated dust and provide for enrichment of undesired materials in the second dust-removing stage. Such an enrichment is also given with the dry electric filters used nowadays. The fine dusts occurring in the last field of electrostatic gas cleaning may be proportionally discharged from the material circuit and thus the composition of the returned dust can be influenced. Meanwhile, after continued structural development of the exhaust gas cleaning for sintering plants, another solution for utilization of oil-containing rolled scale is feasible, which will still be dealt with. The multi-stage exhaust-gas cleaning, however, lately has gained importance with exhaust gas desulphurization for sintering plants.[5]

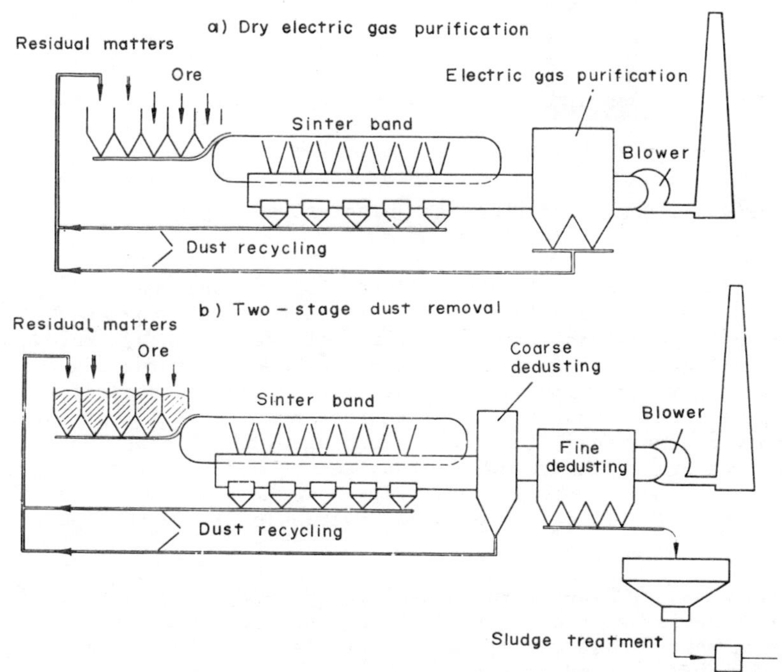

Fig. 4. Sintering plant of material circuit.

BLAST-FURNACE OPERATION

Crude-iron production in blast furnaces implies the generation of blast-furnace gas, which has a temperature of 200°C on the throat and a dust content

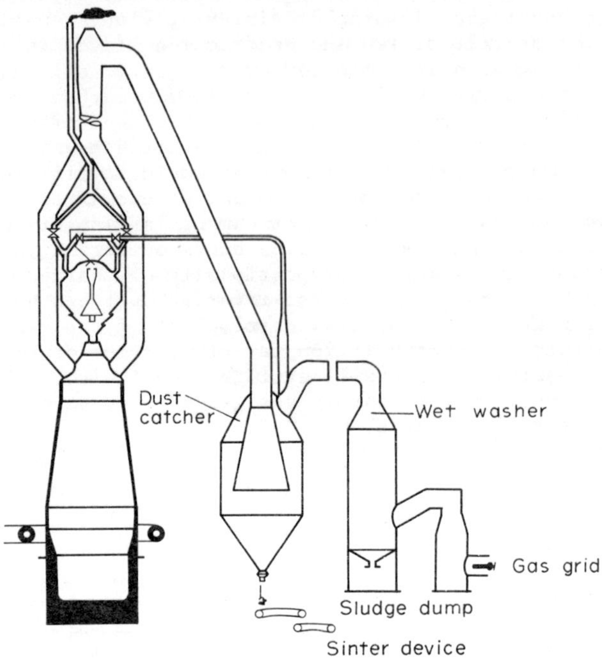

Fig. 5. Dust arresting with blast-furnace gas.

of 10 - 30 $g/m^3{}_n$. Even until 1960, generation of more than 100 kg dust per ton of crude iron was calculated, and this dust generation meanwhile was reduced to average values of 50 kg per ton of crude iron by improvement of the burden preparation. In earlier plants, the blast-furnace gas is cleaned in a mechanical stage primarily consisting of a gravity separator (dust bag) and of a centrifugal separator provided in a downstream low-pressure washing cooler and wet electric filter. Gas cleaning with new blast furnaces of increased pressure at the throat is realized by a dust bag or centrifugal separator and high-pressure washers (Fig. 5). The dust content of the clean gas ranges below 1 - 10 $mg/m^3{}_n$ with the available plants and operating conditions.

The process water of blast-furnace cleaning nowadays usually are of circuit-type. The process water of low- and high-pressure washers need not meet high requirements as to solid materials so that clarification in settling basins provides for sufficient cleaning. With new blast furnaces, however, sludge contact plants have been primarily used in the last decade.

The arrangement of a centrifugal separator does not influence the dust content of clean gas, but a cyclone may offer advantages for dust utilization as a result of the reduced sludge generation and of the changed composition, since the zinc and lead content of the dust increases with decreasing grain size (Fig. 6). Similar results are attained with steelworks dusts, the coarse proportion of which contains 0.2 per cent zinc and the fine proportion of which (less than 200 μm) has zinc contents of 1.5 - 3 per cent. On the other hand, investigations have shown that an increased coarse proportion increases the drainage capacity of the sludge.

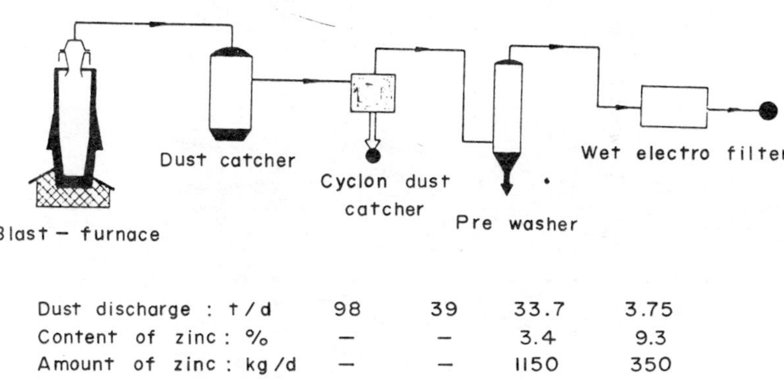

Dust discharge : t/d	98	39	33.7	3.75
Content of zinc : %	–	–	3.4	9.3
Amount of zinc : kg/d	–	–	1150	350

Fig. 6. Dust and zinc occurrence with blast-furnace gas purification.

Flue dust has an iron content of usually 35 - 50 per cent and is completely applied in the sintering plant. Flue gas washing water sludge can only sometimes, i.e. with high zinc content, be distributed to the non-iron metal industry for further processing so that the flue gas washing-water sludge, which occurs, is to be deposited completely on dumps.

Flue gas washing-water sludge nowadays is seldom deposited in the form of thin sludge. The water content of a sludge transported to a deep dump amounts to approximately 95 per cent, and after 1 year a water content of 40 - 50 per cent at a density of 1.7 g/cm^3 is attained. Apart from these exceptions, flue gas washing-water sludge nowadays is drained by means of vacuum-type rotary filters or in filter presses prior to transport to the dump, and filter presses are primarily used in new plants. Comparative tests with flue gas washing-water sludge and practical experience with the use of vacuum-type rotary filters has shown water contents between 30 and 60 per cent, while 25 - 40 per cent were attained with filter presses. Interconnection of additional thickeners or the addition of polyelectrolytes did not bring about any advantages so that such steps are dispensed with. Solid-jacket centrifuges did not prove satisfactory for drainage of flue gas washing-water sludge as the centrifuge outlet contained high solid contents, which were due to the high fine-material content of the centrifuge inlet and to the enrichment of these materials in the flue gas washing-water circuit.

STEELWORKS

In the last decade, steel production in the Federal Republic of Germany rapidly developed towards oxygen blast steel (Fig. 7). It is expected that the production of Thomas steel will be cancelled by 1980 and the proportion

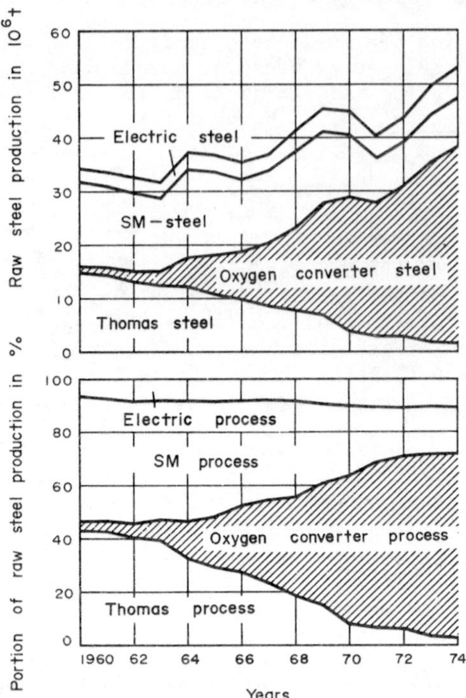

Fig. 7. Development of crude steel production in the Federal Republic of Germany.

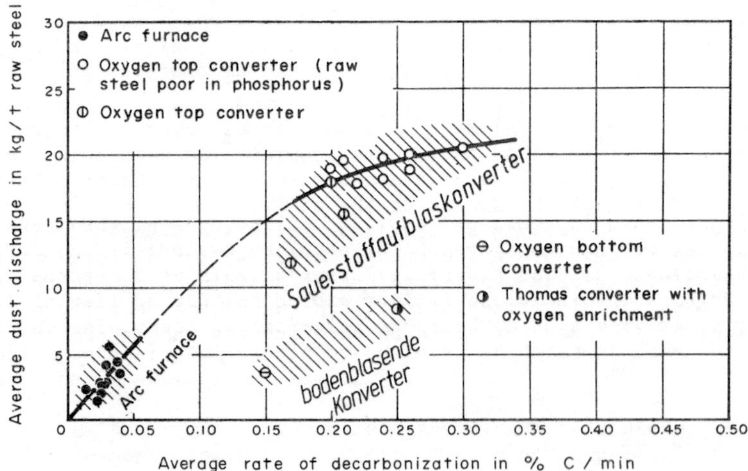

Fig. 8. Dust occurrence with different steel production methods with refining by gaseous oxygen.

of openhearth steel will range between 12 and 15 per cent. From the point of view of waste management and waste removal, therefore, the dusts and sludges of oxygen blast steel production are of particular importance (Fig. 8).[6] The dust contents, which are usual nowadays, can only be attained by two-stage high-power washers or at least two-field electric filters. Some oxygen blast steelworks are equipped with a dry dust-removing system, although only washers were applied in the new constructions of recent years.

The process waters resulting from the wet dust removal and exclusively used in circuit in the Federal Republic of Germany are clarified in two stages. Coarse separation is provided in an upstream hydrocyclone, current classifier or in a gravity separator subject to high surface charges in which the coarse-grain proportion of a grain size of approximately more than 200 μm is separated. Drainage of the separated coarse-grain stage is done, for example, by means of a rake classifier or by a scraper conveyor belt structurally connected to a gravity separator, by which final humidities of 10 per cent are attained.

For drainage of the sludge occurring in the second cleaning stage formed as a thickener, primarily vacuum-type rotary filters of cell structure are applied. The residual moisture of the filter cake usually range between 25 and 40 per cent and even to 60 per cent in special cases. Filter presses should be of greater importance in steelworks in future, since handling of these sludges with reduced water content is facilitated and further preparation generally requires a water content which cannot be attained by vacuum-type rotary filters. Thermal drying is provided in one German mill only (Fig. 9). Granulating driers, the granulate from which can be easily transported and used in sintering plants, are applied in steelworks abroad.

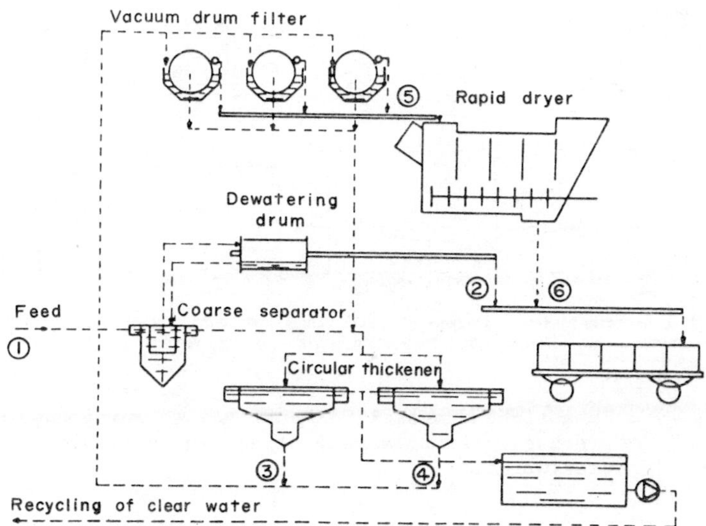

Fig. 9. Treatment of sludges from dust arresting in oxygen blast steelworks.

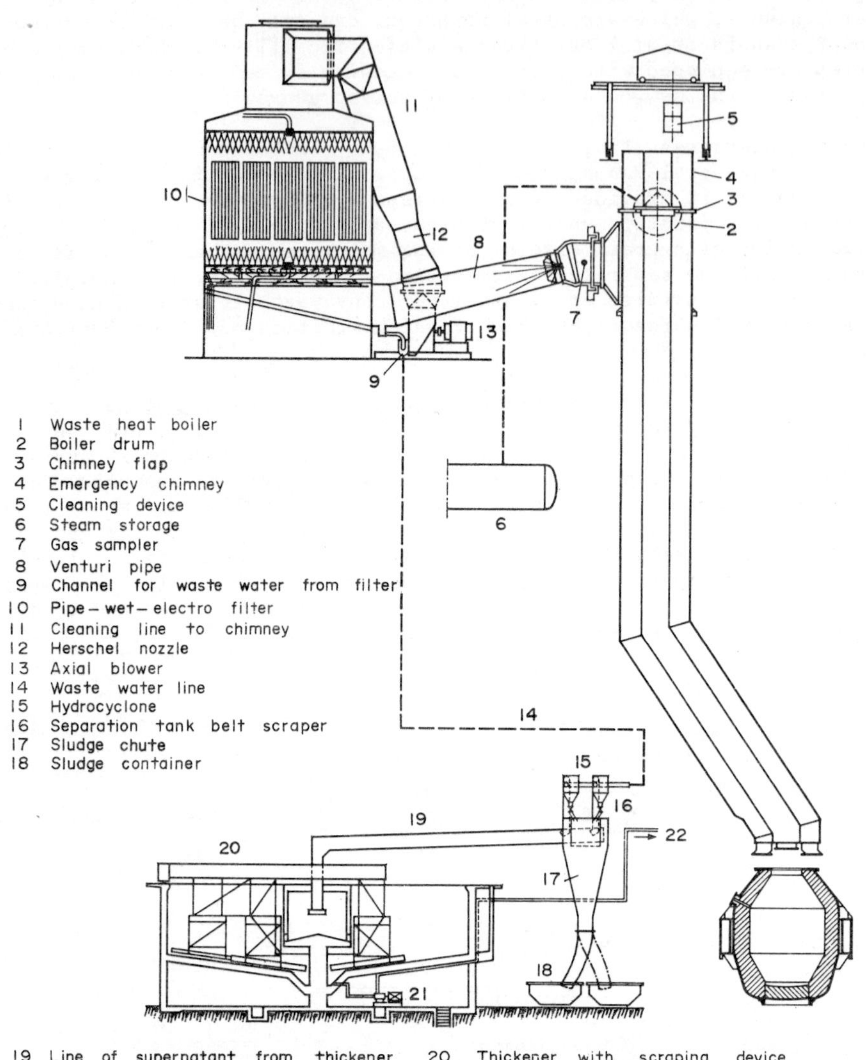

Fig. 10. Exhaust-heat vessel, exhaust gas cleaning and process water treatment of an oxygen blast steelworks.

Eighty-five per cent of the complete dusts and sludges occurring in dust arresting of oxygen blast steelworks again were recycled to the crude-iron production through the sintering plant in the year of generation. In this connection, a method proposed by the Dillinger Hüttenwerke AG should be noted

(Fig. 10). In this factory, the thin sludge is withdrawn by means of an infinitely variable piston diaphragm pump of a solid content of approximately 300 g/l and is fed to this mixing barrel of the sintering plant through a 820-m pipeline. There it is utilized for moistening the starting material. The flow fed by the sludge pump is controlled according to the water requirement of the mixing barrel.[7]

The addition of fine-grain sludges with sintering gave rise to the expectation that there are effects on the sintering output and the properties of the sinter produced if there is a greater proportion of circuit materials. However, investigations made by the Studiengesellschaft für Eisenerzaufbereitung have shown favourable results with respect to the return material occurrence, the sintering output, the barrel stability and the speed of reduction with a 15 per cent proportion of circuit materials and 5 per cent steelworks dusts. Variously, changes in the preparation of the sintering material such as the application of the filter-type sintering method were suggested above all for reducing the influence of fine grains on the gas permeability of the agglomeration layer and the fuel consumption. For sludge drainage, this method employs vacuum-type rotary filters, on which castings of identical size can be produced by a grate rotating together with the filter belt (Fig. 11).[8]

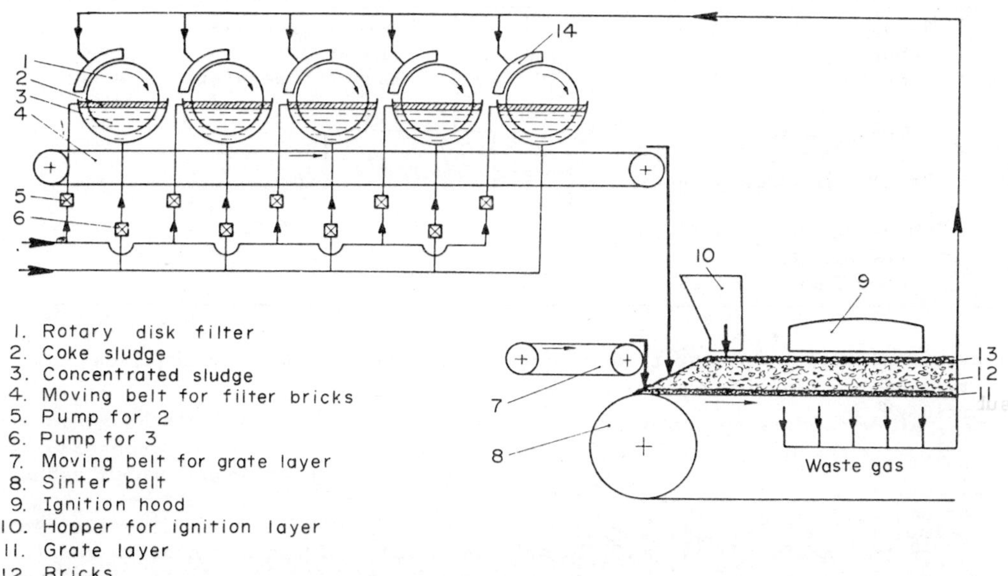

Fig. 11. Schematic structure of a sintering plant after the filtrating and sintering process.

In 1973 approximately 580,000 tons of dusts and sludges from blast-furnace cleaning and from dust-arresting with oxygen blast steelworks were deposited. Utilization of these materials is primarily restricted by their zinc contents, which have a negative effect on the subsequent condensation by the formation of extensions and which favour the chemical wear of the refractory lining. The alkalis contained in the circuit materials have a similar disadvantageous effect. The alkalis entering the furnace together with the burden and the coking coal likewise attack the refractory brick lining of the blast furnaces so that a satisfactory life of the lining for the severely stressed area of the bosh, of the coal bag and of the lower furnace portion cannot be ensured with the conventional refractory materials.[9,10] A further restriction of the repeated use of dusts and sludges may result from their grain distribution.

These interdependences clearly show that the dusts and sludges to be deposited can only be reduced in future if the zinc, lead and alkali-containing waste materials from the metallurgical processes can be successfully separated and utilized. Table 1 lists the methods applicable for this purpose. In the following, all methods, which appeared particularly favourable as a result of a research programme currently carried out by the German steel industry, will be summarized.

TABLE 1

Processes for the preparation of zinc and lead-containing waste materials from steel production

Pyrometallurgical processes	*Hydrometallurgical processes*
Rotary kiln processes	Physical processes
SL/RN process	Flotation
Krupp process	Magnetic separation
Kawasaki process	
SDR process	
SPM process	
Bethlehem-Steel process	
Ferrocab process	
Chlorinating volatilization	
LDK process	
Kowa-Seiko process	

A pyrometallurgical preparation of dusts and sludges generally requires mastery of the following process steps:

- drainage,
- drying,
- bunkering and dosing,
- mixing,
- agglomeration,
- sublimation of non-iron metals,
- generation of an iron-containing furnace discharge and of a discharge of enriched non-iron content, which are suitable for utilization in the crude-iron or steel production and in the non-iron metal production.

The investigations considered a great number of combined methods for mechanical preparation and agglomeration of the residual and waste materials to be prepared. The choice of a process sequence to be applied decisively depends upon the water content of the sludges and upon the requirements placed by the subsequent thermal processing with respect to the agglomerates.

The lower the water content of the starting materials, the simpler the mixture and the greater the choice of possible devices. Mixers with rotating mixing tools, such as baffle mills and Loedige mixers with a downstream crushing mill (rod mill) for homogenization and milling of already formed fine agglomerates, proved particularly useful in the tests.

The production of pellets, sinters and positive castings for forming lumps of the mixed and homogenized waste materials was considered. Investigation of the agglomerates produced in view of strength and thermal decomposition behaviour has shown that pellets of chemical and physical properties sufficient for the subsequent preparation can be produced from the occurring dusts and sludges of flue gas cleaning and trapped steelworks dust. This method, however, is particularly expensive due to the required drying of the sludges. For the time being, it cannot be judged finally, whether pellet production will be substituted by a cheaper method of free or positive shaping. A very simple solution, which has already been tested on large industrial scale for certain mixtures, is the production of a granulate by rolling on a pelletizing plate. The granulate was of irregular shape in the range of 2 - 20 mm and was successfully processed in a rotary drier.[11] Briquetting of the dusts and sludges appears to be impracticable, as briquetting also requires partial drying, as the briquettes could only be separated from the mould with difficulty and as the thermal properties of the briquettes did not meet the requirements. A common feature of all methods is that the water content of the prepared starting materials covers a relatively limited range.

The development of preparation methods for zinc and lead-containing dusts and sludges is focused on rotary-kiln processes derived from the rolling methods for preparation of zinc ores. The differences between preparation of zinc ore and waste and residual materials from steel production reside in a lower non-iron and a higher iron and alkali content as well as in the fine grain size of the waste materials. In the rolling method the zinc contents of 10 - 25 per cent usually contained in the burden are reduced in the rotary kiln with a certain excess of reaction agents (Fig. 12).[12] The zinc reduction is similar to the iron reduction in the temperature range between 900 and 1100°C. The volatalized zinc is oxidized in the oxidizing kiln atmosphere to become zinc oxide and is separated together with lead oxide or lead sulphate which is likewise formed, after cooling in corresponding filter plants, in enriched form as rolling oxide. Until now approximately thirty rotary kilns according to the rolling method of length between 25 and 95 m have been built.[13]

In 1963 Krupp first carried out major investigations with respect to the preparation of flue gas washing-water sludges in a 14-m rotary kiln, at the request of an association consisting of nine ironworks, after preliminary investigations for the Rheinhausen ironworks had been carried out. A total of 150 tons of charging material was processed in those tests and a volatilization of 95 - 99 per cent of zinc and of 97 to more than 99 per cent of lead was attained. The formation of extensions was not overcome by this test series so that further tests for the rolling method for preparation of ironworks dusts and sludges were initially dispensed with in Germany.[14]

After extensive investigations, several test campaigns in 14-m and a 41-m rotary kiln were carried out by Lurgi Chemie and Hüttentechnik GmbH, August Thyssen Hütte AG, Berzelius Metallhütten-Gesellschaft mbH as well as Fried. Krupp Industrie- und Stahlbau GmbH in the years from 1974 to 1976 and mixing ratios between 25 and 50 per cent flue gas washing water sludge and 50 - 75 per cent sludges and dusts from dust trapping in oxygen blast steelworks were checked. The results of the investigations can be summarized in the finding that final contents in iron-containing furnace discharge of less than 0.05 -

0.1 per cent were attained with a zinc volatilization of 95 - 99 per cent. Lead volatilization varied considerably, showing final contents between 0.001 to 0.1 per cent Pb. The grain composition of this discharge only admits a restricted application in blast furnaces due to a proportion of 35 - 70 per cent above 4 mm for the different processes and makes it necessary to apply agglomeration with part of this material. Processing in steelworks is affected by both the grain size and a relatively high sulphur content. The Zn and Pb content of the rolling oxide was determined at between 40 and 55 per cent. The generation of mineral solids from preparation of the iron-containing discharge of the furnace is low and substantially consists of fuel ashes.

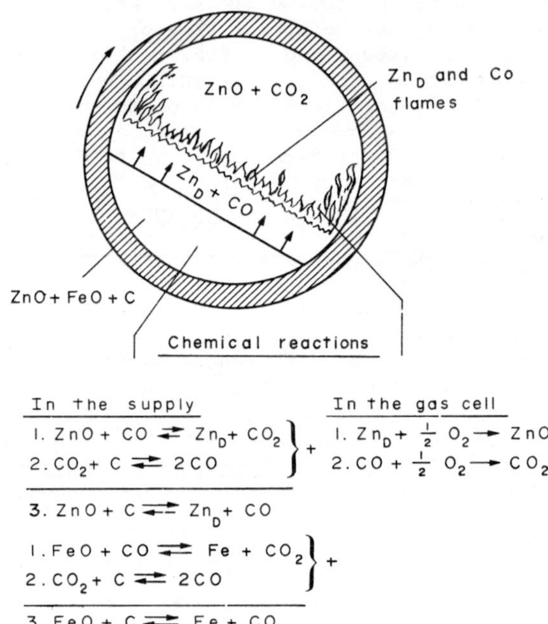

Fig. 12. Zinc volatilization in rotary kiln (according to F. Johannsen).

This report cannot deal with the differences between the investigated processes. Extensive investigations still are necessary to ascertain still unknown technical characteristics. A process flow sheet for carrying out the process is exemplified in Fig. 13.[14]

Approximately 20 per cent of today's crude steel production results from open-hearth furnace smelting processes with high scrap content, which will be subjected to dust arresting by dry electric filters in future. Approximately 70 per cent of the occurring dust is soluble in water and cannot be deposited in non-processed state. Low iron contents as well as relatively high zinc, lead and alkali contents prohibit utilization of this dust in crude iron and steel production. The investigations so far carried out with respect to the application of hydrometallurgical processes for preparation of zinc and lead-containing dusts and sludges have shown that this process, according to the current state of knowledge, can only be reasonably considered for preparation of waste of high zinc content and thus only for preparation of selected dusts and sludges. Hydrometallurgical preparation therefore appeared reasonable for open-hearth dusts, since these dusts cannot be prepared in a rotary kiln. The high alkali contents result in a reduction of the smelting point as well as in an unfavourable fusing behaviour of the throughput and favour an increased formation of extensions. Furthermore, it is to be feared that the major portion of the alkalis enriches in the iron-containing furnace discharge and the applications for this discharge are restricted.

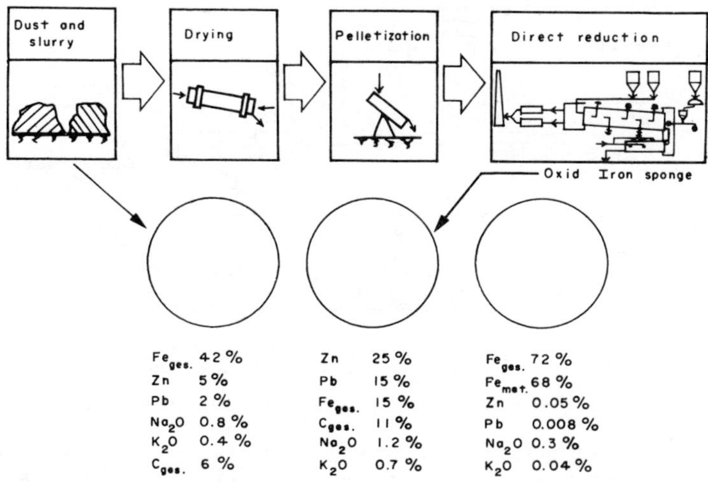

Fig. 13. Preparation of dusts and sludges in the rotary kiln.

The investigations so far carried out in this respect with Hoesch Hüttenwerke AG and Lurgi Chemie und Hüttentechnik resulted in process proposals (Fig. 14). The dust withdrawn from the electric filters is adjusted to a pH of 8.7 - 9 by the addition of water and caustic soda, the washed sludge is thickened up to a solid content of 10 per cent and drained in a filter stage. The results so far attained show that the drained residue of a Zn and Pb content of more than 20 - 25 per cent can be used for non-iron metal production. Further investigations aim at developing a suitable utilization for the dissolved Na-K sulphates occurring with the eluate. A prototype plant on the basis of the elaborated flow diagram is being built by the Hoesch Hüttenwerke AG. The solution for processing these dusts appears urgent because this company already disposes of an operating plant for dust-arresting for an open-hearth furnace and the dust so far must be temporarily stored in bags.

ROLLING MILLS

The process water utilized in hot rolling mills serves to cool the rolls, the roll bearings and the rolled material as well as to descale and to transport the scale to the clarification plant. A step-by-step preparation is usually applied for the process water charged with oil and scale and primarily held in circulation, such preparation being the most secure and economic method for treatment of rolling mill sewage.

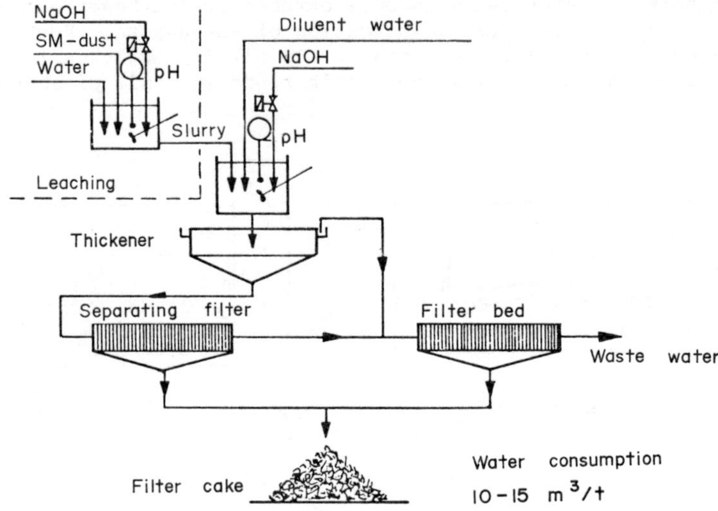

Fig. 14. Preparation of open-hearth furnace dusts.

Mechanical cleaning of rolling mill sewage results in 2 - 7 per cent and in an average of 4.4 per cent of rolled-steel production in the form of scale. In 1973 the rolled-steel production amounted to 36.7 million and the corresponding scale quantity to 1.64 million tons. To treat the rinsing water from rubble filter systems, the addition of flocking agents is often required. In some factories, even sludge-contact plants are applied. Clarification of this sewage involves the generation of oil-containing and fine-grain rolling-mill sludges, which are difficult to treat.

The majority of the scale is returned to crude iron production through the sintering plant. A small amount of very coarse scale is directly applied in the blast-furnace or even in the steelworks. The operational difficulties and damage with exhaust-gas purification with sintering plants if rolling scale containing oil is applied have already been described. The iron content, the amount of scale occurring and the undesired deposition of oil-containing waste materials require investigations so as to secure utilization of oil-containing rolling scale and oil-containing rolling-mill sludges in the future. To this end, combustion of the oils and greases and the application of washing processes as process technologies are being investigated.

Rolling scale is removed from gravity separators by grippers in all German ironworks. Big plants have developed different designs of scale bunkers for drainage, temporary storing, distribution and, to a certain extent, for mixing scale. The first stage of a complete utilization of oil-containing rolling scale in future will, unavoidably, be the separation of foreign matter from the scale and classification of the scale. In an expensive development,

Hoesch Hüttenwerke AG now apply a mechanical preparation for rolling scale. Classification by hand could not be dispensed with. With an output of 1000 t/d, approximately 5 tons of foreign matter/d are separated.

The current research programme for waste management and waste removal in the steel industry includes investigations for application of the fluidized-bed furnaces with sand bed and scale filling for combustion of the oil contained in the rolling scale. The results attained confirm the suitability of the fluidized-bed method for pre-treatment of oil-containing rolling scale, additional oil admission not being necessary with an oil content of more than 6 - 8 per cent. The resulting roasting charge corresponded to a usual concentrate with a magnetite proportion of 66 per cent, 20 per cent of which can be added to ore sintering plants without affecting the sinter quality. If such roasting charge is added, the efficiency reduction resulting therefrom can be compensated by adding hydrate lime.

The fluidized-bed process is in competition with washing processes, if usual rolling scale of low oil content is to be cleaned from oil down to a residual-oil content of approximately 0.2 per cent to be required from aspects of operating safety with dust removal from sintering plants. Details about the performance of such a method by washing-agent regeneration cannot yet be described, but the application of such a plant in a German ironworks is to be expected in 1977. Approximately 2000 tons of oil and greases per annum will be recovered from the annual scale generation of this company, which amounts to approximately 300,000 tons per annum.

In this connection, a process proposal by A. Supp and K. Zimmermann should be noted. They described the preparation of oil-containing rolling-mill sludges at temperatures between 80 and 90°C and the use of alkaline cleaner solutions (Fig. 15). Investigations for the application of oil-containing rolling scale in rotary kilns are planned.[15]

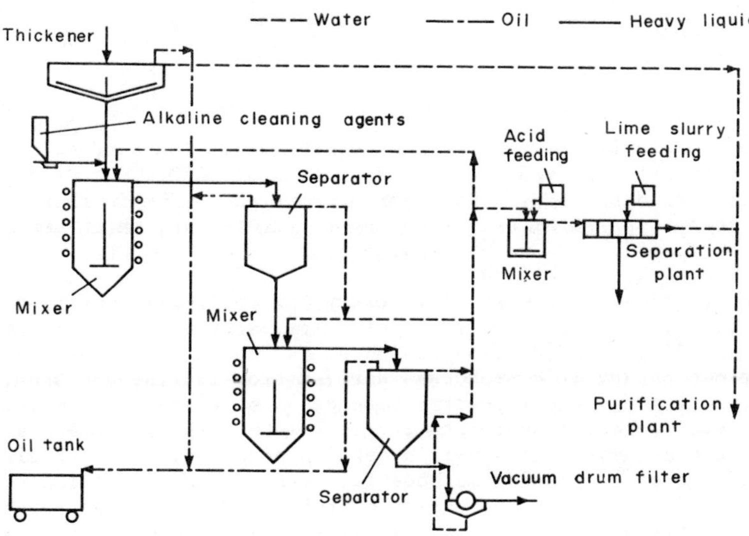

Fig. 15. Plant for removal of oil from rolling-mill sludges.

TABLE 2

Betrieb/ Enttallstoff	Enttall 10^3t/Jahr	Enttall kg/t Rahstahl	Wieder-verwertung kg/t Rahstahl	Deponie kg/t Rahstahl	Haucke Theobald 1976
Kokerei					
Schlammbelebungs anlage*	0.3	0.006	0.004	0.002	
Sonstige Schlomme*	65	1.3	1.3	0.01	
Sinteranlage	550	11.1	11.1	0.01	Schlamm-und Staubentfall bei der Roheisen-und Stahlerzeugung im Jahr 1973
Hochofenwerk					
Gichtstaub	1350	27.5	26.3	1.2	
Gichtschlamm*	437	8.8	-	8.8	
Sauerstoffblas stahlwerk					
Staub	173	17.8	9.3	8.5	
Schlamm*	360	15.1	15.1	-	
Elektrostahlwerk	48	9.2	2.3	6.9	
Walzwerk	63	1.3	0.6	0.7	
Summe					
in 10^3 t/Jahr	3057	-	2407	650	Stahleisen E.W. 762
in kg/t Rahstahl	-	61.7	48.6	13.1	

*bezogen auf Trockensubstanz

SUMMARY AND OUTLOOK

Already today, approximately 80 per cent of the dusts and sludges occurring in steel production are utilized and only approximately 650,000 tons or 13 kg/ton of crude steel are deposited. The major proportion of the deposited dusts and sludges are those types from dust trapping in steelworks, which are produced by cleaning of the flue gas washing water (Table 2).

The research programme of the steel industry obviously will not result in the development of methods for preparation of dusts, sludges and scale occurring in crude iron and steel production, which so far could not be utilized. In this connection, rotary kiln processes derived from rolling and used for the preparation of flue gas washing-water sludge and of dusts and sludges from the oxygen blast steelworks, a hydrometallurgical method for the preparation of open-hearth furnace dusts for non-iron metal production, the mechanical pre-treatment and use of extracting methods for preparation of rolling scale and the fluidized-bed method for combustion of rolling-mill sludges appear to have good prospects.

The results so far attained are not yet sufficient for the design of technical plants so that the operating results of prototype plants often still have to be waited for. This applies, for example, to the agglomeration of the iron-containing furnace discharge, the recovery of the coke excessively added to the rotary kiln process, domination of the alkali and sulphur content and the utilization of the eluates generated by the hydrometallurgical preparation of open-hearth furnace dusts.

Judgments on the possibilities and results of an extended recycling have to take into account that the development is directed towards increased production units and these plants, the quality of the products and environmental protection impose higher requirements on the charging materials and additives. The quantity balance shown in Table 2 meanwhile has become worse as a result of these developments, particularly with respect to utilization of steelworks sludges and dusts. On the other hand, the improved burden preparation and the new flue gas cleaning systems on blast furnaces of increased pressure on the throat result in a reduced production of flue gas washing-water sludge.

In future, waste management will impose its own requirements on construction and operation of plants for protection of water and air. The construction of new plants and improvement of existing plants increasingly have to utilize the possibility of separating the waste materials so as to provide improved conditions for utilizing them.

Utilization of the current waste materials requires high investments and considerable operating costs. These steps must primarily be considered from the point of view of a reduced use of dump surfaces. The possibility of recycling for conserving the natural resources should not be overestimated in this respect. The analytic iron content of the dusts and sludges involved, for example, amounts to approximately 300,000 tons a year, which is very low as compared with scrap, which amounted to approximately 20.6 million tons in 1973. The conditions for recovery of non-iron metals likewise should not be exaggerated, if the current calculations are confirmed. Zinc consumption in the Federal Republic of Germany amounted to 458,000 tons in 1973 with a recycling of approximately 25 per cent and a recovery of 75,000 tons of zinc. On the other hand, it is to be assumed that approximately 18,000 tons of zinc per annum can be recovered from the waste materials of steel production without consideration of the open-hearth furnace dusts.

The results from aspects of conservation of the resources clearly show that decisions for carrying out expensive steps for recovery of waste and residual materials cannot be a function of recycling, but must depend on the actual ecological necessities and economic effects.

The development in the field of air cleaning gives rise to the expectation that the generation of dusts and sludges as a total will continue to increase. This applies particularly to dusts resulting from dust removal from foundries, secondary dust removal, metallurgical main units and locally in view of the exhaust gas desulphurization to be considered for overcharged areas.

REFERENCES

1. W. Theobald and G. Schnegelsberg, *Berg- u. hüttenm. Mh.* 116, 328-340 (1971).

2. D. Drechsel, *Stahl und Eisen*, 91, 509-515 (1971).

3. *Reine Luft für morgen. Utopie oder Wirklichkeit?* Hrsg.: Minister für Arbeit, Gesundheit und Soziales des Landes Nordrhein-Westfalen. Düsseldorf, 1972.

4. M. Haucke, W. Theobald and A. Wutschel, *Stahl und Eisen*, 92, 565-575 (1972).

5. H. Hille and H. Kahnwald, Untersuchung über die Abhangigkeit des Staubgehaltes im Sinterabgas und der physikalischen Eigenschaften des Staubes von den Einsatz- und Betriebsbedingungen einer Sinteranlage. - Abschlussbericht des Forschungsvorhabens der Europäischen Gemeinschaften für Kohle und Stahl, DOK.No. 6252-11/1/218.

6. D. Eickelpasch, H. Kahnwald and H. Tichy, *Stahl und Eisen*, 92, 575-581 (1972).

7. K.H. Bauer, A. Hubert, N. Bock, R. Quinten and N. Weiler, *Stahl und Eisen*, 90, 389-397 (1970).

8. J. Moeljono, Agglomerieren von Feinerzschlämmen durch Vorverformung bei der Filtreirung. Aachen, 1971 (Dr.-Ing.-Diss. Techn. Hochsch. Aachen).

9. Y. Saito, Vortrag anlässlich der 34th Ironmaking Conference des American Institute of Mining, Metallurgical and Petroleum Engineers of 15th April 1975 in Toronto.

10. H. Wysocki, *Stahl und Eisen*, 96, 141-154 (1976).

11. H. Kossek, H. Maczek, H. Rellermeyer and H. Serbent, *Stahl und Eisen*, 96, 482-486 (1976).

12. F. Johannsen, *Metall und Erz*, 26, 4-10 (1929).

13. G. Meyer and R. Wetzel, *Techn. Mitt. Krupp*, 30, 19-28 (1972).

14. G. Meyer, Bearbeitung von Hochofen- und LD-Schlamm im Drehrohrofen. Vorgetragen auf der gemeinsamen Sitzung der Arbeitsausschüsse des Ausschusses für Metallurgische Grundlagen, des Ausschusses für Umweltschutz und des Hochofenausschusses, dated 25th April 1975 in Düsseldorf.

The Application of Material-Saving and Low-Waste Technologies in the Metal Container Industry with Special Reference to Drawn and Wall-Ironed Beverage Cans

Walter Sprenger

Schmalbach-Lubeca GmbH, Braunschweig, Federal Republic of Germany

INTRODUCTION

Even before the energy crisis in 1973/74, questioning the overall economic productivity of entire industries had come into fashion. Environmentalists, consumers and the critical press supporting this trend pay special attention to the packaging industry. They charge this industry with recklessly wasting the resources of our earth and promoting the increase of consumption without restraint. They claim that the industry's consumption of material is unjustifiably high, especially since it fails to exhaust all opportunities for material and energy saving by critical value analysis. As a consequence, the already strained ecological balance is burdened even more.

This global reproach is not only exaggerated but, as can be seen from the following details, completely fails to hit the nail on the head.

One of the major marketing problems of the metal container industry is the rather pronounced homogeneity of its products. Enterprises in this industry can make equal use of practically all existing technologies. And, as in all markets for standardized commodities, competition in the tinplate package industry is in the first place price competition. To maintain its profits, the industry must permanently search for and realize cost savings. Competition from similar products and materials overlaps and is aggravated by competition from other materials (e.g. glass, plastics and paper). In view of the fact that raw materials and supplies account for nearly 55 per cent of the total sales volume of the metal container industry, particular attention should be paid to this factor in the form of continuous and intensive value analysis. For this reason, critical value analysis and realization of material and energy savings account for more than 80 per cent of research and development activities carried out by leading enterprises of the metal container industry. As up to 91 per cent of the energy required for the manufacture of metal packages, i.e. from plate production to the finished can, is consumed in the production of the basic material and only the balance of 9 per cent is consumed by the metal-container industry, any reduction of material consumption is at the same time a substantial reduction of energy consumption. Lower energy consumption, in turn, results in a reduction of the environmental impact.

Starting points for a policy of resource-preserving package design in the metal container industry are the continuous reduction of plate thickness and

the improved utilization of plate area in the manufacture of bodies and top and bottom ends for three-piece cans.

Apart from the substance material "steel plate", the surface finishing material "tin" has always been a subject of particular interest. Tin resources in the earth's crust are limited, if seen from an economic point of view. Remarkable progress has been achieved in the recent past as a result of a policy of economizing the use of this diminishing material.

Without any government intervention and simply as a result of the inherent logic of our free enterprise system we achieved what is considered rational.

1. THE APPLICATION OF LOW-WASTE AND RESOURCE-PRESERVING TECHNOLOGIES IN CLASSICAL FIELDS OF TINPLATE PACKAGING

A means of packaging, which has remained unchanged in its overall dimensions for many decades, is the 1-kilo tinplate food can. It is therefore a particularly good example for demonstrating the success achieved in optimizing the material consumption.

1.1. Plate thickness reduction

The most significant starting point on the long road towards economizing of material consumption was the reduction of the plate thickness. It became feasible as

the cold rolling process was introduced in the supplying industry enabling them to supply tinplate within gradually narrower thickness and hardness tolerances;

the decreasing strength of the can - as a result of the decreased plate thickness - was compensated by beading of the body and by new top and bottom end profiles;

the mechanical stress on the can during processing in the canning industry was reduced, for instance by the introduction of sterilization with overriding pressure.

In the course of this optimization process, which extended over a number of decades, the plate thickness of the 1-kilo food can was reduced from

	1935	1976
bodies	0.28 mm	0.20 mm
top and bottom ends	0.32 mm	0.24 mm

The weight of the finished can decreased from 140 g to 98.5 g in 1976, i.e. by as much as 30 per cent.

1.2. Reduction of waste by improved utilization of the tinplate area

Parallel to the weight reduction of the finished can, the process of optimizing the utilization of the plate area was carried out. The first partial success was achieved by the change from the "German standard sheet size" 530 x 760 mm, which had been the only sheet size available for the manufacture of round top and bottom ends until 1938, to the sheet size 642 - 773 mm. The utilization of area could thus be increased from 74.7 per cent to 78 per cent (see Fig. 1).

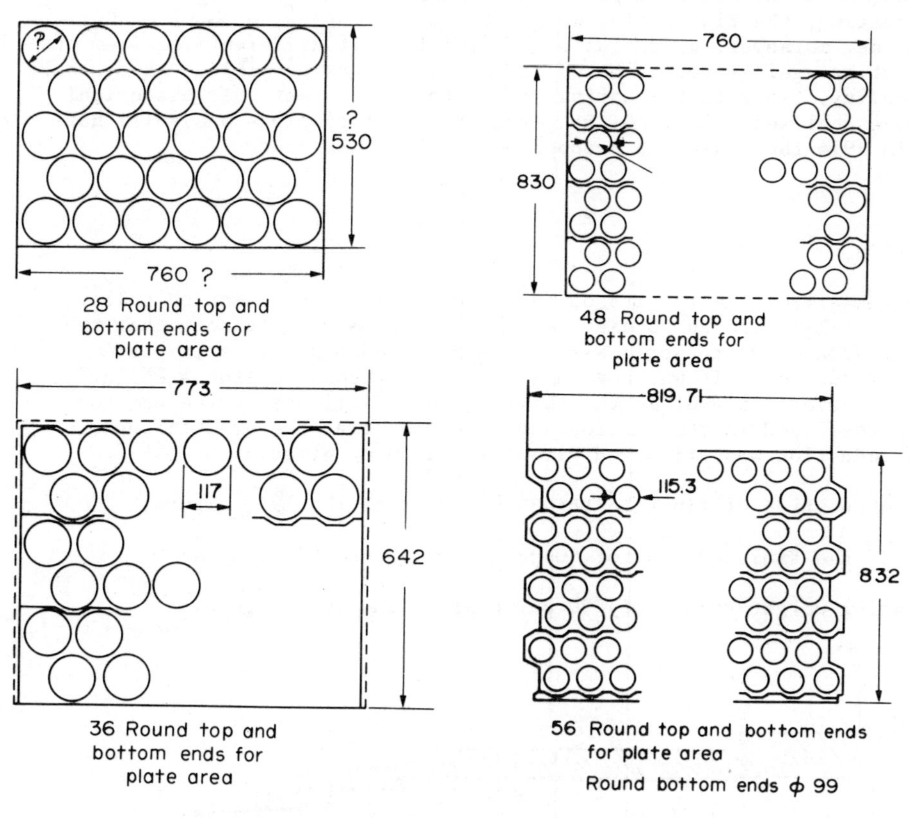

Fig. 1

Fig. 2

Even more remarkable was the progress achieved by the change to so-called double-scrolled sheets. A prerequisite for this step was the installation of larger rolling and tin-coating lines in the rolling mills as well as larger coating lines in the metal-container industry.

The decisive step, however, was the development of the cell shearing process. It resulted in an improved utilization of the sheet area of 85.7 per cent (see Fig. 2).

1.3. Reduction of tin coating weight

In view of the fact that the price of tin exceeds that of the basic substance steel by more than a power of 10 and that, moreover, tin is subject to strong speculative price fluctuations, the steel and tinplate converting industries have always endeavoured to be particularly economical with this material. Tin is used as a means of inside and outside protection against corrosion as well as of side seam soldering. The tin coating weight obtained by the hot-dip tinning process, which was in common use until 1950, was at least 24 g/m^2, i.e. a thickness of about 0.005 mm per side. After the introduction of electrolytic tin coating, the tin coating weight could be reduced to about 0.00015 mm per side and, moreover, differential coating to suit different requirements for protection against corrosion on the two sides was possible. As a result of this policy, tin material savings of up to 80 per cent over historical values were achieved. In 1960, 58 per cent of all tinplate was still hot-dipped, in 1974 the share had dropped to merely 2 per cent.

1.4. Effects

Figure 3 summarizes the results of about 40 years of optimization policy, quoting the 1-kilo can as an example. It is interesting to note that the material shortage due to World War II did not induce progress in plate-thickness reduction. It was not until the market changed from a seller's market to a buyer's market in the fifties and inflationary price and cost situation developed in the sixties and seventies that the breakthrough in material reduction was triggered with the following effects:

total consumption of tinplate to: 64 per cent of the 1930 figure;

weight of the finished can to: 60 per cent of the 1930 figure;

tin coating weight to: hardly 30 per cent of the 1930 figure.

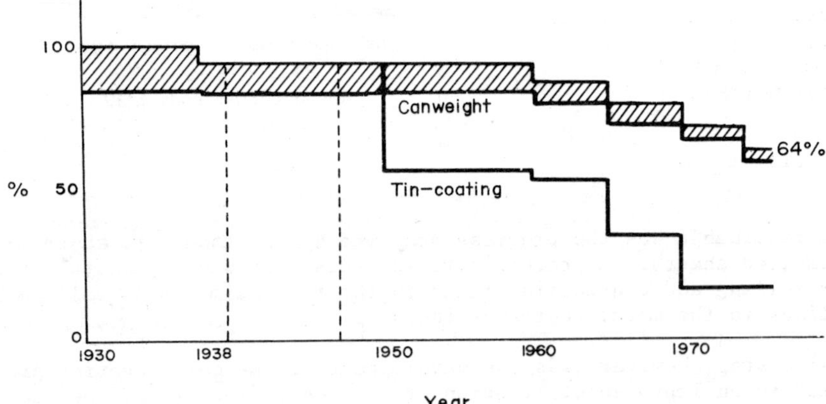

Fig. 3

The continuous price increase for tin has now led to abandoning tin on top and bottom ends wherever this is possible, depending on the corrosivity of the product packed, and using tin-free chromated steel plate instead. Examples are soup cans, pet food cans and drawn fish cans. The share of TFS plate in total plate consumption by the package manufacturing industry rose from 4.7 per cent to 10.4 per cent in the period 1967 - 1975.

To quote another example:

Formerly, 1 ton of tinplate was required to manufacture 25,000 cans for evaporated milk. Today, only 0.63 ton is needed for the same number of cans. Examples of this kind could be quoted without end.

The overall effects of a policy of most economical use of materials applied throughout the entire West German metal-container industry are demonstrated by the following figures:

the share of tinplate between 0.15 and 0.20 mm thickness in total shipments by West German rolling mills to the metal-container industry increased from 20.7 per cent in 1970 to 40 per cent in 1975;

the share of economical tin coatings also increased substantially, as can be seen from Table 1 (in per cent of total).

TABLE 1

	El	Differential coating	Single coating	Total
1965	22.7	1.7	4.0	28.4
1970	22.5	17.4	2.8	42.7
1975	51.2	20.8	2.7	74.7

The average tin coating weight of shipments made by a major West German tinplate manufacturer decreased from 15.02 g/m^2 in 1965 to 10.44 g/m^2 in 1975.

1.5. Other environmental impacts and potential remedies

Apart from the environmental impacts of material production, of waste resulting from container production (see 1.2) and the problems involved in the disposal of household waste (see 2.4) environmental impacts are as inevitable in the manufacture of cans from the basic materials as in any other industrial process. These impacts are closely related with the energy consumption. Although only 10 per cent of the total energy required is consumed in can manufacture proper, as mentioned before, the metal-container industry has always focused its efforts on the elimination or reduction of resulting pollutants.

Pollution can be classified in three types:

water pollution,

noise pollution,

air pollution.

Water pollution. Compared with the technologies for other materials, especially paper and corrugated fibreboard, water pollution in metal-container manufacture is extremely low. In the production of three-piece cans, water is only used for washing of printing plates. Apart from this, it serves as a cooling agent, where it is slightly heated.

Noise pollution. In contrast to this, noise is produced in nearly all operations of can manufacture. It is, however, confined to the production shops. Inevitable noise could in some instances be reduced far below the admissible limits by very expensive measures, such as cabins built around entire machines.

Air pollution. Air pollution is mainly caused by solvent-based coatings and printing inks currently used for surface protection of the packages and by the thermal energy required for drying them. The coatings which are in common use today usually consist of resins which are dissolved in organic solvents at a ratio of 1:2 to 1:5. A very thin coating film is applied and subsequently subjected to a thermal treatment, in the course of which the solvents are expelled and the resins polymerize. Large-scale experimental work carried out to make the solvents innoxious either by condensation or by incineration did not yield satisfactory results which would justify their general introduction industry-wide. Therefore, the industry has tried for some time to find "ecologically safe" coating systems. Some of these systems comprising modified coating materials, modified methods of application and modified drying processes have already been adopted for large-scale industrial use. Modifications were carried out on

materials:

water-based coatings, in which the share of organic solvents is largely replaced by water;

powder coatings which do not contain any organic solvents (especially used for inside coating);

non-varnish printing inks. These improved printing inks are suitable for the production of sufficiently scratch-resistant decorations on cans without need for the hitherto used top coat or varnish. Material and energy consumption are considerably reduced and emissions from the varnishing process are completely eliminated;

the drying process: UV-drying.

This process has already been introduced for drying of printing inks and will rapidly gain importance. The following advantages result from drying of printed sheets with ultra-violet radiation with regard to the environment:

no emission of organic components;

extremely short drying cycle (only seconds instead of 10 - 12 minutes in conventional drying ovens);

reduced energy consumption.

2. THE DEVELOPMENT FROM THREE-PIECE BEVERAGE CANS TO TWO-PIECE DRAWN AND WALL-IRONED CANS FOR BEER AND BEVERAGES

2.1. *The development of the beverage can market*

After having more or less concluded the development of efficient packaging systems in the sixties, lastly for products like wine, milk and milk products, which solve the problems of modern marketing, the only domain left for conventional schemes of scale and packaging is the beer and beverage industry. In view of the problems of material and energy consumption arising in connection with a full-scale change to one-way packages as well as of its impact on the waste volume, it is easy to understand that the public keeps a close eye on this market segment. If the total volume of packed beer and soft drinks (approximately 115 MM hl in 1975) is calculated in units of 0.33 l, approximately 35 billion fills will result.

However, a development analogous to that in the USA cannot be expected for the Federal Republic of Germany. The structural differences between the two countries are so great that a transfer of US American packaging and distribution systems to our market is practically impossible.

To name a few reasons:

In spite of the tremendous size of the USA, compared with the Federal Republic of Germany, its beer and beverage industry is highly concentrated. A small number of companies with few production plants supply areas many times the size of the Federal Republic of Germany. This offers an opportunity for the can to fully display its advantages in distribution. The situation in Germany differs so obviously that it need not be explained by a bundle of figures. Moreover, the conventional distribution system using returnable packages has been firmly established in both the beer and beverage industries in the Federal Republic of Germany, due to the prevailing habit of buying these products in crates in supermarkets or through COD services.

Cans and one-way bottles as an alternative to the conventional means of packaging is therefore no longer at issue. Their function rather is to complement the conventional supplies and to reach the consumer at those places where he will appreciate the convenience and handiness of a can. Moreover, it avails itself for distribution channels where the conventional systems are too expensive. For instance, the can is the package of choice for off-home consumption, whether bought on the road or in the supermarket for consumption during the next weekend trip. The seasonal fluctuations of sales clearly show that trade and consumers accept the can predominantly for such occasions. The limitations of the can are thus defined. Its share in the total output of the beer and soft drink industry - estimated to rise to 165 MM hl in 1980 - may increase to 4.5 - 5 per cent maximum in 1980 according to estimates by the metal-container industry. This share would be equivalent to 2.0 - 2.2 billion units. About 200 MM units of the total canned beer production are likely to be exported. These cans will not add to the waste volume produced in our country. The *per capita* consumption in the Federal Republic of Germany in 1980, based on a population of 60 millions, will amount to about 33 cans per year or 0.6 cans per week. The additional waste produced by an average family of four will be 2.5 to 3 cans per week.

2.2. *Developments for a reduction of material and energy consumption in beer and beverage can manufacture in the past and in the future*

In part 1 of this report, an account was given of the efforts and success of the metal-container industry in pursuing a policy of material and energy cost reducing processes and specifications, taking the 1-kilo food can as an example.

If it were reasonable to pursue such a policy in the conventional fields of application of tinplate packages, this is even more so for beverage cans. Alone the fact that the competitive package "one-way bottle" with accessories costs not more than about 70 - 75 per cent of a can, forces the metal container industry to exhaust all conceivable opportunities for a reduction of material and energy consumption and for an increase in production speeds. The same rule applies as before: about 90 per cent of the energy costs are incurred in the production of the can stock and only the balance of 10 per cent in the production of cans.

2.3. *Economical use of material in the production of beer and beverage cans*

Table 2 gives a very clear picture of the success achieved by the packaging industry on its way towards a permanent reduction of material consumption using the weight of the finished can as a yardstick. Compared with the historical cone top can of the thirties, the can weight could be reduced to about one-half by the soldering process for three-piece cans, while a considerable qualitative improvement could be accomplished at the same time.

TABLE 2

Development of the can weight of 0.35-1 cans from 1935 until 1975

			in % of	
Year	Can type	Weight in g	1935	1951
1935	cone top can with crown cork	93.0	100%	
1951	3 per cent tinplate can with O.T. end	83.0	89.2	100%
1965	3 per cent tinplate can with aluminium easy-open end	70.1	75.4	84.5
1967	" " "	63.6	68.4	76.6
1969	" " "	62.7	67.4	75.5
1972/73	3 per cent necked-in tinplate can with aluminium easy-open end	48.4	52.0	58.3
1975/76	2 per cent drawn and wall-ironed tinplate can with aluminium easy-open end	38.0	40.9	45.8
1976	2 per cent drawn and wall-ironed tinplate can with aluminium easy-open end, lightest US type	34.5	37.1	41.6

Essentially, these savings for the three-piece can were achieved by plate-thickness reductions (see Table 3).

TABLE 3

	1951	1969	1976	1976 in % of 1951 figures
Body	0.30	0.22	0.155	51.7
Bottom end	0.36	0.30	0.26	72.2
Top end		-	-	

Note: ends made of aluminium since 1965.

Activities were comparable for the critical material tin. Specifications developed as shown in Table 4.

TABLE 4

	1960	1970	1975
Body	E2	E1	E1
Bottom end	E2	E2/1	TFS
Top end	E2	Alu	Alu

Note: E1 electrolytic tin coating 2.8 g/m^2 per side
E2 " " " 5.6 g/m^2 " "
E2/1 " " " outside 5.6 g/m^2
inside 2.8 g/m^2

Result:

TABLE 5

Tin required to produce 1 million three piece cans in the years 1960 to 1980

Year	kg Sn	= % of 1960 basis
1960	437 kg	100.0
1970	204 kg	46.7
1975	153 kg	35.0
1975 dwi tinplate can E2	196 kg	44.8
1980 " "	98 kg	22.4

2.4. *The development of the two-piece drawn and wall-ironed beer and beverage can*

About the middle of the sixties, the can-making technology for three-piece soldered or welded cans reached its limits when the body plate thickness had been reduced to 0.155 mm in the USA, the world's largest market for canned beverages. In fact, the steel industry would be capable of rolling even thinner plate, but costs would rise progressively. It was no longer possible to warrant a trouble-free production of cans by the conventional technology (for instance, printing) if a thinner and harder and, hence, more brittle material were to be used. The pressure of the market and the efforts of the aluminium industry to enter into this big market led to the development of the drawing and wall-ironing technology. The first material used for this process was aluminium, in the course of time followed by tinplate, mainly in Europe, e.g. in the UK, the Netherlands and the Federal Republic of Germany.

An essential part of the material-forming operations involved in can-making were shifted from the rolling mill to the can maker by this process. The plate thickness could be reduced from an original thickness of 0.30 mm, which remains constant in the bottom of a two-piece can, to 0.10-0.11 mm in the body.

Thus the weight of the finished can could once more be reduced from the final stage of 48.4 g for the three-piece can to approximately 38.0 g (78.5 per cent). In comparison to the weight of the historical cone-top can of the thirties, this means only 38.0 per cent of the original material weight. The lightest drawn and wall-ironed tinplate can in the United States weighs not more than 34.5 g.

This light-weight can has been the decisive factor in the development trend in Germany since its introduction in 1974. According to the plans of the German can-making industry, the market growth beyond the existing volume of 750 million three-piece cans will fully be covered by cans produced by the drawing and wall-ironing process.

At this stage, where the technology for the manufacture of two-piece tinplate cans from E2 tinplate has not fully matured, no further reductions in tin consumption can be realized which would be comparable with the last stage of development of the three-piece can (specification: body E1, bottom end TFS plate (tin-free)). A switch to E1 for two-piece cans may be expected, however, in the medium term. This would mean another reduction of the tin coating weight of the finished can by another 36 per cent compared with three-piece cans. In other words, the drawn and wall-ironed E1 can will consume only 22 per cent of the tin needed for a three-piece can in 1960.

2.5. *Effects of switch from three-piece to two-piece can-making technology upon household waste volume*

Here are some figures:

The Federal Republic of Germany produced 19 million tons of household waste in 1973. The share of packages was 6.8 million tons (34%). The share of metal packages in the package waste was 0.722 million tons, approximately 22,000 to 25,000 tons of which were beer and beverage can waste.

In Table 6, attempt is made to estimate the metal package waste, based on differentiated growth rates for the entire package waste, and to assess the quantitative effect of the increased use of cans; for reasons of simplification, a complete switch of future beer and beverage can sales to the material-saving drawn and wall-ironed can can be assumed.

The hypothetical figures seen in Table 6 show that - even if a very high growth rate of approximately 22 per cent per year is assumed for beer and beverage cans - the weight share of metal packages in household waste will remain practically constant.

We know that the waste-handling authorities are not so much interested in the weight of the waste, for municipal authorities, the volume of the waste to be disposed is the decisive factor. In this respect the drawn and wall-ironed can, which is gaining ground, has a definite advantage over the conventional can, which is stronger and more rigid. Because of its thin walls the drawn and wall-ironed can may be compressed easily - both by consumer and by the waste-collecting vehicles - and thus requires less space than the conventional can. This results in a better utilization of the waste-collecting vehicles and cost reductions.

The expected plate-thickness reduction (see weight of lightest US can) will result in still easier compression of the drawn and wall-ironed can in the future. Looking at garbage dumps, we can see that dumped two-piece cans have a considerably lower residual volume than conventional three-piece cans.

It may be permissible to make a somewhat provoking statement here. One major problem of national economies with poor raw material and energy resources is the recycling of usable raw materials from household and industrial waste. Recycling of tinplate waste from the metal-container industry has always been practised by recovering the tin in detinning plants for re-use. The recovery of raw materials from household waste is more problematic, because it does not contain usable raw materials to an extent which would easily justify recycling on a commercial basis. Along these lines, a larger share of beverage cans in household waste would improve the chances of recovering the approximately 700,000 tons of metal packages normally contained in the waste by using a magnetic separating process.

CONCLUSION

The above statements show clearly that the research and development activities of the metal-container industry are primarily directed at an economical use of raw materials. Due to the pressure exerted by the market, side effects were obtained which are highly desirable for the national economy. The experience gained in the decades of development activities was transferred to the manufacture of beer and beverage cans. Within approximately 20 years, the weight of a finished can was reduced to less than 50 per cent of its original weight; this was a decisive contribution towards preserving national resources. Our present wall-ironing technology may be expected to bring about further significant progress. In waste management, the thin walls and the flexibility of the drawn and wall-ironed cans permit better reduction of the household waste volume disposal. The progress in magnetic separating of tinplate cans from waste will allow an increasing share of metal package waste to be recycled.

TABLE 6

Effects of switch to 2-piece beer and beverages cans upon share of metal packages in household waste (in tons)

Year	Household waste (1)	Package waste (2)	Metal package waste excl. bev. cans (3)	Market trend bev. cans (in MM cans) (4)	Bev. cans in household waste (5)	Metal package waste, total (6)	Metal package waste share in (2) (7)	Metal package waste share in (1) (8)
1973	19.0 m. t	6.8 m. t	0.7 m. t	500 x 48.7 g	0.022 - 0.025 m. t	0.722 m. t	10.6%	3.8%
Hypothetical growth	2.5%	3%	2%*	approx. 22%		2.7%		
1980	22.6 m. t	8.4 m. t	0.8 m. t	2000 x 38.0 g†	0.076 m. t	0.876 m. t	10.4%	3.9%

* Based on an average real metal package growth rate of approximately 2.5 per cent per year minus effect of further material reduction.

† Based on approximately 4-5 per cent beverage can share in total beer and beverage production.

LITERATURE

Habenicht, Dr. G., *Die Bedeutung von Forschung und Entwicklung für die Verpackungsmittelindustrie im letzten Viertel des 20. Jahrhunderts*, published by Heinz Prast, Ewers Verlag, 1975.

Tangermann, D., Verpackungen aus der Sicht des Herstellers, published in *Neue Verpackung*, 10 (1974).

Tangermann, D., Optimierungsmöglichkeiten bei der Herstellung und dem Einsatz von metallischen Verpackungen, published in *Neue Verpackung*, 5 (1975).

Tangermann, D., Die Weissblechpackung im Rohstoff-Kreislauf, published in *Neue Verpackung*, 8 (1975), Zürich.

Tangermann, D., "Verpackungen und Umwelt", for RGV publication, not yet published.

Walzstahl vereinigung, *Statistik zum Jahresbericht 1975 der Fachgruppe Weissblech*, Düsseldorf, 1976.

Disposal of Ironworks Waste

Rudolf Roth

Mannesmann, AG Huttenwerke, Federal Republic of Germany

Production works are intended for the transformation of raw material to products forwarded either to further treatment or directly to the consumer. By their nature those raw materials are not pure. Moreover accessory substances are needed for any process which do not exist in pure form too. Thus, apart from the end product, waste materials are necessarily produced (by-product, circuit materials, residuary materials, waste products). Today it is a very important task to remove those products in an environmentally desirable way, and to re-use or to recycle them. It would, however, be better to keep the volume of by-products as low as possible from the beginning by suitable methods.

This paper will show how Mannesmann AG Hüttenwerke approached the methodical re-use or recycling of waste products according to their kind, volume and place of occurrence.

In 1975 there were 2.25 million tons of waste products:

1.053 m. t from the blast furnace,

0.962 m. t from the steel plants,

0.054 m. t from the rolling mills,

0.181 m. from auxiliary plants of production ranges.

The waste products are subdivided as follows (in million tons):

Blast furnace slag	0.945
Steel works slag	0.670
Scale	0.148
Refractories	0.011
Vessel slopping	0.134
Sludges and dusts	0.161
Plant rubbish	1.590
Plant refuse	0.020
Oleiferous water	0.002
Oleiferous residuary and refuse material	0.001
Total	2.250

These materials have been disposed of as shown in Fig.1.

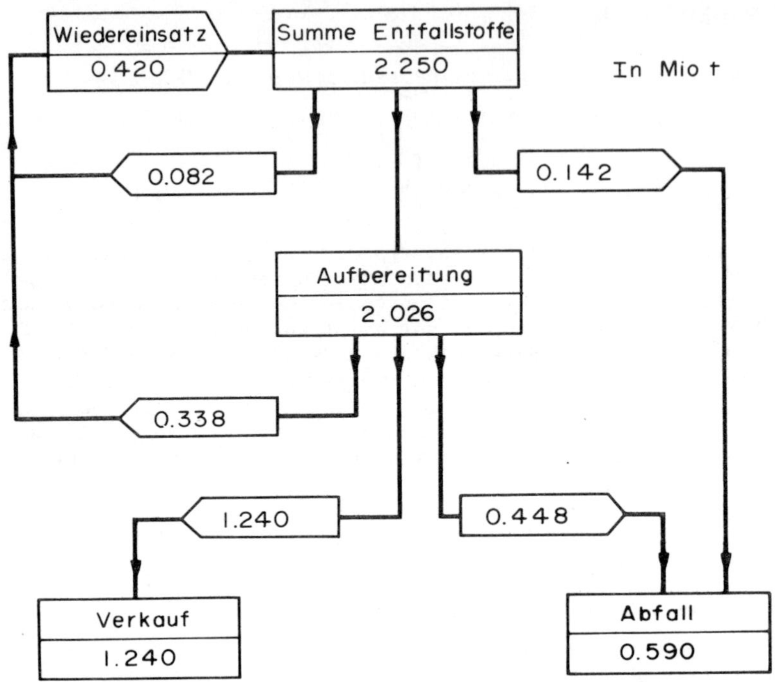

1. *Recycling*. 0.420 m. t = 18.6 per cent of the waste products resulting in the iron works. Of them about 0.053 m. t go to the blast furnace, 0.296 m. t to the sintering plant and 0.070 m. t to the steel plants.

There are flue dusts (40,000 t) as well as filter dusts from the sinter plant supplied to the blast furnace via mixing beds of the sinter plant. Moreover, about 100,000 t = about 75 per cent of vessel slopping, about 32,000 t LD-slags, about 155,000 t ferruginous mill scale and flame scales as well as smaller volumes of water circuit sludges are recycled.

The coarse materials are directly supplied to the blast furnace, the fine particles (< 12mm) are re-used in the sinter plant via mixing beds. Skulls, which are high ferruginous scaffolds, are separated from the steel plant slags and charged in the steel plant instead of scrap.

Recently trials have successfully been completed to recycle a part of the sludge of the vessel dust which is collected separately via the coarse separator. By this means we have succeeded in treating about 10,000-15,000 t = 10-15 per cent of the totally resulting sludges in the sinter plant.

Vessel clearance material is partly used for the production of patching after segregation, crushing, grinding and mixing with tar. Cleared mixer bricks and protective bricks of the OH furnace respectively are conditioned by refractory firms and used as admixtures for refractory mixtures.

2. *Sale*. 1.240 m. t = 55 per cent of waste products; 1150 m. t are slags — especially blast-furnace slags - and 0.09 m. t are phosphorous steel plant slags.

The use of blast-furnace slag is well known as a road building material in splinter grain coarsenings for the production of bituminous covers, binding and bearing layers, and in gravel grades for road foundations as well as its use as an element for the cement industry and for the production of slag wool. The phosphorous slag is sold as fertilizer.

3. Disposal enterprises carry away 0.590 m. t = 26 per cent of waste material. In this respect it must be stressed that these quantities are not all tipped on waste heaps but are partly recycled after preparation by these firms, e.g. for road building, earth dams, reclaimed ground. In fact the quantity to be deposited is much smaller, viz. only about 10 - 15 per cent.

The ratio of the resulting waste material to the material really to be deposited has been reduced to about 7:1 to 10:1.

We shall now report on some measures using special technologies for the reduction of waste materials.

By charging of high ferrous ores (63 - 65 per cent Fe) into the blast furnace the resulting quantity of slags has been reduced. Today it amounts to about 300 kg/t pig iron in comparison with 500 kg/t pig iron about 15 years ago.

Alkalis in the burden result in forming of alkali cyanides in the blast furnace. By charging of ores poor in alkalis into the blast furnace the development of these materials, also known as "cyanic acid", could be reduced to a minimum. Thus, for some years the production of alkali cyanides has been practically nil.

Another way to reduce waste material is to use bonderization sludges resulting from cold reduction, especially in the motor-car industry. Trials have shown that an enrichment of LD slags with phosphor and by that their use as a fertilizer will be possible. An economical utilization according to contemporary thinking is, however, not to be expected, as the resulting quantities of bonderization sludges could only be collected at high cost by reason of the distribution of the sites when they are produced in the Federal Republic of Germany.

Pickling sludges resulting from the neutralization of the residuary pickling acid can practically be eliminated by the introduction of regeneration plants for the pickling acid. Thus instead of the neutralization sludge, e.g. iron sulphate, heptahydrate is arising which can be re-used for the production of sulphuric acid or of iron pigments.

In the production of oxygen steel, a slag suitable for direct production of fertilizers containing phosphate and lime could be attained by using crude phosphate instead of the usual fluxing agents containing fluorine. With this method, moreover, the emmission of fluorine will be reduced.

Of increasing importance is the method of direct reduction of iron ores. The advantage of this process is that only one slag results from the steel production operation.

Moreover, direct reduction is suitable for the conditioning of waste materials, e.g. of scales and iron works dusts. In this regard the reduction of solids in the revolving tubular kiln has already been proved by testing.

The introduction of this method is a decisive step in the reduction of quantities of waste material in the iron and steel industry. A detailed report on the conditioning of sludges and dusts will be given elsewhere.

By 1976 about 6 m. t/year of sponge iron have already been produced by different direct reduction methods. Up to 1977 the capacity will be increased to about 10 m. t/year. Up to 1980 the capacity is estimated to about 23 m. t/year and up to 1958 to about $6\frac{1}{4}$ m. t/year. At present the capacity in the Federal Republic of Germany amounts to about 2 m. t.

The Heye-EPB Process, a Low-Waste Technology

Vollmar Hallensleben

Heye Glas-Fabrik, Obernkirchen, Federal Republic of Germany

Until 10 years ago, the hollow-glass industry had not produced any important technological developments. As a matter of fact some weight reductions had been introduced from time to time, but important successes were not noted since the state of the production technology at that time simply imposed certain limits.

Only in the sixties, when raw-material resources and waste removal were initially considered seriously, consideration had to be given as to how far, on the one hanc careful use of raw materials and thus, on the other hand, reduced waste quantities could be applied to change the previous development.

Apart from the known advantages of chemical strength, gas-impermeability, neutrality as to taste, transparency and thus high-quality image, glass exhibited two disadvantages: high weight and fragility. Therefore it became an aim to combine the many advantages of glass with light weight and high stability.

As is known, glass in neutral state is of higher strength than steel, but this strength is weakened by damage to the surface—even friction on conveyor belts is sufficient. This was an aspect which had to be dealt with and an effort was made to find a corresponding solution. An effective protection is attained by coating glass after it has left the mould and when it is still in hot state, with metal oxides. The "dull" surface resulting therefrom is not suitable for high-speed filling machines. The required antifriction properties are attained by spraying a relieving liquid.

This was the basis for reducing the wall thickness of glass receptacles. Glass distribution, however, was difficult to attain. Each chain is only as strong as its weakest link, and this also applies to glass, which is only as strong as its weakest portion. As bottles principally are still produced as at the beginning, i.e. by blowing (compressed air simply replaced the force of the lungs), the distribution could only be influenced indirectly. Only wide-mouthed articles such as preserving jars were produced by a combined manufacturing method consisting of pressing and blowing. The technical sequence will be explained to create a better understanding:

Hollow-glass articles are produced in two phases. First the glass post is located on the "gathering-mould side". Here the mouth is produced and the bottles are preformed. A so-called "parison" is formed. Then this parison is pivoted to the "finishing-mould side" and the final bottle is blown.

Two production methods, i.e. the blow-and-blow process and the press-and-blow process, can be carried out on the IS (individual section) machine used worldwide. The two processes differ with respect to the step on the gathering-mould side.

In the blow-and-blow process (see Fig.1) the mouth is formed by "blowing down". Then the upper end of the gathering mould is closed and blowing through the mouth is carried out after a small "compressed-air level" has been removed. The resulting hollow space in the parison is created by air. The air searches the easiest path, i.e. "hot" glass portions in the glass post will be deformed in a different way from "cold" ones. The non-uniform wall thicknesses in the "88" process result from the type of process already on the gathering-mould side and can be clearly seen on the finished bottle.

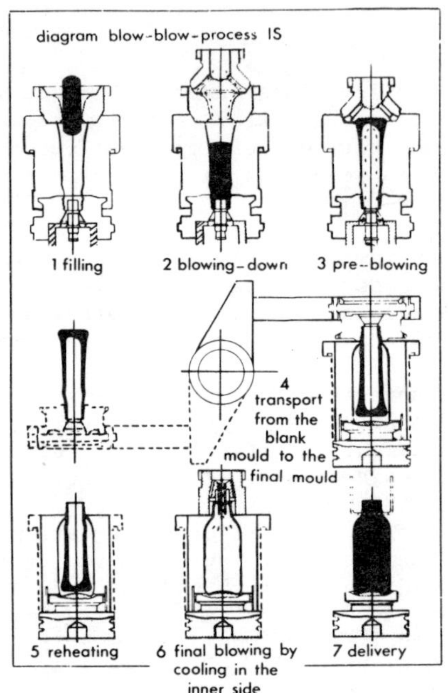

Fig.1. Functional diagram. 1 - Blow-and-blow process diagram. 2 - 1 Filling. 3 - 2 Blowing down. 4 - 3 Preliminary blowing. 5 - 4 Transfer from gathering mould to finishing mould. 6 - 5 Reheating. 7 - 6 Final blowing with internal cooling. 8 - 7 Discharge.

In the press-and-blow process (see Fig.2) the glass is pressed into the gathering mould by "pressing" by means of a metallic die. The mouth is formed during this step. As die and glass post together have the same volume as the hollow space surrounded by the mould, the glass post positively will distribute over the free space due to the pressing action. Uniform bottle wall thicknesses can be produced in this manner.

Hitherto, the press-and-blow process could only be used for the production of wide-mouth articles such as preserve and marmalade glasses, as pressing of narrow-mouth bottles was not yet possible because of the small size of the mouth – particularly with large-volume bottles. The firm Heye, however, continued to develop the press-and-blow process which now can also be used for narrow-mouth articles, i.e. bottles (Heye-EPB process).

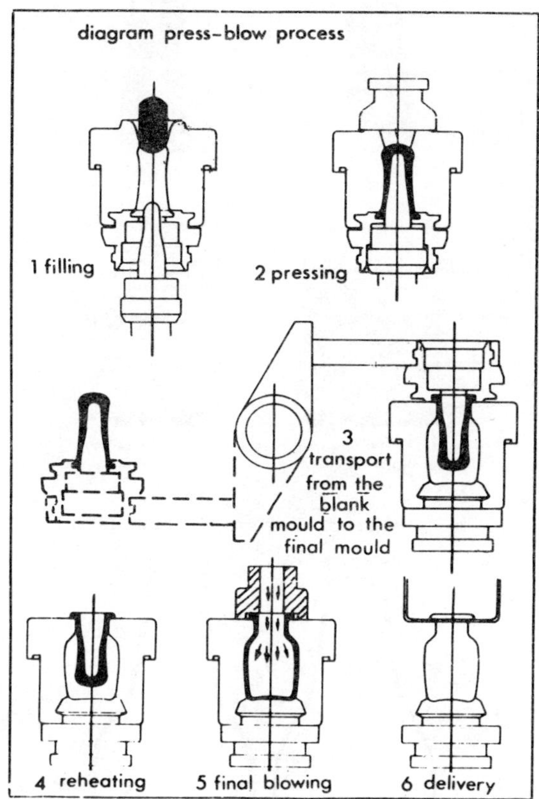

Fig.2. Functional diagram. 1 - Press-and-blow process diagram. 2 - 1 Filling. 3 - 2 Pressing. 4 - 3 Transfer from gathering mould to finishing mould. 5 - 4 Reheating. 6 - 5 Final blowing. 7 - 6 Discharge.

Thus, as a result of the more uniform wall thicknesses, it is possible to produce these articles of same stability and reduced weight.

Starting from a theoretical wall thickness of 100 per cent, the minimum wall thickness, i.e. the wall thickness of the thinnest portion of the bottle with the blow-and-blow process, is approximately 40 per cent, whereas it is approximately 60 per cent with the Heye-EPB process.

If a certain bottle shall have a minimum wall thickness of 1.2 mm, the theoretical wall thickness of the blow-and-blow process is 1.2/0.4 = 3 mm. Producing the same bottle by the Heye-EPB process, the theoretical wall thickness is calculated 1.2/0.6 = 2 mm. This corresponds to a weight reduction of approximately 33 per cent by the Heye-EPB method for a bottle of 1.2 mm minimum wall thickness as compared with the production by the blow-and-blow process.

It can be seen from Figs. 3 and 4 that the glass distribution of a bottle produced according to the Heye-EPB method is much more uniform than that of a bottle produced by the blow-and-blow process. Figure 3 shows a section through a blow-and-blow bottle of 210 g as compared with a Heye-EPB bottle of 185 g.

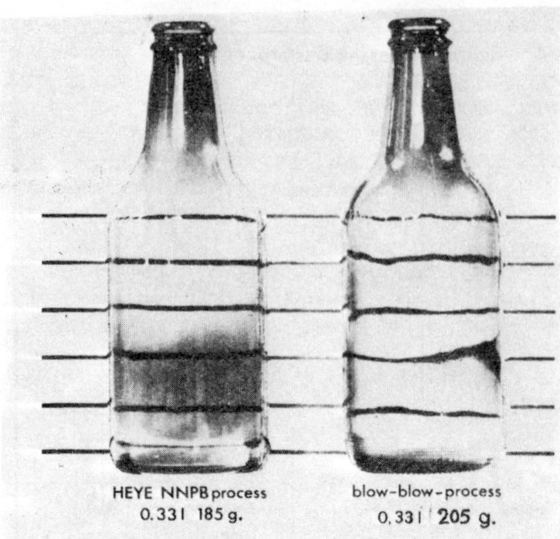

Fig. 3. Heye-EPB process

Fig. 4. Blow-and-blow process

In Fig. 4 a blow-and-blow bottle and a Heye-EPB bottle are arranged in front of a background of parallel lines. The light is difracted by the non-uniform wall thicknesses of the blow-and-blow bottle; the parallel lines are distorted. With the Heye-EPB bottle the lines remain parallel by virtue of the uniform wall thickness.

The frequency of strength values of a blow-and-blow bottle and a Heye-EPB bottle is shown in Gaussian curves (Fig. 5). These curves show that the strength values of the blow-and-blow bottle vary much more - the Gaussian curves are lower and the base is wider than that of the Heye-EPB bottle. Furthermore, the strength values of the Heye-EPB bottle are higher.

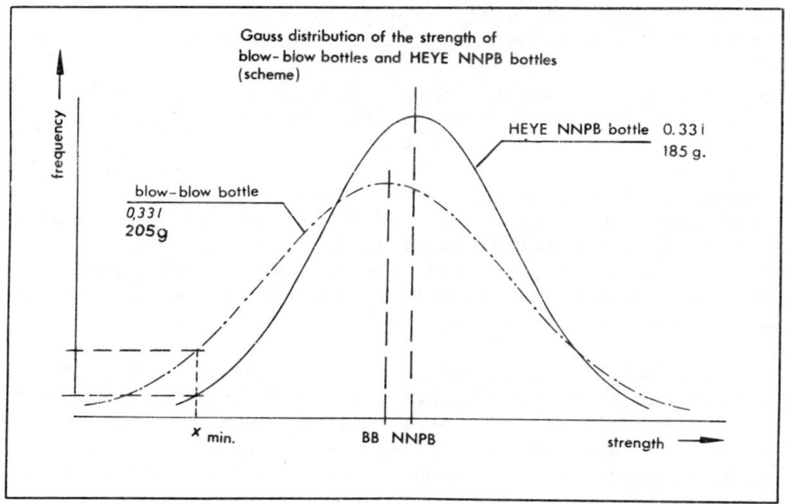

Fig. 5. Gaussian curve. 1. Gaussian distribution of strength of blow-and-blow bottles and Heye-EPB bottles (scheme). 2. Blow-and-blow bottle. 3. Heye-EPB bottle. 4. Frequency. 5. Strength.

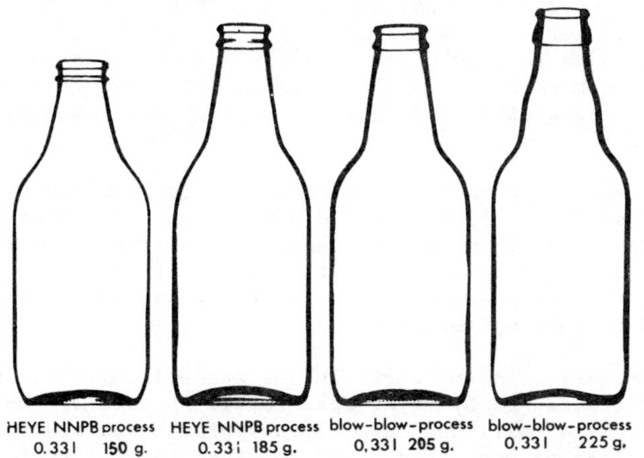

Fig. 6. 1. Heye-EPB process. 2. Heye-EPB process. 3. Blow-and-blow process. 4. Blow-and-blow process.

Figure 6 shows the development of a 0.33-l beer bottle. Initially it had 225 g and was produced by the blow-and-blow method (first from right). Then the shape was somewhat modified and a 210-g bottle could be produced by the blow-and-blow method (second from right). A 185 g bottle has been produced for many years by the Heye-EPB method (third from right). According to the latest development, the weight could even be reduced to 145 g (left bottle).

Before packing the Heye-EPB bottles at the production site, they pass special test machines so as to ensure that the quality of these light containers complies with the customer's requirements and that the consumer's security is guaranteed.

As the weight reduction naturally implied some changes of shape of the bottles, the filling industry initially showed little interest. Only stagflation, generally increasing costs and environmental problems required new considerations. Even companies operating worldwide became ready to subject their product image to certain changes, if such changes resulted in cost reductions. Apart from purchase advantages, the lower weight and volume provide considerable internal savings for both empty and filled bottles. Less pallets (more bottles on one pallet) and less truck loads, reduced freight, less package and hence again and again less material consumption as well as some other factors may add 50 per cent to the reduction of the purchase price.

The use of lighter bottles is useful not only to the filling industry but also to the customers, who save freight, storing space, internal transport costs and store space. Particularly the last aspect is of importance to the retailer, as the shelves for high-sales products at eye level are most important so that the possibility of accommodating more goods in such shelves appears to be very attractive.

Above all, however, it should be borne in mind that a light bottle is more acceptable for environmental reasons. In 1976, approximately 870 million 0.33-1 lost beer bottles will be consumed in Germany. Considering the beer bottle shape shown in Fig. 6, this will result in 195,750 tons of waste for the 225-g bottle, 160,950 tons for the 185-g bottle, which is a reduction by 34,800 tons or 17.8 per cent, while the waste of the 145-g bottle is reduced to 126,150 tons, which is 69,600 tons or 35.6 per cent less than for the 185 g bottle used for the time being. Although the glass industry nowadays disposes of a well-functioning system of recycling, lighter bottles generally favour the waste disposal.

Finally the consumers welcome a weight reduction for the packing. Parking-places are not found adjacent to all stores and housewives are primarily interested in the goods rather than in packages. Above all, refrigerator space is limited. Bottles of thick walls, moreover, require longer refrigerating times and both require space in the refrigerator and more energy so that considerable savings will be achieved in this sector, too.

The lightest beer bottle in the world is the so-called 0.33-1 Snobby bottle of 130 g glass weight produced by Heye. This bottle is used by the Paderborner brewery and hundreds of millions have already proved satisfactory. Moreover, conventional machines do not dispose of technical facilities to realize further weight reductions. As the material proper admits a further reduction, new production machines must be developed, which are able to produce bottles of even lower weight but of unchanged load capacity. In other words, a new bottle generation was created.

In the following, the advantages, which could be attained as compared with the former Heye-EPB method, are enumerated:

1. The new method makes it possible to reduce the wall thickness of the bottles of the new generation to as little as 0.5 mm. Bottles produced according to the Heye-EPB method only admit a calculated minimum wall thickness of approximately 2.5 mm.

2. While the portion of lowest wall thickness amounts to 40 per cent with the blow-and-blow method and to 60 per cent with the Heye-EPB process, the new method permits to increase these limits to approximately 80 per cent. Thus, theoretically a further weight reduction to 25 per cent as compared with the lightest glass bottles existing nowadays is possible.
3. An increased number of main deformation phases of the new process permits a controlled glass distribution and thus a concentrated strengthening of the main load-bearing zones of the bottles.
4. The type of glass transport during the manufacturing process of the new technology improves the precision of the bottle dimensions considerably. With a particular bottle, the tolerance field could be reduced from +2 mm to + 0.2 mm.

A prototype of this new machine, for which meanwhile a German patent and various patents abroad were granted, has been working in our factory for years. It meanwhile proved the theoretical calculations so that the first standard types of this machine are already produced.

Heye received an innovation subsidy of DM 2.26 million from the Federal Government for this modern development.

Referring to Fig. 6, where the lightest 0.33-l beer bottle is a 145 g type, it will be possible by this machine to produce a bottle of same contents and same load capacity and having a weight of 77 g. Considering the present total quantity of lost beer bottles of 870 million, 66,990 tons or 128,760 tons of waste = 65.8 per cent less than with the initial bottle of 225 g would result. Compared with the lately introduced 145 g bottle such new reduction amounts to 59,160 tons or 53.1 per cent.

It should also be noticed that all glass receptacles in the market might be of similarly reduced weight.

Cost/Benefit Aspects of Non-waste Technology

Introductory Report

Charles J. Cicchetti (Rapporteur)

University of Wisconsin, Madison, Wisconsin, U.S.A

I. INTRODUCTION

In these brief introductory remarks I will review four excellent papers, outline my own thoughts on the application of benefit/cost analysis to non-waste technology and suggest some questions that might help to encourage and focus a discussion of these matters.

The papers prepared for Topic III are most important contributions when considered individually. Even more important when taken as a unit I am convinced that they represent significant insight into the conceptual issues associated with the application of economic analyses to the choices related to a non-waste technology. I am confronted with a true dilemma. I could review each paper in great detail and make it possible for the rest of the participants to avoid reading each paper. On the other hand, I could express my views in my own terms, but making sure each of you understand that if there is any insight in such a review it is undoubtedly the result of my careful reading of the four submitted papers.

I will lean towards the latter choice and add in some thoughts of my own in the process. My reason is that I think each of us should take the necessary time to read each of these papers carefully. Additionally, I am not sure I could do any of them justice, if I tried to reduce them to outline form. Before proceeding further let me list the titles and authors of each paper.

(1) "An economy without waste: economic considerations", by Dr. C. Cala and Professor J. Wieckowski of Poland.

(2) "The cost advantages of production techniques without waste", by M. J. Picard of France.

(3) The introduction of non-waste technological processes in the Hungarian Silicate Industry," by Dr. J. Talabér of Hungary.

In addition there is a fourth paper that I found to be an excellent analytical statement of the economic problems under consideration. I am not sure if it was supposed to be formally included in Topic III, but I shall use whatever discretion a rapporteur has and recommend to each of you that you consider this an important statement and that it formally be made part of Topic III.

(4) "Economic aspects of the optimal volume of solid waste", by L. A. Belashov of the USSR.

Despite my decision to avoid summarizing each paper individually, I want to add a short statement about each. I am an economist who teaches beginning University students about the application of economics to environmental problems. My task will be far easier in the future having read the very lucid statement by Dr. Calla and Professor Wieckowski. Their paper summarizes the prospects, problems and need for economic input into the formulation of a coherent environmental programme. Pages 557-570 are such an excellent, easily understood statement of these matters that I now have a ready measure against which to compare my own accomplishments as a teacher. I recommend that each of the participants read this paper in its entirety as it is a direct statement and a most significant contribution.

The second paper that I read was prepared by M. J. Picard. At the outset I felt reluctant to read it because I was so excited by the first paper. Accordingly, I thought the second paper had to be a major let-down by comparison. I was quite wrong! The paper segregates the matter into most useful dimensions. But then the author provides a blueprint or tableau for performing a benefit/cost analysis of the problem. It is a practitioner's guide prepared by one who is obviously an excellent theoretician. It adds conceptual insight. After one recovers from the meaning of its full contribution and wonders why there is still half a paper left to read the author produces an unexpected but most valuable final reward to the reader. Three case studies are outlined that demonstrate the flow from conceptualization to the practitioner's tableau to the policy analyst's reality. This paper is also a must reading for each participant concerned about either the conceptual and/or practical matters associated with the application of benefit/cost analysis to non-waste technology or other environmental matters.

In contrast to the two conceptual papers, the third paper prepared by Dr. J. Talabér is a detailed comprehensive case study. It is completely grounded in the same sort of analytical framework as the others, but because it is a case study of the Hungarian silicate industry in great detail, it brings some of these rather abstract principles to life and therefore transfers considerable meaning, reality and utility to the reader. Whether the reader is interested in the waste problems of the silicate industry or not I highly recommend a careful reading of this case study if one is interested in either the application of benefit/cost analysis or the subject of non-waste technology.

The fourth paper is excellent in two very important ways. It was prepared by L. A. Belashov. It is a complete and lucid mathematical statement of the economics of the optimal volume of solid waste. I also found the paper most useful for a second reason. It demonstrates that the amount of environmental waste that is optimal is not necessarily zero. There are benefits to be had from proceeding in this direction, but so are there costs. This conclusion is something I believe we must keep in mind both in condidering Topic III as well as in discussing the broader context of the entire agenda of the seminar. I recommend that each participant, even the non-mathematically oriented, take the time to read this important paper. At the very least it will serve to remind each of us of the fact that zero or "non" waste may not be our final goal. However, I believe that after reading this excellent paper you will all agree that the insight learned is far greater than this important reminder.

II. SOME QUESTIONS AND IDEAS FOR DISCUSSION

Several thoughts came through to me as I read the four papers and thought further about the subject of non-waste technology. I recognize in advance that some of these matters have undoubtedly been considered in advance, nevertheless I shall still raise them to focus our discussion.

I would like to begin positively and state that I believe we can think of our economic systems from a materials balance approach. This means that we can easily identify the benefits of avoiding waste and the costs of reducing waste to an optimal level.

Let me outline what I think belongs in each category.

A. The benefits from reduced waste and/or pollution include:

(1) Preservation and conservation of our natural resource base.

(2) Protection of the assimilative capacity of the environment thus making it possible to absorb naturally those discharges that continue to be released into the environment in the form of air, water or solid waste pollution.

(3) Increased amenity services from the environment including: life sustenance, biological or genetic information, recreation and aesthetic values.

B. The costs of avoiding waste and/or its negative impacts that we might call pollution include:

(1) Reduce through output of the economy which can be achieved by:

(a) Reduced Gross National Product or output (GNP).
(b) Changed mix of GNP.
(c) Increased production costs.
(d) Changed technological production.
(e) Recycled materials.
(f) Lost jobs.
(g) Increased consumer prices.

(Note: not all the above may be necessary or desirable; additionally some of the above could actually represent net savings (e.g. changed technology) rather than costs.)

(2) Wastes emitted to the environment might be pre-treated so that their negative effects on the above-mentioned benefits could be reduced.

(3) The release of waste into the environment can be controlled in time, place and medium in order to reduce the damages associated with such waste.

(4) The assimilative capacity of the environment can sometimes be augmented so that more waste can be emitted with reduced environmental damages.

The two lists stated above may or may not be all inclusive. But it is apparent that when the costs of abating waste are contrasted with the benefits the optimal level is not likely to be zero as the term "non-waste technology" implies. My first question can be stated as: "Does anyone believe zero waste except for highly toxic substances is an appropriate national or world goal?"

My answer is that I do not. The public policy response is to determine an optimal level of waste by trading off the benefits from avoiding waste with the costs of avoiding waste. There are two immediate problems: measuring benefits and costs and allocating the rights to pollute or reduce waste discharge to the members of society. The second problem means that additional costs must be considered: enforcement and monitoring costs. In selecting one (changed technology) of the many possible ways of avoiding waste or its negative impact called pollution this group omits consideration of many other approaches. Some of them may be less costly. In evaluating the benefits and costs of technological change the opportunity cost of foregoing or ignoring other policies must be considered. My second question can be stated as: "Does this group want to limit their consideration of avoiding waste to technological change and if so does it want to consider the possible opportunity cost associated with other alternative waste-reducing approaches that have been excluded?"

One of the most difficult problems associated with avoiding waste is the way a society implicitly or explicitly delegates responsibility for avoiding its collective problems. We can all agree that when the benefits or costs of avoiding wastes are internalized to a manufacturing entity the problem is more tractable. At this seminar, mixed and centrally planned economies are represented. My third question is related to the way external waste costs are considered in these different social-political contexts. "Does the way an economy is organized make a difference in confronting the external costs of waste discharge?"

My belief is that each nation is more alike than it is different in this regard. I believe capitalists minimize costs and ignore external costs unless required to by government. It also seems likely that a plant manager in a centrally planned industrial economy also minimizes costs or meets a target output without considering external matters unless required to by some other part of the government. I may be quite wrong in this regard. I believe, however, that we should consider the matter carefully and openly in this seminar, if we are to help each other solve our common pollution problems.

Finally, most of the above discussion can be represented by final questions. I believe that benefit/cost analysis is relatively easy to perform but difficult to begin. That rather vague statement means that the most important part of benefit/cost analysis is deciding the objective function. In other words, for whom is the analysis being performed, or more crassly which benefits and costs matter? It is easy to say that our goal is social welfare maximization or economic efficiency. In practice we are more likely to settle for cost minimization or cost effectiveness in achieving a particular target or goal. My last question is: "What objective does this group want to adopt either theoretically or practically in establishing a benefit/cost framework?"

I offer the above questions to stimulate discussion and not to guide or focus it. I hope if the participants feel that my questions are less important or useful than others that they will feel free to correct my judgment. I am enough of a politician to be very resilient. I offer the above only as my personal view of what I would find to be an interesting session.

Cost-Benefit Considerations in Waste-Free Production Methods

J. Picard

Director, Agence Financière de Bassin, Loire-Bretagne, France

Examples of business ventures to establish pollution-free integrated technologies of which we are aware in France up to now often find their basis in an act of faith on the part of their promoters, and they owe their success to a highly empirical approach which is worked out progressively with the development of the project.

This is indeed far from a rational investment committed on the basis of exhaustive preliminary technical and economical analyses.

The absence of any reference, technical uncertainties, the need to resolve numerous problems which only crop up as the development of the venture proceeds, and the inability to recognize numerous costs and other factors at the beginning of work would well seem to render illusory for the moment any study of that type.

Such an impression is further reinforced by a consideration of the fact that the technical solutions which are applied require numerous adjustments over a long period of time after the stage of industrial application has been reached.

These experiences lend themselves for the moment more to *à posteriori* evaluations of results than to feasibility studies whose purpose generally is to compare variables of industrial projects making use of competitive processes, but which are perfectly defined and which have been developed for a long time.

There is, furthermore, a lot to be learned from such *à posteriori* evaluations. They are indispensable when it proves desirable to make the proceedings rational and to form an opinion on the possibilities of encouraging business firms to resort to specialized, integrated pollution-free methods proper to a process.

We now have available in France a sufficient amount of varied experience to allow an initial review of benefit-cost considerations.

After having presented the theoretical context for the analysis, I would then like to discuss a few of the recent representative cases in France which demonstrate the benefits of these methods which are most frequently encountered.

I. THEORETICAL CONTEXT OF THE ANALYSIS

In order to establish a framework for this evaluation, we should refer to the situation in which an industrial polluter finds himself.

The industrialist confronted with a pollution problem has the choice between three possible approaches.

1. He does nothing at all and continues to pay the fee.

   This situation should only be temporary.

   Firstly the fee is likely to go up to a point that it will then considerably encumber the operation of the enterprise. Further, pressure from riverains, associations and the government (classified enterprises, local authorities, Basin Agencies) will continue to make itself felt until the industrialist agrees to reduce the pollution coming from his plant facilities.

2. He reduces the pollution by way of a treatment of effluents coming from the factory, located downstream from the production facility. This involves a treatment plant for water purification, dedusting systems, incinerators and removal of waste products to a community plant approved for such treatment.

3. By calling into question the processes and systems themselves, he then investigates the means to eliminate the pollution itself.

   This could be a question of either making changes to the operations which generate the pollution (a change of methods) or to reintegrate the pollutant into the production cycled (recycling, recovery, upgrading to profitable uses). These are the waste-free technologies still called in France "techniques propres", or specialized, integrated pollution-free methods proper to the process.

These three scenarios are the ones to be taken into consideration and they will be compared in a benefit-cost analysis of production methods which are free of wastes.

For the economic analysis, the first scenario will serve as the situation of reference. This baseline is the *status quo*. The other two scenarios represent the two possible variables in pollution control which it is appropriate to evaluate.

The analysis then consists of a separate evaluation of the additional costs and benefits which are connected with the selection of one or the other system.

In addition, the following need to be defined:

a list of criteria used as a basis of the evaluation;

the various viewpoints which can be taken into account: that of the community at large, that of the government, and that of the industrialist.

After having drawn up a list of criteria which can be used, I would then like to give a few indications about the manner of using such an evaluation to determine a policy for the promotion of technologies which are waste free.

A. *List of Evaluation Criteria (Tables 1(a) and 1(b))*

The first list includes monetary or non-monetary criteria which are characteristic from the viewpoint of the industrialist.

They enable a measurement of the impact of the choice of the method to reduce the pollution either on the management of the industrial enterprise concerned (monetary criteria), or on the general standing of the enterprise and its development (monetary criteria).

The benefits and the costs taken into account are those additional benefits and costs which are incurred as compared to the earlier situation of the *enterprise*.

The second list brings together criteria which are characteristic of the viewpoint of the different parties (taxpayers, government, local authorities) which represent certain aspects of the *community at large*.

The total net benefit to the community will be expressed by the algebraic sum of the benefits and costs for the different, economic agents, a sum which will be obtained by bringing together the two lists of criteria. As a consequence of this aggregation, criteria connected with internal transfers between agents will disappear (fees, government financial aid to industry) which, from the viewpoint of the taxpayer or the government, enter into the matter, but not into that of the community taken in its broadest sense.

B. *The Application of the Results from the Evaluation to a Determination of a Policy for the Promotion of Waste-free Production Methods* (Table 2)

An evaluation from the twofold viewpoint of industry and the community makes it possible to determine the difference which could exist between social costs and cost to the industry, and thus to gain insight into guidelines for a policy to be instituted by the government agencies concerned.

The analysis, if it has been completely carried out, should thus provide an answer to the following questions:

Is the net benefit to the community from waste-free technology greater than the net benefit from classical treatment methods?

Do waste-free technologies lead to a net benefit to industry, and if so, what is it?

Can the net benefit to industry, if it exists, be expressed in money terms?

By taking the position of the case in which waste-free technologies are a net benefit to the community at large which is greater than the one of classical treatment techniques, it is possible, on the basis of answers to the last two questions, to set up a framework which will make it possible to form a policy for the promotion of waste-free methods.

The following frame of reference could be set up (Table 2), which is given as an indication only and should not be assumed to be the policy for assistance which might be adopted in France by the Basin Agencies.

Table 1(a)

SOME CRITERIA FOR THE EVALUATION OF WASTE-FREE PRODUCTION METHODS

Specific criteria from the viewpoint of industry

BENEFITS	COSTS
I. *Benefits Quantifiable in Monetary Terms*	II. *Monetary Costs*
Sale of recovered products	Depreciation of the investment
Development of a new line of activity based on the extracted pollutant	
Operational savings on consumption of intermediaries water power adjuvants	Additional consumption of intermediates necessary adjuvants
Savings on fees	power
Gains in productivity	Other, additional operational expenses
improvements in the use of materials by:	
reduction of losses at the source	
recycling of the pollutant	
elimination or decrease of manpower	
Financial aid according to industry by the community	
II. *Other Benefits*	
Advantages related to the maintainance of the installation at its present location (if its fate was in question)	II. *Non-Monetary Costs*
Technological advances over competitors	Disadvantages related to a possible decline in product quality (for example in the case of total recycling)
Taking out patents	
Limitations of a less desirable location (for a new plant facility)	
Gains on required surface area	
Ease of maintenance	Costs of retraining personnel
Improvement of the corporate image with riverains the government	
Improvement of the corporate image and its impact on its marketing and development strategy.	

Table 1(b)

SOME CRITERIA FOR THE EVALUATION OF WASTE-FREE PRODUCTION METHODS

Specific criteria from the viewpoint of the community

BENEFITS	COSTS
I. *Quantifiable Benefits* (existence of indicators)	II. *Monetary Costs*
Effectiveness in the reduction of nuisances	Community investments in research and development
matter in suspension	Financial aid accorded by the community to industry
organic matter	
toxicity	Decline in fees collected by the Basin Agencies
salinity	
color	
solid wastes	
composition of gaseous effluents	
dust	
Savings of rare resources	
savings in water	
savings in raw materials	
power savings	
savings in space	
II. *Other Benefits*	II. *Non-Monetary Costs*
Quality of the reduction of nuisances	
reduction ensured once and for all	
reduced risk of accidental pollution	
less pollution transfer	
Elimination or limitation of pollution control by the government	
Industrial activity stays in place (if this was arranged)	
employment	
contribution to the local economy	
Improvement of the environment at the work site	
National impact of spin-offs from advanced methods	

Table 2.

Examples of a Promotional Policy to be Set Up in Terms of the Characteristics of the Net Benefit from each Type of Industrial Technique as Compared to the Initial Situation.

NET BENEFIT as compared to the initial situation				EXAMPLES OF WASTE-FREE TECHNIQUES	INCENTIVES	
TO THE COMMUNITY		TO THE INDUSTRY				
Evaluation waste-free method	Evaluation classical treatment	Evaluation waste-free method	Evaluation classical treatment			
		negative	negative (better solution)	change of methods very costly		special financial aid
positive (better solution)	positive	negative (better solution)	negative	change in methods	free technical advice and information	same financial aid as for classical treatment
		positive but not set in money terms	negative	recycling of pollutant keeps the factory active		loans-reduced interest
		positive and set in money terms	negative	development of a new line of activity based on the recovered pollutant	information from technical centers and professional organizations	financial participation in feasibility studies

To make use of a frame of reference of this type, it is of course assumed that there is a capability of classifying the various possible courses of action in terms of the nature of the net benefit which they will provide to the industrialist. This is not done without some difficulty, as already mentioned, which can only then be overcome when more complete and more numerous analyses of such experiences have been performed.

II. EVALUATION OF INTEGRATED WASTE-FREE TECHNOLOGIES

Following these theoretical considerations, I would now like to present a few representative examples of situations most frequently encountered.

These examples could be grouped into two categories according to the economic merits of the operation:

The operation is not profitable for the industry (the net benefit to the industry is negative), but the integrated waste-free method has a better cost/effectiveness ratio than outside treatment, which largely justifies its choice.

The operation is profitable in itself to the industry (net benefit is positive and quantified in money terms), all considerations of pollution abatement aside.

In order not to distort the comparison from the viewpoint of the community at large, we will not take into account in these examples any internal transfers between the government and industry as an incentive to control pollution (fees, subsidies).

It goes without saying that a consideration of a reduction in fees on the one hand, and investment aid on the other, further improve profitability in these operations from the viewpoint of the industry.

A. *Integrated Waste-Free Technologies, an Improved Cost-Effectiveness Ratio*

Compared to external treatment of identical effectiveness, integrated techniques proper to the process always have a more favorable economic status both from the viewpoint of industry and the community.

Moreover, it is often after they have become aware of the high cost of customary treatment methods that industrialists, advised by the Basin Agencies ("Agences de Bassin"), then look into internal methods of pollution reduction.

This explains why the use of these techniques are most frequently encountered in industries which deal with the most extensive amount of pollution (paper-pulp, paper mills, chemical industry) or in industries in which the level of pollution when referred to total sales is particularly high (food industry, potato starch plants, yeast factories).

1. *Recovery of Proteins and Potassium from Unfermented Yeast Solutions.* The

production of yeast is always accompanied with the discharge of wastes heavily laden with oxidizable matter.

At the Strasbourg plant of the "Société Industrielle de Levure FALA" (SILF) (a private company producing yeast) the concentration of this unfermented yeast solution (in French *moût*: the same as wort of beer, must of grapes, and unfermented wine) allows recovery of potassium sulfate which can be used in the manufacture of fertilizers, and of proteins much in demand in cattle feed.

This method provides an almost complete reduction of pollution. For 1 ton of yeast produced, the BOD goes from 127 kg to less than 6 kg, and the COD from 174 kg to less than 10 kg, which for both cases is an efficiency of 97 per cent.

In addition, recycling of water reduces water consumption by 70 per cent.

However, a biological treatment station reducing pollution to the same extent, but without doubt lesser efficiency, would have required twice the investment than those which were necessary.

Table 3 further shows that receipts from the sale of the protein concentrate and the potassium extract balance off operating costs in spite of the high power costs for the evaporation operation.

Considering operating expenditures which would have been those of a biological treatment plant, it is finally an annual savings of more than 1 million francs per year, and a savings of 5 million francs which has been made one the investment at the beginning.

Table 3

Annual Operational Balance Sheets Comparing an Integrated Installation and an Equivalent Treatment Plant

	Integrated Installation	Equivalent Treatment Stn
Investment	5,200,000 F	10,800,000 F
Operating Expenditures	860,000	1,080,000
Receipts		
Sale of "Viprotal" 6 400 tons (at 120 F/t)	768,000	
Sale of potassium extract 1 650 tons (at 150 F/t)	247,500	
Operating balance	+ 155,500 F	- 1,080,000
Gain from the operation (excluding amortization)	1, 235,500 F	

2. *A Non-Polluting Method for the Dehydration of Chloral.* The synthesis of chloral gives a product containing 82.5 per cent chloral, 12.5 per cent water, 2.7 per cent hydrochloric acid and 2.75 per cent of heavy chlorinated compounds.

In the former process which used oleum as a dehydrating agent (concentrated and enriched sulfuric acid), waste water was not treated and it then contained the chlorinated compounds (0.8 t/day), and sulfuric acid (12.5 t/day). The new process features as a replacement of oleum a specific solvent which ensures the extraction of the water and provides for the separation of the chlorinated compounds and the hydrochloric acid. The major amount of toxic pollution which was previously dumped into the river was henceforth eliminated.

The investment for the external detoxification of these effluents (at 25 F per equitox/day) would have amounted to 70 million francs and operating costs for the installation would have irremediably meant the end of chloral production. However, operating costs for the new process are only 1,600,000 francs per year more than those entailed in the dehydration of oleum. The system which was selected, incidently, recovers some 400 tons per year of hydrochloric acid.

Faced with costly treatment, the choice of an "integrated waste-free system" would become indispensable. The principles which are employed have been known for a long time and are part of the basic procedures of the chemical industry. To set up such procedures, however, it would have required a highly skilled labor force due to the complexity of the installations in order to guarantee the proper operation of both the manufacturing process and the treatment process. Since the treatment process is integrated into the preparation the chloral, it gains the benefit of the care given to every stage of the production of this product.

Table 4

A Comparison Between the Former Process, the New Process and External Detoxification

	Former Process Dehydration with Oleum	New Process Distillation
Investment	-	8,492,000 F
(for classical detoxification)		= 70,000,000 F
Operating costs (per 1200 tons of chloral)	4,734,120	6,153,360
Operating balance (excl. depreciation)	1,419,240 F	
Fee due to the Basin Agency for 2,800,000 equitox/day in 1975/76	1,120,000 F	

3. *Recovery and Profitable Upgrading of Organic Matter in Suspension in the Waste Water of a Plant in the Potato Industry.* The processing water for

potato chips and potato puree is heavily laden with organic matter. The "Coopérative Agricole Vico" at Vic-sur Aisne (Farm Cooperative) has successfully undertaken the extraction of organic matter by way of several successive separation steps. The recovered product is marketed in the form of cattle feed.

Due to this operation, the pollution is markedly reduced since the MIS goes from 8 t/day to 0.4 t/day, and the organic matter from 7 t/day to 2 t/day.

The total cost of the installation is 2,660,000 francs which represents about one-quarter of the cost of an equivalent biological treatment plant which it would have been necessary to build had this approach not been selected.

Furthermore, in spite of the high costs for power for dehydration, operating costs are almost balanced out by the sales of animal feed and in any case are less than operating expenditures for the equivalent biological treatment plant (see Table 5).

This organization is the first in Europe to have carried out this investment. Due to an analysis of problems raised by the dehydration installation, it was able to strengthen its technical position over its competitors.

B. *Integrated Waste-Free Methods, A Source of Profit to Industry*

In the examples which we have just looked at together, the advantage of integrated pollution-free methods to the industrialist only existed as compared to external treatment methods which were less efficient at higher cost. In other cases, although certainly less numerous, recourse to integrated waste-free

Table 5

Annual Operating Balance Sheets Comparing the Approach Adopted to an Equivalent Biological Treatment Plant

	Approach Adopted	Equivalent Treatment Plant
Investment	2,660,000 F	11,780,000 F
Operating costs	1,690,000 F	1,040,000 F
Receipts from the sale of cattle feed 2000 t (at 450 F/t)	900,000 F	
Operating balance	-790,000 F	-1.040,000 F
Gain from the operation (excl. depreciation)	250,000 F	

technology is balanced out by an improvement in the operating account of the enterprise combined with total success in the field of pollution control.

This scenario is optimal since there is both a net benefit in money terms for the industry and a net benefit for the community. The common interest thus comes together with the financial interests of the industry.

These operations which are profitable for industry seem to occur mainly in two types of companies : firstly, in business ventures which are too small to have been able, in the past, in the absence of any coercion, to undertake the necessary research, and secondly in companies much larger, but with so little competition that there was no incentive to streamline their operations.

1. *Smoke treatment and dust recycling in a white metal alloy plant.* This first representative case concerns a small factory producing tin alloys based on lead and antimony which is located in an urban area, and in which 2.5 per cent of the production departs in the form of dust in the air, only then to fall out on the houses of its neighbors.

The operation which was carried out consisted of replacing the old, mostly ineffective smoke boxes with highly perfected sleeve filters, for which the fabric and anti-clogging system were specially designed. In addition, dusts of metal oxides which were recovered were formed together and then recycled on a continuous basis in the furnace.

The former filtering systems avoided at best 50 per cent discharge in dust emissions. Were one to assume that this represented 5 per cent of the daily furnace charge, 375 kg of metal oxides were to be recovered every day. However, just as much was being discharged into the air!

Now at present, the sleeve filters with an efficiency of 99.9 per cent are recovering: 15,000 x 0.05 x 0.999 = 749.2 kilograms. Emissions into the air have fallen off to a few hundred grams. Measurements carried out by "CERCHAR" have verified that the smoke no longer contains more than 5 to 7 mg/Nm3/ of metal oxides.

A new installation would have required a relatively heavy investment for a small firm, in the order of 300,000 F.

The results presented in Table 6 show that the new receipts enable an exceptionally rapid amortization. In this case of the manufacture of an alloy with 40 per cent tin and 60 per cent lead, daily savings reach an additional 4300 francs.

By taking the most unfavorable scenario in which the factory only works with lead, with a value ten times higher than that of tin, the basic investment would still be amortized in one year!

This profit picture still only reflects one of the benefits of the new installation. Due to the continuous recycling of dusts, the working conditions have considerably improved. And while at the same time providing major additional receipts due to an approximate 2.5 per cent savings in materials representing about 2.5 per cent of total production, the high level of emission treatment has eliminated any visible discharge from the smoke stack. By putting an end to the justifiable protests of the neighbors, the firm was thus able to continue its activities in an urban area.

2. *Recovery of prune juice.* The second case concerns a small firm in the south-west of France specialized in the processing of prunes.

The prunes are dehydrated for storage, and then rehydrated before marketing. The rehydration baths constitute the main source of pollution, equivalent to a town of 15,000 residents.

Table 6

Daily Operating Balance Sheets Comparing the Former and New System for a Furnace Charge of 15,000 kg per day and The Preparation of an Alloy Containing 40 per cent Tin and 60 per cent Lead

	Former Process: Cyclones and Smoke Boxes		New Process: Sleeve Filters
Investment	-		300,000 F
Operating costs			200 F
Income from recovery of metals			
tin + lead (50% of dusts)	4400		
tin + lead (99.9% of dusts)			
-tin: 259 kg at 30 F/kg			7770
-lead: 388 kg at 3 F/kg			1160
Operating balance	+ 4400		+ 8730
Gain from operation (excl. amortization)		4330 F	

Sorbic acid used in the baths is an inhibitor of biodegradation, which makes the effective treatment of waste water by biological purification impossible.

The profitable upgrading of the pollutant was then looked into, first of all in the form of dried material for cattle feed, and then later, after a market study, in the form of concentrated prune juice to be used in dietetic foods. For this small firm, this then became a real diversification of its product line.

The recovery of all of the matter in suspension, the removal of the water in the form of steam, and the recycling of the acidic distillates now enable the suppression of all of the pollution.

As can be seen in Table 7, a biological treatment plant would have cost three times more than the system set up, and which would have been clearly less efficient considering the presence of the sorbic acid.

Operating expenditures are only about half of the income from the sale of the prune juice in an amount of 150 tons per year per 3000 tons of prunes processed.

The profitability of the operation appears to be beyond question when it is considered that the rate of profitability (gross profits/investment) is about 45 per cent.

3. *Treatment of methionine mother water by evaporation.* Conducting profitable operations is not the exclusive prerogative of small firms.

A major subsidiary of Rhône-Poulenc, the "Société Alimentation Equilibrée"

Table 7

Annual Operating Balance Sheets Comparing the Installed System, Including its Recovery of Prune Juice, with an Equivalent Biological Treatment Plant

	Installed System	Biological Treatment Plant
Investment	235,000 F	768,000 F
Operating costs	140,000 F	77,000 F
Production and sale of 150 tons of prune juice		
expenditures for filtration and packing	52,000 F	
income from sales (at 2 F/kg)	300,000 F	
Operating balance	+ 107,500 F	- 77,000 F
Gains from the operation (excl. amortization)	184,500 F	

(Balanced Foods Corporation, located at Commentry, which is a major producer of methionine, found itself in a position requiring treatment of its effluents. It went into a system for the treatment of mother water using evaporation. In an installation which, due to its size, is unique in the world, the operation extracts alternately methionine and sodium sulfate, which is made possible by the different solubility of the two products.

This treatment brings the effluent down from 800 m^3/day to 100 m^3/day. The COD went from 30 t/day to 8 t/day, while the BOD was reduced to 10 t/day. Recovery of sodium sulfate reduces the salinity of the effluents.

Because of the high value of the recovered products, the operation is clearly a profitable one, as can be seen in Table 8. The concentration of the effluents enables the recovery of 123 tons of sodium sulfate and 20 tons of methionine every day. These products, considering their content, represent 13 million francs of annual receipts. Even though the concentration process is costly in power, operating expenditures do not exceed 10.5 million francs. This gives a gross profit, before amortization, of about 2.5 million francs. For an investment of 7 million francs, the time to recover the capital invested in this installation would be about 3 years, whereas the useful life of such equipment should not be less than 10 years.

As with the two previous cases, this appears to be a rate of profitability greater than those normally sought in the selection of industrial projects.

In such cases, pollution control no longer appears as non-productive. It reveals monetary benefits by contributing to a more rational organization of production and an improved management of resources. It can, of course, be said that if the management of these firms had been optimal, such steps would have been taken without waiting for pollution control to make it necessary.

Nevertheless, in these cases, pollution control is not only a catalytic factor for innovation, but also a factor in productivity.

Table 8

Annual Operating Balance Sheets Comparing the Selected System and an Equivalent Biological Treatment Plant

	System for treatment of mother water	Equivalent biological treatment plant
Investment	7,000,000 F	9,600,000 F
Operating costs	10,500,000 F	960,000 F
Income from sale of methionine and sodium sulfate	13,000,000 F	-
Operating balance	= 2,500,000 F	- 960,000 F
Gain from installed system as compared to an equivalent biological treatment plant (excluding amortization)	3,460,000 F	

The Introduction of Non-Waste Technological Processes in the Hungarian Silicate Industry

József Talabér

Director, Central Research and Designing Institute for the Silicate Industry, Budapest, Hungary

1. INTRODUCTION

First of all we must state that in our following considerations the concept of non-waste technology is understood as including all the processes and measures, by which the by-products obtained during the manufacture of some product, or part of the product, emitted as pollution into the environment may be converted directly or indirectly into a product of the given, or of some other, manufacturing process.

The selection of methods to be applied for utilization is basically modified by whether the involved waste is

(a) the main product of the actual manufacturing process,

(b) a by-product, which may be utilized within the technological process,

(c) a toxic by-product, or

(d) an effluent by-product.

The measures applied under (a) and (b) agree with the economic interests of the given single-purpose process. For those under (c) and (d), external solutions must be found for their conversion. The production of powered materials must be regarded as a necessary concomitant of the processes in the silicate industry, as their application is indispensable for the acceleration of the heterogeneous chemical reactions occurring at the relatively high temperature (100°C).

On the other hand, it may be stated that the main bulk of the silicate industrial powers require measures involved by (a) and (b) and a lesser part of the same belong to those under (c) and (d) requiring other measures appropriate to these conditions.

Thus, in the silicate industry the conditions of waste reduction are evaluated as dependent on the development of the technological equipment and processes.

As the result of technological development, by-products are in general being utilized to an ever-increasing extent. By reprocessing the rock in coal mining, a fuel may be obtained and the rest may be utilized as raw material for producing cement or light concrete aggregates. Blast-furnace slag is used in its original form for the production of building blocks or for road construction: in granulated form as a hydraulic cement additive or as foamed slag for concrete aggregate and for thermal insulation purposes. In certain countries, and this is true of Hungary, blast-furnace slag is definitely in short supply. Dealing with the problems of utilizing silicate-base wastes has been taken up in detail only in the last 20 years. In this paper I shall give a short survey of the present state of their utilization and point out the wider range of poss-

T

ibilities and the tasks involved.

II. UTILIZATION OF SILICATE INDUSTRIAL WASTES

II.1. *The Formed Wastes - the Measure of their Utilization*

Table 1 shows that although the total efficiency of the separation is improving, the emission of contaminants in the cement industry remains nearly on the same level.

Table 1

Year		Cement and lime industry	Raw ceramic and insulation material industry	Stone gravel industry	Building industry prefabrication
1975	1. Dust emission in % of the product	5.1	1.0	0.2	0.5
	2. Average separation efficiency (%)	80.0	80.3	55.2	90.0
1980	1. Dust emission in % of the product	1.7	0.5	0.1	0.1
	2. Average separation efficiency (%)	92.2	86.2	69.0	95.0
1985	1. Dust emission in % of the product	0.4	0.3	0.1	0.1
	2. Average separation efficiency (%)	98.3	89.0	77.2	95.0
1990	1. Dust emission in % of the product	0.3	0.1	0.1	0.1
	2. Average separation efficiency (%)	98.6	95.0	89.2	99.0

Table 2 shows the percentage distribution of separated waste materials by branches of the Hungarian silicate industry.

Wastes, which can be recycled only with difficulty or not at all, must be reckoned with mainly in the wet and semi-dry cement production processes and in the raw ceramics industry. Though the percentage of these cement industrial wastes in the product is decreasing year by year, the problem of these hardly treatable wastes remains unchanged, due to the rising volume of production. The greatest attention from the aspect of waste utilization must be devoted to cement production. The raw ceramic wastes are not powders, but lumpy rubble materials, are generally geographically distributed, and are suitable for recultivation purposes, e.g. for filling up abandoned clay mines.

TABLE 2

	1975	1980	1985	1990
Cement industry	62.5	72.8	78.0	80.2
Lime industry	3.7	4.0	4.9	5.7
Fine ceramics industry	0.1	0.1	-	-
Glass industry	0.4	0.5	0.6	0.6
Raw ceramics industry	23.2	13.0	7.5	4.1
Building industry - prefabrication	2.6	2.3	2.1	2.2
Stone-gravel industry	7.5	7.3	6.9	7.2

II.2. *Utilization of Cement Industrial Dusts in Present Technological Processes*

On studying the utilization of cement factory dusts it is advised to evaluate separately the flue dusts of the dry, semi-dry and wet technological processes.

Dry process. On the basis of foreign experience, it may be stated that in the case of a high efficiency separation, the recycling of the dusts leads - at least periodically - to imbalance in the burning and to shutdowns to be eliminated by external intervention. Yet several processes of an experimental or definitive character are being applied, resulting in the reduction of the alkali and sulphur circulation and thereby in continuous operation (application of cooling ribs, addition of gypsum, charging the dust into the calcination zone).

It must be stated that no generally accepted, applicable method is known, but the leading experts of the world cement industry consider that the problem of dust recirculation (alkali circulation) may be regarded as solved from the aspect of the future of the dry process technology.

The separation and utilization of the flue dust drawn off with the gases emitted from the *wet process* rotary kilns is still unsolved. The main difficulty is that the flue dusts, even if separated, cannot be recycled, or only a very small part of them can, into the technological process. If more than 1-2 per cent of flue dust is mixed with the raw slurry, the handling of the latter is made very much more difficult; it may turn into a jelly-like mass, which can be neither pumped nor mixed. Numerous experiments were carried out to eliminate this effect, by changing the technological process, but with no promising results; no such process is known in other countries. So in the present situation the wet processes must be studied on the basis of two different aspects. One group of questions involves the separation of the mass of dust from the flue gases of the clinker-burning furnaces, and the second, which is graver, than the first, is the handling, deposition and final utilization of the separated dusts.

II.3. *Utilization of the Cement Industrial Dusts Outside the Factory*

As mentioned already, studies and experiments aimed at the utilization of the flue dust of the wet process clinker-burning furnaces have been carried out in several directions and some utilization processes are known from scientific literature, as, for example:

- burning the flue dust to clinker in specifically designed semi-dry, or dry process furnaces;
- utilization for filler material (in road building);
- utilization for other building industrial products (e.g. for producing light concrete aggregates).

According to the results of extensive research work carried on in 1960-62 in our country the flue dusts may be utilized for preparing low-strength concrete in small-size concrete block production lines designed for this purpose and - in a corresponding composition - for the improvement of mortar.

Studies in agriculture showed that the furnace flue dusts enriched in alkalis may be favourably used for amelioration of the soil in vineyards. Also the possibility of using them for road building and civil engineering works, mainly for soil solidification, arose.

The cement factory dusts may be utilized in road building for soil stabilizer binding material.

Preliminary soil stabilization experiments were carried out in Hungary with the dusts of wet process cement factories in 1968. The soil used in the experiments was mainly fine sand. The result of the experiments was relatively favourable, showing that 4-5 per cent no. 500 cement may be replaced by 8-10 per cent dust, if the frost stability of the samples is to meet somewhat reduced requirements. On studying the economy of the utilization of cement factory dusts it must be considered that the present demand for cement may be reduced by reducing the cement requirements for soil stabilization.

II.4. *Utilization of Stone Pit Wastes*

Rock materials may be used for the lower layers of road structures using the following technological means:

- mechanical stabilization,
- stabilization by cement,
- meagre flue-dust concrete,
- mine dirt with granulated blast furnace slag.

Utilization of stone pit wastes is an important national economic interest from the aspects of both environmental protection and overall saving.

The simplest and at the same time cheapest method is the use of stone pit wastes for *mechanical stabilization without binding material*. The bottom layer of the road construction or the protective layer often can be built without considerable special investments. The suitability of the materials in their actual state may be established by preliminary examination, or - in individual cases - the missing finer fraction (mortar) may be added to them.

In the case of greater and continuous demands it is advisable to use mechanical loading on the dumping site. For mechanical stabilization the 0/50 waste may be used best, as on the one hand its equivalence factor is higher (0.7) and on the other hand this fraction cannot be mixed for use in foundation layers with binding material. The stone pit dirt may be economically used for mechanical stabilization, especially in regions, which are poor in granular materials.

As to utilization so far it may be mentioned that the bottom layer of a long section of the concrete motorway M-7 consists of limestone mine dirt.

Roadway foundations with binding material - such as stabilizing layers, concrete, flue-dust-lime type concrete, or foundations from granulated blast-furnace slag - may be prepared from grain wastes and impure crushings of max. 20 to 25 mm diam. These materials could be utilized for mechanical stabilization only with an equivalence factor of 0.5, while that of the granular base layers prepared with binding material in a mixing machine amounts to 1.2 - 1.5.

In the case of relatively fast solidifying base layers mixed with cement binding material the transport distance may be shorter - due to the shorter bonding time. However, the distances can considerably be increased by the solution of transport concrete (mixed during transport). The technologies slow-binding flue dust and granular blast-furnace slag permit the storage of the mixture for a few days, whereby the operation of the mixing plant and the construction of the base layer itself may be made more continuous.

II.5. *The Utilization of Glass Industrial Wastes*

The glass industry produces at present 75 per cent of its products for the packaging industry. The greater part of the bottles remains in the households as household refuse. In developed countries 8 per cent of the household refuse is glass. As glass is practically insoluble in water and solvents, it does not affect the natural waters and nature itself, as a chemical; it can be separated without any problems by systematic collection, or processed easily in incinerators, or in composting processes. In addition the reprocessed and classified glass fragments may be recycled into production, or utilized for road and dam building and as raw material in the building materials industry. In the future, the glass industry will be obliged ever increasingly to utilize the wastes of its own products. This circumstance will favourably affect the waste-disposal cares of big towns and the glass packaging means will probably be used ever more widely due to the increased requirements of environmental protection.

The glass industry is capable of eliminating the lead and fluorine components from the gases poisoning the ambient air.

In the future the modes of elimination will be determined by the following measures:

improved furnace construction,

reduced specified fuel consumption and the spread of electric heating,

the application of briquetted or granulated mixtures. By improving furnace construction - along with improved productivity - the consumed fuel is considerably reduced, so the air will be contaminated less by the flue gases. During the last 10 years fuel consumption has been reduced by about 20 per cent in spite of the fact that production has increased by an average of 25 per cent. The abandonment of coal heating in the glass industry has completely eliminated the phenolic contamination, one of the greatest dangers to natural waters. The introduction of briquetting of the powder-form products eliminated the dust emission of the factories. The introduction of electrical melting in glass production permitted the prevention of the lead and fluorine vapours during production, which are especially harmful to the environment, as the raw material can be charged into these furnaces in such a way that it retains the harmful materials released at high temperatures. Thus, in addition to the advantages from the aspect of environment protection, economic advantages have also been attained, by the saved volatile raw materials. The price of the latter was sufficient to cover the entire costs of electrical heating.

Glass production may be rendered practically a non-waste process. The wastes produced during the manufacturing processes in the factories can be utilized in the plane and hollow glassware factories with advantage, as they accelerate the melting.

In large-scale industrial processing, e.g. of plate glass, the wastes may be collected in orderly deposits and recycled into the production with the advantage mentioned above.

III. UTILIZATION OF THE WASTES OF OTHER INDUSTRIAL BRANCHES IN THE SILICATE INDUSTRY

III.1. *Power Plant Ashes*

III.1.1. *The amount of the obtained ashes and the measures of their utilization.* The energy demand of the world is expected to increase to about 9 times that of the fifties in the second half of the twentieth century. Therefore the production of solid fuels - though their ratio will decrease from the present 52 per cent to about 22 per cent - will develop intensively for a long time to come. The quantity of combustion products (ashes and slag) of the power plants will also increase accordingly. In the United States, for example, this quantity increased between 1966-1974 from 25.2 to 59.5 million tons, 14.6 per cent of which is utilized at present. According to United Nations data, the measure of utilization of ashes and slag in relation to quantities in 1971 was in the countries listed below as follows: United Kingdom 53.6 per cent Federal Republic of Germany 45.6 per cent, France 67.8 per cent.

In the European socialist countries altogether about 140-150 million tons of ashes and slag are produced yearly, 10 per cent of which is utilized in the average. The measure of the utilization is different in the various countries: 28.2 per cent in the Polish People's Republic, 21.2 per cent in the Czechoslovak Socialist Republic, and 19.5 per cent in the German Democratic Republic, while in the other countries it is merely 4 - 6.5 per cent. In the following mainly the problems of the silicate industrial utilization of the ashes amounting to about 85 per cent of the combustion products will be discussed.

As the above data show, the measure of the utilization of the ashes lies well below the possibilities, although useful secondary raw materials can be produced from the useless waste material by appropriate technologies.

III.1.2. *Characterization of the ashes.* Most widespread are the *acidic* ashes containing 45 (– 60 per cent SiO_2, 15 - 30 per cent Al_2O_3, 5 - 15 per cent Fe_2O_3 and 4 (– 10 per cent CaO. These are used in most areas, where a hydraulic bonding capacity of the ash is required (e.g. gas-silicate production, cement completion material, mortar preparation, etc.). The *basic* ashes contain much CaO (20-45 per cent), complicating their utilization, as on the one hand they incline to self-bonding under the effect of air humidity and on the other hand a disadvantageous swelling during the utilization may be caused by the overburnt, slowly dissolving free CaO. As to their mineral composition the ashes consist mainly of aluminum-silicate glass, with 10-30 per cent crystalline components (mainly mullite, quartz, dioxides, etc.).

III.1.3. *Fields of utilization. Gas-silicate production.* Although gas-silicate (gas-concrete, cellular concrete) has been produced for about 40 years, the ash as basic material has been used for this purpose only for 15 to 20 years. The reason is that the technology requires uniform raw material quality and that of the ashes formerly tended to fluctuate.

With the modernization of the power plants the ash quality become more uniform, the unburnt contents were reduced to a minimum, so particularly in the last decade gas-concrete production with ashes as the basic material developed fast. In the United Kingdom 360,000 tons of ashes were used for this purpose in 1966, 467,000 tons in 1971, in the Polish People's Republic about 700,000 tons in 1971 and 1 million tons in 1975.

The ashes of the power plants are utilized in the socialist countries mostly for gas-concrete production (about 25 - 30 per cent of the total utilized quantity), but quite a large quantity of it is utilized for this purpose also in capitalist countries, like the United Kingdom, Sweden, or the USA. The ash-base gas-concrete is used for the production of small bricks, medium blocks of various volume-weights and tensile strengths, for partition wall and thermal insulation plates and, in plants with up-to-date technologies, for big-size reinforced panels. The technology proved correct and a further development of ash-utilization in this way can be expected.

Cement production. In cement production acidic ashes are used nearly exclusively as raw material components (or for their partial replacement) or as a hydraulic additive (as blast-furnace slag, or the puzzolane materials).

Utilizing the ash as raw material is only justified if there is no sufficient clay available, or if it is of an inferior quality, and the ash is nearby and its transport does not cost much more than that of clay. Among the developed capitalist countries the ash is used for this purpose in France, the Federal Republic of Germany and the USA and, from the CMEA countries, in the Polish People's Republic and the Soviet Union.

Utilizing the ash as a hydraulic additive to cement is of far greater significance. This possibility arises from the fact that the acidic ashes have a chemical composition very near to that of the natural puzzolane materials, used for a long time in cements. The advantage of the ash against puzzolane is that it is available in the form of dry, fine powder, while the puzzolane must be mined, dried and ground, i.e. requires more power and is more expensive.

In Europe it is France, where the largest amounts of ash are used as cement additive.

In this field multiple, detailed research has been carried out recently, clarifying the production technology, the properties, the durability and the problems connected with the utilization of ashes. On the basis of all this the spreading of cements with ash additives may be expected. In the Hungarian People's Republic, for example about one-third of the whole cement output is produced with ash additives at present.

This field of application appears to be highly promising, because it offers the greatest economy in cement production with a considerable saving of power, along with the reduction of environmental pollution and of the occupied areas. Simultaneously it increases the capacity of the cement industry allowing reduction of import, and an increase of export of cement, respectively.

But for the wide-range introduction of ash, as a cement additive, a number of problems must be solved. The fine fractions of the ashes must be gained in the dry state, and their storage and transport to the cement factories must be solved (by developing the necessary specialized transport vehicles). Certain concrete designing specifications must also be modified.

Concrete and mortar preparation. The ash may be added as artificial puzzolane material not only to cement, but also directly to concrete. In certain countries and in constructions of great volumes of concrete this method is applied, especially for mass concrete objects (valley barrages, hydraulic power plants, large size concrete foundations, nuclear power plants, etc.), where a part (in certain cases even 30 - 40 per cent of the cement is replaced by ash in the concrete. In this way the cracks caused by harmful thermal effects are avoided, the weight of the object is reduced, and the costs are considerably reduced. But the utilization of ash in this way is only occasional, depending on the nature of the construction.

In mortars, the ash may be used as binding material, additive or to improve the consistency, in finer or coarser form, depending on the end-purpose. Prepackaged cement mortars, or dry mortar with lime hydrate and ash can also be produced. The investigation of the utilization of ash in mortars has been closed; however, this type of utilization is not widespread due to insistence on the conventional methods and partly to the lack of the necessary organizational measures.

Other fields of application. The ashes and slags of power plants can be used for many other purposes; a rich reference literature is available in this field. As the volumes of the utilized ash are generally not great, we list them merely without a detailed evaluation.

Ashes are used, for example, as basic material in combined alum earth and cement production, brick and tile production, for building blocks, as well as for filler material of asphalts, etc.

From the powdery coal-ash obtained in ever-increasing amounts in the power plants, high-quality, high-porosity bricks can be produced with clay bonding in screw presses and big size hollow building elements (300x250x140 mm) in vibropresses. The content of the powdery coal-ash in the raw material may be 30 - 60 per cent, depending on the plasticity of the clay and with vibropress forming it can be increased up to 75 (- 85 per cent. The first production technologies were developed in the United Kingdom and the USA. The production of power coal-ash blocks was started in our country in the sixties. Experiments are being carried on in Turkey for the utilization of power plant ashes in producing bricks of 4-12 per cent of lime content, pressed by 10 (- 400 kp/cm^2 and heat-treated at 48 - 80 C for 1 - 6 days.

III.2. *Utilization of Other Waste Materials*

III.2.1. *Cement production.* The conventional raw materials of the cement industry are limestone and clay. The latter may be replaced partly or completely by industrial by-products or waste materials of similar composition having the required properties. Clay is replaced - mainly in cement production by the dry and semi-dry technology - in many places by blast-furnace slag and in some factories by power plant ashes. A part of the clay is replaced mainly in wet processing plants, by mine dirt, or - as, for example, in Hungary - by coal wash slurry, offering also a calorie saving in the burning. Certain technologies are designed *ab ovo* for a complex raw material utilization, as for example, in processing bauxite, or nepheline-containing rocks for alum earth production, the so-called "belite-sludge" of a high dicalcium silicate content obtained as a residuum after the pyrogenic digestion and leaching is utilized - with further limestone addition - for burning into cement clinker or for cement production.

Pyrite clinker is used in a lower ratio, but in some places in a relatively great volume, as well as iron scale, phosphorous gypsum, red mud and sodium silicofluoride, etc., added to the raw material as mineralizers, respectively, for facilitating the burning. In Hungary only pyrite cinder is used from among these materials in 1-2 per cent, added to the raw materials in all our modern cement factories.

For a hydraulic additive mainly natural puzzolane was added earlier to the cement, but with the development of the industry the ultilization of by-products and industrial wastes for this purpose gained an ever greater role.

The most widely used by-product is the blast-furnace slag, which is a valuable additive of the cement in granulated or ground form. It slows down somewhat the initial solidification and reduces the heat generation of the cement, but it gives a higher final strength and an improved sulphate stability to it. The measures of the blast-furnace slag utilization is highest in the Soviet Union, where 27.3 per cent of the 104.3 million tons of cement produced in 1972 was blast-furnace slag portland cement, for which 18 million tons of blast-furnace slag were used.

The cement industry utilizes, besides blast-furnace slag and ash, other by-products in similar quantities. So, for example, the phosphorous gypsum formed during the production of phosphorous fertilizers (or the much cleaner, so-called chemical gypsum, obtained as a by-product of the more modern processes) may by very well applied to replace natural rock gypsum as a bonding regulator of the cement. Some other industrial wastes and by-products applied in relatively insignificant quantities may also be mentioned, as, for example, the sulphite waste liquor and similar derivates used for facilitating the grinding, or various mineral oil-processing by-products, as, for example, the naphthene soaps, added to the cement for obtaining plastic, or hydrophobic properties.

III.2.2. *Raw ceramics industry*. *Brick production by fuel, mixed with clay* (coal dirt, powder coal ash, domestic heating coal wastes). The low-caloric value *coal containing barren materials* may well be utilized by mixing them with the basic material of bricks. In this way about 50 - 70 per cent of the number of calories required for burning the bricks may be ensured, making the combustion process also more uniform and faster. A further advantage is the reduction of the ash and slag quantities by mixing these coal-containing wastes into the basic material to be disposed of and the lower specific calorie consumption. Mixing coal-containing wastes in considerable amounts is practised in our country even at present.

Brick production from inorganic wastes. Trials to utilize inorganic wastes as building materials were performed in the United States, by adding chemical additives, pressing and air-drying the mixture.

From among the industrial wastes used for fuel the high calorific value *paraffine-containing press-clay* and the *oily filter clay* /4000-8000 kcal/kg/ may be mentioned. Both materials are used in circular kiln operation, mixed with the heating coal. Their utilization is limited by the spread of gas- and oil-fired tunnel-furnaces.

III.2.3. *The production of thermal insulation material*. Basic material: acidic blast furnace slag:

$$M_K = \frac{SiO_2\% + Al_2O_3\%}{CaO\% + MgO\%} \quad 1$$

with the condition that the sulphur content bonded in the form of sulphides is below 0.5 per cent. The cooled blast-furnace slag is cut up, crushed and classified. For producing slag wool the fraction d=30 to 80 mm is the most suitable.

Processing. The lumpy blast-furnace slag is melted in a cupola furnace at about 1300-1400°C with coke fuel. The melt tapped in a continuous flow from the cupola is formed by a vapour current, or a centrifuge to a bulk mass of elementary, diam. 4.8 μm, fibres. The slag wool is processed by manual or mechanical methods into thermal insulation building elements and structures.

Thermal conduction factor: about 0.04 Kcal/m.h.°C density: 75-200 kg/m^3.

Economic background. In Europe about 1 to 1.2 million t/year blast-furnace slag is utilized for producing slag wool and mineral wool respectively, mainly in the east European countries. This trend is diminishing, as the considerably higher-quality rock and glass-wool products are gradually gaining more and more ground in this field. The utilization is still viable in the less developed countries.

IV. CONCLUSIONS AND PROPOSITIONS

The above comments show that wide possibilities exist, but still much remains to be done in the field of creating non-waste technologies and a widespread utilization of the industrial waste materials.

We have shown that the main mass of dusts in the silicate industry consists of dust materials of the cement factories. The amount of these is increasing in proportion with the improvement of separation. The most obvious solution seems to be the introduction of the modern dry process technology with floating heat-exchange system, which saves heat-power. The reconstruction of the old factories in this sense permits a direct utilization of 90-95 per cent of the dust materials in the technological process. The remaining 5 per cent may be utilized in road building, or as cement additive, in vine culture amelioration and in the production of light concrete aggregates. The utilization outside of the factory is limited mostly by transport costs to great distances, and therefore processes sited close to the involved factories are to be preferred.

For the utilization of stone pit dirt mainly the mechanical and bonding stabilization processes in road building should be taken into consideration. In the mentioned fields of power plants ash utilization 40 - 50 per cent of the obtained ashes could be utilized instead of the present 10 per cent in our countries, to the considerable benefit of our national economies. Therefore the amounts of the ashes, their composition and utilization possibilities must be surveyed with respect to the given raw material and power supply situation of the individual countries.

Technical specifications or standards are to be elaborated for the requirements of the ashes to be used in the various fields.

The concepts of the development of the building and building material industries can and must be harmonized with the utilization possibilities of the available ashes.

Economic Aspects of Non-Waste Management

C. Cala* and J. Wieckowski**

**Ministry of Science, Higher Education and Technology, Warsaw, Poland*
***Professor, Director of Institute of Management, Warsaw University, Warsaw, Poland*

ECONOMIC PROCESSES IN NON-WASTE MANAGEMENT

The concept of non-waste management and its implementation on a large scale in production processes is a practical step towards taking into account th ecological aspects of the contemporary economy. In this respect, the Paris symposium may be treated as an attempt to develop the ideas and issues presented during the earlier ECE seminar in Rotterdam which dealt with the ecological aspects of economic development planning. The development and implementation of the main issues of non-waste management may also be considered as an attempt to draw practical conclusions from the warnings of the Club of Rome. The reports of the Club of Rome suggest the possibility of collapse of social-economic development, unless certain spontaneous phenomena are brought under control, the phenomena which are present in the five fundamental variables of this development - increase of population, limited possibilities of production and nutrition, use of natural resources (especially of those which are not renewable), industrial production and gradually rising pollution of the natural environment. The concept of non-waste economy is a proposal for counteraction against these phenomena and their spontaneous character, concentrating especially on three variables - rational use of natural resources, new methods of industrial production and issues of environmental design and protection.

Much has recently been said about the need for *new socio-economic models*, especially in view of the uneconomical processing of raw materials and energy and attitudes to consumption in general. In the process of creating such models the concept of non-waste economy, including non-waste technologies, should be treated as a basis.

Why is the role of non-waste economy in socio-economic development being stressed so much, and what are the possibilities of practical application of these concepts? A simplified answer may be based on two fundamental premises:

Technological progress in the member countries of the Economic Commission for Europe shows that it is possible to introduce production processes by means of which the conflicts between the spheres of production, consumption and protection of the natural environment may become less acute. Both the symposium of the Comecon countries in Dresden and the ECE seminar in Paris show that there exists an enormous amount of technological capability in this field. The main problem then is in its implementation on a large scale and in the exchange of experience between different countries. Simultaneously, in every country intensive scientific research is being conducted on the development of technologies of "clean" production.

It is to be hoped that more and wider activities will be undertaken in the sphere of organization and that *economic means and stimulants* will be developed in order to apply technologies based on wasteless economy in a more popular way.

Economic issues are becoming enormously important in the analysis of the entire concept of non-waste economy. Existing economic systems search for solutions which minimalize the costs of production and exploitation and the results of this are the following phenomena and dilemmas:

unsatisfactory use of all goods in the process of exploitation (only the most economical solutions are chosen); non-consumed goods and wastes remain, which increase costs by the need of utilization;

counter-action against pollution of the environment calls for additional investments in order to restore the environmental equilibrium;

non-waste technologies are costly, hence somebody has to cover the additional expenses connected with their application.

One general conclusion is easy to arrive at. The above phenomena appear in analyses done up to now as separate, not interconnected elements, not linked within one, comprehensive economic calculation. Thus there is a conflict of interests on two levels - micro and macro. Interests on the micro-scale are those of an individual producer or consumer, or even of productive branches and consumer groups, whereas the macro-scale concerns the social and economic interests of the whole country. For what can be profitable from the point of view of one branch or consumer group is not necessarily good for the national economy as a whole. In other cases, what is non-profitable on the micro-scale may be desirable or even necessary for the national economy in spite of increased costs.

The problem of economical calculation in regard to the concept of non-waste economy should, of course, not be restricted to one country. Wide mutual interconnections between countries in the field of social and economic affairs require the inclusion in this calculation of such elements as raw materials prices on world markets or environment pollution covering certain regions consisting of several countries.

The creation of a homogeneous and coherent system of economical calculation with regard to non-waste economy is a difficult task, so that we may only speak about a gradual approach to a certain optimum for a specific situation, and the process remains a continuous one. The aim of the present statement is to indicate some elements of this calculation and some possible ways of treating phenomena within economic categories. At the same time its intention is to demonstrate some experiences and proposals for maintaining a non-waste economy.

THE MEASURES OF EFFECTIVENESS OF NON-WASTE ECONOMY

There are technical, organizational and legal measures which influence the increase of effectiveness of non-waste economy. These apart, certain economic means may and ought to be applied to influence rationalization in this field. With the help of economic instruments it is possible to stimulate the macro-economic systems: branch industrial organizations, administrative districts (urban and rural agglomerations) or separate economic units (industrial firms, commerce or transport). Nevertheless in each particular case different forms of intensification of non-waste economy should be applied.

In the case of macro-economic decisions undertaken in a great economic or administrative system a complex calculation of the effectiveness of non-waste economy could be applied. In such cases a wide-ranging analysis of total benefits to, and costs connected with, rational material economy should be provided, with special attention to scarce natural resources. The calculation should take into consideration the issues of biological balance of the natural environment, since big industrial centres and urban agglomerations are the potential source of destruction of the present balance. Industrial progress, and the separation of production plants from the sources of raw materials and power, create enormous needs for means of transportation (transport of material goods and people, transfer of information) - all this results in new threats to the natural environment. Actions leading to utilization of new sources of materials and power as well as of agricultural products are often connected with losses in other parts of the national or world economy. Even organized leisure or tourism may create serious hazards to natural assets.

Thus, large-scale decisions require a complex examination of the problem, with special emphasis on the following:

new industrial investments and their sites;

new urban agglomerations or development of existing towns;

building or rebuilding of new land, water and air communication routes;

introduction of waterflow systems and watering or de-watering of certain territories;

changes in the structure of cultivation, fertilization and agricultural production.

This type of decision requires the application of direct expenditure calculations and calculations connected with keeping waste to the minimum in the whole complex of economic activities. The appreciation of different solutions in this field is the most difficult task. Partial information on expenditures and their results can be developed by technologists, physiologists or natural scientists, but presentation of the same issues in the form of benefit or non-benefit, in other words in economic terms, remains the domain of economists. Our experiences in this field are often unsatisfactory, for it happens frequently that the immediate benefit to particular organizations are calculated, and it turns out that when action is undertaken, the results are less than were expected. Then the need arises to cover unpredictable social expenditures, far exceeding the scope of activities of a given firm.

This is the reason for losses in the national or even the world economy. It is fully justified then to make the management of big industrial or administrative organizations responsible for the misleading results of their calculations. Losses caused by imperfect analysis of all costs and results should be covered by the funds of these organizations, thus providing a stimulus for intensification of non-waste economy.

The results of non-waste economy are better understood in terms of units acting in conditions of economic independence, for in that case a traditional cost/ benefit yardstick may be applied. The first thing to do is then to decide what elements of non-waste economy can be connected with the activities of those units and how to estimate the costs and benefits of such an economy.

The second is to state the means of including the expenditures and profits of non-waste economy in the general profit calculation. The main issue is to stimulate the managements of particular enterprises to rationalize their non-waste economy by linking it with the profit calculation; a fruitful exercise only if the enterprise's profitability is treated as generally acknowledged criterion of effectiveness.

There are, among others, the following methods of combining the different phenomena of non-waste economy with the calculation of profits which may be applied:

Correct estimation of raw materials or other scarce resources, as well as energy materials. Prices of scarce materials can be raised if the industrial processes concerned involve waste. An enterprise sustains an economic loss as a result of waste, whereas the use of the wastes can be profitable. High taxes can be imposed on investments related to the use or stocking of scarce materials but the taxes should only concern wastage. Such taxes would increase the value of the wastes. The conclusion therefore is that taxation decreases the profitability of production.

Determination of higher prices for construction or production technologies guaranteeing a fully efficient non-waste economy. This phenomenon may be translated into profitability calculations in several ways. One - identifying this type of production by high selling prices; second - application of tax reductions by a system of rebates and thus increasing profitability. It is also possible to raise taxes on technologies or constructions which do not guarantee the efficient use of wastes, and in case of importing - to apply preventive customs charges to licences or finished goods bought abroad.

Estimate of the consequences of neglecting the principles of non-waste economy and thereby upsetting the environmental balance. This concerns a situation where wastes from a technological process, from the operation of equipment, or from the utilization of chemical products, either in the form of destruction of packaging or from the products as such, constitute a threat to the natural environment. In such cases it is necessary to remove the causes of the threat on the one hand, and on the other, to link the damage to the environment with the unit guilty of not following the rules of non-waste economy. The size of the penalty or compensation which decreases the production can be the measure of the threat or real damage.

Support for action to deal with the causes of threats to the natural environment. Expenditure connected with the rationalization of a non-waste economy may sometimes be very high and thus exceed the financial possibilities of particular enterprise and industrial or administrative units. In that situation it could be more convenient to pay a prescribed fine than, for instance, to invest in a sewage-treatment plant. Similarly, the complex utilization of wastes seems to be very costly. Hence, financial means for these aims should not be treated as part of the current expenditures of a given firm, thus lowering its profitability to a great extent. Such means should be provided from special funds in the hands of administrative organizations or state bodies.

These are some examples of linking the phenomena of non-waste economy with the profitability calculations of industrial or administrative organizations. Inclusion of these phenomena in the economic calculation aims at stimulating these units to intensify action in the field of non-waste economy. This presupposes that administrative or state bodies are endowed with powers to take action with precise economic consequences (e.g. taxation, application of fines and penalties, creation of special funds in an administrative district etc.).

NON-WASTE ECONOMY AS AN ELEMENT IN THE CALCULATION OF EFFICIENCY

The system of non-waste economy concerns many fields of industrial activity, above all, the production and consumption, as well as the functioning of the natural environment in which these processes are taking place. The scope and results of non-waste economy can be measured in technological, economic and social categories. The forms and scope of organizational and technological activities aimed at improving non-waste production and consumption may be similarly determined.

Non-waste systems can be examined in a number of ways:

(i) as a measure of social loss resulting from failure to follow the principles of non-waste economy.

(ii) as a means of calculating the efficiency of investments and results in the maximum reduction of waste in production and consumption.

(iii) by means of economic stimuli it is possible to influence the managements of enterprices and administrations to reduce waste to the minimum. As non-waste system requires the support of legal rules which define:

methods of measuring social inputs connected with the system,
methods of calculating the effectiveness of improvement in non-waste systems,

the economic instruments influencing the managers in the sphere of non-waste production and consumption,

forms and scope of natural resources protection (those which are rare in nature),

form and scope of environmental protection against the consequences of not following the principles of non-waste production and consumption.

This paper deals mainly with the economic issues concerning the functioning of non-waste systems. Nevertheless, in some cases the need to use legal means to back up economic action is recognized. In some cases, international rules should be established.

Let us now discuss briefly the phases of the production and consumption processes and environment protection, within which the non-waste system should be made the subject of economic calculations.

New Products Preparation Phase

In this phase, economic calculations have a very important place, above all for an estimation of actions improving the scope of non-waste production, and at the same time for the application of stimuli in this phase of the productive process.

I. It is possible to develop and put into practice the methods of calculating the effectiveness of expenditures concerning a manufactured product. The main issue here is to estimate the price of manufacturing in relation to the inputs proposed and from the point of view of natural limits on the availability of raw materials, energy resources, etc. Inputs of natural resources can be weighted, using for example:

high prices of such inputs,

tax charges on such inputs.

In both cases the higher estimates should be introduced into the calculations of effectiveness, of different methods of fabrication. The second solution is even more convenient in making economic calculations, for tax rates are easier to manipulate than prices. Depending on the economic situation, change of the latter may require a longer period of preparation and provoke wider economic repercussions.

II. The effectiveness of various manufacturing methods should be analysed from the point of view of the amount of waste which may be created in the production process. This means that the manufacturer has to include proposed technological solutions at the conceptual stage. Therefore any waste-reducing action should be included in the cost of the future product. For instance:

not to include wastes in production costs in cases where the comprehensive use of wastes is included in the concept;

to include in production costs all wastes expected in the course of production by a given method if the manner of their utilization has not been previously decided;

to include all wastes expected in the production costs, with additional taxation, in case when they could destroy the natural environmental balance or create additional nuisances in the process of being used.

Undoubtedly, estimation of the amount of wastes expected in the phase of production adds to the work load of the entrepreneur or designer, an additional burden related to the technological process and to fabrication.

III. The analysis of the efficiency of particular fabrication methods should also include the costs of raw materials and energy necessary in the further exploitation of product, as well as the wastes created by those inputs. As in the previous case, it is possible to introduce various forms of charge on construction costs in the form of taxes on a particular method. These would add to the costs, depending on:

improvements in packing (easily-obtainable, non-polluting materials);

improvement in exploitation (not requiring expensive or scarce energy sources; wastes of materials and energy not harmful to the environment in the forms of toxic smokes, gasses, liquids, etc.);

improvement of transportation and handling (similarly to the above cases, raw materials and energy needs have to be considered, as well as environmental threats posed by improvement of the goods produced);

improvement of conservation and repairing of a product (similarly, the raw material and energy demand and harmfulness for environment connected with necessary conservation and repairs are to be considered).

In course of analysis of effectiveness of different forms of fabrication, attention should also be paid to the last phase of the product's life, i.e. to its disposal. The designer could be assigned to the task of devising a method of disposal of a product. The following forms of calculation can be applied in this phase:

to decrease the fabrication costs of a product by the value of inputs which can be reintroduced in the process of production;

levying of taxes on the cost of products, in cases where special storage is needed or the environmental balance is influenced in any other way; the rates would be diversified according to the degree of harmfulness of the disposed product to the environment.

The calculation of effectiveness of different technological solutions is, as a rule, based on hypothetical data. Therefore it is not a real calculation but an assumption based on future, imperfectly known situations. The second main difficulty is to estimate the investments needed for future production if non-waste principles are not observed. The main issue here is to state the basic cost and the tax to be imposed (distinguishing by price or by taxes).

This paper does not pretend to offer any definite proposal for a solution but merely to outline the problem of establishing a link between an estimate of design costs and respect of non-waste requirements in the course of their production and use. Developing of principles for the calculation of effectiveness which would include the above propositions is, in the authors' opinion, quite possible but nevertheless requiring action by administrative or state bodies.

Whatever way one wishes to accommodate the degree of rationalization of fabrication in the classical profit calculations (so as to respect no-waste principles), other means of pinpointing the entrepreneur's activities in this field may be applied. All the phenomena mentioned can be detailed point by point and after they are added up various versions may be allotted to different means of fabrication classified according to their non-waste capacities. The allocation of products to a given category can then influence the price being paid the constructor for his design. Similarly, the income earned by the producer in the later stage may also be differentiated from the non-waste point of view. This is possible by means of tax concessions which would increase the profitability of producer.

Phase of production of goods. In this phase the non-waste character of production is directly connected with the economic activity of the producer and with the functioning of the natural environment in which this activity takes place. It is thus the phase of processes with specific economic results, which sometimes are long-lasting and irreversible. Making up for mistakes made in this phase is in most cases very costly and labour-consuming, therefore economic instruments must exist which will make it possible to rationalize non-waste production and to stimulate the level of effectiveness in that field.

Final decisions on the technology of manufacture are taken at the production stage and so one establishes material inputs; raw materials and equipment are selected and the quantities are determined; energy requirements are also established, etc.; the flow of materials through different phases of processing; methods of employment of materials and energy, the quantity of necessary supplies, of reserve stocks, and the volume and method of waste utilization; decisions on the exploitation of the product, consumption of materials and energy during exploitation, spare parts storage, scope of repairs, material wastes resulting from exploitation, etc.; the degree of relationship between the production process and the natural environment; supply of water, pollution from smoke, gases, dust, sewage, etc.; possibilities of utilizing materials existing in exploited products; methods of utilization of used packing.

These elements should be related to the calculation of efficiency. The material inputs and labour are taken into consideration in the traditional calculation of original costs, whereas the problems of non-waste production are represented in that calculation to a very much lesser extent, which primarily concerns the following elements; providing of complex utilization of material and energy resources (without losses and wastes), economical methods of utilization of limited or scarce raw materials, a necessary minimum storage of materials, rationalization of packing (reclamation and storage), preparatory conditions for rational exploitation of a product and its non-waste disposal.

These processes can be included in the calculation of efficiency, estimating them in terms of cost and profitability. It is thus possible:

to increase the prices of scarce materials and sources of energy, or to impose taxes on such inputs;

to state prices for natural resources not presented in terms of value, such as: water, land, materials not processed, etc. (such prices would be applicable merely for the measurement of rationalization),*

to distinguish complex organization of wastes utilization (i.e. to include non-utilized wastes in the category of losses with a higher price, exceeding their nominal value, or to grant tax privileges);

to impose a high rate of tax on stored material resources which would decrease the profitability of production;

to oblige producers to pay fines or compensation in cases where they do not follow the principles of protection of the natural environment;

to impose additional taxation in cases of production of goods and packing which cannot be utilized or disposed of after exploitation according to non-waste principles;

to introduce high preventive customs duties in cases where materials are imported which do not correspond to non-waste rules in their use and

---

* In the socialist economies goods not exploited are as a rule not estimated from the point of view of their value.

post-use disposal.

Phase of exploitation and disposal of products. In this phase, stimuli to the rational application of non-waste technologies are introduced. Nonetheless, it has to be admitted that the exploitation of a product is to a great extent determined by its designer and producer. There are possibilities, though, in this field to influence the actions of the consumer-user. Of these we can select:

application of increased prices or preventive customs duties on goods which do not correspond to non-waste norms. This can react unfavourably on producers and importers who do not follow non-waste principles;

imposing fines on users who disturb the natural environment balance (air, water) by pollution with wastes or the used parts of products or packing, etc.;

encouraging with rebates and allowances for purchases of new parts or energy equipment on the return of used parts or packing;

imposing of penalties for leaving products derelict contrary to the regulations;

oblige the user to repair damage to the natural environment caused by improper exploitation of plant.

All these methods are designed to retrieve the costs from the user's budget in case of failure to respect the principles of non-waste economy.

IMPACT OF SITING OF INDUSTRY ON THE VOLUME OF NON-WASTE PRODUCTION

The analysis hitherto has been dealing with economic instruments and stimuli which can be applied in a gradual move towards non-waste economy. There are also other forms of economic undertaking that stimulate an introduction of "clean" technological processes in the national economy. Among these we should include the proper siting of economic organizations, especially industrial plants. This is a matter of regional planning.

Attention has been paid to the problems of siting of industrial plants since the end of the nineteenth century; and theoretical scientific research in that field has been largely developed. The main stress, however, up to the period of World War II was put upon the optimal choice of site for a model plant depending on the interplay of forces between the availability of resources and the selling markets, with the factor of transportation taken into account. The influence of production methods on environmental protection had not been raised in research at all, and played a small role even in the classical works of Alfred Weber* or August Losch.†

"Pre-ecological technologies", using a term applied by Ananichev,‡ had pride of place in the methods of production, and resulted in widely noted degeneration of the environment caused by industry.

The choice of site for a plant always depends on many factors such as: distance

*Alfred Weber, Ueber den Standort der Industrien, 1909.
†August Losch, Die raumliche Ordnung der Wirtschaft, Jena, 1940.
‡K.W. Ananichev, EEC. ENV/AC.4/R.3, 1974.

of material resources, selling prices and markets, conditions of technological infrastructure (energy, labour force, housing, schools, etc.). Let us now assume that all these factors are similar for different variants of siting of industrial plant (which is practically impossible) and concentrate our attention on the issue of the damage caused by production methods in relation to the natural environment. The question of siting can then be discussed in relation to three forms of organization:

single plant;

simple agglomeration of industrial plants;

plant agglomeration with co-ordinated production processes on the principles of non-waste technology.

A single plant may be sited anywhere, under the above assumptions, provided that such methods of production are applied in its activities which do not damage the balance of the surrounding environment. From the point of view of heavy industry, such conditions are economically difficult to satisfy. They mean that every plant has to operate separate equipment for fumes and sewage treatment, handle their wastes in a way which would take care of the environment, or to assure "clean" production, which is more expensive than the ordinary method. It means that only those plants can be freely sited which as a rule have a "clean" process of production, as for instance the textile industry, food industries, etc. The image of possibilities of choice of site is simplified, for in many cases damage to the environment may result not from direct productive operations but from indirect causes such as an increased transportation demand.

Simple agglomeration of industrial plants is a form of industrial organization commonly found in all industrialized countries. The proximity of plants grouped in a geographically defined place gives great economic advantages. First of all they have joint use of the infra-structure of the area (roads, means of transportation, loading services, power network, etc.). Such plants are not generally linked by common production processes or co-operation, being a set of independent units acting separately. With regard to environment protection such agglomerations may give *additional advantages through the design and use of common equipment for water and sewage treatment*, and to a lesser extent, *gas and dust treatment*.

It should be said, however, that introduction of common waste treatment plant in an old industrial area is often coupled with serious obstacles, firstly because of difficulties of finding the necessary space in a densely built-up area, and secondly because of the reluctance (for psychological reasons) shown by managers of individual units when the problem of joint investments is brought up.

It can be assumed, though, that in planning and building new industrial plant complexes, joint multi-purpose water and sewage-treatment equipment will act as an important factor stimulating siting and investment in a defined area. Equipment of this kind may be treated as necessary technological infra-structure (just like the road network) which in many cases may be implemented earlier than full development of the area. The best economic effect can be obtained in *agglomerations of interlocked industrial plants*. The plants operate within a defined system creating one productive complex, whereby one of the branches leads and others co-operate. This type of complex can apply on a full scale all varieties of non-waste technology; one may even say that without the application of non-waste technologies the organization and operation of such complexes would be impossible. This special role of concepts of wasteless technologies makes it possible to intensify many economic effects which are difficult to obtain in other forms of siting. Among those effects, according to Soviet economist A. Probst,* the following may be included:

Rational combination, in near or immediate neighbourhood, of different kinds of production successively processing the same raw material, thus allowing for vertical linking of various stages of a technological process which gives a substantial economy in investment and exploitation costs.

The costs of transferring the object of production from one place to another are *minimalized* as a result of placing all the stages of production within a small area thus leading to economizing of labour and lowering of transportation costs. It should be remembered at the same time that in case of dispersed siting of industrial plants these costs in some branches of industry reach 20-40 percent of the original costs of the final product (e.g. in metallurgy, chemicals or building materials).

Providing more developed forms of specialization and co-operation, and easier solving of management and organization problems.

Effective utilization of wastes produced by one plant in another (recirculation). This also concerns the utilization of secondary energy sources (waste gases with fuel properties, hot water, used steam, etc.).

Economies are obtained as a result of a varied time-table (daily or seasonal) of heat and electric power consumption by plants of different production schemes, depending on the time of day.

The effect of joint use of large-scale equipment for water and sewage treatment or atmospheric protection may also be obtained in case of simple industrial agglomerations, as was stated earlier.

To these results of non-waste technologies others may be added, such as the economy in the use of labour, staff qualifications, common social services, etc. A. Probst states that the general economic improvements obtained by implementation of such production complexes may be estimated as 20-30 per cent of investment costs, so the role of non-waste technologies in this area is undoubtedly of great importance.

NON-WASTE TECHNOLOGY - CONDITIONS OF IMPLEMENTATION

The effectiveness and speed of implementation of non-waste economy requires the fulfilling of a number of conditions, both on a country scale and on the wider, international scale. This concerns the model of consumption and the means of fulfilling the consumer's needs.

Firstly, the possibilities of prolonging the life of goods should be analysed. Recent marketing activities aim *inter alia* to stimulate new needs and fulfil them in a diversified way, depending on the wishes of chosen consumer groups. This kind of activity, characteristic of countries of high technological

*A. Probst, Vaprosy razmestsheniya socyalistitsheskoi promyshlennosti, *Izdatelstvo Nauka,* Moskva, 1971.

development, tends to evoke apparently new needs, but is, as a matter of fact, merely an exchange of one product for another alleged to be of better quality. This increases the degree of wastefulness of labour and inputs connected with:

design of new products;

switch of a production process to new processing or technology;

disposal of apparently useless (technologically old) products;

disposal of spare parts and repair service developed for products withdrawn from exploitation.

A very limited increase in the value of goods to the consumer sometimes involves efforts and means, disproportionate to the results obtained. Although this improvement of quality is often purely apparent, there nevertheless exists the phenomenon of forcing the consumer to purchase and accept a new product with the simultaneous abandoning of an old but fully usable one.

It often happens, too, that both consumers and producers act irrationally in both production and consumption, being aware of the forced necessity of changing a product irrespective of the extent to which it has been used. Accelerated use is commonly encountered in cases of expensive products involving high material inputs (aircrafts, cars, washing machines, fridges, TV and radio sets, etc.).

Not only consumption goods are governed by these rules, but also technological equipment used in industry, agriculture, communications, information or services.

This situation primarily arises from producers wishing to increase their incomes, whereby consumption models created by countries of high technological standards are transferred by analogy to other countries. This leads to the tendency of unjustified "keeping the pace" in the field of quality of exploited products which goes beyond the economic possibilities of those countries. Thus there is considerable prodigality of material and labour resources.

It appears then that the apparent diversification in fulfilling needs created by the consumption models of highly developed countries should be limited. There is a need for standardization of production and a considerable decrease in the speed of exchange of products which accordingly would have a longer life.

An important condition of making non-waste technology effective is the need to modernize the methods of estimating the consumer values of new products (patents, licences). One of the fundamental criteria of estimation should be the scope of non-waste production and exploitation of a new product instead of, or rather coupled with, the presently applied calculation of production and exploitation costs. A low-cost level is often a determining factor, although the design of a new product requires considerable use of rare (naturally limited) resources, or a negative influence on the natural environment, both in the process of production and consumption. For implementation of this condition it is necessary to introduce proper legal regulations on the national and international scales.

Similarly, another important factor in implementing non-waste technology would be to introduce limitations on the variety of technological processes and economical norms for production and exploitation processes. Implementation of these proposals is fully feasible, both on the national and international scales, which was proved by different limitations introduced during the fuel and energy crisis period. It appears that it is possible to spread these limitations to other spheres of the world economy (especially with regard to raw materials) in which certain deficiencies are expected in the near future. Limitations of use could be linked with disseminating the real possibilities of use of plastics and utilisation of secondary products. This would often require creation of social funds (state or economic associations) which would enable conditions for adaptation (in most economic way) of substitute goods not utilized until now. Proper regulations in this field are necessary though, both on the national and international scales.

This problem concerns also the utilization of exploited products and used packing. These issues are passed over to the consumer who neither is aware of his responsibility nor has the proper means at his disposal to utilize surplus goods or to guard the environment from pollution. Purchasing of used products and packing is in most cases uneconomical (in the traditional understanding of the subject), unless special social funds are introduced. A situation of profitability is not created until the crises occur, e.g. in war conditions or by a reduction of supplies on the part of the raw materials producers.

The economy of used products and packing also requires the introduction of norms of wide international scope.

It appears then that the implementation of non-waste technology calls for a calculation of effectiveness, both on micro- and macro-scale of the economy in the fields of:

organisation and siting of industry;

prolongation of life of goods;

limitation of the variety needed to fulfil consumption needs;

organization of secondary utilization of products and packing after first use;

organization of utilization of substitute goods (secondary or low-profit);

organization of environmental protection and biological balance.

Apart from the above, possibilities and condition of stimulation in the area of rationalization of non-waste technologies should be analysed, with special consideration given to the economic instruments discussed in the paper which influence the attitudes and actions of producers and consumers.

Action to set non-waste technology in motion should be accompanied by decisions made at the level of producers' and consumers' association and regional or state management as well. These actions often require long-range planning accompanied by executive power. Centralized planning in the area of non-waste production should take into account long-term prognoses of production and consumption. Actions in the fields listed below should be undertaken in the widest international co-operation:

- exchange of information and experience on introducing non-waste technology;
- estimation of macro- and micro-economic effectiveness in the process of non-waste economy improvement;
- support for non-waste technology (estimation criteria of patents, licence, etc.)
- design of patterns and models of consumption in order to rationalise the exploitation of world resources;
- support for and execution of conditions providing non-waste production and exploitation (preventive customs payments, fines, taxes);
- research on rationalization of non-waste production and exploitation;
- providing of necessary means of environmental protection and redressing biological balance.

All the problems analysed in the present paper could be used as a foundation for further discussion of the economic aspects of wasteless management and technology.

Ways and Means of Implementing Non-waste Technology

Introductory Report

M. Schubert (Rapporteur)

Technical University, Dresden, German Democratic Republic

Regarding the preceding Topics I - III of the seminar it seems to be suitable to consider ways and means of implementing non-waste technology on a functional basis, including the sphere of production and consumption. Taking into account the present reports, there may be technological, economic, juridical and social possibilities of attaining this objective. Moreover, some requirements concerning education and training as well as international co-operation have been mentioned. Due to the fact that all contributions came late and only four of them were finished when this report was composed, other available documents were used.

Implementing non-waste technology enables

a reduction of non-used wastes and thus a reduction of environmental pollution, and on the other hand,

a reduction of the specific consumption of natural resources to secure human needs.

Zygankov* shows in his report that the Communist Party and the Soviet State pay much attention to the problem of environmental protection as an important factor to guarantee prosperity to the present and future generations of the country. He stresses that such a complex and scientific-technologically complicated object as the introduction of waste-reducing technologies can be realized only by the combined efforts of all industrial ministries, the academy of science and the institutes of technology in planning and co-operation of their work.

An Interministerial Scientific-technological Council, which is responsible for the complex problems of environmental protection and the economic use of natural resources, planning the implementation of non-waste technology and the use of wastes as secondary raw material, is an example of how to solve the problems.

The author recalls the success of the council mentioned above, in connection with the introduction of new technologies from the Socialist Republic of Azerbaidjan. Here, by means of a manual about the use of waste products, a profit of approximately 4 million roubles was arrived at.

Comparing economic variants it is essential to take into account the lower environmental pollution and use of natural resources with the implementation of non-waste technology.

The law concerning the limitation of wastes by emission aims at protection from wastes as far as possible. The alternative solutions are reduction of

---

* A. Zygankov, Discussion paper ENV/SEM.6/R.6/Com.6, "Activity of interagency scientific council for complex problems of preservation of the environment and rational use of natural resources in the field of non-waste technology". (Available in Russian from ECE Secretariat, Palais des Nations, Geneva, Switzerland).

quantity of emitted wastes and lower density, for example by means of tall chimneys. There is an additional limitation for the maximum wastes emittable by a plant or factory. This is an effective stimulation for the implementation of non-waste technology because reducing the wastes or using them is more advantageous than converting them into material, which can be dumped without danger. For the time being the limitation of emission is regulated only according to the technological state of the art of reduction methods.* Eventual damage and annoyance in spite of the regulations (injury to the health of human beings or animals, reduced growth of plants, limitation of recreation facilities, reduction of agricultural land, increasing corrosion) is insufficiently taken into account when comparing different technological procedures from the politico-economical point of view, the more so, as they arise from the complex action of several emittants and reach critical values only after exceeding a certain threshold.

Therefore Hofman* suggests to add "costs for pollution" to the costs for production and distribution. These costs are costs for

(1) reduction or prevention of wastes,

(2) compensation for negative social effects,

(3) compensation for losses of raw material.

The costs for pollution correspond to the increased expense in the productive and non-productive system of the economy. The itemized costs are explained in detail and subdivided. The importance of costs for compensation of losses of raw material is stressed. Though the cost-benefit relation of non-waste technology belongs to Topic III, the importance of the real, i.e. comprehensive valuation of the overall social expenses — often not calculated at present, must be emphasized as an effective means for non-waste technology to succeed.

Huissoud† reports in detail how stimuli for the advancement of non-waste technology are combined effectively in France: legislation concerning limitation of emission, financing selected research work and model plants. Who pollutes - pays. The incoming financial means are used to develop non-waste technology. For better use of wastes the installation of a National Office for Wastes is planned.

In my opinion state-controlled and planned economic regulations and means are useful instruments for the advancement of waste-reducing and non-waste technology. Thus, in socialist countries costs and fees for the economic use of natural resources and sanctions to firms polluting the environment have turned out to be very effective.

In the German Democratic Republic the economic regulations in force stimulate planned measures of environmental protection. Firms which do not observe the limits of emission into water or air or do not set aside the planned investments for environmental protection have to pay sanctions to the state. Moreover, they are obliged to pay for the damage and to finance the restoration.

To secure economical use of water, fees are to be paid for the use of groundwater and surface-water, according to the maximum quantity that may be used.

---

* K. Hofman, Discussion paper ENV/SEM.6/R.6/Com.7, "Economic and social efficiency of non-waste and low-waste technologies".

† R. Huissoud, Discussion paper ENV/SEM.6/R.6/Com.4, "Statutory and financial provisions for the establishment of manufacturing methods free of waste products".

For the economical use of resources and minimizing wastes the following economic regulations seem to be practicable to me:

High prices for rare raw materials should be fixed, stimulating economical use of the material, and prices for products made from secondary raw material should be kept in a reasonable relation to primary raw material. Prices for secondary raw material should stimulate firms to use it.

Fees for the natural resources (soil and water), should be territorially and quantitatively different for greater effect.

Compensations and sanctions imposed on firms when the law and regulations for keeping the biosphere clean are not observed.

Financial aid (lower taxes, depreciation) to firms which observe environmental protection. Additional taxes for products and plants which pollute the environment more than is allowed by regulations.

Financial support and credits for environmentally protective measures.

The application of economic measures depends on the socio-economic structure of the social system and on the conditions for economic development of each country.

Industrial plants came into being, using only certain portions of a raw material for production. Non-waste technology, however, demands complex and exhaustive use of natural resources and therefore the common involvement of materials and energy.

The main premise for the economical use of raw materials and environmental protection is a planned and long-term programme for the use of natural resources and of the available funds. In the socialist countries nation-wide plans for the development and implementation of low-waste technologies and the use of wastes have been effective. The state plans for science and technology contain research and development tasks for better use of resources and for keeping the biosphere clean. The five-year plans and annual plans contain obligatory tasks for decreasing the emission of wastes into the environment as well as for the economical use of wastes for secondary raw material. In the German Democratic Republic, for example, for the economical use of material special plans have been worked out balancing the output of wastes like brown coal ash, cinder, plastic wastes, sulphite liquor and others with the degree of their eventual use or their elimination, respectively. Standardizing the maximum use of material aims at the economical utilization of material.

Investigations are to be made in order to decrease wastes either by using material giving less wastes or by utilizing the resulting wastes. In this connection the report by Gutt *et al.** is relevant, giving a survey of the locations and use of most of the industrial by-products and wastes of the United Kingdom building trade. The necessity of planning future production, eventual substitutions and programming transport are pointed out.

We will have non-waste technology by

(a) reduction of wastes per unit of time, and

(b) an increase of the re-use of wastes.

* W. Gutt, Discussion paper ENV/SEM.6/R.6/Com.3, "A survey of the location, disposal and prospective uses of the major industrial by-products and waste materials".

Lötzsch and Schubert* compute the wastes $\dot{A}$ per unit of time as

$$\dot{A} = \Sigma(a_i+b_i) + \dot{E}_i + \Sigma c_j\dot{V}_j + \Sigma d_k G_k + \Sigma e_l \frac{G_l}{t_{N,l}}$$

with

- a_i wastes per product E_i depending on the raw material,
- b_i wastes per product E_i depending on the technological procedure,
- c_j wastes which arise from the use of a product j per quantity of the product V_j,
- d_k wastes which arise from the use of a product k per unit of time depending on the used quantity of the product,
- e_l quantity of wastes which arise after the use of the product l per product G_l,
- $t_{N,l}$ average life-time of the product.

Reducing a_i and b_i by utilization of all components of the raw material, improvement of separating devices, implementation of new technological procedures (for instance, chipless shaping) and utilization of technological cycles is the main object of technological research. As the discussions of Topics I - III showed, this object is aimed at more and more.

It is important that non-waste technology will be involved in new technological procedures from the very beginning to make it part of the procedure, not to work it out subsequently. In that regard, research and development must be done in a new way.

It should be stressed that changes of a_i and b_i can be realized only with huge economical units in the future and that they demand planned development over a long period of time to become economically effective.

There are only limited possibilities to reduce the specific wastes c_j for a product. Ballast could be cut, for instance, by using less unnecessary packaging or less no-return packaging. As to specific wastes when using products (d_k), diminishing is possible only by improving the efficiency and/or the specific indexes. The amount of wastes after using the product corresponds to the mass of the product itself, disregarding low losses of material by wear. Reduction of e_l is possible only by re-using parts of the used product.

There are still possibilities to diminish the specific mass of products and the average life-time of products, i.e. an increase of the ratio $G_l/t_{N,l}$.

At present there is a trend to shorten the physical life-time of high-fashion consumer goods corresponding to their fashion life-time. In connection with non-waste technology it should be vice versa.

The use of wastes is not yet satisfactory because the following conditions are not always realized:

(1) there must be a need for the refined wastes;

---

* P. Lötzsch and M. Schubert (GDR), "Technical and technico-economic aspects of the application of non-waste technology", ENV/SEM.6/R.6/Com. 2

(2) technological procedures and plants for the refinement of wastes must be available;

(3) the law should prescribe the use of wastes as the best method to eliminate wastes in a harmless way;

(4) refinement of wastes should economically be more effective than emitting or dumping them into the biosphere, provided the prescribed limits are kept.

The use of industrial wastes has increased owing to information and "waste-exchanges" in recent years. There are still problems in connection with the use of compact wastes to be transported from distant areas when material of the same quality can be mined at nearer places.

The use of wastes is difficult when they are low-concentrated (SO_2 in exhaust gases) or mixed with other parts (household refuse). Processing wastes of one sort (waste paper, scrap-iron) allows raw material to be economized in every ECE country today, but the refinement of used textiles is usually impracticable, because textiles consist of different fibres which cannot be refined economically if at all.

In the future it will be necessary to develop and manufacture products by certain technology from the point of view of using wastes in a cycle of production-consumption.

The necessity of a complex consideration of the relationship production sphere - consumption sphere - biosphere is pointed out by Neddens.* He attributes the problems caused by the adoption of non-waste technologies largely to an inadequate recognizing and understanding of the interdisciplinary character of environmental problems. Moreover, he arrives at the conclusion that in future the state has to exert an increased influence through planning in order to overcome such difficulties. Neddens refers also to the problems arising thereby in free market economies, where, in his view, the balance between individual freedom and social demands has to be held. In this connection, the limits and problems are indicated which should be taken into consideration, if the state influences upon an economic system based on the principle of a free market economy. The principles of the environmental policy adopted in the FRG which were mentioned by Neddens, i.e.

the principle of prevention,

the principle of causation,

the principle of co-operation,

and the instruments of management (penalties, supporting actions, provisions, prohibitions) treated in connection with them cover and characterize all the problems caused under the conditions of a free market economy in contrast to the planned economy of the socialist countries, but by this interpretation both the different possibilities and the limitations in the respective economic systems become evident.

* M. Neddens (FRG), Discussion paper ENV/SEM.6/R.6/Com. 9. "Administrative way and means of implementing non-waste technology".

This is revealed as well in the papers of Reid* and Smith†. Reid quite correctly states that there is a basic conflict between the optimization of individual net benefits and the optimization of societal net benefits. The measures indicated in the paper by Reid which may be taken to implement non-waste technologies with regard to society's preference functions were already mentioned in the introductory report. Furthermore, Reid mentions nationalization and/or direct government control as possible measures aimed at improving the utilization of resources. Proceeding from a consideration of the risks involved in the analytical registration and assessment of the increase in benefit or the marginal impact of Federal support for research and development, Reid set up criteria to be used as aids for decision-making in project selection. In this connection, benefit and risk are assessed for the private sector and related to the expenditure on research and development. Reid states that a prerequisite for maximizing the efficiency of the federal programme is that research should be in conformity with both development and the private sector's investment criteria.

By estimating the solid waste associated costs to about $29 per ton of solid waste Smith makes a valuable contribution to an estimation, by means of which an economic stimulation is to be achieved, aimed at reducing the municipal waste produced by containers, paper and packaging. What is not discussed, however, is that part of the desired recycling which concerns the collection of recycled wastes. In connection with the problem of how to use the additional financial means transferred to local governments from solid waste product charges we could also speak of an economic closing of the recycling circuit if these means were directly spent on the collection of the recycled wastes.

Starting with the definition of non-waste technology Kroonenberg discusses three interrelated main problems of the non-waste technology concept relating to the use of natural resources. The author holds the view that the non-waste technology concept has to be emphasized in the training of engineers and technologists to make sure that the future designers are susceptible to this field. Proceeding from four points which consider both manufacturing and use of products, flows of material, energy and wastes, both designing and the designed product, the attempt is made to give a methodical concept for a systematic approach of non-waste technology in the design process. Furthermore, the attempt is made to formulate mathematical expressions for the material, energy and the influencing parameters. Showing the possibilities for reducing wastes and minimizing material losses by such parameters as the number of different products, life-time and wear-out, amount of material per product, number of parts per product, implemented manufacturing processes, use of new materials and production processes, installed power, energy losses, energy content and recyclability of products Kroonenberg‡ draws up a list of the fields where the designer, i.e. the engineer and the technologist, has to work today and in future to implement non-waste technology. The detailed programme for an enginners' training course given in Appendix II of this article is of certain interest to all of us because it substantially integrates the whole field of environmental policy which is effected by the engineering sciences. In addition, the author proposes the application of a computer model which simulates the influence of design decisions on the use of material, energy costs and the environmental impact of a product.

---

* R. Reid (USA), Discussion paper ENV/SEM.6/R.6/Com.10, "Non-waste technologies: ways and means of implementation".

†F.L. Smith (USA), Discussion paper ENV/SEM.6/R.6/Com.8, "An overview of solid waste product charges".

‡ H.H. van den Kroonenberg, Discussion paper ENV/SEM.6/R.6/Com. "The role of design education in non-waste technology".

From the standpoint of both the central administration and the special industries, Nunn* discusses the usefulness of the material flow analysis in the post-consumer waste treatment in Norway, e.g. for newsprint, thermoplastics and foodstuffs. By means of such an analysis the waste-paper industry got, for example, the information necessary for planning the expansion of its collection capacity for household waste-paper. Another material flow analysis included a detailed study of the production and consumption of the five most important thermoplastics (polyethylene h.d., polythylene l.d., polypropylene, polystyrene and polyvinyl chloride), which altogether amount to 93 per cent of the Norwegian thermoplastics (polyethylene h.d., polyethylene l.d., polypropylene, polystyrene and on the supposed average product lifetime were used as a basis to estimate the development to be expected in plastic waste until 1980. Furthermore, a material-flow analysis of several foodstuffs was carried out in order to investigate the amount and composition of all waste products and to assess their efficient recycling potential. These examples showed that the material-flow analysis is a useful tool to be applied in the initial stages of research aimed at a better utilization of natural resources. This method may be most profitably applied for marketing waste products, when the industry utilizing these wastes does not produce them itself. This method, however, does not include the effect on the utilization of other natural resources.

More could be done for the development and implementation of non-waste technologies if engineers and managers were better informed.

A common but untrue opinion is that the use of non-waste technology is too expensive in general, that there is a lack of good solutions for such technologies and that the costs for research and development are too high. We must overcome this prejudice by comprehensive information about the general significance of non-waste technology, possibilities for its use. Every engineer, scientist, economist, socio-economist and the general public must be encouraged to help.

The results of international conferences like the symposium of the states of the Council of Mutual Economic Assistance, held in Dresden in March 1976, as well as the present seminar, must be made public. Of great importance could be brochures with technical and economic indexes like that published by the Ministère de la qualité de la vie.

In educating scientists and engineers we must teach the students that non-waste technology is the strategic aim to be reached within a short period of time. Their basic and special knowledge will be a sufficient tool to reach that aim. To become a specialist can only be a part of our education.

There is still a lack of knowledge about non-waste technology, mostly on the part of engineers already working for a long period of time. They must learn about it by special training, increasingly sponsored by the organizations of engineers and scientists as is done in several socialist countries.

The far-reaching and difficult tasks of the development of non-waste technologies and their implementation need further international co-operation. The impetus given by the Conference on Security and Co-operation in Europe and the initiative of the ECE, to place this task in its work programme give the premise that practicable solutions as recommendations to governments may be worked out.

*David W. Nunn, Discussion paper ENV/SEM.6/R.6/Com. 5, "Applications of material flow analysis in resource management".

In our opinion, the results of the symposium of the states of the Council of Mutual Economic Assistance about theoretical and technological-economic problems concerning non-waste technology, for which my country had the honour to be host in March 1976 in Dresden, and the conclusions drawn there will also be useful for our work. We consider the elaboration of economic methods and indexes for the evaluation of low-waste technologies an important matter also in connection with the exchange of information about economic stimuli and the technological problems within the ECE.

Suggestions for Problems and Questions to be Discussed Concerning Topic IV

1. How can technological research be applied in the most effective way for the implementation of non-waste technology with new technological procedures and products; which topics are most suitable for international co-operation?
2. To what extent is the estimation of social damage or other indexes of environmental damage suitable, together with other economical and juridical stimuli, to encourage the introduction of non-waste technology?
3. How can exact planning and forecasting, on the basis of large economic units, help to implement non-waste technology more rapidly?
4. What are the demands concerning the education of engineers, scientists and social scientists (dimensions, intensity, participants)?
5. Which kind of information and co-operation with social organizations should be developed?

The Role of Design Education in Non-Waste Technology

H.H. van den Kroonenberg

Twente University of Technology, Enschede, Netherlands

1. INTRODUCTION

Non-waste technology is defined as "the practical application of knowledge, methods and means so as within the needs of Man to provide the most rational use of natural resources and energy and to protect the environment".

With this definition as a starting-point it is possible to investigate how this desired situation can be reached. It is of importance to state that the present situation has arisen without the concept of non-waste technology in mind.

Therefore the implementation of non-waste technology demands at present measures different from those that will be necessary in the future. The three interrelated problems within the non-waste technology concept are :

the limited amount of raw materials,

the menace of energy shortage,

the rapid pollution of the environment.

To master these problems three different kinds of remedy can be conceived:

postponing,

curing,

preventing.

With the two first types of remedy, one has to be sure that the present situation is not already leading to a catastrophe. An example of delaying or postponing action in face of the rapid depletion of present energy and rew material resources, to hope that alternatives will be available in time.

In fact this can only have short-term effects. An example of the curing method is to try to recycle the present waste materials as well as possible. This is a middle-term remedy.

A real long-term solution can only be achieved by ensuring that problems in relation to materials, energy and environment do not arise.

In Fig 1 a review is given of possible means to handle the material, energy and environmental problems according to the three mentioned approaches.

Problem-solving approach			Field
Postponing	Curing	Preventing	
New resources (ocean) Quick depletion existing reserves	Recycling	Designing for material conservation Designing for long life	Material
New energy sources (ocean) Quick depletion existing reserves	Total energy systems	Increasing energetic efficiencies Isolation	Energy
Remote location of industry (artificial island)	Selective collection of waste	Non-polluting processes	Environ-ment
Short-term	Middle-term	Long-term	effect

Fig. 1. Examples of problem-solving approaches

Because the "preventing" method takes all the future impacts into account this problem-solving method is equal to the design-method.

Most products, devices, machines, etc., to be used in the future still have to be designed and this places the designer in a key position, especially for the implementation of non-waste technology.

During the education of engineering designers the non-waste technology concept has to be emphasized to make sure that our future designers are fully aware of this problem field.

Non-waste technology, however, is such a broad field, including so many and so varied activities that it is impossible in a design curriculum to give attention to all these items.

The "methodical design", however, makes it possible to build the philosophy of non-waste technology into the design activities rather than going into details of all possible aspects.

Methodical design enables the designer in the various phases of the design process to pay attention to certain aspects of non-waste technology, which are specific to that part of the design process.

We therefore start with a short review of methodical design and the related non-waste technology aspects. It will show that already a great deal of applicable knowledge is available in this field.

2. STARTING-POINTS

Before further consideration is given to non-waste technology, four starting-points are proposed to serve as guidance .

(a) As a first starting-point for non-waste technology the location of design in the total life cycle of products is of interest. This is depicted schematically in Fig. 2. In the design phase of products the designer has to take into consideration both the manufacturing and the use of these products.

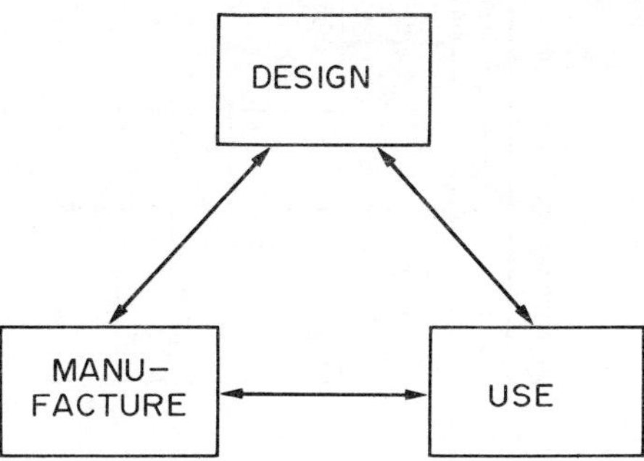

Fig. 2. Life cycle

(b) The second starting-point from which to arrive at a clear statement of the problem is the materials flow necessary for the realization of designed constructions, as is shown in Fig. 3. The accompanying energy flows and waste flows are also indicated in the diagram. Consideration of this diagram in combination with Fig. 2 shows clearly that the designer is situated in a very strategic position in this problem field.

(c) The third starting-point is the design process itself. In the design process three different phases can be distinguished as is shown in Fig. 4. In the first phase the design problem is clearly defined. Here the question can be asked why the product to be designed is desired. In case this question leads to the answer that the product is not desired. Then it is shown that the designer, with the help of other persons and institutions that have to have a say in the decision process, can have a clearly motivated say as to whether a product will be made or not.

The decision to realize the product means that this product just by its presence will have an influence on the environment. The first phase of the design process is therefore mainly related to the environmental problems.

In the second phase of the design process, the working action of the construction based on one or more physical principles is established.

Here, the manner in which the desired function of the construction to be designed is fulfilled is important. The working principle of the desired processes to be performed has a great influence on the energy consumption of the construction to be designed. Therefore the second phase of the design process is mainly related to the energy problems. This second phase is finalized by

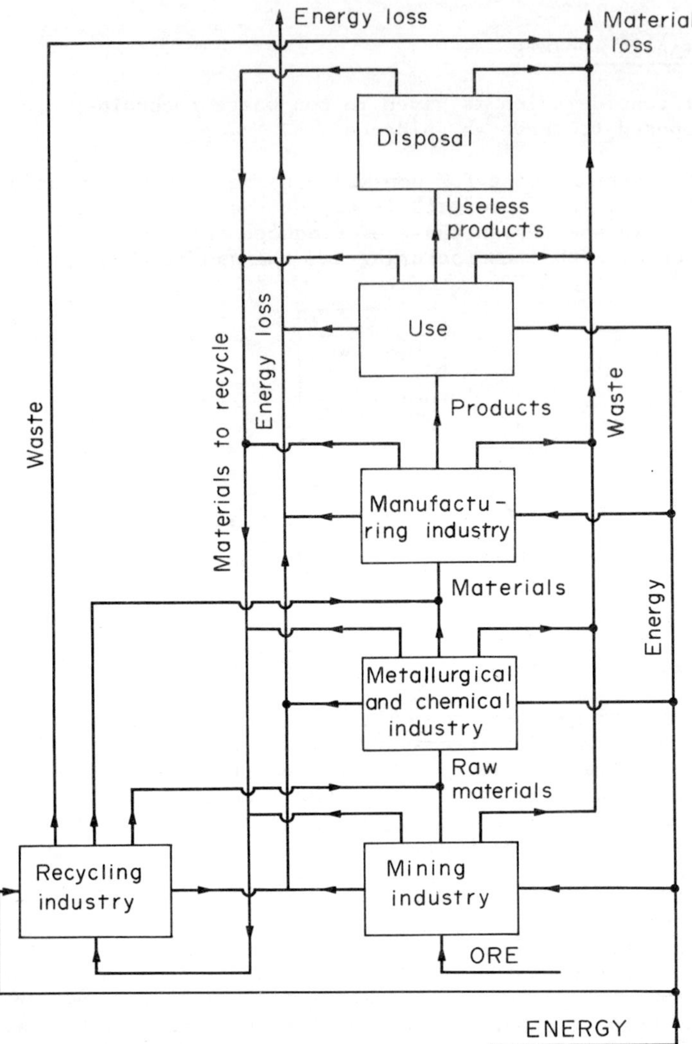

Fig. 3. Materials cycle

the so-called "structure" which is in fact a skeleton sketch of the working principle. In the third phase the final form and dimension of the construction is established. In this phase the materials to be used and the manufacturing methods are selected. It will be clear that this third phase of the design process is mainly related to the raw material problems.

(d) A fourth starting-point is the design-subject itself. In the field of mechanical engineering a division in design levels is possible according to Fig. 5. 2. The design level shows the measure of complexity of the design-subject. From Fig. 5 it follows that questions of raw materials are mainly related to parts and components. Energy problems play an important role at the machine level, whereas environmental problems are connected with the technical systems at the highest design level.

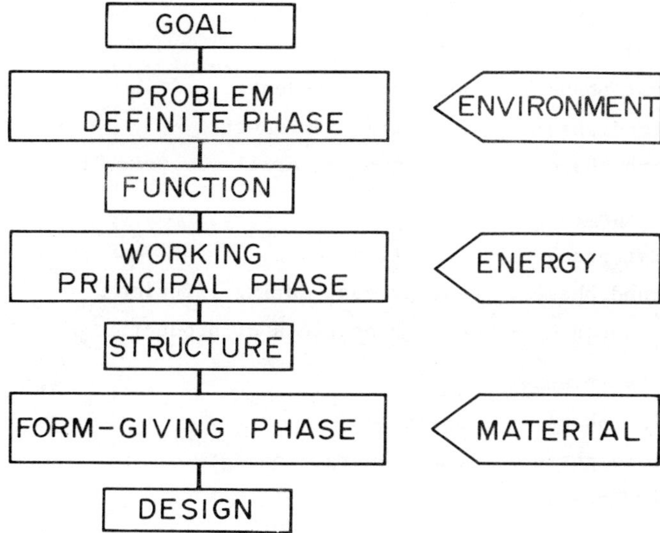

Fig. 4. Design process.

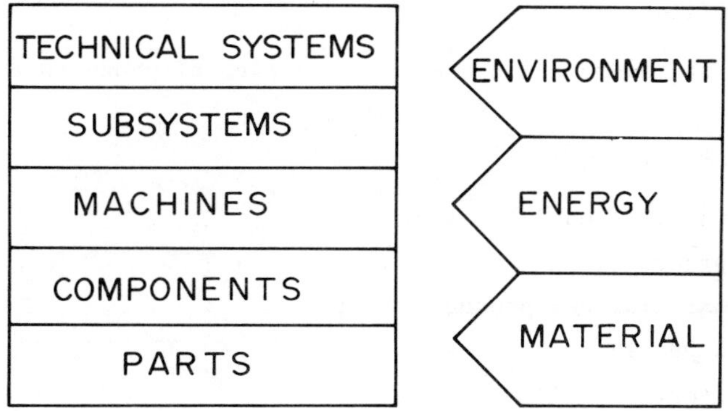

Fig. 5. Design levels.

These four starting-points will be used for a systematic approach to non-waste technology during the design process.

3. NON-WASTE TECHNOLOGY AND ENGINEERING DESIGN

In order to minimize the amount of materials used for a product and in order to keep the energy consumption and the environmental pollution as low as possible, the designer can take into account a number of important factors, which can be arranged according to the three phases of the design process as follows :

A. *Materials*

first phase	number of products
second phase	lifetime of product
third phase	amount of material per product

B. *Energy*

first phase	installed power
second phase	energy cost during lifetime
third phase	energy contents of product

C. *Environment*

first phase	pollution
second phase	total energy concept
third phase	recyclability

A. *Materials*

The total amount of raw-material used for a series of products can be expressed as follows :

$$W = P.\ M.\ (1-R),$$

W is weight of material used per year in kg,

P the number of products,

M the weight of material per product in kg,

L the lifetime in years,

R the recycle factor.

Via the recycle factor the material problems are related to the environmental problems.

The influence of the number of products, the lifetime and the materials amount per product are treated respectively in the three successive design phases.

The number of products. To keep the material consumption low, the first thing that can be done is to keep the number of different classes of products and the number of products per class as low as possible. This is something that the designer cannot control by himself, but it is a mission for the public to refuse useless and undesired products. In the first phase of the design process or in a feasibility study previous to the design process the assessment of the need plays an important role.

The lifetime. In the second phase of the design process the choice of the working principle has an important influence on the lifetime of the construction. Lifetime can easily be demonstrated by considering the reliability curve as shown in Fig. 6.[3] The reliability of a construction is the estimate that this construction will perform its function during a certain time and under well-defined circumstances. The mathematical expression is :

$$R = e^{-\lambda t}$$

in which R- stand for reliability, λ is the failure rate and t is the time.

With the help of the failure rate the mean time before failure (MTBF) can be expressed as :

$$\text{MTBF} = \frac{1}{\lambda} .$$

Because the first failure of a construction can lead to the necessity to dismantle the construction it is important to increase the MTBF as much as possible by means of a low failure rate λ.

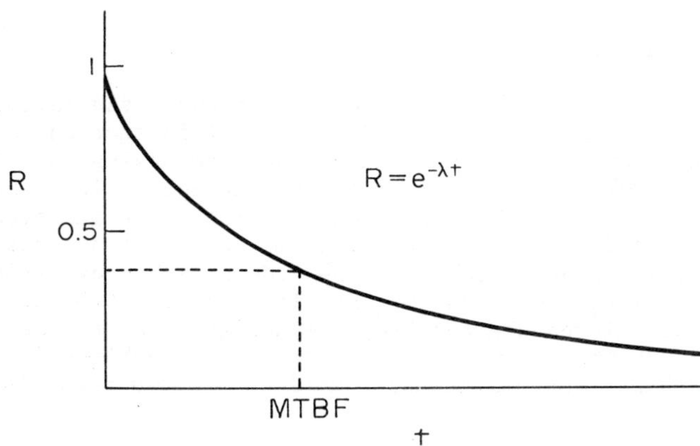

Fig. 6. Reliability.

The failure rate versus the time can be depicted as a so-called bath-tub curve according to Fig. 7.

In this curve three sections can be distinguished :

Infant mortality. This is the period during which the faults due to poor design and production become apparent. The material loss related to infant mortality can be reduced by increasing the quality of production.

Normal life. In this section of the bathtub curve a constant failure rate is assumed, because during normal ultilization the products fail due to chance alone. By increasing the quality of the product, the failure rate can be lowered, but it is also possible to lower the failure rate without changing the product just by good directions for use.

The designer has to anticipate the consumer's behaviour by making his design less sensitive to misuse and by making sure that the user receives the right information to avoid failure as much as possible.

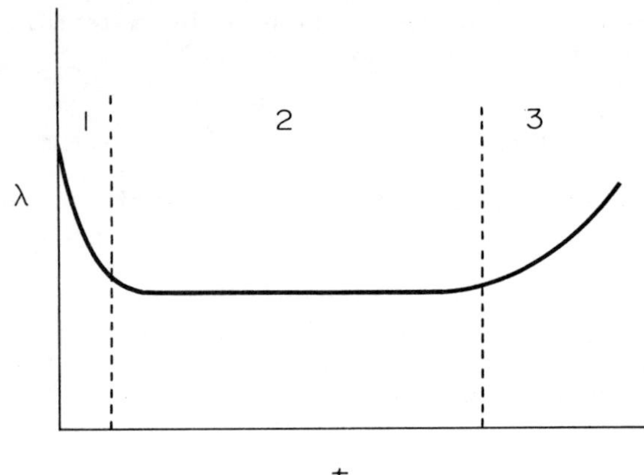

Fig. 7. Bath-tub curve.

Wearout period. In this section of the bathtub curve the failure rate increases due to wear, corrosion and fatigue. This section can be shifted to the right side by good maintenance thus increasing the lifetime of products. The designer can during the design activities make allowance for maintenance aspects.[4] By preventive maintenance and eventually replacement or repair of parts, the lifetime of the construction can be prolonged. This maintenance has to be made possible by paying attention to easy assembling and disassembling. If a construction has to be disassembled for maintenance purpose, the designer has to take care that this can be achieved very easily. In case disassembling is difficult there is a chance that maintenance will be avoided thus risking a short lifetime.

The designer has to take care that the accessibility to maintenance points is good. This can be achieved by mounting parts that have to be replaced from time to time preferably outside the casings instead of inside the casing.

To keep the consequences of wear for the material consumption of a construction below certain limits the designer can take care that wearing parts contain little material.

By making these parts easily accessible and easy to remove, the designer has fixed the places where wear occurs. By replacement of these parts only little material is lost while due to the efficient replacement an important time saving can be achieved. If due to long replacement times or repair times the construction has to be taken out of service too long, there is a danger that for economic reasons the whole construction has to be replaced by a better one, which is a bad thing from a lifetime point of view.

Amount of material per product. This is the total amount of material to make a product including the material removed during manufacturing processes. This question is typical of the third design phase. Dyson[5] has given a number of lines of conduct for the designer to save material. Among others some typical lines of conduct for non-waste technology are the following :

(a) minimizing the number of parts per product;

(b) maximum use of material by minimal material loss during manufacturing ;

(c) the use of new materials and production processes.

(a) Minimizing the number of parts per product. Striving to reduce the number of parts in a construction will result in material saving because each connection between two parts requires extra material. Besides material saving this has a favourable influence on the lifetime of the construction, because this can result in higher reliability and better maintainability and repairability.

In case the reduction of the number of parts concerns moving parts this will also result in less wear which also has a favourable effect on the lifetime.

(b) Minimizing material loss during manufacturing. From the point of view of material conservation, metal forming is better than metal cutting. Dyson reports the material losses dependent on the manufacturing process. This is shown in Table 1.

TABLE 1

Process	Net use %
Extrusion	88
Thermal forming	82
Cutting (plates, profiles)	67
Drilling	52

(c) Use of new materials and production processes. By applying new materials by which new processes are made possible, substantial material saving can be achieved. The use of material by precision casting, sintering, forcing and rolling is maximal. By using suitable materials for these processes much waste can be avoided. The designer can influence this development in a favourable sense.

The above lines of conduct (a), (b) and (c) hold good for small, relatively light loaded constructions. For heavy loaded constructions, additional design rules have to be added :

1. Full use of the strength of materials by striving for tensile and pressure stresses.
2. Providing the material in the working lines of the acting forces.
3. Avoidance of bending.
4. Keeping the product of force and working length minimal.

A low weight of the product is, besides the direct material saving, also of importance because mass forces are smaller, handling and transport costs are lower and the load capabilities are better.

To arrive at a low weight of the product several combinations of material characteristics are important. In Table 2 these combinations are shown. The values of these combinations have to be low in case a light construction is required.

TABLE 2

Load	Strength	Stiffness
Tensile/pressure	$\frac{\gamma}{\bar{\sigma}_t}$	$\frac{\gamma}{E}$
Bending	$\frac{\gamma}{\sqrt[3]{\bar{\sigma}_b^2}}$	$\frac{\gamma}{\sqrt{E}}$
Torsion	$\frac{\gamma}{\sqrt[3]{Z^2}}$	$\frac{\gamma}{\sqrt{G}}$

Nomenclature : γ specific gravity,
$\bar{\sigma}_t$ max. tensile strength,
$\bar{\sigma}_b$ max. bending strength,
Z max. shear strength,
E modulus of elasticity,
G modulus of rigidity.

B. Energy

The total energy consumption of a product is the sum of the energy in the materials, the energy used during manufacturing and the energy consumed by the product during its lifetime.

Installed power. In the first phase of the design process the designer can ask himself whether it is necessary to install much horsepower in a certain product or construction. It is often possible to meet the functional requirements with much less power than installed because the installed power is often used as a selling argument or as a status symbol (for example vacuum cleaners and automobiles). By reducing the installed horsepower purposely the total energy consumption during the lifetime of the product can be reduced.

Energy cost during lifetime. In the second phase of the design process in which the working principle is established, it is possible to select working principles with high energetic efficiencies. For machines which can be composed of an energy source, a transmission and a load this can be done with regard to energy converters, transmissions and loads. Energy converters can be divided into three groups. Natural energy converters such as windmills and waterwheels. Thermal energy converters such as steam intallations or diesel engines and direct conversion such as thermo-elements and photocells.

In Fig. 8 the average energetic efficiency of these three types of conversion is given. The transmissions can be divided into mechanical, hydraulic, pneumatic and electric transmission. In general the transmissions have a much higher efficiency than energy convertors.

The efficiency of loads cannot be given because the variety of possible loads is too great.

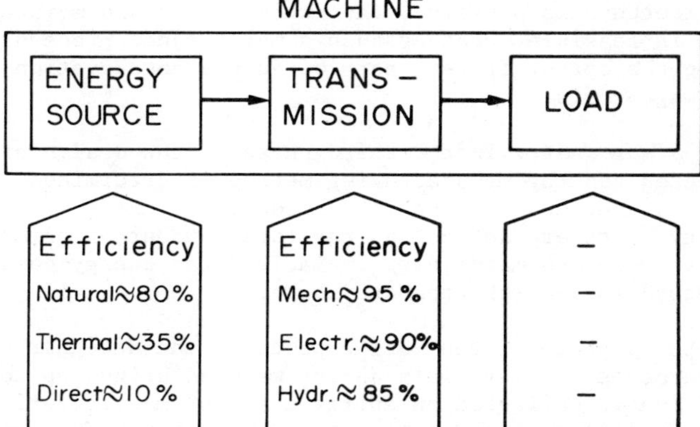

Fig. 8. Machines.

An example is the horizontal transport of a mass which can occur under dry Coulomb-friction, viscous friction by lubrication and rolling friction by providing rollers or wheels.

The friction coefficient which is a measure for the transport losses are respectively about 0,2, 0,02 and 0,002. This illustrates that in general by choosing a certain working principle also the maximal possible efficiency is fixed.

A high efficiency is often desirable to attain a long lifetime, because in case the efficiency is reduced due to friction the wear involved will cause a short lifetime.[6]

During use, distinction can be made between energy costs for fuel or electric consumption of the construction to be designed and the energy costs of maintenance and repair.

The fuel and electric energy consumption is dependent on the efficiency of the construction to be designed. During design the designer can use his influence on efficiency especially by choosing favourable working principles as is shown above.

Also the energy costs of repair, replacement and maintenance during the lifetime can be taken into account by the designer. Maintenance procedures such as painting, cleaning, lubrication and energy costs of replacements can be quantified already.

Intensive maintenance and repair can result in a longer lifetime. Prolongation of lifetime means a modification of the energy costs per year according to :

$$\frac{E}{L_1} - \frac{E + \Delta E}{L_2} \quad \text{KWh/year,}$$

E = total energy costs of the product,

ΔE = energy costs of extra repair and maintenance,

L_1 = normal lifetime,

L_2 = lifetime with extra repair and maintenance.

In case this difference is positive, the extra repair and maintenance based on energy costs is feasible. Whether such a maintenance procedure will take place depends on the costs. It is a form of energy conservation that often involves extra wage costs.

Energy contents of products. In the third phase of the design process the materials are selected and the manufacturing method is determined. By selecting materials, the designer has to take into account the energy costs of these materials. Energy costs are defined as the total amount of primary energy (thermal) used to make the materials. In Table 3 the energy costs for several materials expressed in kWh (th)/ton are shown.[7]

Further in the third phase of the design process, the designer has to add the energy used to process the materials during manufacturing. Up to now less systematic information is collected on energy costs of manufacturing processes. Processes consuming the most energy are those whereby the materials have to be heated, such as casting, hotmilling, forging, heat treating and (for example) the drying of paint.

With regard to the environmental aspects it can be stated that the impact on the environment greatly depends on the policy followed in relation with energy and material problems.

Pollution. In the first phase of the design process, the designer has to make a list of requirements concerning the total pollution which can occur after realizing of the construction of the product.

This depends on the location in which the designed construction or plant is being used. In cases where the plant is used on an offshore artificial island, codes can be different codes from those in a residential area.

The pollution referred to covers all aspects such as direct pollution, waste, noise, heat disposal and scenic pollution.

Total energy concept. In the second phase of the design process the working principle is established. It depends on this working principle whether environmental problems can be minimized by the application of the total energy concept. In case this concept cannot be applied the possible produced waste heat has to be released into the environment.

Recyclability. In the third phase of the design process, the choice of the materials to be applied has a great influence on the possible recycling of these materials.

By measuring recyclability not only can the environmental impact be reduced, but also material and energy conservation achieved. In Table 3 the energy required for preparing and recycling various materials is shown.

By legal measurements the increase of the recycling-factor can be stimulated. As long as legal directions do not exist the designer can influence the material conservation from his own responsibility by taking care that the materials used in his products can be recycled.

This can be done by taking into account the following aspects:

(a) The designer can take care that not too many different kind of materials

TABLE 3

Energy required for the preparation of 1 ton metal in kWh,

	From ore	From scrap	Energy profit
Steel	13,200	10,600	2,600
Copper	20,000	8,000	12,000
Aluminium	90,000	40,000	50,000

are used in one product becuase this makes the recovery difficult at the end of its life. Moreover, he can select materials which can easily be recycled.

(b) The products in which different material combinations are applied have to be designed in such a way that for recycling purposes different materials can easily be separated by dismantling.

(c) In case material combinations cannot be separated these combinations preferably must be used as raw materials for special alloys.

The decisions on which the material selection is based are mainly taken in the third phase, the form-giving phase of the design process. It is therefore of great importance that the designer can dispose of information on the recycling properties of different materials and material combinations.

This holds good especially for the scarce materials, such as copper, lead, tin, zinc and nickel. In some cases these materials can be replaced by other newly developed materials. In case replacement is not possible, miniaturization can be very effective. Therefore it is also of importance that data on scarce materials and possible replacement materials are made available to the designer.

CONCLUSION

In the methodical design process many relevant aspects of non-waste-technology can systematically be arranged.

For the designer of mechanical equipment and products some knowledge necessary for the implementation of non-waste-technology proves already to be available.

Methodical design, however, is more than only a systematic approach to design problems; it can also be used to stimulate creativity.

For the implementation of non-waste technology creative thinking is indispensable, especially to solve existing problems according to the non-waste technology concept. In Appendix I two new concepts for complex technical systems are described, in which non-waste technology aspects play an important role.

Although the existing knowledge in the field of non-waste technology is still limited it is possible to use it for the education of engineers in an engineering design course in the university. In Appendix II a proposal for a graduate design course, related to non-waste technology, is given. It is of great importance for the implementation of non-waste technology in society that an adequate number of well-motivated engineers accept, execute and propagate the non-waste technology concept.

APPENDIX I

Examples of Designs with Non-Waste Technology Aspects

A. Capsule transport. In general pipelining is a good example of non-waste technology in the field of bulk transportation. Capsule pipelining, however, can be considered as the best example of non-waste technology in comparison with the transportation of liquid and slurry by pipeline. This concept is not yet operational, but many feasibility studies show that there is a fair chance that the technological problems of capsule transport can be solved. Capsule transport consumes less energy than slurry and even less than liquid pumping.

Slurry transport can cause severe pollution because the slurry preparation requires the breaking and grinding of the material. At the end of the slurry pipeline dewatering of the slurry can cause environmental pollution. If the material to be transported is encapsulated these problems will not occur. The capsules can be returned by a parallel pipeline and can carry any kind of return load. In single pipeline systems it is also possible to dispose of the capsules at the end of the line, because the capsule material itself is used for other purposes. An example is the transport of grain in aluminium capsules. After arrival at the end of the pipeline capsules can be transformed into aluminium sheets or bars and used for construction work.

In the department of Mechanical Engineering of the Twente University of Technology students work during the design course on a capsule transport project to apply their design knowledge to the non-waste technology concept.

B. Manganese nodule mining. Manganese nodule mining in extremely deep water (3000-5000 m) is not yet operational. Advanced design of mining equipment, however, is under way. The capacity of these floating mining plants, working thousands of miles out in the ocean, has to be very high to make possible the economic extraction and transport of nodules. The required power for these operations is rather high thus necessitating a complex supply system for fuel with a fair risk of oil-spillage and sea-pollution.

As mining in tropical areas is likely, it is in this case possible to extract energy from the temperature difference between surface water and deep ocean water. This makes the remote offshore installation self-supporting. As in nodule mining the mining plant has to be moved continuously, there is no danger that the surface water around the plant will be cooled too much.

If the beneficiation of the nodules takes place on the offshore mining plant itself, no environmental damage to residential areas will result.

As all these aspects are within the non-waste technology concept, nodule-mining is a good example of applied non-waste technology.

In the Department of Mechanical Engineering of the Twente University of Technology students can work during the design course on an ocean mining project to apply their design knowledge and creative thinking to the non-waste technology concept.

APPENDIX II

Non-Waste Technology Design Courses for Graduate Students

In order to make students more conscious of the potential of non-waste technology and of the important role that can be played by the designer the graduate students have to study related literature on different subjects in this field and to report on certain non-waste technology aspects that are essential for engineering design.

Typical examples of the treated subjects are given below:

(a) Designing for material selection in view of shortages:
  - shortages of resources,
  - economical aspects,
  - recycling.

(b) Designing for recovery and re-use:
  - recovery processes,
  - quality constraints,
  - waste-handling problems.

(c) Designing for low energy costs;
  - energy costs of fuels and materials,
  - energy costs of manufacturing,
  - energy costs during use and dimantling.

(d) Designing for maintenance:
  - accessibility,
  - repairability,
  - economical aspects.

(e) Designing for durability:
  - fatique, wear, corrosion,
  - reliability,
  - prevention and protection.

(f) Designing for non-waste manufacturing:
  - improving process efficiency,
  - energy-saving processes,
  - process innovation.

It is very important not only that the students be active in the field of non-waste technology, but also that the staff is involved. Especially, the long-term-activities can be done by staff-members, while the short-term activities, which have to fit into the long-term plans, can be executed by students.

Therefore at this moment a computer model is being developed in order to determine the influence of design decisions on the use of material, the energy cost and the environmental impact of a product. This model includes the complete life cycle of the product or construction from manufacturing up to the recovery after disposal or dismantling.

The information needed as input is :

the quantities of material required for manufacturing the parts of the product;

the manufacturing of the different parts;

data on heat treatments;

the failure rates of the components;

the average number of hours per year that the product is used;

data concerning recycling of materials;

the economic lifetime of the product.

It will be evident that much research has to be done before all this information is readily available.

The output of the programme is :

(a) For each part,

energy costs of materials,

energy costs of manufacturing processes,

energy costs of heat treatments,

material losses due to manufacturing processes.

(b) For each material used :

total amount of material in the product;

total amount of material required to make the product;

material losses during manufacturing processes;

material losses due to replacement of defective parts during lifetime;

material losses during recovery after disposal.

(c) For the total product:

the technical lifetime;

total energy costs of the materials used;

total energy costs of manufacturing the product;

energy costs for replacing part of the product;

energy costs of using the product;

total energy costs of recovering the material of the product;

total energy costs of the product.

With this programme it is possible to examine different alternatives already during the design phase on the non-waste technology characteristics. As soon as this computer programme is operational it is possible to find guidelines for the designer how to design products from a non-waste technology point of view.

REFERENCES

1. H.H.v.d. Kroonenberg, Methodisch Ontwerpen. *De Ingenieur*, *47*, 915-923 (1974).
2. H.H.v.d. Kroonenberg. Een bijdrage voor een Algemene Ontwerpmethode (1). *De Constructeur*, pp 51-59 (Sept: 1975).
3. Kivenson, *Durability and Reliability in Engineering Design*, Pitman Publishing, London, 1972.
4. D.J. Smith. *Maintainability Engineering*. Pitman Publishing, London, 1973.
5. B.H. Dyson, Efficient utilisation of materials. *Proceedings of the Conference on the Conservation of Materials*, Harwell, 1974.
6. Ch. Lipson, *Wear Consideration in Design*, Prentice Hall, Englewood Cliffs, N.J., 1967.
7. P.F. Chapman, *The Energy Costs of Materials, Energy Policy*, vol.3, no. 1, p. 47, 1975.

A Survey of the Location, Disposal and Prospective Uses of the Major Industrial By-Products and Waste Materials

W. Gutt

Building Research Establishment, Department of the Environment, United Kingdom

The construction industry in the UK uses large quantities of low value materials, particularly aggregates, over 200 million tonnes of which are now consumed per year. The quarrying of this material and its transport inevitably have some adverse effects on the environment. Some 80-90 per cent of all aggregates are moved by road and aggregates make up the biggest single category of road freight. These factors have led to some unpopularity of quarrying operations and tighter control of the granting of planning permission for new workings. At the same time urban development has sterilized some resources, particularly of sand and gravel. It must be emphasized at this stage that the quarrying industries have made great efforts to minimize damage to the environment.

While natural materials are being quarried, other industries are producing over 100 million tonnes of solid wastes per year some of which, if not most, could potentially be used in construction. In order to assess the possibilities for utilization BRE has made a survey of the quantities, location and disposal of the major industrial by-products and waste materials in Great Britain. This was carried out on behalf of the Department of the Environment Working Group on Aggregates and Waste Materials, which in turn presented evidence to the Advisory Committee on Aggregates. The results of the survey and the conclusions of the working group have been published as BRE Current Papers[1,2] while the Advisory Committee have issued a preliminary report.[3] The quantitative results of the survey are summarized in Table 1. It can be seen that in the UK the most important by-product in terms of both current production and stockpile is colliery spoil, followed by china clay wastes. There are also large stockpiles of slate waste, and spent oil shale originating from past industry although little is now produced. From the findings of the survey and the deliberations of the Working Group three research topics emerged with the objective of making the best use of available resources of both natural and by-product materials.

The first of these topics is to assess the additional potential of waste materials for use as aggregates. Because colliery spoil occurs in by far the largest quantities and could therefore make the most significant contribution to aggregate supplies, work has concentrated on this. Colliery spoil is already being used on an increasing scale to produce sintered clinker-like aggregates for blockmaking. BRE is therefore investigating the possibility of producing an aggregate for general purpose use in concrete including structural work and has taken out a patent on a process for the manufacture of such an aggregate.[4] The carbon present in the spoil not only contributes to the fuel needed to fire the product, but may also through reaction with other components of the shale, produce a bloating action. It is therefore necessary in seeking to produce a dense or non-bloated aggregate to devise a process which will allow the carbonaceous matter in the shale to be burnt out at

TABLE 1

Major waste materials in Great Britain - m tonnes, 1974-75

Material	Current annual production	Stockpile	Quantity used
Colliery spoil	56	3000	7–8
China clay waste	22	280	1
Slate waste	1	300	0.03
Pulverized fuel ash including furnace bottom ash	10	-	6
Blast-furnace slag	6	-	6
Steelmaking slag	4	-	4
Phosphogypsum	2	-	-
Furnace clinker	1	-	1
Incinerator ash	1	-	-
Oil shale	-	200	1.5

a temperature below that at which bloating will take place, before finally firing the material at a higher temperature. This process needs to be such as to enable the heat generated in the burning of the carbon to be beneficially used. There are a number of ways in which this might be done and BRE has not yet settled on any particular method. BRE work also includes a fundamental characterization of spoils in an attempt to relate the mineralogical and other characteristics with their firing behaviour.[5]

The second topic, concerned specifically with the use of a waste material, deals with the use of fly ash, the waste from coal-burning power stations as a pozzolanic additive to cement. The current objective is to provide data for an improved Standard Specification for fly ash covering its use as a pozzolanic material and in the long term for any new Specifications for blended Portland/fly ash cements. Samples of ash have been taken from a number of power stations. A range of concretes are being made with these ashes in order to correlate the properties of the concrete with the physical and chemical properties of particular ashes.

Much attention is being given in this investigation to the durability of concrete incorporating fly ash and in particular to the sulphate resistance of the concrete.

A further response to increasing pressure on aggregate supply is to take greater care over matching aggregate quality to the end-use to which it is put. For example, it is wasteful of resources to use a high-grade aggregate such as gravel in situations such as ground floor slabs and foundations where great strength is not needed and which consume about one-half of the aggregates used in new building construction. Therefore an examination is being made of the possible use of sources of rock which have up to now been thought unsuitable because they were, for example, too soft. In the south-east an assessment is being made of the Jurassic limestone while in the north-west sandstones and in the north-east Magnesian limestones are being examined.

The durability of concrete containing these weaker aggregates is, of course, vitally important and this will be assessed as part of the programme of research.

Studies such as those described above will establish the technical feasibility of using a waste material substitute. Even in cases where the technical feasibility of using a waste material is well established, however, the extent of utilization may be inhibited by economic considerations. For example, the technology of expanding slate in a rotary kiln to make a lightweight aggregate is well known and there has been commercial production in the recent past. There are ample reserves of raw material and the product had a good reputation, yet there is no present production of expanded slate in the UK. The costs of production, particularly of fuel, and of transport to areas where there is a sufficient market make the process uneconomic. These two factors are often decisive in preventing or restricting the use of many waste material substitutes which are competing with relatively low cost natural materials.

In assessing the economic feasibility of the colliery spoil aggregate the production costs appear to be the most important factor. In order to keep the running costs of an aggregate plant as low as possible, it is important to make maximum use of the fuel inherent in the spoil but equally the capital cost of the plant must be kept reasonably low.

On the other hand, the use of fly ash as a partial substitute for Portland cement appears to have the potential to provide direct economic benefits. The main method of use of this material in the UK at the present time is at the concrete batching plant and because it is cheaper than Portland cement its use can lead to reductions in the cost of concrete, although the need in some circumstances to provide additional storage silos may offset some of these savings. Blended Portland/fly ash cements are produced only on a very small scale in the UK at present. Their production requires significantly less energy than that of Portland cements[6] but their introduction on a wider scale is unlikely at a time when as at present the cement industry is working below its full capacity.

REFERENCES

1. W Gutt, P.J. Nixon, W.H. Harrison, M.A. Smith and A.D. Russell, A survey of the locations, disposal and prospective uses of the major industrial byproducts and waste materials. BRE Current Paper CP 19/74.

2. Report of Aggregates and Waste Materials Working Group, BRE Current Paper CP31/73.

3. *Aggregates - The Issues*. Preliminary report of the Advisory Committee on Aggregates, HMSO, 1973.

4. UK Patent Application 1979/74.

5. R.J. Collins, A method of measuring the mineralogical variation of spoils from British collieries. To be published in *Clay Minerals*.

6. M.A. Smith, The economic and environmental benefits of increased use of pfa and granulated slag. CP41/75.

Statutory and Financial Provisions for the Establishment of Manufacturing Methods Free of Waste Products

R. Huissoud

Conseil National du Patronat Francais, Paris, France

Policies which have been formulated by the Government in France in the area of antipollution measures for industrial nuisances aim at a definition of objectives and, at the same time, the availability of resources.

These policies consist of the following :

restrictions which must be respected with respect to the environment are clearly posted at every industrial facility;

these restrictions are set up within the context of a production sector, if necessary in a contractual manner;

while, at the same time, making available to industry the necessary financial incentives.

I would like to speak to you first of all about statutory proceedings and their general context.

I will not be presenting here a list or the substance of various pieces of legislation. These hold those perpetrators at the source of pollution-among whom are industrial plants-liable for the consequences of any such pollution. They provide public authorities with the means for regulatory action as concerns factory locations and their methods of operation.

From among these texts, nevertheless, I would like to give particular attention to the Law of 19 December, 1917, as amended, which deals with installations classified as dangerous, harmful to health, and noxious. This Law, whose scope of application should be broadened, for which a new draft bill is presently under study in Parlement, specifically applies to industrial facilities.

(a) Those establishments which are concerned are those whose activities are included in a list set up by Decree. Two categories are to be distinguished according to the type of products, and the "dangerous, harmful to health, or noxious nature" of such products.

Establishments in the second category are only obliged to carry out the formalities of making a prior declaration before going into operation, or prior to any expansion of plant capacity.

Establishments in the first category can only commence operations when they have obtained an authorization for the project by Ministerial Order. This also applies to any expansion of production.

(b) Such Ministerial Orders granting authorization set up, amongst other things, limitations on the discharge of wastes. It is through such Ministerial Orders that direct protection of the environment against pollution is ensured.

By way of example, French regulatory provisions set for each category of paper the maximum DBO_5 flux (biochemical oxygen demand after 5 days), and for matter in suspension, which may not be exceeded.

It is the comments which describe and recommend installations for recycling water and the recovery of pollutants, and which enable under the best of conditions to conclude successfully in these results.

In the area of solid wastes, I would also like to make mention of Section 16 of the Law of 15 July, 1975 relative to the elimination of wastes and to the recovery of matter contained therein, which stipulates that :

> "Decrees eminating from the Conseil d'Etat (Privy Council) may regulate the methods by which certain materials, elements, or forms of energy are to be used in order to facilitate their recovery, or those of materials or elements which are combined with them in certain manufacturing processes".
>
> "Regulatory measures may specifically involve the prohibition of certain processes, mixtures, or combinations with other materials, or they may entail the obligation to be in conformance with certain methods of manufacture."

As can be seen from this, the French Government has progressively endowed itself with regulatory measures which enable it to promote waste-free production techniques. Still, it is necessary to make available to industry the necessary financial incentives.

The general rule is to make the perpetrator of the pollution pay for the costs involved with it, and especially when it is a matter of reducing the pollution. It is within the context of this "the polluter pays up" principle that action programmes have been organized which, on the economic side, accompany the regulatory measures and which are designed to bolster up its effectiveness.

This is an equitable manner of proceeding. It is furthermore an assurance of effective action. In actual fact, it is on the basis of being liable for the economic consequences of his acts that the polluter will then select, as he can better than any other, the means with the best performance and the least cost to meet with the objectives assigned him by regulation, benefiting in the process the community at large with the savings which are entailed.

It would appear to me to be of interest to speak to you first of all about the very first, original application of this principle carried out by the Agences de Bassin (the Basin Agencies) in the case of water, and further to this, about the exceptions to this principle motivated by the establishment of specialized technology proper to the process.

Created by the enabling legislation in the Law of 16 December, 1964, the six large French hydrographic basins constitute the normal and effective framework for a policy of optimal water resources management. As public agencies responsible to facilitate this management with their participation, the Agences Financières de Bassin (Basin Financial Agencies) are in particular empowered to collect fees (which do not have the character of a tax, a fine, nor of a payment in full discharge of adherence to regulatory restrictions) from industries which pollute the waterways, and to grant financial aid to those who make investment specifically to reduce their pollution of water.

The Basin Agencies, very much convinced about the merits to be found in specialized technology proper to a process, whose technical and financial advantages will be dealt with extensively during this congress, have made arrangements for provisions designed to promote them. Any investment in a method of waste-free production may, the same as any other investment in pollution control, receive aid from the Basin Agencies. Furthermore, some of the Agencies, taking into account the reliability aspect of these techniques as compared to customary treatment solutions, have given these operations the benefit of participation rates higher than those accorded to other antipollution investments, such as treatment plants.

Finally, there is a tendency on the part of the Basin Agencies not to grant financial aid to treatment plants built by industry unless these firms have earlier made the necessary effort entailed in installing existing waste-free production methods.

As concerns government aid for the promotion of manufacturing methods free of wastes, it is of several types.

These technologies benefit from special efforts undertaken in the area of research and development. Funding from the "Enveloppe de la Recherche Public" (Budget Allocations to Public Research) is granted to scientific committees concerned respectively with water, air and wastes which operate within the Ministry de la Qualité de la Vie. These then distribute allocations to research teams working on projects which meet with the objectives and subject matter announced by these committees. The matter of waste-free technology is considered to be a priority by all of these committees.

In this manner, research on enzymatic depilation was financed, as was, for example, research on blanching of foodstuffs in fluidized beds, on dry peeling of potatos, on new agents used in mixing wood shavings in the manufacture of paper pulp, on methods for scalding poultry with steam, and still many more.

Upon the successful conclusion of research efforts, predevelopment aid is then usually granted to the first industrial firm which agrees to put the newly discovered technological breakthrough into operation.

Finally, when any new procedure is exceptionally well suited to the protection of the natural environment, it can benefit from a subsidy from the "Fond Interministériel d'Action pour la Nature et de l'Environnement" (Interministerial Action Committee for Nature and the Environment).

Technology in the field of manufacturing methods free of wastes has been able to develop in France during the course of the last 5 years due to the benefits provided by these combined regulatory and financial provisions. These efforts which have now been committed will be continued under the VII Plan, this covering the period 1976-1980.

The establishment of an "Agence Nationale des Déchets" (National Agency for Wastes) as early as 1977, and then subsequently of an "Agence Nationale de l'Air" (National Agency for the Atmosphere) should in these two areas further intensify the promotion of waste-free production methods.

The limitations on the discharge of wastes set a maximum beyond which emissions of any kind may not go. Such restrictions depend both on the activity at the factory and the level of quality which is to be maintained in the natural environment. They correspond to that which will provide the best anti-pollution techniques presently available, or in other words, manufacturing methods free of wastes when such methods exist. The selection of equipment is still left to the industrial concern so as to leave the field open to innovation.

Within this general context, a specific strategy has been developed : "the sector policy".

Pollution problems are in actual fact specific to the various sectors of activity. For example, a particularly dangerous type of pollutant will only be found in the effluents of certain kinds of factories, and will then require only those control measures appropriate to that kind of facility. In the same manner, methods for the reduction of pollution will give results which will vary according to the different types of factories in which they are applied. The extreme case here is specialized technology proper to the process, which in general is to be applied to a well-defined industrial activity.

The Minister of "de la Qualité de la Vie" forwards instructions to the Prefects (the chief administrators in each of the provinces in France) which must determine, for each of the most polluting industrial sectors, the technical framework in which a case by case examination of establishments in that sector will be conducted. These instructions will furnish precise indications on the pollution levels which the best pollution-control techniques presently available make it possible to achieve. These levels, which serve as guidelines for the entire industrial sector, are called "Emission Standards".

Such instructions have already been issued for, among others, metallurgical surface treatments, sugar-beet factories, pulp paper factories, cement plants, some shop areas in the iron and steel industry, distilleries, potato starch plants, and yeast factories.

In order to be effective, this "sector policy" must be realistic, in that it must consider the limitations which are peculiar to the parties involved. This policy, therefore, is set up in co-operation with representatives of the branch of industry concerned.

When it appears that an objective can only be achieved by putting into operation a production method free of wastes, this is then imposed in the instructions sent to the Prefects.

It is in this manner that instructions concerning sugar factories require them to dry clarified foams, and those for potato starch factories to recover at these plants the proteins in potato plant water. Instructions formulated with respect to distilleries require the recovery of yeasts and the wash from second wines. Other examples could also be mentioned.

Nevertheless, in order to leave the field open to innovation, most often it is only the goal which is regulated, and the comments which accompany the instructions decribe and recommend methods of waste-free production which are known at the time of drafting and which make it possible to adhere to the objectives.

Applications of Material Flow Analysis in Resource Management

David W. Nunn

Christian Michelsen Institute, Department of Applied Physics, Bergen, Norway

BACKGROUND

An overview of the way in which we utilize our resources and of the amount and nature of the waste-products generated is a precondition for the formulation of successful resource management or pollution-control policies.

The *material flow analysis method* employs the principle of the conservation of mass to map the utilization of a particular resource from its initial production stage, through the processing industries to the consumer. A picture of the adherent waste-product streams is thereby generated.

In the autumn of 1972, a committee was appointed by the Royal Norwegian Council for Scientific and Industrial Research to coordinate research in the field of solid waste management. The committee has concentrated its efforts on post-consumer waste treatment and the solution of problems which the municipal sanitary authorities have. Besides sponsoring research into waste-treatment technology, a great deal of emphasis has been placed on the need to analyse the flow.

This paper discusses the usefulness of material flow analysis

- from the point of view of a central agency which is also responsible for coordinating research in the field and
- from the point of view of the industries which have been the subject of the investigations.

Analyses of newsprint, thermoplastics and foodstuffs are used to illustrate the issues raised. [1-3]

Newsprint, a Waste-product with a Ready Market

Newsprint is the major component of the paper fraction in Norwegian domestic refuse and comprises II.1 per cent of the total amount of household waste. It is an easily identifiable commodity and the prime object of recycling efforts in many countries.

A detailed analysis of the pattern of newsprint consumption was carried out. The results showed that 7.5 per cent of total newsprint consumption became waste-paper at the printers, 7.1 per cent ended up as waste in the form of unsold newspapers and 77.5 per cent of the total tonnage was sold to the consumer as newspapers. The remaining 7.9 per cent was for diverse purposes such as making up stocks, for archives and for the production of sundry printed matter.

The analysis was conducted on a regional basis thus showing the potential amount of industrial and post-consumer waste for the individual municipalities.

The information provided the waste-paper industry with the overview it needed to plan expansion of its collection capacity for household waste-paper. The industry has since carried out a study on the generation of boxboard waste.

Thermoplastics, a Group of Materials with a Rapid Expansion in Consumption

Consumption of thermoplastics has expanded very rapidly over the past two decades. Their properties are such that they have increasingly been substituted for traditional materials and are now utilized over the entire product spectrum, from drainage pipes to packaging film—products with greatly differing lifetimes.

Analyses of the composition of Norwegian domestic refuse have shown that it contains on the average 5.7 per cent by weight of plastics. The problem addressed in this study centered around how the rapid expansion in consumption and diversity of product lifetime would affect the amount and composition of plastic waste-products in the near future. Plastics waste has a high enthalpy of combustion (approx. 44 MJ/kg) compared with mixed domestic refuse (9,4 MJ/kg) such that minor changes in the plastics fraction can lead to relatively major changes in the combustion enthalpy of household waste. This can be an important factor influencing the choice of treatment method.

In this study, then, the material flow analysis was expanded to encompass the time dimension. A detailed study of the production and consumption patterns of the five most important thermoplastics (polyethylene—high density and low density—polypropylene, polystyrene and polyvinyl chloride) which together account for 93 per cent of the Norwegian thermoplastic consumption, was carried out for one specific year. This information was combined with knowledge of the historical growth in consumption, industrial forecasts of future consumption and assumptions on average product lifetime to estimate the expected development in plastic waste through to 1980.

The analysis showed that polyethylene (low density) packaging will continue to be the predominant source of post-consumer plastic waste in 1980 and that plastics would then constitute 6 per cent of refuse. Apart from giving important information on the composition of post-consumer waste, the analysis highlighted several areas which are of potential interest for recycling. These include shrink film which is being used increasingly as an outer packaging instead of corrugated cartons. It should be possible to coordinate the collection of this scrap material from industry, wholesalers and retail outlets with the collection of boxboard which is already organized.

Although the degree of "in-house" recycling of plastic scrap is very high in the plastics industry, problems often arise when plastic semifinished products are converted to finished products at plants which are located far from the primary processing plant. An example of the latter is the welding and thermoforming of PVC-film and foil into file-covers, packaging, etc. Scrap can amount to 8 - 10 per cent of the amount of film bought. These converting activities are carried out in many small factories spread over the whole country.

The nature of the PVC-scrap is such that it is difficult to utilize in the production of new film. One film manufacturer does, however, buy back scrap from the converters he supplies and uses this scrap in the production of flooring which he also manufactures. In this way he effectively gives his customers (the converters) a rebate on the semi-finished product they buy, thus securing his market and reducing the total amount of system waste.

Other film manufacturers who do not also produce flooring cannot offer this facility. Further use of the converters' scrap is inhibited because its utilization presupposes a knowledge of the scrap's product formula in order to be able to adapt it to one's own product specification. Film manufacturers are understandably reluctant that their product formulae become common knowledge.

To What Extent Can Recycling of Food-waste Improve Self-sufficiency in Food Supplies?

Concern has been expressed in Norway over the low degree of self-sufficiency in food supplies. This has been calculated to 38 per cent expressed as the percentage of the calories of food energy consumed (corrected for the import of animal fodder) which were indigenous.

The estimates of food consumption based on production statistics exceed estimates of food consumption based on the population's physiological food requirements by 16 per cent. This potential wastage represents an amount of food energy equivalent to 30 per cent of Norwegian animal fodder imports. A direct comparison of the energy content of food waste with that of fodder gives an oversimplified impression of the situation.

A material-flow analysis of several foodstuffs was undertaken in order to investigate the amount and composition of any waste products and to assess the potential of recycling them productively.

The material balance approach proved to be adequate for analysing the flow of food through the food-processing industry and highlighted several areas where utilization of our food resources could be improved. For example, the utilization of meat by-products to human consumption was shown to be lower than in the other Scandinavian countries.

A combination of the material-flow method and results of surveys of the composition of post-consumer waste was used to analyse the nature of post-consumer food wastes. This showed that only one-third of post-consumer food waste could be explained as inedible parts of the food supplied - meat bones, potato peelings, etc. The most important sources of food wastes which are not fully utilized at present are summarised in Table 1.

The analysis thus showed that the largest amounts of waste which are not utilized at the moment come from the consumer. Kitchen waste can be used as a supplement in pig feed. Complete utilization of all the kitchen wastes from both private consumers and institutions would reduce the requirement for food concentrates (which are mostly imported in pig farming) by 18 per cent. Utilization

TABLE 1

The amount and composition of some food wastes

Source of waste	Description of waste	Amount (Tonn/year)	Fodder value (1000 FU*/year)
Private households	mixed waste	270,000	70,000
Hotels and restaurants	mixed waste	15,000	4,000
Hospitals	mixed waste	10,000	2,600
Bakers	bread/cakes	6,700 (as flour)	6,700
Slaughterhouses	blood	2,700	600
Dairies			
reception stations	wash water	10,000 (as milk)	2,000
cheese making	whey effluent	29,000	2,000
	dumped whey	185,000	12,500

*FU = fodder unit, equivalent to the fodder value of 1 kg barley.

of waste from institutions is economic at today's prices and moves are being planned to increase the supply of kitchen wastes to farmers.

Although research is now being directed towards finding rational ways of coll-ecting food wastes from private households, use of these wastes as animal feed will make only a marginal contribution to the resource situation. As food wastes constitute about one-third of domestic refuse, any recycling effort here would, of course, have a major impact on the amount of refuse to be treated. For a consumer who is concerned about the food-resource situation though, the best advice would be for him to change his diet and eat more grains and less meat instead of trying to recycle his waste.

General Impressions on the Utility of Material-flow Analysis in Resource Management

These examples have shown that material-flow analysis is a useful tool to be applied in the initial stages of research concerning the better utilization of our resources.

As we have seen in the case of paper and to a lesser extent plastics, the method can be fruitfully employed to map the amounts of residuals produced when a ready market exists for the waste products. It is particularly useful when the industry which can utilize the waste does not itself produce the waste. The information gained from these studies was also important in directing further research effort.

The method does not consider what effect the increased utilization of the resource in focus will have on the consumption of other resources (e.g. manpower and energy). As such it can do no more than provide a pointer to areas of resource use with a potential for improvement.

REFERENCES

1. Carl Christian Hauge, Material flow analysis of newsprint. (Original title "Materialstrømanalyse av avispapir".) CMI report No. 74015-2/CCH. (In Norwegian).

2. David Nunn and Arve Strøm-Erichsen. Material flow analysis of plastics. Energy aspects of recycling. (Original title "Materialstrømsanalyse av plastavfall. Energiaspekter ved gjenvinning".) CMI report No. 75093-I/DWN/AS-E. (In Norwegian).

3. David Nunn and Arve Strøm-Erichsen. The utililization of waste from the production and consumption of food. (Original title "Utnyttelse av avfall fra matproduksjon og -konsum".) CMI report No. 74176-1/DWN/AS-E. (In Norwegian).

An Overview of Solid Waste Product Charges

Fred Lee Smith, Jr.

US Environmental Protection Agency

Non-waste technology has been defined as the planning of human activities to minimize the waste of materials and energy. In a very real sense, non-waste technology may be viewed as a way of ensuring that the material and energy stocks of our spaceship Earth remain adequate for our ever-growing demands.

Several strategies exist to implement non-waste technology. One can proscribe wasteful activities via regulations—for example, efficiency standards for products or processes. Or subsidies may be provided to encourage innovation and implementation of non-waste technologies—for example, tax credits for anti-corrosion research. Alternatively, programs may be undertaken to encourage less wasteful lifestyles—the energy-conservation publicity programs of recent years. Or, one may introduce a direct economic disincentive for wasteful material or energy use and thereby create a broad waste-reduction incentive. These strategies are interdependent and all will undoubtedly play some role as the world moves to economize on its use of materials and energy.

This paper will discuss one economic disincentive strategy—the introduction of solid waste product charges. This proposal would require that a charge be collected from the producer of products that enter municipal waste to cover their eventual collection and disposal costs. For example, a bottle or can manufacturer would pay a fee for each container manufactured equal to the costs of collecting and disposing of the container when it is discarded. The consumer, in turn, would pay a share of these costs via higher product prices.

Product charges were discussed in the June 1975 ECE Seminar on the Collection, Disposal, Treatment and Recycling of Solid Wastes. Mrs. J. Aloisi de Lauderel, in the session on Reduction and Treatment of Municipal Waste at the Source, noted that such charges would encourage industry to consider not only the costs of manufacture, but also the costs of waste disposal. Product charges ensure that waste-related costs of products are reflected in the price of the product.

Charges internalize the external costs of waste in production and consumption decisions—hence, in resource allocation decisions—and, thereby, increase the efficiency of resource use.

Over the last 4 years the US EPA has been investigating this solid waste product charge approach. In our view, the two primary purposes of product charges are:

1. to provide an incentive for producers and consumers to alter their jointly determined market decisions effecting the overall efficiency of resource consumption and, hence, solid waste; and
2. to provide a means of raising revenues for solid waste management or other social purposes. This paper summarizes our findings including a review

of the rationale of the concept, a discussion of design issues, and data on the effectiveness and impacts of product charges. The paper concludes with a brief history of the waste charge concept as a non-waste technology policy option.*

THE CONCEPT

Product charges are a way of "pricing" solid waste collection and disposal services and, therefore, of reflecting more accurately the real social cost of the product. The principal purpose of such a waste charge is to ensure that those whose production and consumption decisions determine the uses of resources will directly bear the costs resulting from their choices (in this case, the costs of collection and disposal).

In the United States today, collection and disposal costs are typically paid for indirectly and collectively through either general real estate taxes or through fixed periodic lump-sum payments by each discarder. This failure to price solid waste services—that is, to charge each waste generator the amount required to collect and process his solid waste—is both inequitable and inefficient. The lack of a fee proportional to the actual amount of waste each discarder generates is inequitable in that solid waste costs are thereby borne equally by all citizens and taxpayers. The individual who generates more waste pays the same as his more conservation-oriented neighbor. The inefficiency arises from the fact that when solid waste services are not priced, there is little incentive for producers and consumers to take action to improve the efficiency of resource consumption; as a result, too much material is used and too little of it is recovered. If collection and disposal costs were directly experienced by the producer, products might be redesigned to reduce waste or to improve recyclability, overall material consumption might be lowered, and secondary material markets might be improved (e.g. by increased recycling). In the jargon of economics, product charges would "internalize" the cost of solid waste management.

DESIGN CONSIDERATIONS

All product charge policy initiatives proposed in the United States to date have included three elements:

1. a charge on products commonly discarded into municipal waste;
2. a reduction in the charge for actions, such as recycling, which improve the overall efficiency of material use and therefore reduce solid waste;
3. the transfer of funds raised by the charge to local jurisdictions (in the US, this would mean a transfer direct to cities and counties).

In addition to specify a product charge proposal, the following issues also had to be addressed: what products should be charged; how high should the charges be; at what production stage should the charge be levied; what procedures should be used to credit producers recycling efforts; and what use should be made of the revenues resulting from the charge? A decision is also necessary on whether (and if so how) a gradual implementation of charges is

---

* A more detailed summary of this work will appear in the forthcoming *Fourth Report to Congress on Resource Recovery and Waste Reduction*, US Environmental Protection Agency.

desirable.

What Products should be Covered?

In principle, all products entering the municipal waste stream should be charged; in practice, the administrative complexity of charging certain products may justify exceptions. To date EPA analyses have evaluated charge proposals including only paper and packaging. The rationale for this limitation is that these categories comprise a major fraction of municipal waste (in the US about 80 per cent of product waste) and include most short-lived products. Moreover, a charge levied only on these product categories would be relatively easy to administer since only some 5000 manufacturing establishments would be involved.

Amount of the Charges

In the United States the cost of collecting and disposing of municipal solid waste now averages around $26 per ton based on actual accounting data. The costs of the environmental damage resulting from waste disposal have been estimated at about $3 per ton, although this estimate is extremely crude. In total, solid waste associated costs in the United States now amount to about $29 per ton of solid waste.

In most cases, weight seems an appropriate basis for allocating these costs among specific product categories. An important exception is the rigid-container product category where the competing products (glass bottles and aluminum or steel cans) have very different weight-to-volume ratios. A flat weight fee would penalize the heavier products even though these might occupy the same volume—and hence involve roughly the same costs—in the collection and disposal process. Thus a different allocation procedure—one related to the volume of each container—seems appropriate here. Also, product categories that require special handling such as tires, furniture, appliances or other bulky discards should be charged at a rate that would reflect their specific handling costs.

In the charge proposal now under study, only the direct costs of collection and disposal are included. Environmental costs were not included since the accuracy of the damage estimate is questionable. Paper and flexible packaging are charged on a weight basis at $26 per ton; while rigid packaging items are charged on a volume basis at a rate of $5.00 per thousand containers.

Adjustments for Recycling

The charge would be reduced for products made from recycling materials, since recycling reduces disposal requirements and provides a market for material recovered from municipal recovery programs. Moreover, the availability of a reduction would stimulate the demand for recovered materials. The reduction for which a particular producer would be entitled would reflect the extent to which that producer incorporates recycled materials into his products. This adjustment would provide a direct and significant incentive to develop and implement non-waste technologies.

Use of the Revenues

In principle, the revenues collected from the charge can be allocated to the general fund, rebated to individual households, or distributed to local governments. All US proposals to date, however, have endorsed a transfer of funds to local governments, where waste collection and disposal actually occur. This selection would reduce the extent to which the consumer might be subject to double-charging (paying once in the price of the product, and once again via the indirect or lump-sum payments now in effect). The funds transferred to local governments would result in either reduced local taxes or increased governmental services. And, in either case, the effect would be to partially offset any initial double charge.

Gradual Implementation

The shift to non-waste technology need not take place overnight. It is only necessary that we begin to move in this direction. Moreover, only sudden change in the price of materials or energy might seriously disrupt a modern economy. For these reasons, a product charge measure might best be gradually introduced over a period sufficient to avoid such disruptions. The length of this phasing-in period is subject to debate; however, the selected design would phase-in the charge uniformly over a 10-year period.

Design Summary

To conclude, the design selected for detailed evaluation would cover only paper and packaging, would be based on direct solid waste costs with the charge allocated on a weight-basis for flexible packaging and paper and a volume or unit basis for rigid containers, and would reduce the base charge for any use of post-consumer waste. Revenues would be distributed to local governments. Finally, the charge would be gradually introduced over a 10-year period. This product charge measure is evaluated below.

IMPACT AND EFFECTIVENESS

Evaluating a product waste charge proposal is a complicated task, since it involves determining the impacts of a complicated series of price shifts, as well as the impact of a relatively large transfer of revenues to local governments. Such a measure would affect both the production levels and prices of products as well as provide an incentive for increased use of secondary materials. Impacts estimated to date include the administrative costs, the recycling and waste reduction effects and consumer price shifts. The distributional effects of a charge measure and the anticipated tax payments of each product sector have also been estimated but are not discussed here.

Administrative Costs

The degree of complexity and potential administrative overhead costs involved in implementing and managing a waste-charge scheme of this type are comparable to those experienced in managing existing national excise taxes in the US. Based on an analytic comparison, such costs should average less than one-half of 1 per cent of charge revenues.

Effectiveness of Product Charges

US EPA studies indicate that a charge will have significant long-term impacts on the rate of recycling for paper and some of the other important solid waste components. The rate of paper recycling in the US, for example, would approach 27 per cent (based on production) by the late 1980s. The direct waste reduction impact is estimated at about 2 to 4 per cent of total production.

Consumer Costs

At the point of consumer purchase, the price increases resulting from a waste charge expressed as a per cent of total cost would be uniformly small; no significant consumer price increases would be anticipated.* The total increased costs of goods and services per year is estimated to average around $10 per capita. Moreover, to the extent that redistributed revenues are used to reduce taxes or expand services, this direct financial burden will be offset.

STATUS OF PRODUCT CHARGE POLICIES

During the late 1960s, a committee of the US Congress held hearings focused specifically on financial incentives for recycling. At those hearings, Leonard Wegman, an engineering consultant, proposed that a "penny-a-pound" tax be placed on all items entering municipal waste. Current policy initiatives are refined versions of this original initiative. The US Congress considered the idea from time to time during the next several years, and in 1975 national solid waste legislation was proposed which included product charges. Comments were received on the idea but no further action was taken at that time.

More recently (20 May 1976), a congressional panel exploring material policy issues held public hearings on the waste-charge idea. And in June 1976, action was taken by Congress to establish a group comprised of various Federal executive agencies headed by EPA to develop and evaluate product-charge policy initiatives. (Final action on this move is pending.) US EPA is already working with the US Department of the Treasury and OECD to resolve other outstanding issues. Progress reports on our evaluations of this non-waste technology implementation policy will be published during the coming year.

*The impact on the producer (and hence on his incentive to improve the efficiency of material use) is much greater, since selling and distribution costs are not yet included at that stage of production.

Administrative Ways and Means of Implementing Non-Waste Technology

Martin Neddens

Rat von Sachverständigen für Umweltfragen, Wiesbaden, Federal Republic of Germany

Political processes of planning and control can be roughly divided into three phases: Programme Development, Implementation and Effect Analysis. Past studies devoted to these processes have mainly dealt with aspects of programme development. Since work on the basic aims and programmes of environment policies have made substantial advances in recent years, problems of implementation are now assuming a position at the forefront of interest.

Experience has shown that environmentally acceptable deployment of existing technologies and the development and implementation of new anti-pollution technologies lead to a wide variety of very different problems in the field of administration. Empirical observations have revealed for example delays in the development of legal regulations; apart from political conflicts of interests, this is evidently due to lack of information, to restricted labour capacities and to coordination problems. Wherever adequate administrative regulations already exist, a lack of supervision is frequently observed because of the general high level of supervision required and also, however, because of a lack of readiness on the part of those concerned to provide supervision. Resistance from those concerned is not in principle directed against administrative measures but more frequently against the associated deadlines. Recently there have also been signs of reluctance in connection with administrative planning measures. Beside the relatively clear obstacles such as inadequate provision of information and existing gaps in research, this leads to such far-reaching problems as, for example, the delineation of administration regions. An unsuitable pattern of dividing the planning region often prevents the external effects of pollution-generating activities within the area from being internalized. As a consequence of this, positive motivation of the parties concerned is difficult.

The field of implementation of anti-pollution technology compared with other areas of politics is extremely heterogeneous. Systematic treatment requires on the one hand an analysis of the structures of administrative implementation itself and on the other hand an examination of the instruments utilized by the administration for implementation of political programmes.

2. PROBLEMS OF COMPLEX IMPLEMENTATION STRUCTURE

2.1. *The Cross-section Character of Implementation Principles*

A large number of implementation problems arise on account of the fact that the cross-section character of environmental problems has in the past only been mastered to an inadequate extent. The necessity to consider administrative measures in future to a far greater extent as laterally related to specialist subjects and as overlapping in media is implicit in the three basic principles which form the basis of the environment policies of the policies of the Federal Republic of Germany and which assume practical significance particularly for implementation of anti-pollution technologies.

The "Precaution Principle" requires advance and formative planning. To this end it is necessary to follow the interactions between the ecological systems in water, soil and air during implementation so that improvements in one sector do not simply lead to uncontrolled pollution in other sectors.

According to the "Instigator Principle" the costs arising for environmentally acceptable technologies are to be internalized in each case at the point in the chain of waste-instigation at which this is most suitable from both economical and administrative points of view. To this end attention must be given both to the interrelated effects existing between the ecological systems and in particular to their relations with the individual sociological spheres of activity in production and consumer sectors.

On the basis of the "Co-operation Principle" individual liberties and social necessities have to be offset against each other in a balanced relationship. The joint responsibility and co-operation of those concerned required for this purpose can only be successful in implementation of anti-pollution technologies if in turn the interdependencies between ecological and social requirements are clearly visible for those concerned.

2.2. *Elements and Context of the Implementation Structure*

The implementation structure considered as something extracted from the political planning and control processes cannot be seen independent of the preceding step of programme development and the follow-up step of effect analysis. Distinction can be made between vertical and horizontal features.

In the vertical direction there exist in the Federal Republic three major implementation steps. The supreme Federal level deals primarily with overlapping legislative preparations,development of legal stipulations and measures involving research and development of technologies which are not harmful to the environment. The Federal State level bears the major part of the responsibility for working out plans and implementation of programmes. The lowest step of community self-administration deals primarily with the application of standards and financing; the major emphasis of direct public services is also to be found here (waste disposal, waste-water purification, etc.) and to an extent supervision duties. In practice there is particular close linkage between the Federal State level and the community level whereas implementation difficulties arising from political conflicts of interest tend more to be evident between the Federal Government and the individual State levels.

The horizontal breakdown of the implementation structure between the administrative departments and the associated individual social groups and fields of activity is much more complicated. This reflects the problems of environmental conservation which have repercussions primarily in the implementation phase, i.e. the fact that environmental conservation when seen in connection with the persuance of almost all other social aims arises as a restriction so that conflicts of aims arise right from the start. The necessity of social division of labour prevents a situation from arising in which the implementation problems arising on the basis of complicated horizontal coordination questions can be reduced by grouping together all environmental needs in one special department. The only solution is to represent as clearly as possible the interrelationships between the individual spheres of activity and their effects on the environment in order to permit rational debate of conflicts of aims.

The danger of shifts occurring in the pollution to which the ecological media, i.e. water, soil and atmosphere, are exposed and the danger of at best sub-optimization requires an additional explanation of the relationship existing between these media of the environment. Examples of these requirements are supplied, for instance, by problems of protecting the water table against pollution arising from dumping of solid waste materials or the arisal of sludge on improved waste-water purification, disposal of which turns out to be particularly expensive. Similar difficulties are encountered in the limitation of gaseous and particle-bearing effluents; the residues arising from air-purifying systems produce pollution in either water or soil depending on which the process is employed.

Contextual representation of the implementation structures does not imply that central control of implementation is intended. This can take place just as well by decentralized means. However, once the implementation structures have been described, structure-generated implementation problems can be seen more clearly and the implementation instruments are easier to analyse. Formal aids are recommended for the description procedure. Figures 1 - 4 show the rough structures arising on implementation of anti-pollution technologies. The phase of political programme development within the process of development of political desires is only slightly touched upon. The requirements of effect analysis in the ecosphere, production and consumer fields is on the other hand worked out in much greater detail because this is particularly important for implementation. This applies all the more so since development and application of environmentally acceptable technologies is basically considered to be the duty of enterprises acting in the production and consumer fields.

The starting point for this representation is the Basic Model shown in Fig. 1. This is to be understood as an overall structure of order for further partial models based thereon. Traditional models of political and social science have primarily considered solely the relationships between the production and consumer sides. To permit systematic account to be taken of ecological requirements, the production and consumer processes must be connected with the ecosphere by raw material and waste flows. The ecosphere comprises the natural water, land and atmosphere-borne ecosystems. From here we can derive the ecological limiting values within which artificial man-made systems have to move. The

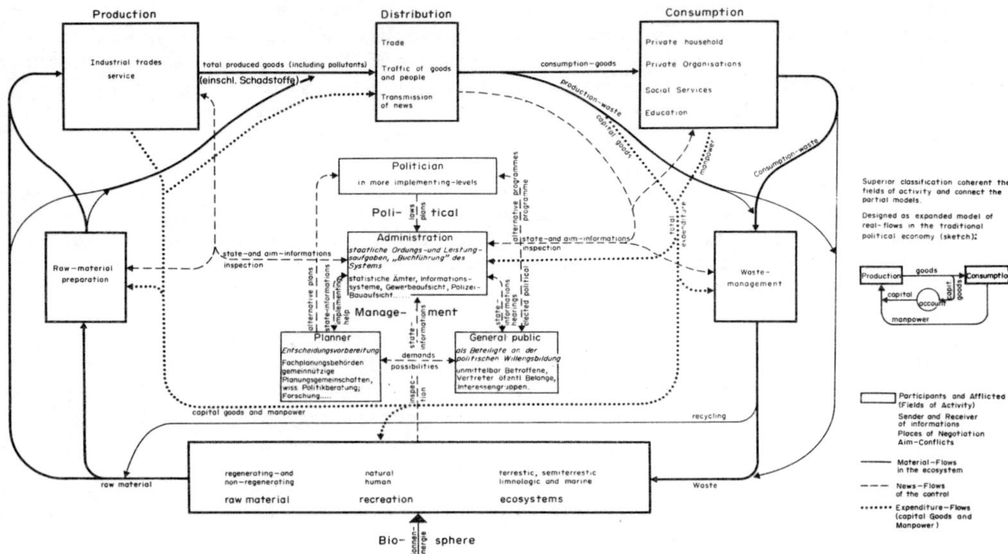

Fig. 1. Functional model of the environmental system.

production and consumer sides comprise technologies and organization forms by means of which the ecosystems are influenced.

At the centre of the implementation model is the control function of the politico-administrative system. This is embedded in the ecological and economic systems by communications flow lines which show points of interaction in the implementation. The individual part systems of the political control system are in turn interconnected by communications flow lines. The relative emphasis of planning which is gaining increasing importance in implementation is associated with considerations relating to participational methods of establishing aims.

The material flows describe the functional relations existing between the ecological and economic systems. They are to be considered as "benefit" flows which are to be contrasted with the "outlay" flows. This takes into account the fact that the implementation structure is directed towards actions and thus on principle towards benefit/outlay considerations.

The implementation of environmentally beneficial technologies must influence the material flows in such a manner that the function of the ecological part-systems is improved and their interaction as a system in ecological equilibrium is retained. The points of interaction are spread over the whole of the system. This reveals the cross-section character of the implementation problems in environmental politics.

The basic model permits individual partial problems to be split off and grouped together in part models. Figures 2 - 4 demonstrate this for solids (waste disposal) liquids (water-purification) and gaseous effluents (air conservation). The political control system here contains the institutions relevant for the particular part sectors. The subsystems of the part models in turn are found spread over the whole system of the basic model. Thus the cross-section character of part problems of environmental conservation is also revealed. The part models are open; their relationship arises from superordinate fields of activity indicated in the basic model.

Delineation of the part models one from the other constitutes division of labour. The interlinkage points between the part models are in each case represented in the drawings at those points at which the related superordinate field of activity is arranged in the basic model - e.g. the air conservation model at the top left shows "Waste water from wet cleaning" which goes into the model for water purification as an input parameter "industrial waste water purification". Likewise, "Community purification sludge" in the case of water purification is found as an interlinkage point with the model for waste disposal.

Successful implementation presupposes sufficient operably defined targets of activity. The discussion of aims or targets is made easier if the part aims are divided hierarchically corresponding to the functional context of the implementation structure. Figure 5 shows a weighted aim system for waste disposal. The debate on conflicts of aims with participation and joint responsibility of those concerned will be rationalized only to the extent to which the objective basic functional relationships are made clearer.

The models show that implementation of anti-pollution technologies cannot only follow one direction via programme development, implementation and effect. Knowledge of existing implementation structures vice versa also influences programme development and limits the choice of promising implementation instruments.

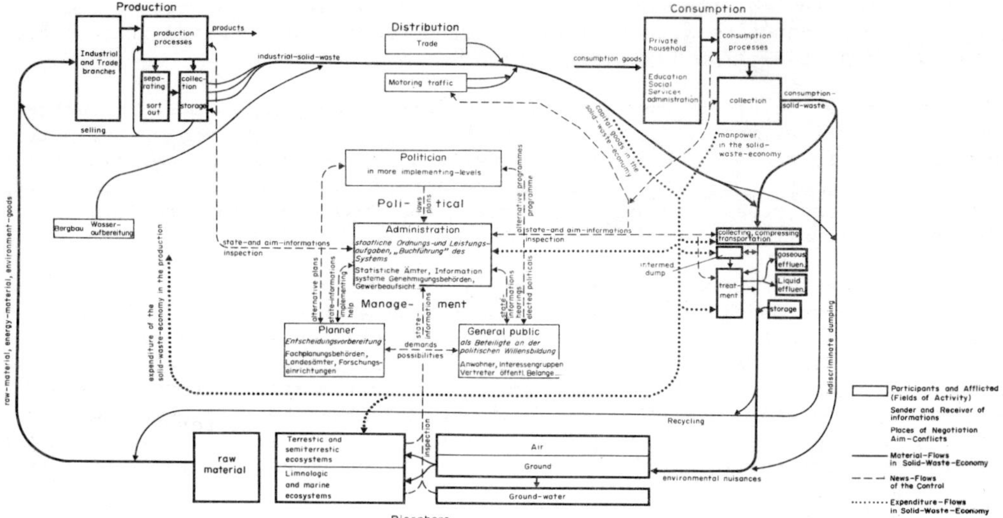

Fig. 2. Functional model of the solid-waste-economy.

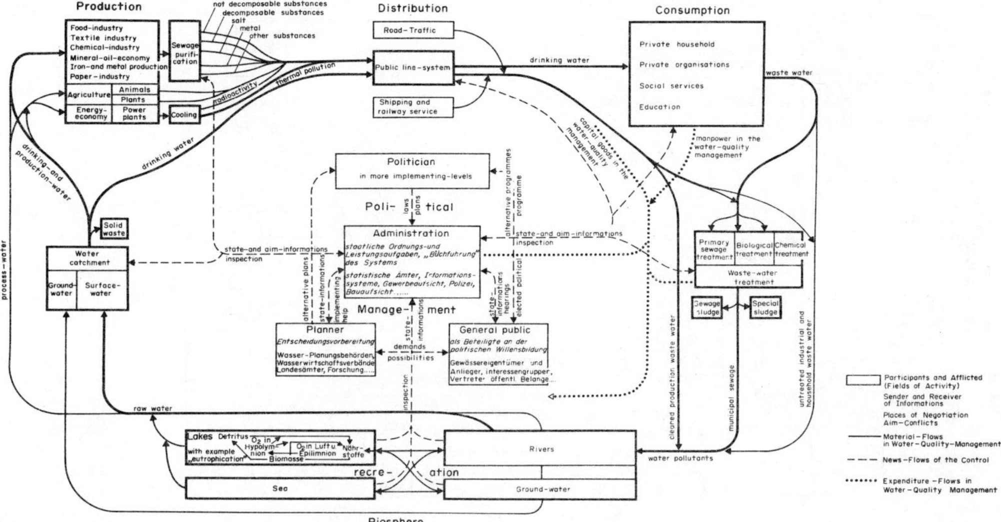

Fig. 3. Functional model of the water-quality management.

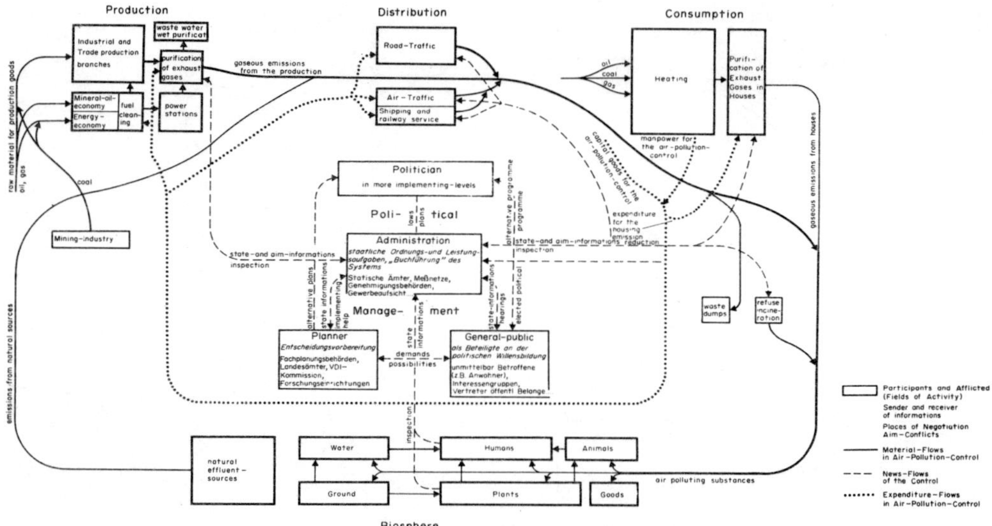

Fig. 4. Functional model of the air-pollution control.

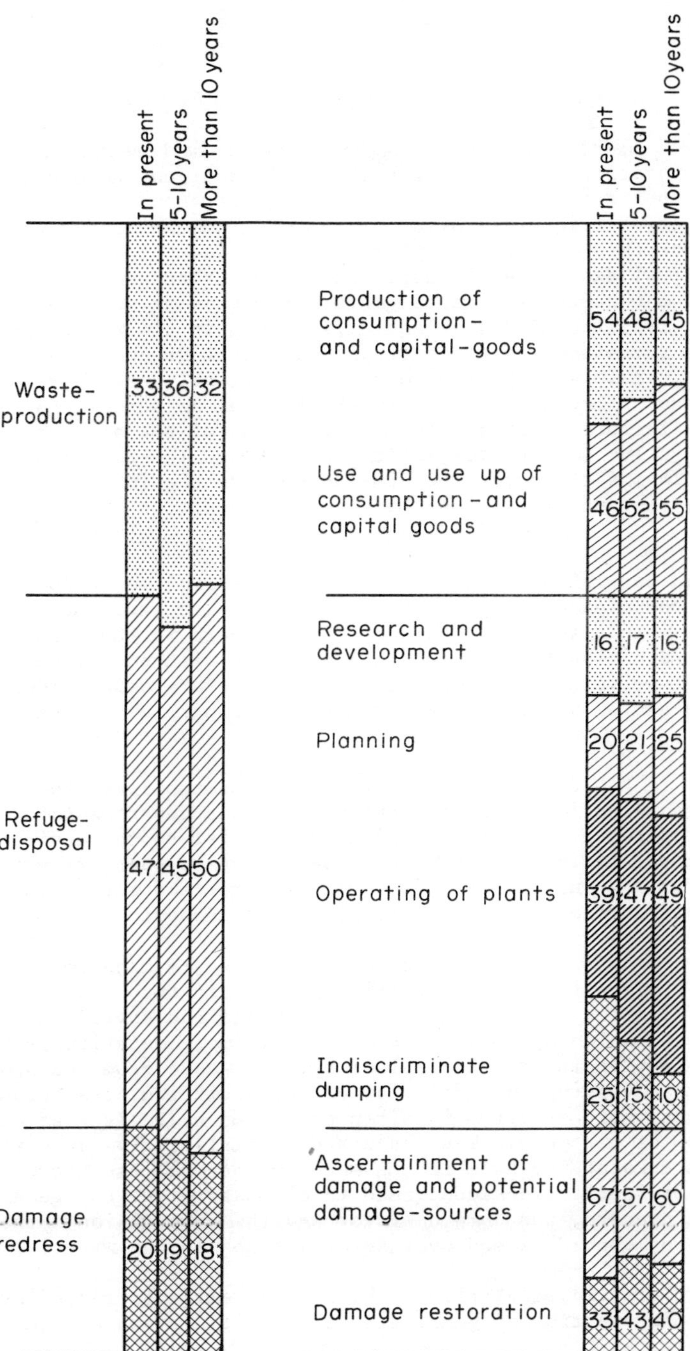

Fig. 5. Weighting of partial objectives in the solid-waste economy. (July 1975 German Council on Environmental Quality).

2.3. *Weak Points in the Implementation Structure*

2.3.1. *General problems.* A multi-stage vertical implementation structure, in particular separation between programme development on the one hand and implementation and application of standards on the other hand, is basically associated with the danger of reinterpretation of the programme aims during implementation. The reasons for this lie on the one hand in the lack of capacity and training amongst low level executive officers and on the other hand in the specific motivation of the executive agents. This motivation of such agents is in turn determined by traditional values and differences in their own institutional interest. And resistance on the part of the social groups affected by the implementation measures may also lead to deviational behaviour on the part of lower executive officers; possibly in the form of an attempt to save a programme from complete failure by adaptation and lowering of the aims or prevention of unforeseen negative side-effects of the programme. The degree of possible reinterpretation of the programme aims depends on the intensity with which the superordinate instances exercise their control function over the lower executive officials.

The pronounced horizontal differentiation on the implementation structure leads primarily to coordination and matching problems. The large number of institutions concerned, in itself tends to lead to delays. There is the tendency towards a danger of coordination arising too late and thus with purely defensive orientation to ensure that private department interests alone are safeguarded. Positive, far-sighted planning based on the "Precaution Principle" is not evident in this connection.

Improvements are hoped here from environment compatability studies as an institutionalized coordination mechanism between various departments. Such a mechanism must contain criteria for effects which are harmful to the environment in the individual social fields of activity. Development of such criteria has hitherto only been very inadequate. Again it would have to take into account a wide variety of interactions from the ecological and economic systems.

2.3.2. *Medium-specific problems.* The aim within waste management is to regulate the material cycle within the industrial process in such a way that generation of waste, utilization of waste, disposal of waste and environment pollution are interrelated in an optimum manner. Special implementation problems exist in the Federal Republic in the disposal of special waste from the production sector, these problems involving in particular capacity difficulties encountered by disposal plants whose erection is often prevented by resistance on the part of these sections of the population concerned. Administrative supervision problems as used to be encountered have in the meantime become less grave even though the large number of institutions concerned in disposal (i.e. the generating concern, the transport concern, the dump operator and the supervision authorities) demand that supervision should extend over several steps in the material flow.

In the longer term implementation problems will shift to recycling rather than disposal or to prevention of generation of specific types and quantities of waste materials.

Within water conservation, structure-generated implementation problems arise from the fact that the purification services provided by a works never serve the company itself but are only beneficial to others. Thus special problems of motivating those responsible for introduction arise. In addition infringements of standards in individual cases can only be proved with substantial administrative outlay for supervision so that mere introduction of standards is inadequate in its effect. Thus instruments are required here which beyond restrictions create economic incentives for purification, such as waste-water levies, quality standards differentiated in accordance with benefit and integrated operation of interlinked river regions.

Gaseous effluents must in all cases be restrained at the institutions where they are emitted. The choice of applying implementation measures at some other point in the chain of generation as arises in the case of solid and liquid waste products does not exist in this case. For these structure-dependent reasons instruments involving decrees and prohibitive statutes predominate in atmospheric conservation, these being accordingly associated with high administrative supervision outlay.

3. PROBLEMS OF INSTRUMENTS IN THE ADMINISTRATIVE FIELD

3.1. *Limits of the Instigator Principle*

Application of the Instigator Principle does not permit implementation of environmentally favourable technologies at economically favourable costs. The Instigator Principle provides for inclusion of the costs for prevention, elimination and offsetting of environmental pollution in the internal cost accounts of the particular instigator. The limits of the Instigator Principle have become evident in information and evaluation problems relating to determination of the costs of environmental pollution. Here ecological targets have to be orientated around approximation solutions as well as around environmental quality standards. Conflicts of aims in this connection often only become fully evident on the lowest level of implementation. Too rigid and too exacting targets then act as obstacles to implementation. In addition to bearing in mind the required conditions in the ecological system, targets must also always take in account the economic and technical requirements (technological state of the art) in the production and consumer fields. However, excessive orientation around the technological state of developments may cause delays due to the fact that the incentive for further development of technology may be lacking as a result of which the technological state of the art may become "frozen" for the time being.

Wherever the allocation effect of the instigator principle leads to too pronounced structural shifts between branches of industry and regions, supplementary measures may become necessary on the basis of the joint burden principle so that if necessary a politically tenable compromise may be reached. In principle exceptions to the Instigator Principle are only economically sensible in circumstances where the instigator cannot be established or where acute emergencies have to be overcome. The administrative means in such cases include granting of lengthy periods for adaptation to the requirements of the Instigator Principle and public financing aid for environmentally beneficial investments.

The decision to be arrived at between alternative strategies in the implementation of anti-pollution technology should be made in accordance with benefit/outlay criteria. A comparison of the instruments for implementation of the Instigator Principle in accordance with cost factors presupposes comparable delineation of alternative costs for environmental measures. Since only very vague success has hitherto been achieved in this connection cost factors can only be included in instrument comparisons to an inadequate extent.

3.2. *Problems of Individual Instruments*

Decrees and restrictive statutes (process and product standards) in general require a high amount of supervision particularly when positive motivation to observe standards is lacking. Process standards can be laid down at uniform levels for an economic area or may be fixed at differeing levels to meet regionally different environmental requirements. On the other hand, product standards must be uniform in order to guarantee the commercial trafficability of the product within an economic area. In this connection only the end-product is of interest. Implementation difficulties arise in connection with process standards as a result of delays in approval procedures arising due to insufficient legal regulations and also due to resistance on the part of persons concerned. *Financial incentives* (levies and subsidies) tend more to encourage further development of environmentally beneficial technologies than decrees and restrictive statutes. The central problem of suitability in the case of financial incentives involves the ultimate motivation effect which can only be estimated against the background of the interest structure on the part of those concerned. For the construction of water-purification plants, for example, motivation cannot necessarily be created with subsidy programmes since the benefit is only enjoyed by others. Beyond purely economic considerations further socio-psychological aspects play a certain role in motivation.

Levies in general only allow a certain environmental quality to be achieved at more favourable costs than by application of process standards if the costs arising for environmentally beneficial technologies vary amongst individual instigators. Levies can also compensate for the remaining environmental pollution which cannot be prevented by product and process standards. Levy systems, however, result in differentiated implementations problems; for instance, in waste management its application has proved to be more difficult than expected several years ago.

Provision of goods and services by the administration has the advantage of directness. If these are not cost-covering then they are contrary to the instigator principle. These funds will thus only be employed in exceptional circumstances and for cross-section related purposes of environmental policies, possibly for promotion of research and development in furtherance of training and in the provision of information.

3.3. *Tendencies Towards Change in the Development of Instruments*

Further development of instruments is directed towards achieving a more efficient effect from financial incentives. In addition, instruments are directed towards greater complexity and closer orientation to the aims. They are also directed more towards prevention measures within the implementation process. In the Federal Republic these tendencies are manifest in newly introduced plan for sub-systems of environmental politics such as supra-regional waste-disposal plans, management plans for above-ground waters, air purification in heavy-pollution areas, and countryside conservation programmes, etc.

In the administrative implementation apparatus this implies a shift in accent from intervention and law-and-order enforcement to planning administration.

The instruments used hitherto for these purposes will have to be combined and their combined effect investigated. In this connection there arises a greater need for information and greater demands for control. This in turn makes further overall clarification of the relationships existing between the general environmental system and the related implementation structures necessary.

Non-Waste Technologies: Ways and Means of Implementation

Robert Reid

Vice President, Energy and Environmental Analysis, Inc., Arlington, Va., U.S.A.

OVERVIEW

The concept of non-waste technologies is not foreign to the industrial sector. As manufacturing processes have become more sophisticated and the use of automated technologies has increased, the recovery and recycling of product residuals has improved. Tons of residuals once vented to the atmosphere, discharged to the waterways or left to accumulate on the land are presently being processed to recover valuable components. Of course, as defined, the concept of non-waste technologies is considerably broader than simply the recovery of waste by-products. The question of how we can most rationally use our natural resources is a subject of mutual concern across all segments of society and the international community.

IMPLEMENTATION STRATEGIES

In a free-enterprise society the use of our national resources has generally been left to be resolved by economic forces. Even the structure of many of our regulatory programs is geared to work through the market system. Environmental programs may be the best example, where the thrust of existing legislation is to internalize the external costs imposed on society from various economic activities. In effect we are saying that the most rational use of our resources can be translated into the most economically efficient use of these resources. Unfortunately, the two are not necessarily synonymous.

Economists have long recognized that basic conflicts exist between optimizing individual net benefits and optimizing societal net benefits. What is good for General Motors is not necessarily good for society. This concept may be best illustrated in the context of two individuals, A and B, bartering for a commodity. If A and B reach agreement then, presuming they are both rational, both have gained by their negotiating. However, from society's perspective a one-sided transfer from A to B may have increased society's net benefits, over the gain achieved by A and B negotiating independently.

This concept can be transferred to the utilization of our national resources. In a free enterprise system, resource use patterns will be modified if:

the value-in-use of the resource has changed sufficiently to require reoptimization of production and consumption patterns, and technologies and/or substitute resources exist that would enable new use patterns to be pursued.

While these may not represent sufficient conditions, they are certainly necessary in the absences of government intervention. The question of whether society's net benefits would be improved in this context relates to the value placed on these resources. Whereas the private sector will view the commodity in terms of its present value-in-use, the actual benefits to society may far exceed private sector benefits. Reliance, therefore, solely on the operation of the free market will not necessarily maximize societal benefits, especially if the future value of these resources are considered.

Methods of implementing non-waste technologies must include the imposition of society's preference functions on the use of natural resources. Techniques would include:

value pricing of existing resources;

technology transfer and consumer education programs;

subsidization of research and development of non-waste technologies to improve economic viability;

direct government intervention through regulations;

the use of economic incentives such as taxes, subsidies, price floors, capital grants, etc.;

expropriation and/or direct control.

Value pricing is generally what exists in most free enterprise economies today. It fails to consider the future value of resources or the external costs that are perhaps imposed on society. The next five options all involve some degree of external intervention. From the perspective of the United States, the most frequently employed implementation strategies include technology transfer and consumer education programs, subsidization of RD&D, and direct regulation. The use of economic incentives, although possibly more efficient, has not been pursued as frequently and the last option has only been used in extreme circumstances.

The question of which of these options should be chosen to stimulate the use of non-waste technologies probably can only be addressed on a case specific basis. One area that is receiving greater attention in the United States is the use of public sector funds to subsidize private sector research and development expenditures. While it may be too early to reach a judgement as to the efficiency of this approach it is possible to develop a fairly strong justification for public-sector support in this area.

GOVERNMENT SUPPORT OF RD&D

We recently completed a study* for the Office of Conservation, Energy Research and Development Administration focusing on the issue of the role of the Federal government in assisting the private sector to develop energy-conserving technologies. The objective of the analysis was to develop an analytical tool for

*Risk/Rate of Return Tradeoff Analysis: A Case for Federal Support, prepared for the Office of Conservation, ERDA, prepared by Energy and Environmental Analysis, Inc., 30 August 1976.

assessing the incremental or marginal impact of Federal support for energy-conservation RD&D. This information was then employed in the project selection process. The discussion centered on two parameters:

potential net economic benefits to the private sector, and

risks inherent to successful commercial introduction of subject technologies.

Conceptually, these issues are not difficult to grasp. Any time an individual, corporation or, for that matter, the Federal government makes an investment, there is a perceived benefit and perceived risk, the product of which must exceed costs if the project is to compete successfully for scarce RD&D resources and investment capital. Benefits, risks and costs are all dynamic elements which must be assessed prior to reaching a decision.

Federal support for private sector RD&D can affect all three of these parameters. The question therefore becomes: what is the impact of a successful RD&D program in the context of the private sector's investment criteria, and how can this information be used to maximize the efficiency of the Federal program?

CONCEPTUAL FRAMEWORK

Figure 1 illustrates the problem. The vertical axis measures the potential net economic benefit to the innovator and the horizontal axis measures the probability that the innovator will actually realize these benefits.

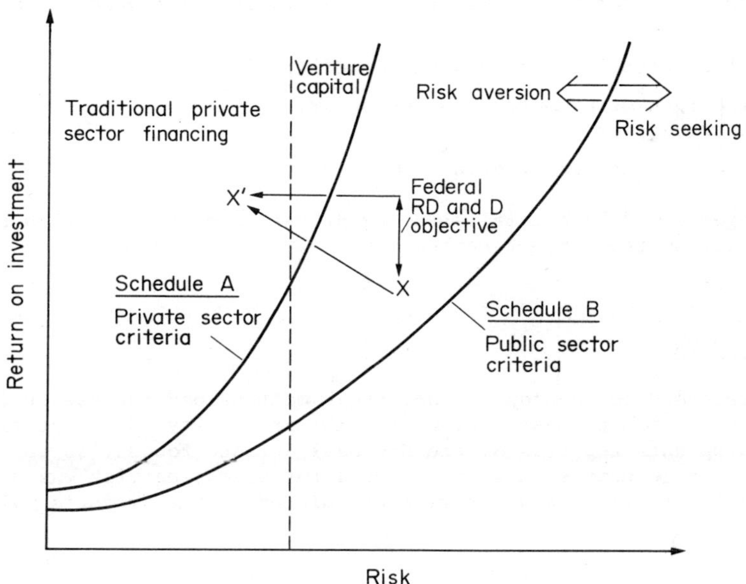

Fig. 1. Benefit/risk tradeoff criteria.

Schedules A and B may be viewed as indifference curves. Projects falling on one of these lines are regarded as equally attractive to the innovator, i.e. he is indifferent as to which to pursue. The difference between schedules A and B, which have been correspondingly labelled private and public sector, relate to the propensity to accept risk. A functional relationship which captures all the important relationships may be defined as:

$$r_{min} = r_{nom}\,(1 - P)^{1/\alpha}$$

Where: r_{min} = minimum required rate of return,

r_{nom} = nominal rate of return,

P = probability of failure,

α = propensity to accept risk $[\alpha > 0]$.

The propensity to accept risk is a function of the individual investor. The private sector tradeoff schedule reflects risk avoidance (i.e. $0<\alpha<1$), the public sector schedule—risk acceptance (i.e. $\alpha=1$) and any schedule to the right of the public sector's—risk seeking (i.e. $\alpha>1$).

The thesis developed was that any project falling above the private sector's or below the public sector's investment criteria represents an inappropriate use of public funds. In the first instance, the private sector, given expected benefits and risks, should be willing to pursue the project independent of public sector assistance. In the second instance, the project is tantamount to gambling with negative expected returns and therefore an equally inappropriate use of public-sector resources. The objective of the RD&D program may therefore be illustrated as altering the position of a project from point X to X' either by:

improving the potential net economic benefit;

reducing the risk of commercial success; or, as illustrated; or

a combination of the two effects.

If such changes could be accomplished the private sector would then be justified in commercializing the subject technology.

RESULTS

The study proceeded to develop an analytical methodology for use in implementing the project selection process. Values for the propensity to accept risk were estimated using data reported by the National Science Foundation and Monte Carlo simulation technique. A scoring model for assessing risk was developed and net benefits to the innovator were calculated at the products value-in-use.

APPLICABILITY

Although this system was developed for use by the Office of Conservation its transferability to other areas of non-waste technology may be possible. The important point to recognize is the divergence between the public and private sector, and the resulting support for government intervention.

Methodological and Strategic Aspects of Non-waste Technology

Introductory Report

J.-F. Saglio (Rapporteur)

Directeur de la Prévention des Pollutions et Nuisances,
Neuilly-s/Seine, France

I. PROTECTION AND MANAGEMENT OF THE ENVIRONMENT

In comparison with the clock of natural evolution, very little time has passed since the *Homo sapiens* colonized this planet and since his development in numbers, the degree of his social organization and the amplitude of his predations reached their present levels, i.e. 100,000 years for the first period and barely a few thousand years for the second. And, when we stop to consider that the most radical changes have taken place within the last century, we can better measure the importance for Man's very future of the thought he is now giving to his relations with his environment.

For, indeed, the major problem posed for our species is just this: Is our activity compatible with maintaining of the fundamental biological balances which condition life on the planet? This is not a simple question and, despite the relatively advanced state of our knowledge, it is not certain that we are now capable of giving a clear overall answer.

The steps currently being taken in this direction make use of the rational methods of scientific analysis.

First of all we must draw up a statement of the various "critical" situations or those assumed to be such, review them and describe them in detail.

Secondly, we must evaluate the consequences of one or more classes of parameters (especially those called pollutions and nuisances) on the entire biosphere which is then represented in practice by its most sensitive constituent elements.

The thresholds for the risk of danger thus graded and quantified make it possible to define in operational terms what Zygankov *et al.* (USSR) so justly termed as the critical ecological stress, beyond which a certain number of damages become irreversible.

Finally, starting from these thresholds - these probabilities of risk-we can engage preventive and curative action.

The research for and the development of waste-free production techniques are placed in the framework of preventive action, and they constitute the privileged method and means.

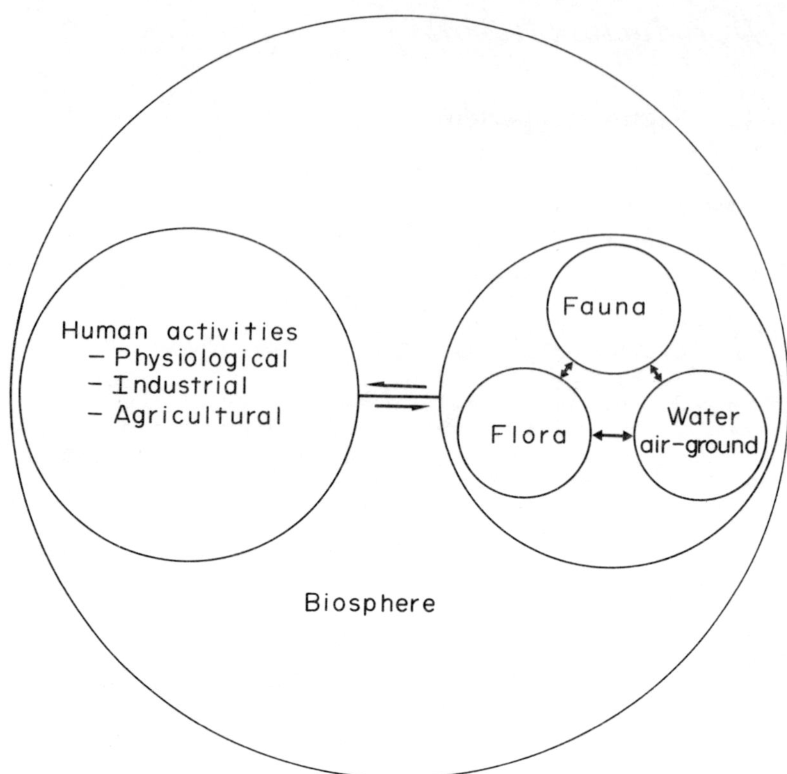

Fig. 1. Man and the biosphere.

II. METHODOLOGY FOR THE INTRODUCTION OF A WASTE-FREE PRODUCTION TECHNIQUE

To really and effectively introduce the notion of waste-free production techniques in a large-scale human activity, especially in the field of industrial activities, several factors are to be considered which intervene at different levels in the mechanism for the development of a waste-free production technique.

First of all there is the technological feasibility and then the internal and external economic factors (costs of energy and raw materials). The socio-economic factors also come into play (standards of living and consumer habits). In addition, there are legislative and regulation constraints. Finally, we should not forget that the type and volume of the activities and the exchanges to which they give rise are situated in an international context.

Currently, there is a trend toward the development of waste-free production techniques in industrialized countries, provided they do not, for example, result in undue reductions in the profit margins. Moreover, a clear distinction should be made between consumer goods and capital goods.

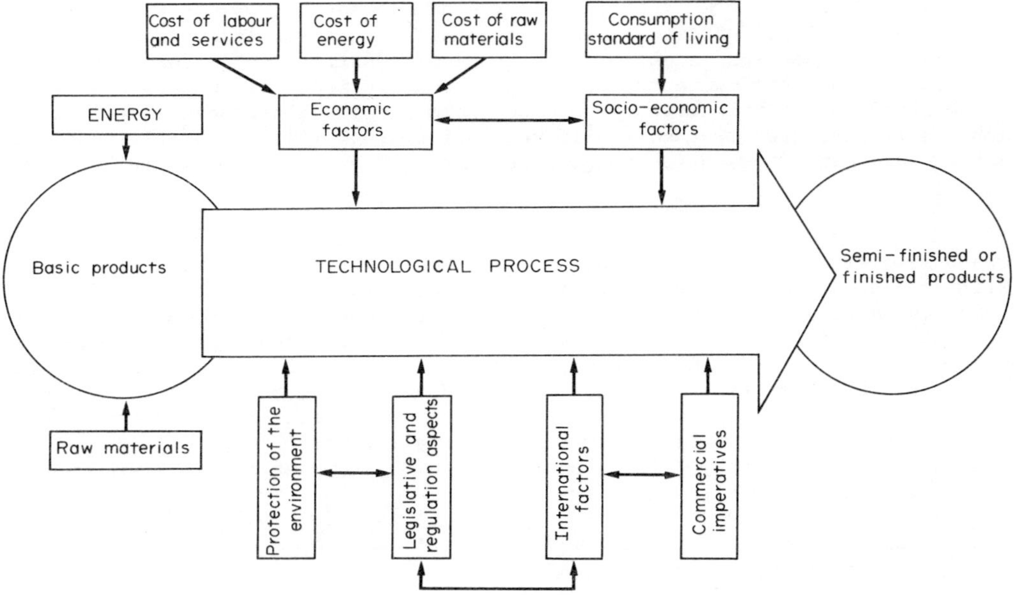

Fig. 2. Factors influencing a processing or production cycle.

Let us take the example of the use of two metals - aluminium and copper - in capital goods, which are subject to intensive price competition and in which these two metals represent a substantial portion of the total price. We can note that the substitution of expensive materials like copper by inexpensive materials such as aluminium, as well as more efficient use of materials which cannot be substituted, are well-established practices.

As far as consumer goods are concerned, there is a less pronounced trend due to the action of various forces: a more compact and refined presentation and standards of living which result in the consumption of more and more energy and goods (household appliances, country homes, several cars per family, etc.), which do not advance toward the ideal image of a post-industrial society in which the fight against waste and the durability of the goods would be the essential characteristics.

The legislative and regulation constraints make up a framework which reflects the needs and requirements of human collectivities faced with the problems of protecting and managing the environment. The costs of the measures taken to apply these laws and regulations, be they punishments or incentives, are becoming a non-negligible part of the total cost price of a product, in the same way as are the labour and service costs (marketing, distribution, etc).

Indeed, in most industrialized countries, on the average it is not abnormal for the costs of protection of the environment to reach 10 per cent of the total cost price of a product.

The facts and trends are essential for the development of waste-free production techniques for, even if we manage to converge the regulation and sociological factors toward an incentive for reducing industrial pollution at the very source, we must still have the sufficient economic means at our disposal and these means must fit in with dynamic development of the production systems, that is to say, they must basically remain inciting, i.e. favourable to change. Here we are dealing with the major problem of the application of the "polluter pays" principle.

Indeed, the taxes and fines can play a dissuasive role but, unless they reach exaggerated levels, they do not, in themselves, have any dynamic effect. Without a doubt, one of the most effective means is the obligation for a polluting industrialist to install the equipment required to treat the pollution he generates. But a better method is the incentive to change the technological processes used within a manufacturing or processing cycle so as to prevent the generation of pollution, nuisances and residual wastes. Moreover, to the extent at which the fluxes of pollution let off in a production cycle decrease, this results in a reduction in the treatment installations (Laskorin *et al.*, USSR), without, of course, taking into consideration the non-negligible savings of energy and raw materials.

These considerations quite naturally lead us to develop and implement an optimization mechanism by economic sub-systems.

Using the criteria for technological choices based on overall "cost-advantage" analyses, we can build production or processing systems in which each unit module is described in physical terms and assessed in economic terms, giving due consideration to the costs of energy and raw materials on the one hand and those of labour and services on the other hand (T.V. Long, USA).

It is not easy to give a description of the physical units, since it mandatorily calls upon three classes of parameters:

- quantifiable parameters (of a technological nature);
- parameters which, by definition, are not quantifiable or difficult to convert into objective criteria (of an ecological nature);
- trend or forecast parameters (A.Zygankov, USSR).

After economic assessment of these various parameters, we proceed to optimize the sub-system in question, by minimizing both the use of resources in the production circuit and the impact of the pollution, nuisances and residual wastes on the environment.

T.V. Long (USA) has perfectly described an example of such optimization with respect to car manufacture. Not only the manufacturing cycle itself, but also the problems which arise upstream (ore production, component manufacture) and downstream (collection of waste productions and recovery for reinsertion into the cycle or another cycle) were included in this example.

Finally, we must point out the importance and the effect of the "international dimension" variable in the development and application of a waste-free production technique policy. Depending on the degree of accuracy of the economic assessments, in particular those which are conditioned by the availability and cost of the resources, and by the international trade rules, we will find ourselves confronted with conditions which are more or less favourable to the promotion and development of waste-free production techniques. In this respect, we should emphasize the particularly beneficial effect of the international agreements on this subject, for they distribute the efforts to made over all the parties concerned and make it possible, in some European communities, to reduce the competitive distortions and thus the elements which hinder exchange.

III. PROMOTION AND DEVELOPMENT OF A WASTE-FREE PRODUCTION TECHNIQUE

First of all, we should emphasize the fundamental and essential role played by "Research and Development" work in the promotion of waste-free production techniques. In West Germany and France in particular, this work is partially financed by public funds and partially by the industrialists themselves. To a certain extent, the financial aid from the State serves as a technological insurance policy against the risks taken, and basically remains as an incentive. Since I97I and I972 in West Germany and France, the long-term environmental research programmes have been oriented around this finality of waste-free production, in particular around the rational use and savings of energy and around the exploitation of renewable energy resources.

Moreover, it is obvious that the privileged field for application of waste-free production techniques is the industrial sector in the largest sense of the meaning, i.e. especially including certain agricultural activities which, due to their volume and the intensive methods they use, do not unfortunately always remain without consequences on the environment.

It is possible to give a very schematic representation of how and where the actions arise which, in an industrial process, lead to production without or with a minimum of waste. The essential steps in this diagram are those in which the three mediums (water, air and ground) are in contact with the flows of energy and of material processed, either as a simple support or as a major factor in the process itself, and where the pollution arises which results in deterioration of the quality of the water, emission of noxious gases or the production of more or less harmful solid wastes and effluents.

These steps or "key points" for the appearance of pollution being isolated in the process, the action to be taken then consists in:

as far as the water is concerned

economizing the water and raw materials (the losses of which result in immediate pollution) by implementing partial or total recycling systems;

as far as the air is concerned

changing, if possible, the reaction which condition the emissions so as to render them less harmful (more complete combustion, elimination of certain gases, etc.) and, in any case, trap the flying particles or the particles in suspension which are likely to propagate in the surrounding atmosphere.

as far as the ground is concerned

avoid the production of abnormal or excessive quantities of solid wastes or effluents which cannot be recycled or reinserted into another process.

At this point we should make a distinction between two classes of wastes or effluents: the primary wastes which result from the phases of the production process and the secondary wastes resulting from the waste products.

Usually, the recycling rates for primary wastes can be quite high, e.g. 85 per cent in the case of copper. The situation is less satisfactory for the secondary wastes. A simple model described by Mr. Fischer (West Germany) makes it possible to calculate the ideal values for the recycling rates of the secondary wastes as a function of the rates of growth in consumption and the proportions of products with short and long service lives.

Thus, for example, in the case of copper, the maximum recycling rate in relation to the total flow is 33 per cent. In 1974 this rate was only 20 per cent. In the case of aliminium, the ideal rate is 40 per cent and the recorded rate is 10 per cent. Theoretically, therefore, progress can be made. However, and in the case of aluminium, for example, recovery comes up against the nature of the consumption network; a considerable part of the metal is converted into elements characterized by a high entropy, which implies high financial and energy costs for recycling.

Moreover, we should not neglect the fact that, in addition to complete technological innovations in which the ideality of the action described above it achieved (zero pollution), and despite the advantages obtained by this type of technological modifications, there will still be a need for some stations for purifying the unavoidable pollution and that they must be taken into consideration for optimization of the system.

Fortunately, in this respect, when proof is given of imagination, it is possible to treat the residual wastes which cannot be recycled to find other profitable uses for them. This type of action should be particularly sought.

However, as for the final remainder of the wastes, we should make allowance for and install collection and elimination systems using processes which are strictly controlled from the environmental point of view, for, in this ultimate link in the chain, we should not allow the transfer of pollution, e.g. from the ground to the atmosphere, through incineration of the wastes. It should be pointed out here that an excellent solution consists in constructing regional, harmful waste-treatment centres (Laskorin *et al.*, USSR).

IV. FORECASTS AND PROSPECTS

The major portion of the human population lives in developing countries (especially as far as industrial development is concerned). Moreover, all the natural resources as concerns both the energy potential and the raw materials are not evenly distributed, and the industrialized countries, i.e. those very countries which have the technological and economic infrastructures necessary for the production of consumer and capital goods are, on the whole and in the present situation, greatly deficient in this respect.

If this situation were to stand still, it would *a priori* render a strict policy for protection of the environment quite difficult. However, and fortunately, a new world economic order which brings about a redistribution of the wealth is beginning to spring up. For their part, the industrialized countries are already making an attempt to meet their part of the challenge. In a highly rationalized economic context for the resources, increased productivity and search for products resulting from sophisticated techniques, the development of waste-free production techniques can be marvellously fitted in. In my opinion, this is what we can reasonably expect on the level of general development.

As far as the rate of this development is concerned, it obviously closely depends on the economic growth rate (Mr. Fischer, West Germany) and on the limits of availability and exploitation of the world resources.

With a basically expansionist hypothesis and considering the resources as being unlimited, the maintenance of a high growth rate required proportional investments, the potential necessary for the development of waste-free production techniques would be put on short allowance. With a zero growth hypothesis, the waste-free production techniques become necessary and mandatory. Analysis of the data concerning the present situation shows that we are most likely headed toward a median path.

By making greater use of renewable resources and by stabilizing the consumptions of energy and raw materials, with the present population density, the implementation of new forms of energy, a chemical industry based on coal and wood and recycling of the phosphates from waste waters, an economic system could be indefinitely maintained without limiting the productivity.

This is our challenge for the end of this century and, without a doubt, the waste-free production techniques will have a major role to play in meeting this challenge.

V. CONCLUSION

The improvement of the quality of life is a line of force which is clearing itself a sometimes difficult path through factors which, among others, are called politics, economics, society, technology, etc., but it is fundamental, for it is the only one whose ultimate finality is the search for and the maintenance of the balances necessary to sustain life itself.

If, one day, we had to say to which science protection of the environment belonged, we could give an answer without hesitation: it belongs to thermodynamics - thermodynamics of life.

And, if we also had to say what our bearing point is, our major dynamic element, that which enables us to make the move from intent to action, from desire to implementation, I would say that it is imagination. Remember what Pascal once wrote: "Imagination has everything at its disposal; it creates beauty, justice and happiness, which is the world's all."

General Aspects of the Development of Chemical Production Systems in Regions with a Complicated State of Environment

A. Zygankov and V. Senin

State Committee for Science and Technology, Moscow, USSR

SUMMARY

Nowadays the development of all sectors of the economy depends more and more on ecological factors, and first of all on the degree of environmental pollution by production wastes and by the products of their decomposition.

The estimation of the interaction of wastes deals with too great a number of varieties. The utilization of traditional mathematical methods implies the formalization and "adjustment" of a complex mechanism of environment to a certain model, which satisfies only one of many states of environment. The discharge of a new waste requires basic changes in the model.

Generally, in the course of the development of a production system, the following concepts should be taken into account:

1. Environmental pollution by any uninterrupted technological process is connected with the wrong discharge and treatment of wastes.
2. Any pollutant will disseminate to a certain limit in a part of the environment after its discharge and simultaneously infiltrate into other parts of it and interact with them.
3. Anthropogenic pollution will stop when perfect production systems are developed.

Natural values can be subdivided into the following basic categories:

1. economically measurable values;
2. economically immeasurable values;
3. values which can be understood by a society only in the future.

As it was agreed at the First International Symposium of CMEA countries of non-waste technology (German Democratic Republic, 1976) optimum in the development of the economy may be reached by combining the use of economic, social and legal measures, having in mind the maximum development of non-waste processes and schemes.

The ecological parameters should be taken into account to the full extent in the prognoses for the development of branches of industry. In such prognoses the establishment of local branch ties and complexes should take into account the fact that the final economic effect may be negative if a closed system of waste treatment is not envisaged.

Discharge of wastes into the biosphere by modern production units is caused by the fact that only economically measurable values are taken into account. Discharges by one unit are made without taking into account the discharges by another. Further increasing of discharges may create a critical situation in a region.

There are certain regions where a *sensitive* natural mechanism was established historically which determines unique conditions for fauna and flora. Such regions can be described with a term of ecological tension, which implies a set of parameters further changing of which will lead to a radical change of fauna and flora and make human life impossible.

In such conditions the problem of the tension reduction to a critical level should be solved, keeping stable necessary rates of the production growth.

Quantitative estimation of the ecological tension is a very difficult task and can be seen as a number of parameters laid down in certain hierarchical order.

These ideas are illustrated by describing the situation in the area of the Crimean peninsular where the idea of creation of a closed system without environment as part of it was exploited to a great extent.

The example leads to a conclusion that the chemical production in a region with a suspected ecological tension should be based on an optimum choice of the main plant which uses natural resources. This plant should be in the centre of the closed system, the elements of which may be different in different cases.

The problem of a complex production system can be solved mathematically in the form of the linear or dynamic programming. The work in this direction is underway in the USSR.

(The complete text of this paper, in Russian, will be supplied on request by the Environment and Human Settlements Division, ECE secretariat, Palais des Nations, CH - 1211 Geneva 10, Switzerland.)

Perspectives for the Development of Non-Waste Technological Processes in Various Branches of Industry

B. Laskorin, A. Zygankov, B. Gromov and V. Senin

State Committee for Science and Technology, Moscow, USSR

SUMMARY

Main branches of industry responsible for the major share of pollution- energy, chemical, petro-chemical, iron and steel, non-ferrous, pulp and paper, industries require billions of roubles for construction and maintenance of the pollution combating installations and equipment.

But even the gigantic scale of a planned construction of these installations cannot solve the problem of the prevention of the biosphere from the negative influence of economic development. The problem can be solved radically only by introducing low- and non-waste technologies.

It seems obvious that at this stage the main tactical task is to introduce non-waste technology into those productions where the wastes are the most dangerous.

The aim of this report is to analyse the achievement in the main branches of industry from the point of view of finding the tendencies in developing certain new technological processes excluding the environmental pollution. The modern level of science and technology is taken as the point of departure. Possible radical changes of technology as a consequence of the development of new methods in principle are not considered.

The subject is considered under the following headings:

1. Creation of productions excluding pollution of water bodies by industrial wastes.
2. Preservation of the air basin against industrial wastes.
3. Complex treatment of mineral raw materials - the basis for the creation of non-waste productions in the future.

Under each heading the achievements and perspectives in the field of non-waste technology in the main industries are analysed. Many illustrating examples are given and discussed. On the basis of these, it is concluded that now each branch of industry has a solid basis for development and introduction of the processes excluding the pollution of the environment.

The development of such processes goes along the following major directions:

1. Creation of the various non-effluent technologies with the extraction of the valuable components and returning the purified water into the cycle.
2. Creation and introduction of the systems for the treatment of wastes from the production and consumption spheres.

3. Creation of regional industrial complexes with links between the different enterprises for mutual treatment of wastes.

A major condition for the acceleration of the introduction of non-waste technology is considered to be the development of new economic principles in the designing of industrial production systems with the ecological aspects taken into account. At this stage, when ecological parameters are not properly represented in economic studies and estimates, the main responsibility for the preservation of the environment in the next 5 - 7 years will still rest with water- and air-purifying systems. In this time the stabilization of pollution can be achieved but not more. In order to achieve the main goal, i.e. to reduce considerably the industrial pollution, it is required to establish the necessary scientific, technical, social and economic conditions for the development of non-waste technological processes.

With this in mind, a well-determined strategy should be developed for research and development activities. The strategic policy should cover a certain number of the hierarchical level from institutions to inter-state programmes.

These measures could allow the introduction of non-waste productions into the main branches of industry before 1990. It is considered feasible to create a non-waste economy covering the production and consumption spheres at the national level in this century.

(The complete text of this paper, in Russian, will be supplied on request by the Environment and Human Settlements Division, ECE secretariat, CH-1211 Geneva 10, Switzerland).

A Method of Assessing Non-Waste Technology and Production

Thomas Veach Long II and S. Ellis

Resource Analysis Group, University of Chicago, 5735 S. Ellis, Chicago, Illinois, USA

It is generally accepted that societal decisions between alternative technologies should be based on a comparison of the time-discounted social benefits net of social costs associated with the technologies. However, in market economies, a significant proportion of the social benefits and costs may not be captured by the prices that result from supply-demand equilibria in the private sector. For example, environmental cleanup and protection costs were largely not reflected in US prices of goods and services until laws were passed that provided a regulatory basis for internalization of these costs. In order to accurately evaluate externalized costs and the services of public goods, we first generate detailed technological descriptions of alternative technologies. The technological descriptions, given in terms of marginal or average physical products of fuels, electricity and materials, can then be used in determining one part of the economic cost — that attributable to energy goods and other natural resources. For a total economic analysis, this cost information must be coupled with that of labor costs and whatever costs are ascribed to the services of machines, physical facilities and other forms of invested capital.

There may be several advantages, however, in focusing initially on a description in physical units and deferring consideration of economic costs. First, accurate, disaggregated, physical data is often much easier to obtain than economic cost information of equal quality. Second, the physical assessment of technologies transcends economic systems and the method could presumably be applied with equal facility in planned and market economies. Third, we avoid the difficulties, such as the proper choice of exchange rates, often associated with international economic comparisons.[1] Finally, for some classes of decisions, the result of system sub-optimization may be of interest, where we concentrate on minimizing the use of resources in the productive process, or, closely related, on minimizing the impact of the residuals of production on the environment.[2] These two fuel-material minimizations may often turn out to be the head and the tail of the same dragon, although this is not necessarily the case.

Over the past 7 years, a method for carrying out these technological evaluations, which we refer to as resource analysis, has been developed at the University of Chicago. The essentials of this method are as follows. The initial step is the choice of a system for study. A system for automobile production, use and/or recycling is shown in Fig.1.[3] In general, we include within the system boundary not only the fabrication step, but also mining, beneficiation and production of all component materials and the possible eventual fates of the commodity after use — either reprocessing or discard.

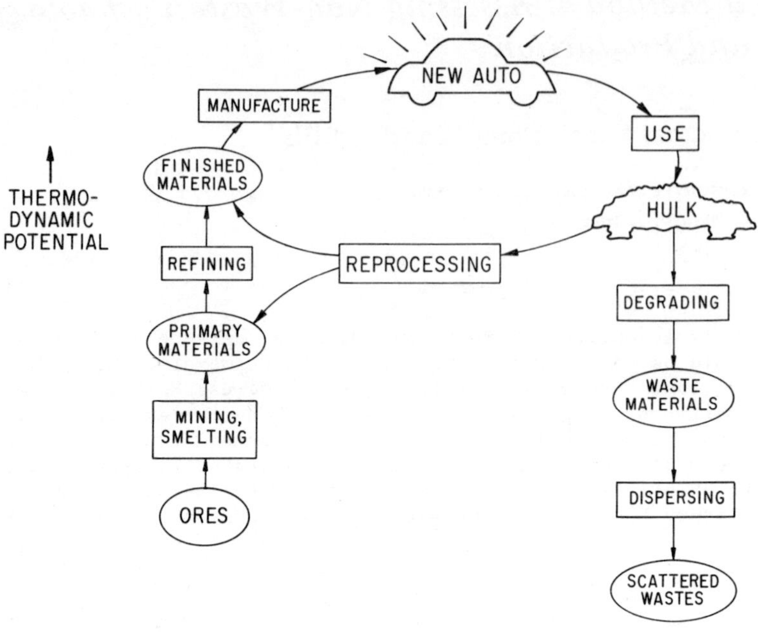

Fig. 1. Automobile production, use, and recycle or discard.

Next, we set up a complete trace of materials flows in the system, a materials balance. For diagrammatic presentation, the names and quantities of materials are shown in an oval, as in Fig.2, with process steps given in rectangles. The quantities are usually normalized to an output of 1 metric tonne or 1 unit (e.g. one automobile) of commodity delivered to final demand. Third, the energy requirement of each process step is assessed. The initial studies have focused on energy requirements, under the assumption that at least a short-term constraint on energy use exists. For instance, the automobile study showed that 135 gigajoules of energy are required for production and that about one-third of this would be saved by recycling. Symbols for fuel and other energy inputs are also given in Fig.2. The possible substitutions for fuel and electrical use are carefully evaluated, but in most investigations, those materials that are present in small quantities or that are not energy intensive have been neglected. Environmental impacts and the material and energy requirements for cleaning up residuals are often ignored, but should be included.

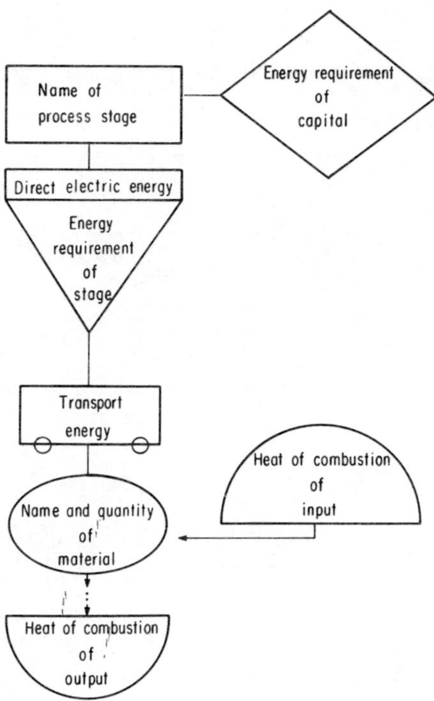

Fig. 2. Symbols for resource analysis.

In Figure 3 an analysis of coal-mining and electrical generation based on coal combustion is shown. For this analysis, we have attempted to accurately evaluate both the requirements for reclamation of strip-mined land and for sulfur removal on combustion by scrubbing. Our initial assumption was that the energy required for reclamation (0.26 GJ/I.37 tonne mined) was equal to that for stripping. More accurate values as a function of maximum grade permitted after reclamation are given in Fig.4 for Midwestern US mines. A more detailed diagram would show additional inputs that have actually been included by the incorporation of their "embodied" energies in the energy requirements given in the triangles. For example, in underground mining a significant quantity of steel is used in roofbolts and plates. The energy required to produce this steel (starting from ore extraction) is included. An important omission in Fig.3 is the failure to show water inputs, particularly in electrical generation. We believe that constraints on water use in specific regions are and will continue to be severe. Thus water as well as energy and other major material flows should be singled out to be explicitly traced.

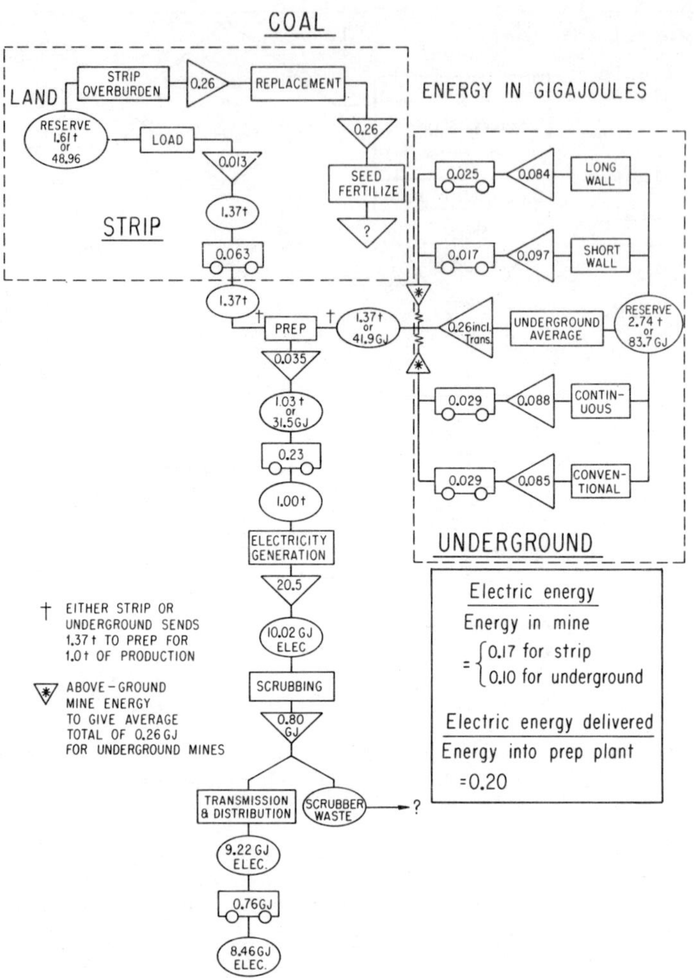

Fig.3 Resource analysis of coal mining and consumption.

In summary, the rudiments of a comprehensive description of technologies that can be used in assessing the opportunities for decreasing waste has been developed. This method begins with a complete materials balance and overlays this with an evaluation of energy inputs based on the average technology or on a specific technology within the industry. In this way, we determine the average (or marginal) physical products for fuels and materials. Maximizing these products in concert — perhaps through minimizing the energy (thermodynamic potential) requirements over the total system — would lead to non-wasteful technologies. However, this information generally must be incorporated into an economic apparatus for decision making. Otherwise, the pursuit of technological efficiency may result in non-optimal utilization of other scarce resources — labor and capital.

Units: MJ/tonne 6500 tonnes/acre	Maximum Grade		
	33⅓ %	25%	8%
Spoil reclamation	17	27	41
Tipple refuse disposal	3	4	9
Sub-total	20	31	50
Highwall	38	45	62
Total	58	76	112

Fig. 4. Energy requirements for reclamation.

REFERENCES

1. For examples, see: R.S. Berry, T.V. Long, II and H. Makino, Energy budgets 5. An international comparison of polymers and their alternatives, *Energy Policy*, 3, 144 (June 1975); and T.V. Long, II, International comparisons of industrial energy use. *Proceedings of the National Science Foundation-Mitre Workshop on Long-Run Energy Demand* (June 1976).

2. A discussion of the method described here and economics is contained in *Workshop Report No. 9*: International Federation of Institutes of Advanced Study Workshop on Energy Analysis and Economics (Lidingo, Sweden; June 1975).

3. An interest in the advantages of recycling solid wastes stimulated one of the first studies in this area of research. See R.S. Berry and M.F. Fels, The energy cost of automobiles, *Science and Public Affairs*, p.11 (Dec.1973), from which Fig.1 is taken.

Non-Waste Technology and the Materials Flow in an Economy: Facts and Perspectives

M. Fischer

Vice-Director, Institute for Systems Engineering and Innovation Research, Federal Republic of Germany

1. INTRODUCTION

The transformation of raw materials into products and wastes is governed by economic forces and closely related to our welfare. However, the range and direction of such transformations is limited by technical, ecological and geological constraints, which have become more obvious in the recent past. A technology which achieves this process within these limits is called a Non-waste Technology (NWT). The transition from today's technology into the state of NWT cannot be made unless the political and economic forces are taken into proper account; it is the aim of this paper to contribute to their understanding by analysing past experience and to proceed from this to a broad forecast of the possible future of NWT.

2. MATERIALS HANDLING IN THE PAST

Besides iron and steel, copper and aluminium are the most widely used metals in industrial economies. They are intimately related to the use of electricity and therefore absolutely necessary for technological progress and economic growth. The extraction of metals from ores is energy intensive (cf. Figs 1 and 2) and implies the handling of a host of accompanying materials which must be converted into by-products if they are not to be emitted as wastes. In order to keep profits at a reasonable level the mining industry continuously tries to reduce the unit operating costs* by raising the efficiency of labour through mechanization, by reducing energy inputs, and by increasing the scale of plants and the yield of the processes.

In the case of copper the exploitation of poorer and poorer ores has become unavoidable. This trend cannot be balanced by technological improvements, mainly since a tremendous increase in the energy input would be necessary (Lovering, 1969). The unit costs (in fixed dollars) are therefore increased, but the unit costs multiplied by the ore grade curve is constant (Fig. 1). The roasting of sulphide ores produces SO_2, which can be converted into sulfuric acid. The higher yield of the modern double contact-process has at the same time reduced the SO_2 emission. The improvements achieved in West Germany are also shown in Fig. 1. In this regard, a completely different situation seems to prevail in

---

* As the quality of goods such as refined aluminium or electrolytically produced copper is fixed (only secondary improvements in quick delivery or technical advice to the buyers are possible), technical progress in the mining industry must concentrate on cost reduction.

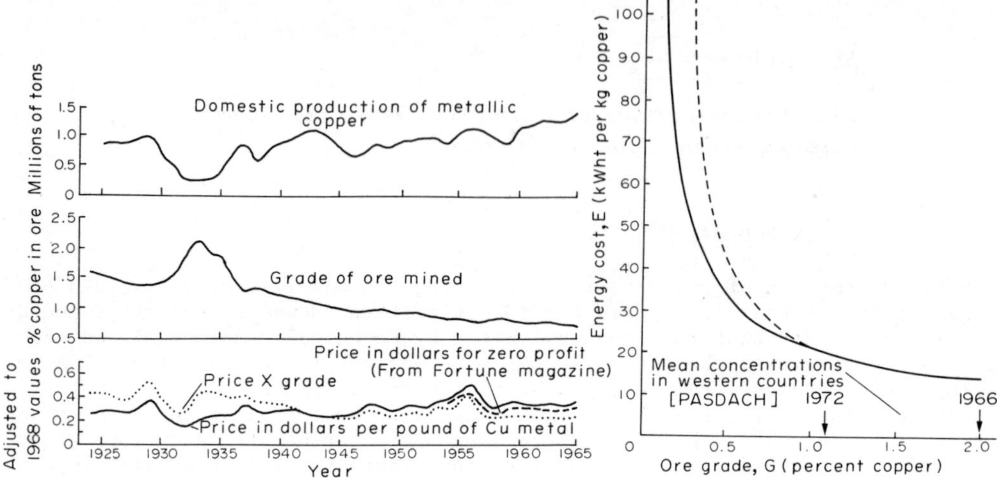

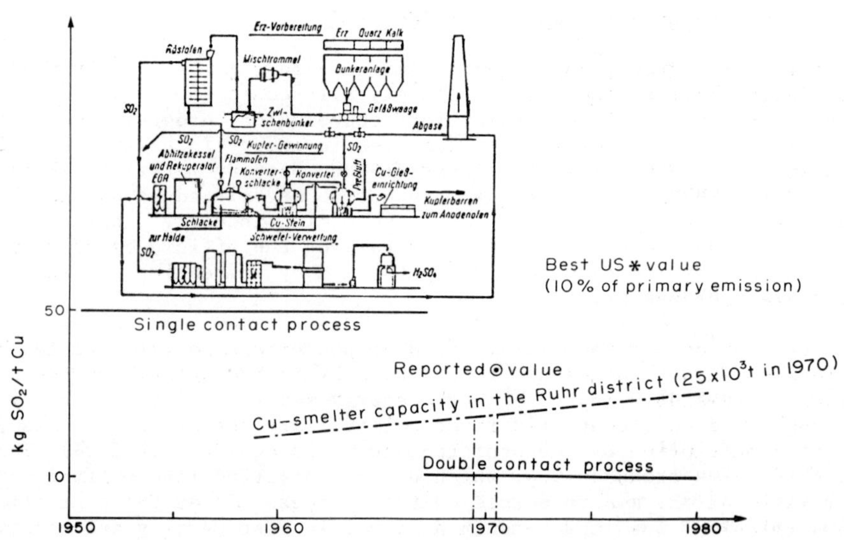

Fig. 1. Specific SO_2-emission from copper smelters in West Germany (for primary material containing equal amounts of S and Cu)

the US (Bridgstock, G.: Sulfur dioxide—special problems in the smelter industry, in Barrekette (Ed.): *Pollution*, New York, London, 1973, pp. 215 ff.): sulphuric acid from copper smelter off-gases would not be competitive because of the great distances between smelters and chemical plants.

I cite this as a reminder of the fact that NWT is often encouraged by a concentrated but diversified industrial structure. In such areas a positive feedback loop works towards NWT: Industrial concentration means dense population which enforces environmental regulations. This promotes the conversion of wastes to by-products which thereby become more readily available and cheaper, thus enhancing the attractiveness of the region to different industries.

Since the 1890s, when aluminium became an industrial metal, it has been produced from bauxit, an ore of roughly constant grade (45-60 per cent Al_2O_3). The production process has always involved the combination of a purification stage (Bayer-Process) with a final electrolytic stage. Reduction of the energy input (Fig. 2) increases in the scale of operation and other measures have steadily improved the process economics, resulting in a price reduction more pronounced than for other metals (see insert in Fig. 2).

Today is is not known how to turn the gaseous emissions of aluminium refineries (specific primary emission of HF: 5 kg/t_{Al}) into useful products, and due to the heavy environmental impacts of great plants ($\approx 2 \times 10^5$ t/y) the allowable emissions have been continuously reduced in West Germany (Fig. 2). This and the great demand for electric energy already limits the further expansion of refining capacity. For every ton of aluminium which is reduced 1-2 tons of "Rotschlamm" must be disposed of, as only minor success has been made in using this material in the building industry or in recovering its iron content.

Summing up the findings of our short survey it turns out that in the basic industries economic self-interest and environmental legislation already established a trend towards NWT clearly without an undue reduction of profits. A reservation must, however, be made with respect to energy: dwindling resources have already resulted in rising energy inputs for Cu-mining and this will certainly be the case also for Al and other raw materials.

When turning to the final production stages of the economy one has to differentiate between investment and consumer goods. As pointed out by, for example, Landsberg and Huddle in the "Materials" issue of *Science* (AAAS Publication 76 - 4) our knowledge of the materials flow in an economy is still in its infancy. Actual and detailed data can be found in a recent study for the West German Government,* but time series are still seldom available. Thus the following statements are still preliminary.

Most of the two metals Al and Cu are used in investment goods, which are subject to an intensive price competition and where the materials input accounts for a substantial percentage of the total price. Therefore the substitution of cheaper materials (e.g. Al) for scarce, expensive ones (e.g. Cu) and the more effective use of those which are not substitutable is a continuous trend†‡

* H. Stodiek (1976), Blume (1977).

† It is hoped that at the time of presentation of this paper relevant data from an ongoing study will become available.

‡ Blume (1977) and Fig. 3.

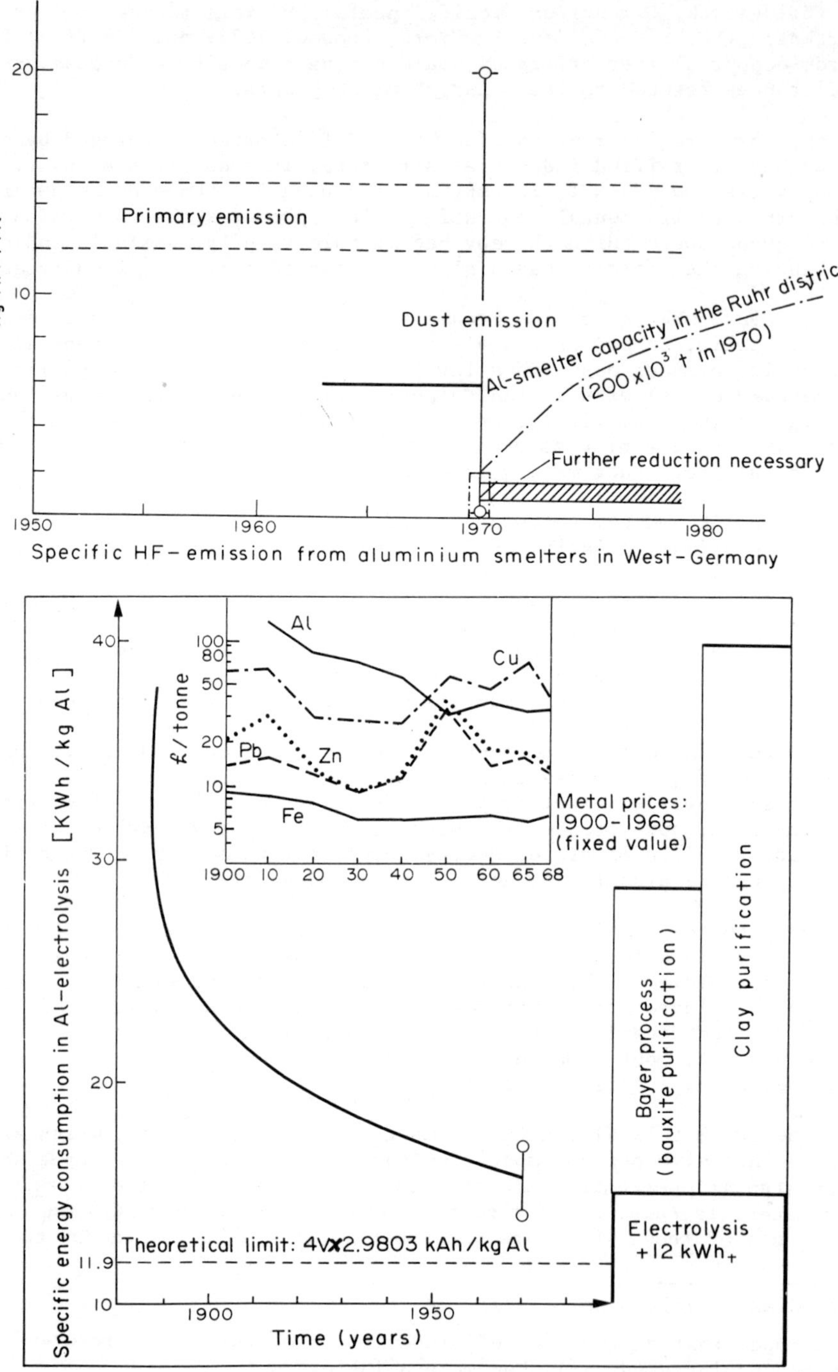

Fig. 2. Mining and refining of aluminium.

(Al-cable for high-power transmission, increasing power per weight in electric-motors). In addition modern technology tends to be more know-how intensive, which leads to a decrease in the input of Cu or Al per unit of investment (whereas the price of 1 kg of heavy industrial equipment is in the order of 15 DM, the price of 1 kg computer hardware is of the order of 100 DM).

The trends in consumer goods are less pronounced. Although there are the same technical tendencies as for investment goods, there are definitely counteracting forces: more package, more comfort by using electricity and household machines, two houses, three cars per family and so on. If the so-called "post-industrial society" is ever to come and if so, what the materials' flow in such a society will look like, is difficult to say (Landsberg, loc. cit.).

If one looks at the economy as a whole the specific consumption of any material as a function of income *per capita* seems to pass through a definite maximum. The US may already be on the declining branches of the Cu-, Al-curves. But others (as plastics) may take over and so specific energy consumption has not declined let alone been saturated.

The study by Stodiek (1976) provides an actual picture of the recycling of non-ferrous metals in West Germany. As in other countries the re-use of primary scrap from industrial operations is quite high (compare also the figures in Figs. 3 and 4) in the case of copper (about 85 per cent), but less for aluminium. But even if the loss of material in the production process is small, there is still a considerable energetic waste which motivates further reduction of primary scrap generation.

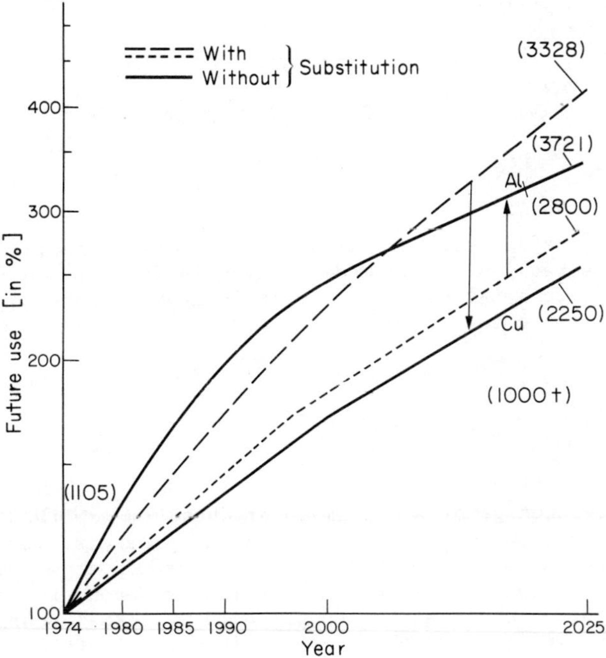

Fig. 3. Projection of future use of Cu/Al showing the effect of substitution. Figures from Blume (1977).

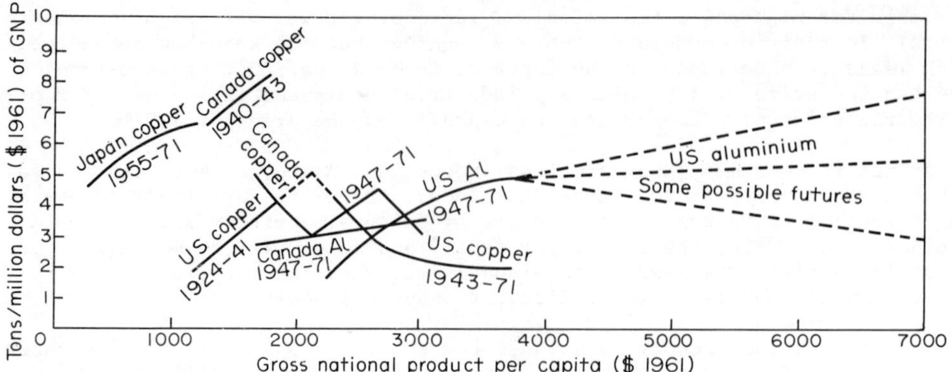

Intensity of use of copper and aluminium: examples from Canada, Japan and the United States

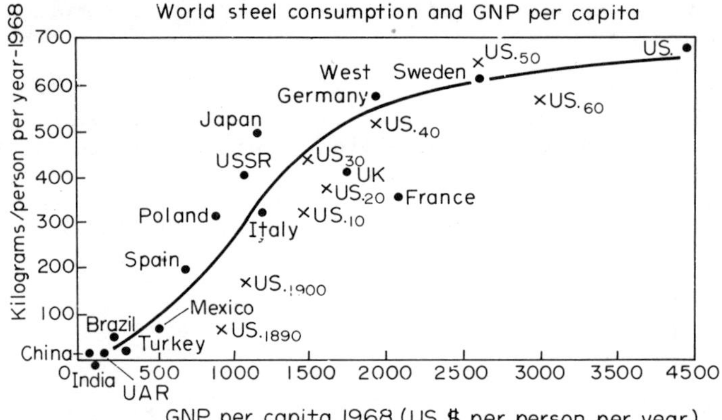

Fig. 4. Specific use of materials.

The situation is less satisfactory is we consider the recycling of outworn products (secondary scrap). A simple model, which is illustrated in Fig. 5, shows that for any specific metal with exponentially growing consumption ($P_{tot} \alpha e^{\gamma t}$) in the stationary state the maximal contribution of secondary scrap to the total flow is

$$\frac{\Sigma \, Si/_{Ti}}{P_{tot}} = u_o + \frac{1}{2} u_1 + \frac{u_\infty}{\gamma T_\infty} \quad ,$$

where u_o is the percentage of short-lived products ($\gamma T_o \ll 1$), u_1 that of goods with lifetime $\gamma T_1 \approx 1$ and u_{oo} that of long-lived products ($\gamma T_\infty \gg 1$). Export and import of final products have been neglected. For copper we may put $u_o^{Cu} = o_1 u$ $Cu \approx 2/3$; this would give a maximum contribution of secondary scrap to the total copper flow of 1/3, which should be compared to an actual contribution of 20 per cent in 1974 according to Stodiek *et al.* This means that 30 - 40 per cent of the available secondary scrap is lost, mainly in household wastes. For aluminium we assume $u_o^{Al} \approx 1/3$ and $u_1^{Al} \approx 2/3$ resulting in a maximal contribution of 67 per cent, which must be corrected by 25 per cent for net export and dissipational losses.

This theoretical value of 40 per cent Al-production from secondary scrap is much higher than the value of 10 per cent given by Stodiek. The figures are plausible* if one remembers that the short-lived products are mainly consumer goods and that there is no recycling of aluminium from municipal wastes in West Germany. Aluminium is a good example of a consumption pattern which is such that a considerable part of the material is converted into a high entropy waste (low concentration, widely dispersed). Therefore recycling would be quite expensive both in financial and energetic terms. The same holds true for one of its most serious competitors: plastics. Furthermore, neither poses any serious problem to the common methods of solid waste disposal, so it is difficult to imagine any change in the present trends.

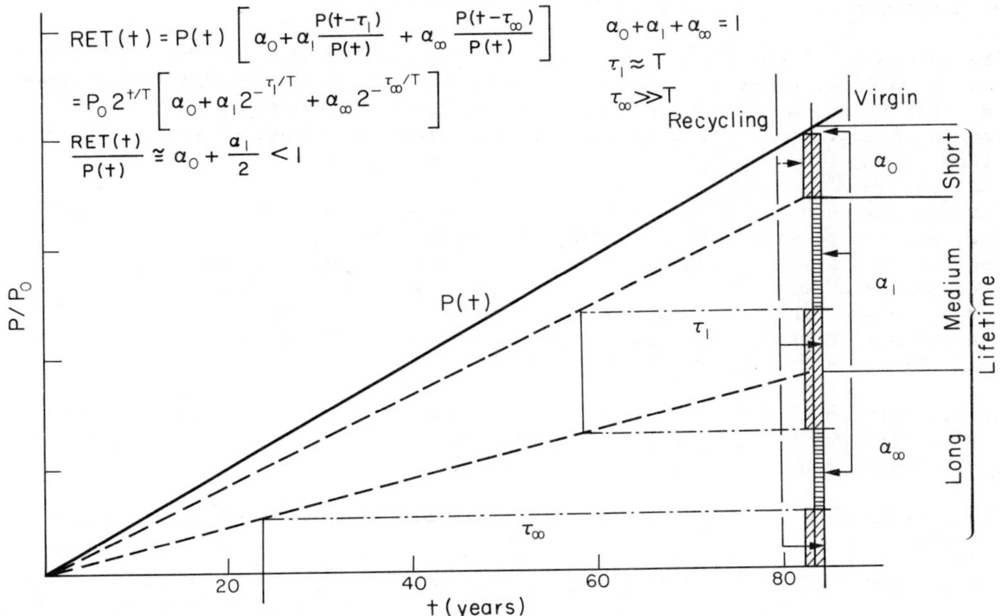

Fig. 5. Recycling potential in an exponentially growing economy.

3. PERSPECTIVES OF NWT

The majority of the human population live in countries which still have to be industrialized. This will greatly enhance the pressure both on resources and on the environment. Furthermore, the developing countries claim a "new economic order", a redistribution of wealth among nations. Both these trends pose serious problems to the industrialized countries, especially to those which have few raw material resources of their own. There seems to be general agreement that at least the export oriented industrialized countries have to meet this challenge by further increasing their productivity and by continuously turning their production to more and more sophisticated technology. The problem is, of course, how this could be achieved within the limitations of resources and the environment and—as far as I can see—the strategies recommended are mixtures from

* Compare Reichert, loc.cit.

two extreme options. The supporters of the high-energy option point out that given plenty of (nuclear) energy, bare rock and seawater become resources. Goeller and Weinberg ("Materials", loc.cit., pp. 68 ff.) have recently sketched a stable technology based on virtually unlimited resources (as iron and aluminium) and shown that the specific energy consumption for metal extraction will be not greater than 2-3 times its present value. However, the stability of their world rests critically upon the assumption that world population will stabilize at 10^{10} people and a *per capita* energy consumption of 7.5 kW*. But as neither population nor energy consumption are limited by natural laws, the availability of raw materials may well stimulate further growth. As growth needs high investment rates, the capital provided for NWT will possibly be kept on the hygienic minimum.

The advocates of the "stop growth now" alternative recommend a halt of the *per capita* material consumption at the present US level. To them NWT is necessary in order to sustain a constant consumption level eventually with decreasing inputs of energy and raw materials and therefore with less environmental pollution.

Governments usually try to muddle through somewhere in between the extremes. They are interested in social stability which, among others, depends on growing *per capita* income and preserved environmental quality. They therefore look upon NWT as a means "to ensure the most rational use of natural resources and energy to provide for the needs of man and to protect the environment" (from the definition of NWT adopted at the preparatory meeting in January 1975, ECE Dok. ENV/AC. 412, p.2 from 15 Jan. 1975).

The transition to NWT will not be just brought to us by anonymous economic forces, it must be made an explicit goal of politics and an issue of primary governmental concern. There are several reasons for this:

- the transition to NWT is often not profitable from the company's point of view;
- environmental regulations must be formulated by the government and enforced by public authorities. In many cases NWT is only brought about by adapting technology to such regulations;
- laws and executive practice have evolved in periods when NWT was unknown or unimportant, and NWT is often not compatible with these traditional patterns;
- although there are clear trends towards NWT, past experience shows that the urgent need for such transitions actualizes through sudden "crises" (environment, oil, food);
- even if NWT can provide a solution it comes too late if R&D does not start long before the crisis is about us. The long-range development of NWT, financed from public funds, functions as a "technological insurance against excessive price demands from foreign suppliers" (ABELSON/HAMMOND in "Materials", loc.cit., p. viii) and dangerous violations of environmental limits.

In the Federal Republic of Germany there are several government programmes which cover different aspects of NWT. The environmental programme from 1971

* Twice this value for the US (because they are in God's own country?).

includes a major sub-program on NWT, some of the results having been presented during this seminar. A program for the economy of solid waste handling has been initiated in 1975, and the energy R&D program is partially devoted to conservation and rational use of energy and the exploitation of renewable energy resources. These programmes are backed up by economical and institutional measures to stimulate the introduction of NWT. In the last summer a R&D programme in the materials sector has been published which aims at providing a technological insurance in the sense mentioned above. Besides measures to enlarge the range of exploitable resources (e.g. deep ocean mining), the programme concentrates on what we call NWT (see Fig. 6 for a table of goals).

As usual in our country the necessary R&D will to the greatest possible extent be conducted in the industry and only partly financed by the government.

Such a programme cannot solely be oriented toward actual problems but must identify future needs. The problem is that nobody can forecast the pattern of consumption, the acceptable price level, and the political preferences 50 or even 30 years from now, but yet this is the technologically necessary time horizon. But it is also clear that if we left these problems to the future, suddenly there would be too many to be solved at all. Our theoretical analysis of this is still going on but a discussion of our preliminary findings may enhance our chances of adequately tackling these problems.

A. Broadening the basis for raw materials' supply

B. Improved use of raw materials

1. Substitution
2. Use of couple-/by-products
3. Decreased material intensity
4. Reduced production losses
5. Improved output
6. More flexibility in the application of materials
7. Increased product quality, lifetime
8. Multiple use of products

C. Recycling

Fig. 6. Goals of the raw-materials programme in the FRG.

Physically a strictly *Non*-waste Technology is impossible. But the concept is a useful idealization, as the "ideal gas" in physics or the "perfect competition" in economics. It designates a state of technology which might be asymptotically achieved. But we must know the price. Although we are not able to forecast its future level we can give the conditions for its stability. In the highly industrialized areas of the ECE countries the capacity of the biosphere to absorb high entropy wastes (waste heat, CO_2, highly diluted toxic substances) is already fully exploited. With further exponential growth of energy conversion and material consumption NWT means exponentially decreasing specific emission or equally increasing efficiencies of energy conversion. But in most cases the present stage of technology is already such that extensive capital investments will be necessary for further improvements due to the law of diminishing returns. These investments do not increase productivity. In fact emission-control equipment induces extra, often highly qualified personnel in supervisory bodies. As an example, the capital needs of the nuclear power program in West Germany - deemed necessary to keep the GNP growth rate on its present level - grow faster than the GNP and are of such a magnitude as to seriously interfere with the needs of other industries. Also there are at present more people working in government on nuclear power regulation and in non-military nuclear research financed by the taxpayers than there are employees in nuclear power stations. This picture will only gradually be changed by a transition to breeders or fusion. Therefore with this type of technology, economic growth in highly industrialized regions comes to a halt in a foreseeable future but on a level of minimum environmental quality and of productivity so low as to make these regions no longer competitive with industries in developing countries.

A completely different approach to NWT would be the reliance on renewable resources and a stabilization of energy and material consumption. With the present population density a solar-based hydrogen economy with a chemical industry based on coal and wood and a recycling of phosphate from waste waters could be sustained indefinitely without limitations on its productivity. Improvements in NWT would really improve environmental quality.

If today's challenge to us is to achieve this transition from present technology in, say, the next 30 years, the immediate steps certainly include:

ongoing efficiency improvements in the use of materials and energy which must be implemented already in the design of processes and goods;

the recycling of non-homogeneous wastes, mainly from final consumption;

a thorough exploitation of the potential of renewable resources.

REFERENCES

Abelson/Hammond (Eds.): Renewable and nonrenewable resources. *Science Compedia* No.4, Washington, D.C., 1976.

Blume, H. (Ed.): *Technologien der Rohstoffnutzung*, Köln, 1977. (*See also* the papers by Herz, Mentzel and Reichert in Nos. 5/6 of *Metall*, Vol. 31, 1977, pp. 471-477).

Chapman, P.F.: *The Energy Costs of Producing Copper and Aluminium from Primary Sources*, Open University, 1973.

Jngenjörsvetenskapsakademien: Materialomsättningen i samhället, IVA-Meddelande, 182.

Lenhart, K.: *Die Emission von Schwefelverbindungen*, VDI-Berichte 186, Düsseldorf, 1972.

Lovering, T.S.: Mineral resources from the land. In *Resources and Man*, San Francisco, 1969, pp. 109-135.

NRW 72: Reine Luft für Morgen, Minister of Labour, Health and Social Affairs, North Rhine-Westfalia, Düsseldorf, 1972.

Sittig, M.: *Environmental Sources and Emissions Handbook*, Park Ridge, N.Y./ London, 1975.

Stodiek, H.: *Substitution und Rückgewinnug von NE-Metallen in der Bundesrepublik Deutschland*, Hamburg, 1976.

Winnacker-Küchler: *Chemische Technologie*, Vols. 2 and 6, München, 1970 ff.

PROGRAMS OF THE GOVERNMENT OF THE FEDERAL REPUBLIC OF GERMANY RELATED TO NWT

Umweltprogramm der Bundesregierung, 1971

Attached to this programme there is a Materialienband (zu Bundestagsdrucksache VI/2710) which contains a wealth of material on the state of the environment and of the technology of its protection

Abfallwirtschaftsprogramm

The programme coordinates all measurements towards an economy of solid-waste handling.

Rahmenprogramm Rohstofforschung 1976-1979
The Fed. Minister of Research and Technology

Research and Development to ensure the raw materials supply in the FRG.

Annex

Inaugural Addresses

Vincent Ansquer

Ministre de la Qualité de la Vie du Gouvernement Français

APRES LES TECHNIQUES DE DÉPOLLUTION, PRIORITE AUX TECHNIQUES ANTI-POLLUTION

Ce séminaire marque une étape essentielle dans la réflexion internationale sur le thème des techniques de production sans déchets; il est l'aboutissement des travaux engagés au sein de la Commission économique pour l'Europe depuis la première réunion, en avril 1973, des Conseillers des Gouvernements pour les Problèmes de l'Environnement, mais il doit être aussi, et surtout, le point de départ d'une concertation internationale féconde entre tous ceux dont le rôle et le devoir sont de veiller à la protection de l'environnement et à la lutte contre les nuisances.

La lutte contre les pollutions d'origine industrielle constitue dans tous les pays une part importante de la politique de l'environnement; l'importance des nuisances engendrées par le fonctionnement des installations industrielles, qu'il s'agisse des eaux résiduaires rejetées dans les cours d'eau, des gaz et fumées évacués dans l'atmosphère, des déchets de fabrication à éliminer, nécessite impérativement la mise en oeuvre de moyens techniques adaptés pour résorber les pollutions; il en va de l'agrément du cadre de la vie quotidienne, de la qualité de l'environnement, voire de la santé des populations. L'activité industrielle, principal moteur de la vie économique, indispensable à la satisfaction des besoins des consommateurs, ne doit plus ignorer l'environnement, tout au contraire. La lutte contre la pollution, de servitude subalterne qu'elle était, est passée aujourd'hui au rang des préoccupations majeures de l'industrie.

Si la lutte contre les pollutions constitue aujord'hui un impératif unanimement reconnu, les approches du problème posé à l'industrie restent diverses. Au risque de schématiser à l'excès, on peut discerner deux attitudes radicalement différentes.

Une première démarche intellectuelle consiste à séparer l'activité industrielle de production, considérée comme noble, d'une activité secondaire d'épuration des rejets polluants, indispensables certes mais jugée accessoire. C'est cette approche qui a surtout prévalu jusqu'à présent: le processus technologique de production est considéré comme un acquis intangible, la production de déchets comme un inconvénient inéluctable. Cette attitude conduit à ajouter, à l'aval des installations de production, des équipements spécialisés destinés à remédier à une pollution jugée fatale. C'est ainsi qu'ont été le plus souvent conçues les stations de traitement d'eaux usées, les filtres d'épuration des fumées, les installations d'élimination des déchets. Les techniques correspondantes sont maintenant bien connues et des résultats satisfaisants peuvent être obtenus pour la plupart des effluents et déchets d'origine industrielle.

Mais il existe une autre voie qui n'a commencé à être explorée que beaucoup plus récemment. Au lieu de prendre la pollution comme un fait inéluctable, de juxtaposer l'usine polluante et son antidote - la station d'épuration - il est aussi possible de prendre le problème à la racine et de s'attaquer à l'origine même de la nuisance; en un mot, il s'agit de produire sans engendrer de pollution. Une telle approche passe par la mise en question du processus de production lui-même; ce n'est plus la recherche d'une solution technique à un problème posé par la technique, c'est un véritable changement de mentalité. Il n'est donc pas étonnant que cette démarche n'ait commencé à être suivie que tout récemment, que les techniques de production sans déchets constituent encore un domaine largement inexploré.

Peut-on à ce stade de la réflexion trancher entre les deux solutions? Vaut-il mieux mettre l'accent sur la *dé-pollution*, le traitement des déchets produits, ou sur l'*anti-pollution*, la lutte contre les nuisances à la source? Mieux vaut prévenir que guérir affirme la sagesse populaire. Tel est bien le principe des techniques de production sans déchets. *Encore faut-il que ces solutions, théoriquement satisfaisantes soient techniquement possibles et économiquement acceptables.*

Il est sans doute encore trop tôt pour faire un bilan définitif des avantages et des inconvénients respectifs de chacune de ces deux solutions. Une évaluation soigneuse doit être effectuée et les travaux de votre séminaire revêtent donc à mes yeux une importance primordiale. Mais force est de constater, sur la base des exemples dont nous disposons, que les techniques de production sans déchets apparaissent déjà très séduisantes.

Car les exemples ne manquent pas: de la fonderie d'alliages métalliques qui filtre le plomb contenu dans ses fumées, à la papeterie qui *recycle en fabrication* les fibres autrefois rejetées dans la rivière; de la féculerie qui récupère dans ses effluents des protéines utilisables pour l'alimentation animale à la teinturerie qui adopte un procédé par solvants sans eaux résiduaires; de la carrière qui récupère du sable dans ses eaux de lavage à l'usine de production de chlore qui sépare et réutilise le mercure contenu dans ses effluents. Dans de très nombreuses branches industrielles des réalisations concrètes sont là pour montrer que des techniques de production sans déchets existent et qu'elles présentent un intérêt majeur.

Les bilans économiques sont également très encourageants. Si l'on compare le coût des technologies sans nuisances - dont il faut déduire la valeur des produits récupérés - aux charges qu'impliquent la construction et le fonctionnement d'installations d'épuration, la balance penche très généralement en faveur de la prévention des pollutions à la source.

Les techniques de production sans déchets débouchent également sur de substantielles économies de matières premières et d'énergie. Est-il besoin de souligner le gaspillage que constitue le rejet dans l'environnement, non seulement des produits valorisables contenus dans les effluents mais encore des réactifs utilisés pour épurer? Et que dire des dépenses d'énergie que requiert le fonctionnement des stations de traitement des effluents.

Il n'est pas douteux que les technologies sans nuisances constituent une voie nouvelle pleine de promesses qu'il convient d'explorer sans tarder. Le gouvernement français, pour sa part, est résolu à faire du développement des techniques de production sans déchets une des priorités de sa politique en matière d'environnement. J'ai décidé d'accentuer de façon très significative l'effort de recherche et de développement engagé depuis quelques années dans ce domaine.

Je demande à trois personnalités du monde scientifique d'animer sur ce point une réflexion et de me faire dans le mois qui viennent de nouvelles propositions afin d'accélérer le développement des technologies propres, et supprimer les nuisances à leur source.

Le séminaire organisé par la Commission économique pour l'Europe des Nations Unies vient donc à point nommé car, dans un domaine aussi neuf et aussi important, la coopération internationale, les échanges d'information et de connaissances sont à l'évidence primordiaux. Je me réjouis donc tout particulièrement de l'heureuse initiative de la Commission économique pour l'Europe et tiens à saluer le rôle éminent de M. Stanovnik, son Secrétaire exécutif. Les conclusions auxquelles aboutira votre séminaire revêtent à mes yeux la plus grande importance.

Sans prétendre influencer vos travaux, je souhaiterais que ceux-ci permettent de faire progresser nos connaissances sur un certain nombre de questions fondamentales.

Quels sont les secteurs d'activité industrielle dans lesquels le développement des techniques de production sans nuisances est le plus souhaitable?

Selon quelles modalités doivent être effectués les choix économiques entre antipollution et dépollution?

Quels sont les domaines dans lesquels la recherche technologique doit être engagée en priorité?

Quel impact le recours aux technologies propres peut-il avoir dans une politique industrielle, dans le choix des modes de production et des localisations des usines?

En terminant, je voudrais vous dire que l'avènement des techniques de production sans déchets me paraît constituer une étape majeure dans l'évolution de nos conceptions; à l'époque de la production a succédé celle de la production et du traitement des déchets; voici venu le temps de la production sans déchets. Votre réflexion au cours de cette semaine est capitale; elle nécessite à la fois une très grande compétence technique et le courage intellectuel du refus des idées reçues; c'est une tâche difficile mais je ne doute pas que vous y réussissiez.

Janez Stanovnik

Executive Secretary, Economic Commission for Europe

What we are essentially faced with, in today's world, is the basic problem of growth. During this century, and particularly after the Second World War, we have witnessed an explosion of growth. In just the past 45 years, the human race has doubled. In this same period, roughly half a century, man's output has more than trebled and so has his use of energy and other materials. One of the famous authorities of France, Pierre Auger, tells us that there are about one million scientists alive today and that this number is more than the total number of scientists in the history of mankind. But our tremendous technological and economic progress has engendered its own problems and difficulties. The very nature of modern technology has resulted not only in the production of more and more usable goods but also in the production of more and more unused waste.

This is not the only problem created by the pattern of growth that we have followed this past half century. One of the biggest problems related to our technology is the fact that it has been very highly concentrated and centralized. Growth has not been as evenly distributed throughout the world as would be desirable. No wonder that critics have been heard from many quarters and on many scores. There have been those, particularly the young, who have criticized our pattern of growth as being the result of a consumer-oriented society. There have been sociologists who have even written books about "wastemakers", thus stigmatizing the pattern of growth we have followed. But none of these critiques has been as strong and powerful as the recent criticism by the developing countries, launched through the United Nations General Assembly in the form of an action programme for a new international economic order. Underlying this demand for a new world order was a revolt by the majority of mankind against the type of growth that the industrialized nations have followed for the past 50 years. These people feel that the existing world economic order does not transmit growth impulses equally around the globe. Moreover, they feel that the high concentration of growth is not accidental, but rather an intrinsic consequence of the kind of development we have been pursuing.

It is often pointed out that the developed world represents only 25 per cent of mankind and that its population inhabits only 25 per cent of the land surface; but that its people are using 80 per cent of the world's energy and resources in producing between 70 and 80 per cent of the total world product.

The demand for a new international economic order is therefore a logical and understandable reaction to an unbalanced world economy. What in my view is most significant in this appeal for a new world order is that it does not primarily focus on a certain rate of growth, but rather on a demand for equality and justice. I feel personally that equality, like peace and prosperity, is indivisible. We cannot be half-free and half-slaves, half-rich and half-poor. Equality is one of the very basic elements in a human approach to human behaviour. If equality is to become the cornerstone of international relations and world order, then we must reconsider carefully the type of economic development we have been pursuing. If the developed world is to contribute something essential to the construction of a new international relationship, now is the moment for us to examine critically our past and present patterns of economic behaviour.

The Economic Commission for Europe has been engaged for quite some time in the search for a new model or a new pattern. From the Stockholm seminar, by way of the Rotterdam seminar (and through much relevant work within the Principal Subsidiary Bodies of the Commission), we have come to this seminar today - a seminar that, in my view, represents the most important step taken so far towards the evolution of a new model for growth. Up to now we have been developing economically as though everything were unlimited. There has been no serious problem of manpower. Capital has been amply available. The technology was there. And, most of all, energy and raw materials were abundant and accessible. It took the energy crisis to remind us of the limitations that can affect policies and patterns of growth. During our period of affluence, we were just running forward, not looking at what we were leaving behind for the next generation to clean up. We have been rushing ahead, believing we were thus fulfilling our historic mission and improving our civilization.

Now the moment of truth has come. And now that we are living through it, I firmly believe no stone should be left unturned. We should question every single assumption on which we have heretofore been building our palace of growth. We must search for new foundations that better correspond to the needs of humanity and that will place our developed world on a more fair and equitable footing in the international community of nations. It is clear, therefore, that the search for a revised model of growth, to which this seminar will contribute one of the most important elements, is the foremost task to which the affluent nations must address themselves today: a positive effort to build something new and more just.

Many people, particularly outside this conference room, may consider that the demand for non-waste technology and development is a utopian idea, but I am convinced that many changes in man's history have occurred only through such utopian visions. Perhaps non-waste production seems a fanciful idea *today*, but may I suggest that we will never arrive at non-waste technology if we do not at least start now with low-waste technology. No one expects to accomplish miracles immediately; but we must start, and so far we have not even made a start.

If we look at what statistics - and they are very scarce and scattered in this field - can tell us, the figures are startling. Each year between 10 to 15 tons of waste are thrown away, *per capita*, in our part of the world (and I have seen these figures confirmed by somewhat contradictory sources). Of course some of this waste comes from agriculture and about one-third from mining and mineral production, but there are still 1 to 2 tons per person which come from households, where, I think, the possibilities for reduction are the greatest. I was very impressed when I saw the US figures: American cities pay four billion dollars annually for *collecting* household waste, and then pay another half billion dollars for its *disposal*. These studies also show that one billion dollars worth of various materials is hidden in this household waste. In 1 ton of aggregate waste there are 200 dollars worth of aluminium, about 17 dollars worth of paper, and some 15 dollars worth of other useful materials like glass. Now, if these figures are at all close to the truth, I believe it is a cause for serious concern. Surely we will never be able to recuperate from the waste all the aluminium, all the copper, all the glass, all the iron; but we must also recognize the fact that today we are hardly recuperating anything (only between 5 and 10 per cent) and that we already have at hand the technology for achieving a much higher degree of recuperation. Of course we are then immediately faced with the main problem: at what cost? This is a worry which is my daily bread - namely, the problems of economics, efficiency and cost. My question is, therefore, to what extent does the actual cost of recuperation reflect reality? For example, when the cost of energy was four to five times lower than today, we considered that normal; but when it was suddenly raised four to five times in a few months we adjusted to the new normalcy, and now we are arguing that the present cost is quite normal and that any increase is abnormal.

The question of the normal cost of resources has not yet been resolved by economic science. Instead it is left to political bargaining, which often involves a show of power, force and strength. But there is one consideration which could very easily guide us, and that is that there is not only a private cost, but also a social cost. The private cost of a certain production may be considered lower than the social costs, which include the nuisance and the waste problems left in the wake of such private production. There is, accordingly, a basic contradiction between private cost and social costs. This is, accordingly, a basic contradiction between private costs and social costs. This is a complicated problem which you will not be able to solve at this symposium; but any contributions you can make will be invaluable, since nobody has as yet provided a satisfactory answer to it.

There is still another contradiction: that between the present and the future. When I discuss these matters with my children, they naturally reproach me and my generation for simply having a good time and leaving the problems for them to solve. What is the proper discount for the future? What should we take on ourselves; what can we leave for the generations to come? Again, an extremely difficult question. But there is one thing I will not hesitate to say publicly at this symposium. I simply do not think that we, in one generation of 25 years, have the right to spend resources which have accumulated in the earth's crust over millenia. There are generations to come. We are just brief inhabitants of this earth and just as our ancestors have given life to us, we have given life to those who will be living after us. We belong to one human race - with its past and its future. I believe, therefore, that technological solutions must be found which will make it possible to bridge the gap between past, present and future.

Similarly, as I said before, I do not think that we can live on as a wealthy island within a sea of poverty; not because I think we will be threatened militarily, or in any other way, but that we simply will not be able to be human if we remain so deeply divided into rich and poor. It will be a problem of internal corrosion that will not permit us to go on in this way. For this reason, I consider the challenge with which you are confronted today as far from being merely a technical matter; it is a profoundly human matter.

Economics has been thought to be a dismal science for too long. We should not forget that economics was born from moral science. Questions of morality and distribution and equality were the foundation of economic science, and we must go back to them again. I am therefore profoundly grateful, Mr. Minister, that we are meeting for this truly historic conference in this country which is the cradle of humanism. We must start thinking of economics again in humanistic terms. It is only in this way that we can contribute to the construction of the better world that mankind is now demanding.